Materials Science and Engineering: An Integrated Approach

Materials Science and Engineering: An Integrated Approach

Editor: Lily Chen

NY RESEARCH PRESS
New York

Published by NY Research Press
118-35 Queens Blvd., Suite 400,
Forest Hills, NY 11375, USA
www.nyresearchpress.com

Materials Science and Engineering: An Integrated Approach
Edited by Lily Chen

International Standard Book Number: 978-1-63238-572-7 (Hardback)

Cataloging-in-Publication Data

Materials science and engineering : an integrated approach / edited by Lily Chen.
p. cm.
Includes bibliographical references and index.
ISBN 978-1-63238-572-7
1. Materials science. 2. Materials. 3. Engineering. I. Chen, Lily.
TA403 .M38 2018
620.11--dc23

Printed in China.

Contents

Preface

This book was inspired by the evolution of our times; to answer the curiosity of inquisitive minds. Many developments have occurred across the globe in the recent past which has transformed the progress in the field.

Materials engineering is a discipline that integrates several fields such as chemistry, physics and engineering. It helps in the discovery of new materials, mainly in the form of solids. The engineering of materials aims to enhance their mechanical, electrical, optical, magnetic and chemical properties. The aim of this book is to present researches that have transformed this discipline and aided its advancement. The various sub-fields of material science along with technological progress that have future implications have been glanced at. Scientists and students actively engaged in this field will find this book full of crucial and unexplored concepts.

This book was developed from a mere concept to drafts to chapters and finally compiled together as a complete text to benefit the readers across all nations. To ensure the quality of the content we instilled two significant steps in our procedure. The first was to appoint an editorial team that would verify the data and statistics provided in the book and also select the most appropriate and valuable contributions from the plentiful contributions we received from authors worldwide. The next step was to appoint an expert of the topic as the Editor-in-Chief, who would head the project and finally make the necessary amendments and modifications to make the text reader-friendly. I was then commissioned to examine all the material to present the topics in the most comprehensible and productive format.

I would like to take this opportunity to thank all the contributing authors who were supportive enough to contribute their time and knowledge to this project. I also wish to convey my regards to my family who have been extremely supportive during the entire project.

Editor

1

Cytocompatible and multifunctional polymeric nanoparticles for transportation of bioactive molecules into and within cells

Kazuhiko Ishihara[a,b], Weixin Chen[a], Yihua Liu[b], Yuriko Tsukamoto[a] and Yuuki Inoue[a]

[a]Department of Materials Engineering, School of Engineering, The University of Tokyo, Tokyo, Japan
[b]Department of Bioengineering, School of Engineering, The University of Tokyo, Tokyo, Japan

ABSTRACT
Multifunctional polymeric nanoparticles are materials with great potential for a wide range of biomedical applications. For progression in this area of research, unfavorable interactions of these nanoparticles with proteins and cells must be avoided in biological environments, for example, through treatment of the nanoparticle surfaces. Construction of an artificial cell membrane structure based on polymers bearing the zwitterionic phosphorylcholine group can prevent biological reactions at the surface effectively. In addition, certain bioactive molecules can be immobilized on the surface of the polymer to generate enough affinity to capture target biomolecules. Furthermore, entrapment of inorganic nanoparticles inside polymeric matrices enhances the nanoparticle functionality significantly. This review summarizes the preparation and characterization of cytocompatible and multifunctional polymeric nanoparticles; it analyzes the efficiency of their fluorescence function, the nature of the artificial cell membrane structure, and their performance as in-cell devices; and finally, it evaluates both their chemical reactivity and effects in cells.

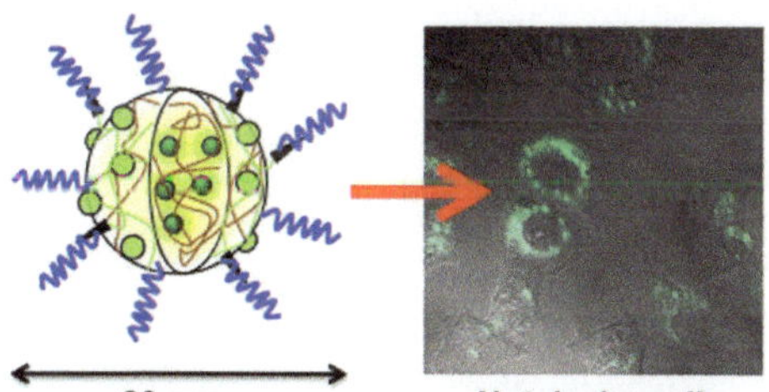

Cells can uptake polymer nanoparticles, embedding quantum dots covered with cytocompatible phospholipid polymers and cell-penetrating peptides.

KEYWORDS
Nanomedicine molecular science; zwitterionic polymer; fluorescence property; cytocompatibility; surface modification; in-cell device

CLASSIFICATION
30 Bio-inspired and biomedical materials; 600 Biomaterials, Nano materials

1. Introduction

Nanoparticles have remarkable potential and find broad applications industrially as well as academically due to their various functionalities.[1–6] They are obtained from metals, metal oxides, metallic alloy, and polymers and they show unique physical and chemical characteristics compared with those of bulk materials. Nanoparticles can be dispersed in solvents to provide stable suspensions due to interactions between the nanoparticle surfaces and the surrounding solvent molecules. The large surface-to-volume ratio of nanoparticles induces high efficiency when applied as catalysts. Depending on the size of the nanoparticles, certain metal nanoparticles will display surface plasmon resonance and magnetic nanoparticles will display superparamagnetism. Moreover, assemblies of nanoparticles provide unique structures with various functions.[7–13] The nanometer scale of these structures makes them suitable for analysis of microscopic areas, such as targeted intracellular regions. Nanoparticles are actively used in pharmaceutical and biomedical applications as carriers for biomolecules to facilitate efficient delivery into cells and tissues.

However, the use of nanoparticles in biological systems is often limited by their specific structures. Nanoparticles possess high-energy surfaces and can therefore lead to aggregation of colloidal particles by

CONTACT Kazuhiko Ishihara ishihara@mpc.t.u-tokyo.ac.jp

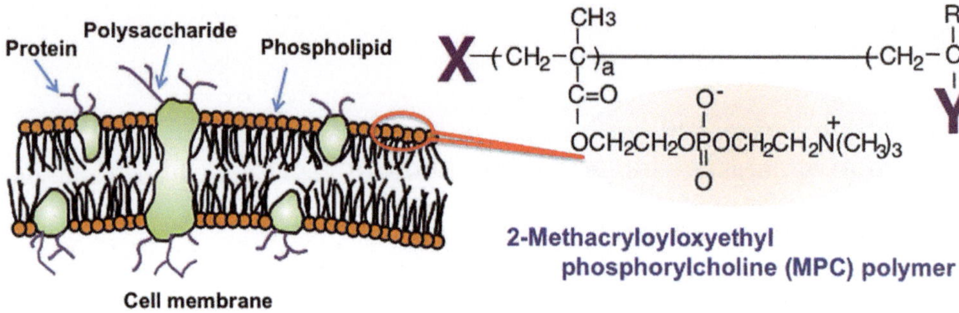

Figure 1. Cell membrane-inspired molecular design and chemical structure of MPC polymer.

attractive van der Waals forces or electrostatic interactions. Random adsorption of proteins on nanoparticle surfaces reduces the functionality according to the size effect. In a cell culture medium, nanoparticles are adsorbed onto the cell membrane and are internalized by the cell through endocytosis. Under biological conditions, regardless of the nature of the application, nanoparticles occasionally induce unexpected cellular responses with disrupted functionalities because of nonspecific protein adsorption and the corresponding biological responses. To overcome these problems, nanoparticle surfaces are engineered to render them more biocompatible and bioinert.[14–19]

Several methodologies are available to obtain bioinert surfaces on nanoparticles, including surface modification with natural macromolecules [20–23] or water-soluble synthetic polymers. Water-soluble synthetic polymers, such as poly(ethylene glycol) (PEG) and its derivatives, are used for grafting or coating the nanoparticle surface.[24,25] These polymer chains expand into the aqueous medium, generating a highly mobile hydrated layer that surrounds the nanoparticles. The steric hindrance generated by the polymer chains disturbs aggregation of the nanoparticles and prevents undesirable biological reactions. Recently, however, some articles have reported accelerated blood clearance of PEG-modified nanocarriers, and immunoresponses such as production of immunoglobulin M by PEG. Thus, it is necessary to note this point when PEG is used as a surface modification reagent for nanoparticles.[26–28]

Other methodologies have been investigated for surface modifications using phospholipid derivatives such as phosphatidylcholine derivatives. Although they are used as pharmaceutical carriers of bioactive reagents in the bloodstream, their stability is insufficient to enable long-term circulation in the bloodstream. Hence, a polymerization technique is a promising method to improve the stability of the phospholipid assembly. Phospholipid derivatives with polymerizable groups have been synthesized.[29–32] One of the most effective of these is 2-methacryloyloxyethyl phosphorylcholine (MPC), which possesses both a methacrylate group and a monomer, and accordingly, various kinds of MPC polymers have been used to synthesize the zwitterionic phosphorylcholine group.[33–35] MPC is polymerized easily by radical polymerization, even in the presence of other monomers (Figure 1). The MPC polymers are hydrophilic and are electrically neutral due to their unique zwitterionic structure. They have good stability under biological circumstances, namely, they are biologically inert at pH 7.4 and under high ionic strength. They have excellent biocompatibility, specifically resistance against protein adsorption, cell adhesion, and tissue immunoreactions.[36–42] Thus, based on these fundamental characteristics, selected MPC polymers are used as surface modification materials in medical devices, such as implantable artificial hearts and artificial hip joints.[43–48]

Considering the functional properties of MPC polymers and their usefulness in providing a bioinert platform on the nanoparticle surface, the present review focuses on elucidating precisely designed MPC polymers and adding insight into the recent promising results related to the fabrication of nanoparticles for in-cell imaging.

2. Surface modification of nanoparticles with MPC polymers

Substantial research data exists related to modification of nanoparticle surfaces by MPC polymers. Table 1 summarizes the various nanoparticle systems using MPC polymers. In such preparations, the components used as core materials include biologically active compounds, inorganic nanoparticles, and polymer nanoparticles. The nanoparticle diameters range from a few nanometers to a few hundred nanometers. The amphiphilic nature of MPC polymers induces formation of polymer aggregates in an aqueous medium, along with the generation of a hydrophobic region, thereby facilitating entrapment of hydrophobic bioactive compounds including anti-cancer compounds.

Table 1. List of nanoparticles covered with MPC polymers

Core material	MPC polymer	Diameter (nm)	Functions	References
Bioactive molecules				
Pacritaxel	PMBN-EGF	50–75	Anti-tumor effect	49
Pacritaxel	PMBN-per S1	50	Anti-tumor effect	50
Doxorubicin	P(MPC-*co*-EtOEMA)	—	Anti-tumor effect	51
Griseofulvin	PMB30	20–50	Pharmaceutical activity	52
Tranilast	PMB30	*c.*100	Solid dispersion	53
Cyclosporine A	PMB30	*c.*180	Solid dispersion	54
Plasmid DNA	PMB50	190 ± 3	Transfection effect	55
DNA/β-CD complex	PMPC-adamantyl	110–150	Gene carrier	56
DNA	PMD-folic acid	100–165	Gene carrier	57
CHP	PMPC	59.7 ± 27.4	Self-aggregation	58
Inorganic nanoparticles				
$CaCO_3$, $BaCO_3$	PMPC	1,000–3000	Biomineralization	59
Graphene oxide	PMPC	500–1500	Biomedical application	60
Magnetic particle	PMPC	*c.*400	Immuno-reaction	61
Fe_3O_4 nanoparticle	PMPC	18.7 ± 1.5	Contrast agent of MRI	62
Fe_3O_4 nanoparticle	PMPC	8.9 ± 0.1	Contrast agent of MRI	63
Fe_3O_4/SiO_2	P(MPC-*co*-AAm)	630–670	Molecular imprinting	64
SiO_2	PMPC	12	Stabilizer	65
Mesoporous SiO_2	PMPC	<1100	In cell carrier	66
TiO_2 nanoparticle	PMPC	100	Suspension in aqueous media	67
Ag nanoparticle	P(MPC-*co*-HEMA)	130–160	Thin film formation	68
Ag nanoparticle	P(MPC-*co*-HEMA)	100–500 (Film)	Anti-bacterial activity	69
Au nanoparticles	PMPC	100–300	Protein detection	70
Au nanorod	P(MPC-*co*-DHLA)	17.8 ± 1.8 × 53.1 ± 3.5	Imaging in cell	71
Pd nanoparticle	PMPC-*block*-PDiPAMA	*c.*20–30	Catalyst for coupling	72
CdSe/ZnS QD	PMPC	12	Imaging in cell	73
CdSe/ZnS QD	PDbNbM	20–30	Tracking in cell	74
Polymer nanoparticles				
Polystyrene	$PMPC-NH_2$	333–611	Template for Au nanoparticle	75
Poly(lactic acid)	PMBN	250–300	Immobilization of IgG	76
Poly(lactic acid)	PMB30	400–500	Cell-based assay	77
Poly(lactic acid)	PMB-PL	127	Photoinduced release of protein	78
Poly(amido amine)	PMPC	115–156	Dendorimer dispersion	79
Poly(BMA)	PMPC	100	Near-IR imaging	80

PMPC: poly(MPC), PMB50: water-soluble poly(MPC-*co*-BMA) MPC unit content 50 mol%, PMBN: poly(MPC-*co*-BMA-*co*-MEONP), CHP: cholesteryl groups-bearing pullulan, P(MPC-*co*-HEMA): poly(MPC-*co*-hydroxyethyl methacrylate), P(MPC-*co*-DHLA): poly(MPC-*co*-methacryloyloxyethyl dihydrolipoic acid), P(MPC-*co*-AAm): poly(MPC-*co*-acrylamide), P(MPC-*co*-EtOEMA): poly(MPC-*co*-2-ethoxy-2-oxyethyl methacrylate), PMB30: water-soluble poly(MPC-*co*-BMA) MPC unit content 30 mol%, molecular weight 5.0×10^4, PMD: poly(MPC-*co*-N,N-dimethylaminoethyl methacrylate (DMAEMA)), PDbNbM: poly (DMAEMA)-*block*-poly(MEONP)-*block*-poly(MPC), PMB-PL: poly(MPC-*co*- BMA-*co*-4-(4-(1-methacryloyloxyethyl)-2-methoxy-5-nitrophenoxy)butyric acid), PMPC-*block*-PDiPAMA: poly(MPC-*block*-2-(N,N-diisopropylamino)ethyl methacrylate).

[81,82] MPC polymer aggregates are stable when injected into the bloodstream and concentrate in cancer tissue, triggering a dramatic anti-cancer effect. This delivery system is ideal for transporting poorly water-soluble anti-cancer compounds into living systems.[49–52] Additionally, inorganic nanoparticles can largely be dispersed in an aqueous medium by surface modification with water-soluble MPC polymers. Contemporary progress in the field of polymer synthesis has established the surface-initiated living radical polymerization procedure, which could be applied to MPC associated surface modifications of nanoparticles.[83,84] During the polymerization process, the living radical initiator is immobilized onto the surface of inorganic nanoparticles and living radical polymerization is carried out to generate a poly(MPC) grafted layer.[60,63,66,70,73] The poly(MPC)-grafted chains have nearly the same degree of polymerization and hence the same chain length. This indicates that the diameter distribution of the poly(MPC)-grafted nanoparticles depends on the original core nanoparticles, and therefore uniform functionalities are expected. In addition, modified nanoparticles could efficiently disperse in an aqueous medium.

3. Surface modification of quantum dots with MPC polymers

Semiconductor nanoparticles such as quantum dots (QDs) have received strong interest as a promising material in biological imaging research, and can serve as alternatives to organic fluorescent dyes.[85–89] QDs are typically between 2 and 6 nm and have unique optical properties depending on the size, such as fluorescent multi-color emission spectra, high quantum yields, and excellent resistance to photobleaching by excitation light. These optical properties are suitable in most fluorescence applications, particularly for long-term monitoring of labeled substances, an area in which QDs have a singular advantage over conventional fluorescent organic dyes.

In vitro biological applications of QDs include cell labeling and tracking cell migration, and *in vivo* they are applied as contrast agents in tumor-tissue sections.[90–93] The QD surface is generally covered by trioctylphosphine oxide (TOPO) and *n*-octylamine derivatives to form a stable hydrophobic layer.[94] QDs are difficult to disperse in aqueous medium themselves and sometimes they induce serious damage to the target cells and tissues. Thus, the surfaces of QDs must be modified

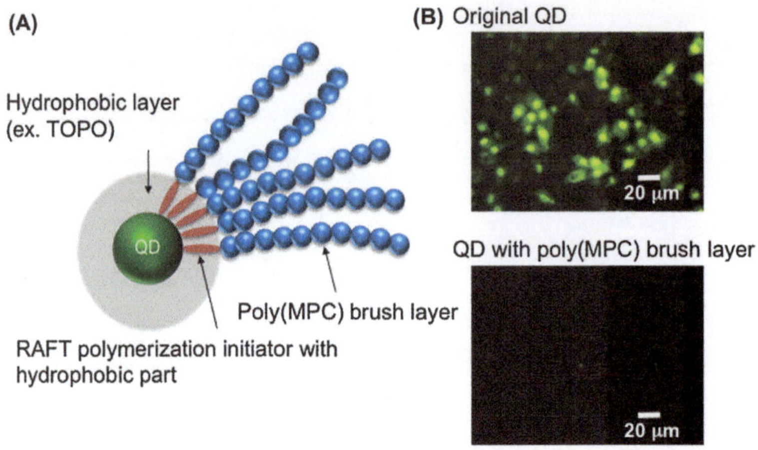

Figure 2. Schematic representation of QD modified with poly(MPC) (A) and fluorescence microscopy image of QD internalization into cells (B).

with hydrophilic or water-soluble compounds to achieve stable and highly sensitive bioimaging without cytotoxicity in cell culture medium.

Preparation of QDs modified with the MPC polymer was reported initially by Matsuno and coworkers [73]. They successfully grafted poly(MPC) chains onto QD (CdSe/ZnS) surfaces using reversible addition-fragmentation chain transfer (RAFT) polymerization (Figure 2). In this process, an amphiphilic RAFT polymerization initiator forms a micelle-like structure in an aqueous solution and solubilizes the QDs that are covered with a TOPO layer. Subsequently, at the surfaces of QDs treated with the RAFT polymerization initiator, surface-initiated RAFT polymerization of MPC can occur under normal RAFT polymerization conditions. The poly(MPC) chains formed are then immobilized stably on the surface. The QDs modified with poly(MPC) chains show good cytocompatibility and are able to inhibit uptake by HeLa cells, although the QDs modified with poly(MPC) chains have very small diameters (~12 nm).

Liu et al. [74] reported the preparation of a fluorescent probe for evaluating pH conditions in cells, which showed a combination of pH response and changes in fluorescence spectra (Figure 3). The nanoparticles had a core-shell type structure with a QD core and a block-type water-soluble polymer composed of a poly(*N*,*N*-diethylaminoethyl methacrylate (DEAEMA)) segment, a poly(ω-(*p*-nitrophenyloxycarbonyl oligo(ethylene glycol)) methacrylate (MEONP)) segment, and a poly(MPC) segment as the shell. The poly(DEAEMA) segment could induce a stretch-shrink transformation with a change in the pH of the medium due to protonation of the DEAEMA units. An organic fluorescent dye, Alexa 594 cadaverine, was immobilized in the poly(MEONP) segment in the block-type polymer. The nanoparticles could disperse well in an aqueous medium and fluorescence resonance energy transfer (FRET) between the QD core and fluorescent dye was observed in the media at pH values of 7.4 and 5.0. This change in pH corresponded to an endocytosis process known as the proton sponge effect. The fluorescence spectrum was significantly altered between pH 7.4 and 5.0 because the distance between the QDs and fluorescent dye changed according to the pKa (7.1) of the pH-responsive poly(DEAEMA) segment. Thus, when the distance between the QDs and fluorescent dye was within several nanometers at pH 7.4, FRET was induced from the QDs (the donor) to the fluorescent dye (the acceptor). This produced an increase in the fluorescence intensity of the red fluorescent dye. Alternatively, when the distance increased at pH 5.0 due to protonation of the polymer DMAEMA units, independent fluorescence of the QDs and fluorescent dye were observed. To enhance the cellular uptake of the MPC polymer-covered QDs, one of the cell-penetrating peptides (CPPs),[95,96] octaarginine (R8), was immobilized at the terminal of the poly(MPC) segment. After addition to the cell culture medium followed by 35 min of incubation, the nanoparticles were internalized into the cells by endocytosis, and fluorescence from both the QDs and fluorescent dye was observed due to the formation of FRET. However, after 70 min of incubation, the fluorescence from the fluorescent dye disappeared in response to the decrease in the pH value. Henceforth, after 105 min of incubation, the fluorescence from the fluorescent dye was recovered owing to escape of the nanoparticles from the endosomes. These nanoparticles find a promising application to monitor the live transportation process of molecules into cells, along with the possibility to track their pathway inside the cells continuously.

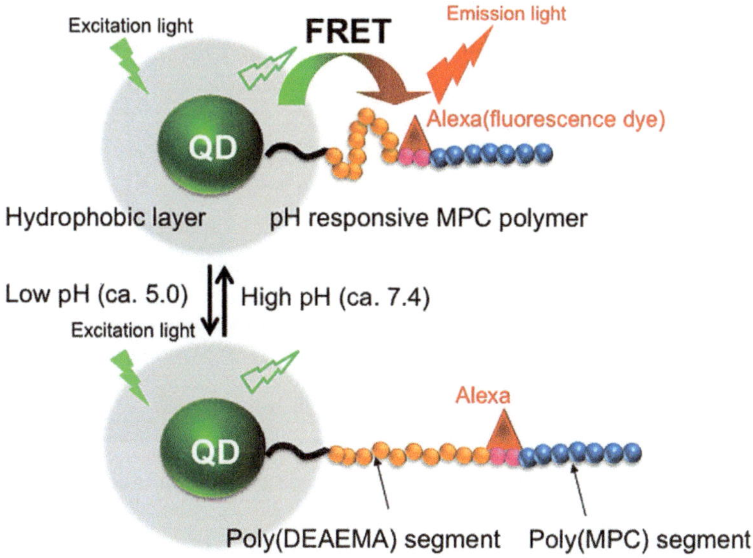

Figure 3. Schematic representation of QD modified with pH-responsive triblock-type MPC polymer and its FRET generation in response to pH change.

4. Entrapment of QDs in polymeric nanoparticles covered with MPC polymers

Previously, we prepared polymeric nanoparticles covered with MPC polymers in an aqueous solution using a simple solvent evaporation method, wherein amphiphilic MPC polymers were used as a suspension stabilizer. A hydrophobic polymer dissolved in an immiscible organic solvent was then added to the MPC polymer solution. Dichloromethane was used as the solvent due to its low boiling point and the hydrophobic polymers used were poly(L-lactic acid) (PLA) and polystyrene (PSt). The fabrication of polymer nanoparticles containing QDs was carried out using the solvent evaporation technique, which was simple and inexpensive (Figure 4).[76–78,97–100] Hydrophobic interactions among the materials play an important role in the mechanism by which the amphiphilic MPC polymers, PLA, and QDs form nanoparticles. The core of the nanoparticles composed of PLA and QDs, which are insoluble in water, was solubilized in dichloromethane. During the solvent evaporation process, evaporation of dichloromethane resulted in the precipitation of hydrophobic PLA chains with QDs at the interface of the aqueous medium. Additionally, at this interface, the MPC polymer chains formed entanglements with PLA chains, resulting in the formation of a stable MPC polymer layer on the surface of the PLA core.

Instead of MPC polymer, the use of poly(MPC-*co*-*n*-butyl methacrylate (BMA)-*co*-NEONP) (PMBN) was analyzed.[97,101] Dissolution of PMBN in aqueous medium indicated an amphiphilic nature and, due to the presence of hydrophobic BMA units, tended to form polymer aggregates at a PMBN concentration of 0.10 g dl^{-1}. Dichloroethylene droplets containing PLA and QDs were dispersed in the PMBN solution by sonication. During the preparation process, the sonication time was controlled to maintain the fluorescence intensity of the QDs. The PMBN/PLA/QD nanoparticles showed good dispersibility in aqueous medium, such as phosphate buffered saline and cell culture medium. The ultraviolet absorption and emission fluorescence spectra of QDs in the polymer nanoparticles were the same as those of commercially available QDs suspended in toluene (Figure 5). It was confirmed that the entrapment of QDs provided good long-term stability with no change in either solubility or fluorescence performance, even beyond six months of storage at 4°C. The PMBN/PLA/QD nanoparticles functioned well between pH 4.0 and 9.0. Photobleaching examination revealed that PMBN/PLA/QD nanoparticles were more resistant to continuous excitation photoirradiation compared with an organic fluorescent dye, fluorescein isothiocyanate. The entrapment of QDs using PLA and PMBN had no influence on the optical properties of the QDs. The hydrodynamic diameters of the PMBN/PLA/QD nanoparticles were around 10–20 nm, as measured by dynamic light scattering (DLS), and the size distribution was found to be narrow. Atomic force microscopy (AFM) showed that the morphology of PMBN/PLA/QD nanoparticles was spherical. According to AFM observations, the size measurements were consistent with the hydrodynamic diameter determined by DLS measurements. Moreover, transmission electron microscopy (TEM) revealed several groups of QDs appearing as dark spherical objects, a result of the electron-dense QDs being entrapped within a single polymer nanoparticle (Figure 6). X-ray

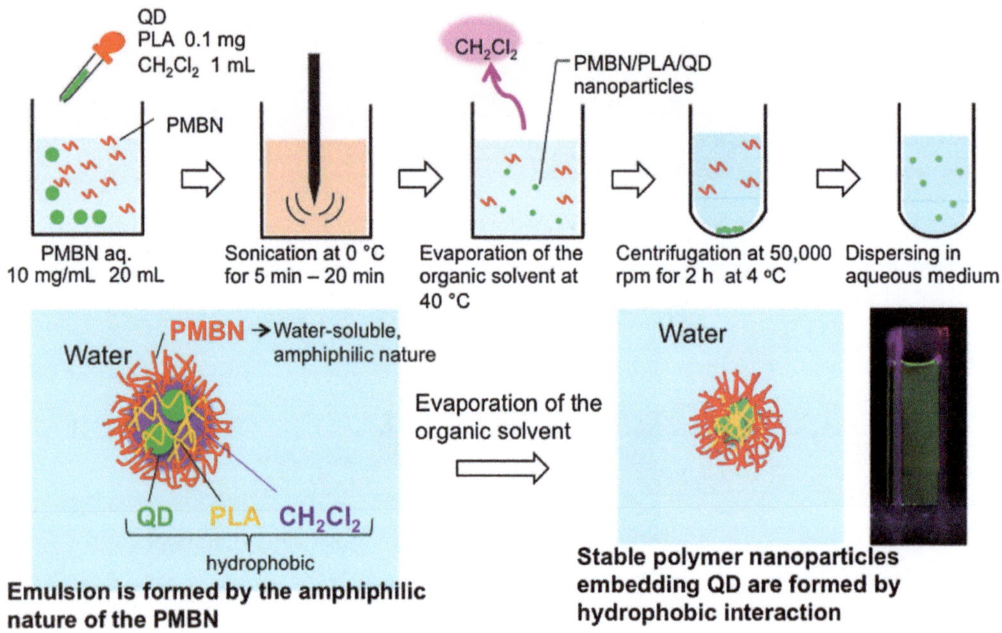

Figure 4. Schematic representation of the entrapment of QDs by PLA nanoparticles and coating with MPC polymers.

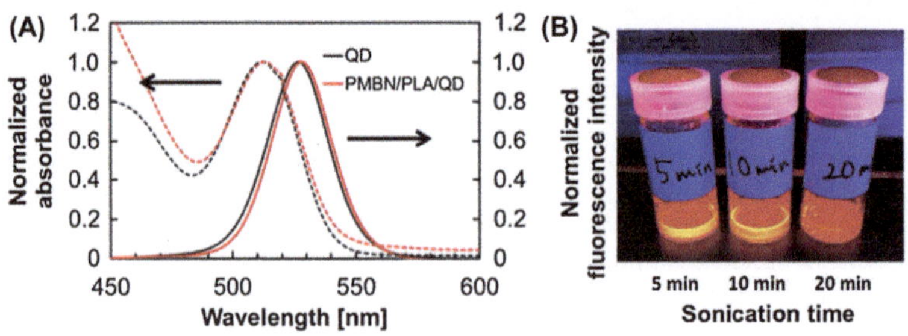

Figure 5. Fluorescence spectra of the QD and PMBN/PLA/QD nanoparticles (A), and fluorescence image of the PMBN/PLA/QD nanoparticles prepared with various sonication times (B).

photoelectron spectroscopy (XPS) analysis indicated that the PMBN/PLA/QD nanoparticles had specific XPS signals at 133 eV and 403 eV, which corresponded to the phosphorus atom in the phosphate group and the nitrogen atom in the ammonium group of the MPC unit, respectively. Thus, from these findings, the surface of PMBN/PLA/QD nanoparticles was covered with phosphorylcholine groups in the MPC polymer.

5. Transport of MPC polymer nanoparticles with biomolecules into cells

Immobilization of biomolecules on the PNBN/PLA/QD nanoparticles is important for successfully penetrating the cell membrane. R8, a well-known CPP,[95,96] and octaglycine (G8) were independently immobilized onto the surfaces of nanoparticles and applied to cell culture medium of cultured HeLa cells. As shown in Figure 7, G8-immobilized PMBN/PLA/QD nanoparticles (G8-PMBN/PLA/QD) were not internalized in the HeLa cells, even after incubation for 24 h. However, the R8-immobilized PMBN/PLA/QD nanoparticles (R8-PMBN/PLA/QD) were internalized effectively into the cells. The difference between these two nanoparticles indicated that the PMBN/PLA/QD nanoparticles without signal oligopeptides could provide the highest signal to noise ratio among all existing imaging probes, as there was no background fluorescence due to nonspecific uptake of imaging probes by the cells. This is due to the excellent cytocompatibility of the MPC polymers covering the nanoparticles.[36–42]

The internalization process of R8-PMBN/PLA/QD nanoparticles by the cells was examined. The R8-PMBN/PLA/QD nanoparticles adhered to the cell membrane within the first 5 min. Then, internalization of the

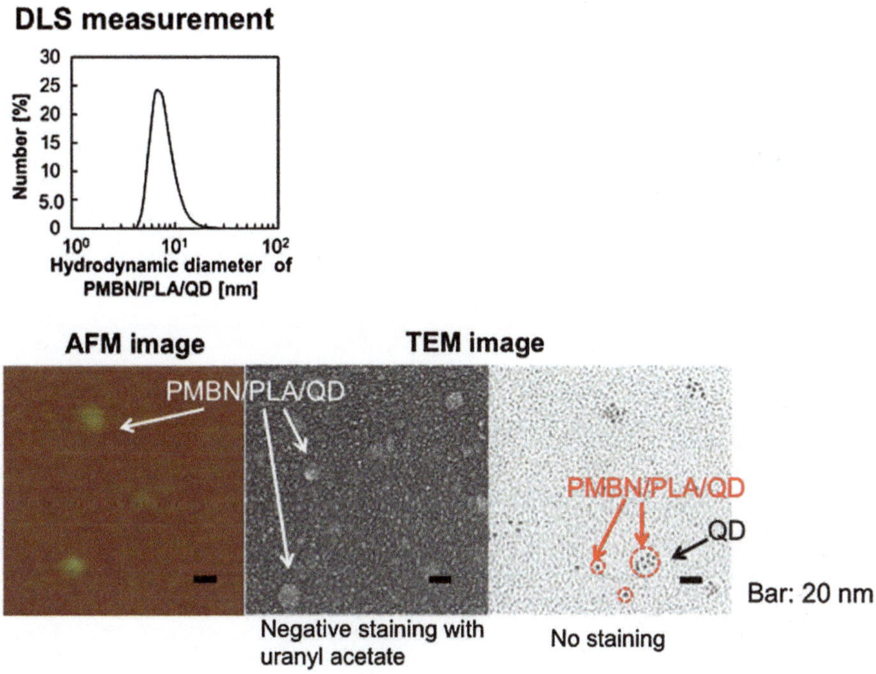

Figure 6. Morphological observation of PMBN/PLA/QD nanoparticles.

nanoparticles into endosomes was initiated over the next 15–30 min. The amount of internalized R8-PMBN/PLA/QD nanoparticles increased from 1–3 h. All R8-PMBN/PLA/QD nanoparticles adhered to the cell membrane were internalized by the endosomes within 5 h. These results provided the kinetic information of the R8-mediated internalization of nanoparticles into the cell.

Cytotoxicity analysis indicated that PMBN/PLA/QD nanoparticles had no cytotoxicity three days post internalization into the HeLa cells. However, another possibility was induction of an inflammatory response. Further studies were required with respect to the inflammatory responses. This was analyzed by detecting the relative expression of TNF-α mRNA to GAPDH mRNA in RAW264.7 cells after incubation with R8-PMBN/PLA/QD nanoparticles or G8-PMBN/PLA/QD nanoparticles for a day. No significant difference was observed. This suggested that the MPC polymers on the surface of the PMBN/PLA/QD nanoparticles suppressed the inflammatory response in RAW264.7 cells during the CPP-mediated internalization into cells. These findings clearly showed that any undesirable interactions between PMBN/PLA/QD nanoparticles and the cells, such as non-specific uptake, cytotoxic effects, or induced inflammatory responses, were eliminated.

Arginine-rich peptides are typically used as CPPs. The effects of other oligopeptides on permeation through the cell membrane are not well understood. We evaluated the role of various octapeptides immobilized on the PMBN/PLA/QD nanoparticles (Figure 8).[102] The selected octapeptide had simple sequences of only one kind of amino acid, hydrophobic and neutral tyrosine (Y8), hydrophilic and neutral asparagine (N8), anionic glutamic acid (E8), weakly cationic histidine (H8), cationic lysine (K8), and mostly hydrophilic and cationic arginine (R8). Of the octapeptides tested, only K8- and R8-immobilized PMBN/PLA/QD nanoparticles could be internalized into HeLa cells. These findings were supported by the report of another research group arriving at the same result that R8 and K8 octapeptides could function as CPPs. These results suggest that molecules functioning as CPPs should possess hydrophilic and electron donating properties.

To understand the effect of the sequence of octapeptides composed of inert and active amino acid residues for the process of internalization into cells, the following series of octapeptides were prepared using glycine (G) and arginine (R) and analyzed: G8, GGGGGGRR (G6R2), GGGGRRRR (G4R4), GGRRRRRR (G2R6), and R8. Figure 9 represents images of cells examined by fluorescence microscopy. Addition of the respective octapeptide-coated nanoparticles to cell culture medium initiated cellular contact between the nanoparticles and the cultured cells. The results indicated that only the R8-immobilized nanoparticles could be internalized into cells. Next, the effect of the surface density of R8-mediated internalization of nanoparticles into cells was evaluated. To achieve this, the ratio

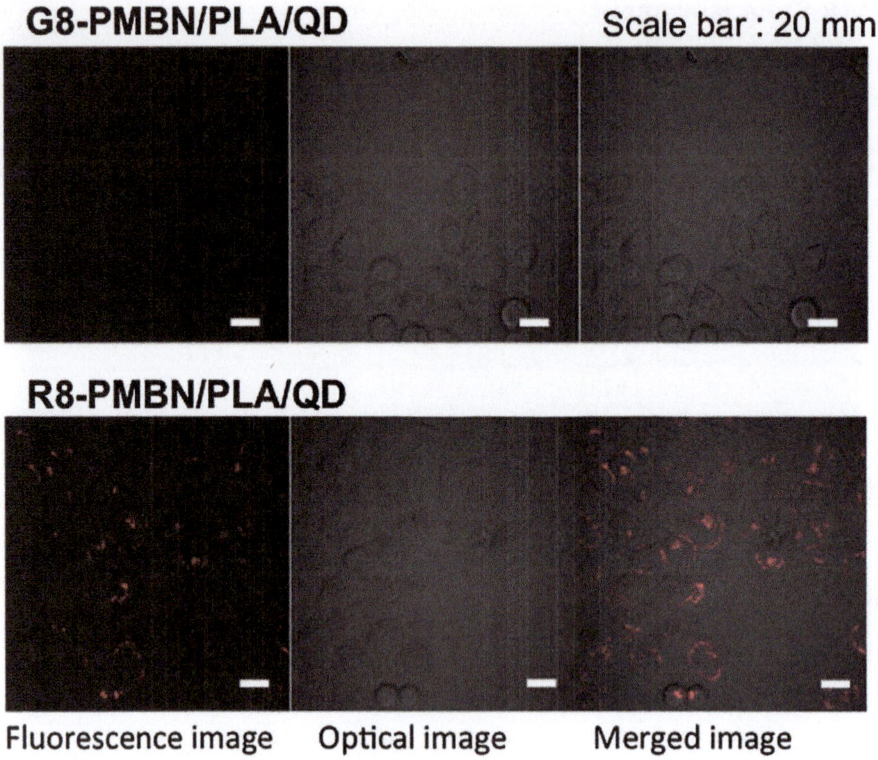

Figure 7. Uptake of the octapeptide-PMBN/PLA/QD nanoparticles by HeLa cells.

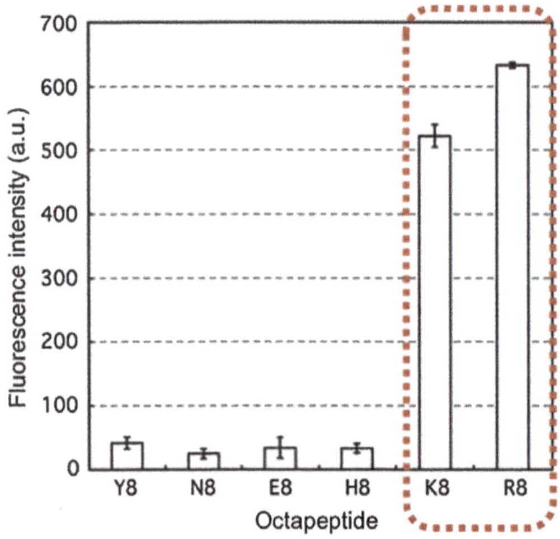

Amino acid		Hydrop hobic Index	PI	Charge
Arginine	R	-4.5	10.76	Cationic
Lysine	K	-3.9	9.74	Cationic
Histidine	H	-3.2	7.59	weakly cationic
Asparagine	N	-3.5	5.41	Neutral
Glutamic acid	E	-3.5	3.22	Anionic
Tyrosine	Y	-1.3	5.66	Neutral

Figure 8. Effect of the chemical structure of octapeptide immobilized on PMBN/PLA/QD nanoparticles on cellular internalization. (Reproduced from [102] with copyright permission, Wiley-VCH)

of the concentration of R8 to G8 in the feed solution during immobilization was altered to control the surface density of R8. The results of the internalization of the nanoparticles are shown in Figure 10. Although the surface ζ-potential of G8-PNBN/PLA/QD nanoparticles was slightly negative (−11 mV), it increased correspondingly with an increase in the R8 fraction until a ratio of 0.4. However, up to the R8/G8 ratio of 0.4, the surface ζ-potential was constant around 0 mV. On the other hand, the amount of nanoparticles internalized into the cells increased around the R8/G8 ratio of 0.4 and a significant increase was observed up to the R8/G8 ratio of 0.6. These results were interesting, as they suggested that surface charge was not an essential

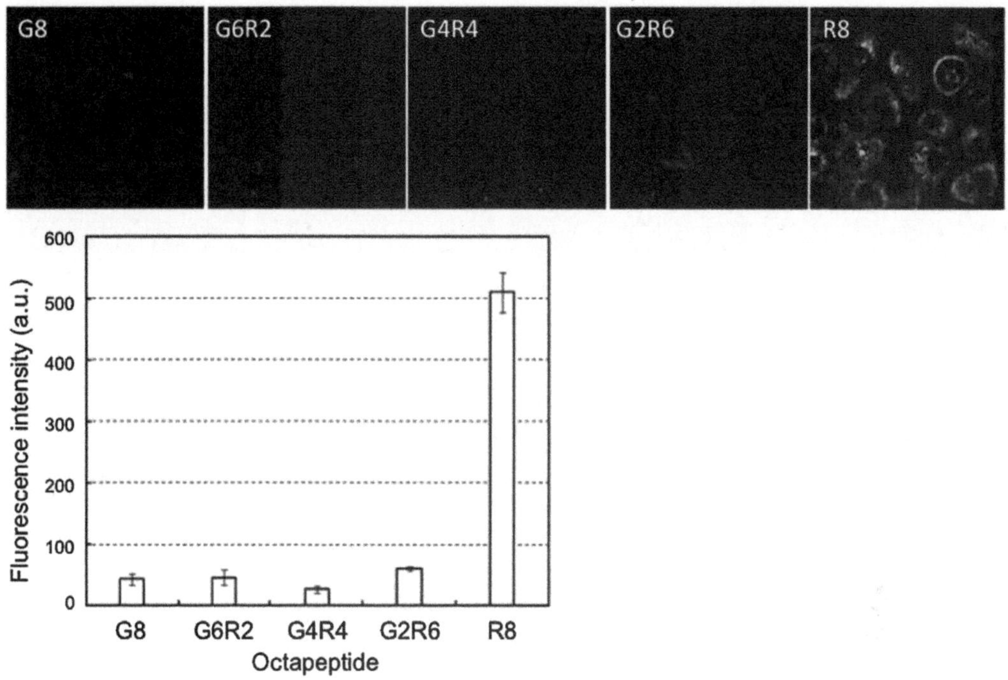

Figure 9. Effect of the sequence of octapeptide immobilized on PMBN/PLA/QD nanoparticles on cellular internalization. Internalization of nanoparticles was evaluated by fluorescence microscopy (upper panel), and the fluorescence intensity was determined for each nanoparticles (lower panel). (Reproduced from [101].)

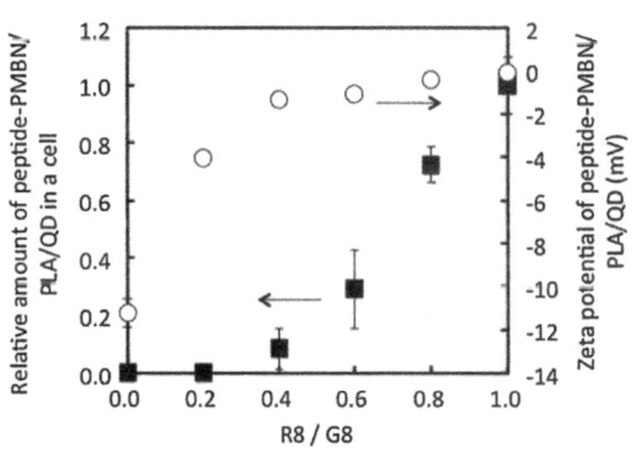

Figure 10. Effect of composition of R8 and G8 immobilized on PMBN/PLA/QD nanoparticles on cellular internalization.

factor for internalization of nanoparticles into the cells. This opens up the scope for further research and consideration to analyze possible molecular interactions between the oligopeptide unit and the cell membrane surface. PMBN/PLA/QD nanoparticles not only inhibit nonspecific cellular uptake by mammalian cells, but also provide insight about specific interactions between biomolecules and cells when bioactive molecules like R8 are immobilized on them. From these results, it is concluded that R8-PMBN/PLA/QD nanoparticles are a promising probe for high sensitivity and non-inflictive cell and tissue imaging technology.

6. Behavior of R8-immobilized MPC polymer nanoparticles *in vivo*

Observation of the dynamics of nanoparticles in cells is important for understanding the typical endocytosis pathways of internalization. The R8-PMBN/PLA/QD nanoparticles associated with HeLa cells significantly.[97,102] No change in cell proliferation was observed compared with proliferation of the original cells for more than 30 h of cell culture. The distribution of nanoparticles was observed by fluorescence microscopy during cell proliferation as shown in Figure 11. During the proliferation process, the R8-PMBN/PLA/QD nanoparticles in the cells were distributed to two daughter cells by cell proliferation, and the amount of R8-PMBN/PLA/QD nanoparticles in each cell decreased. Correspondingly, the fluorescence intensity of the R8-PMBN/PLA/QD nanoparticles in all the cells did not change for about 30 h.[102] This was the first report indicating long-term retention of nanoparticles inside cells.

7. Future perspectives

Useful applications of nanobioengineering would not be achievable without the suitable design of nanostructures for materials and devices. Continued progress in biomedical nanotechnology requires additional systematic approaches for the chemical and physical design of

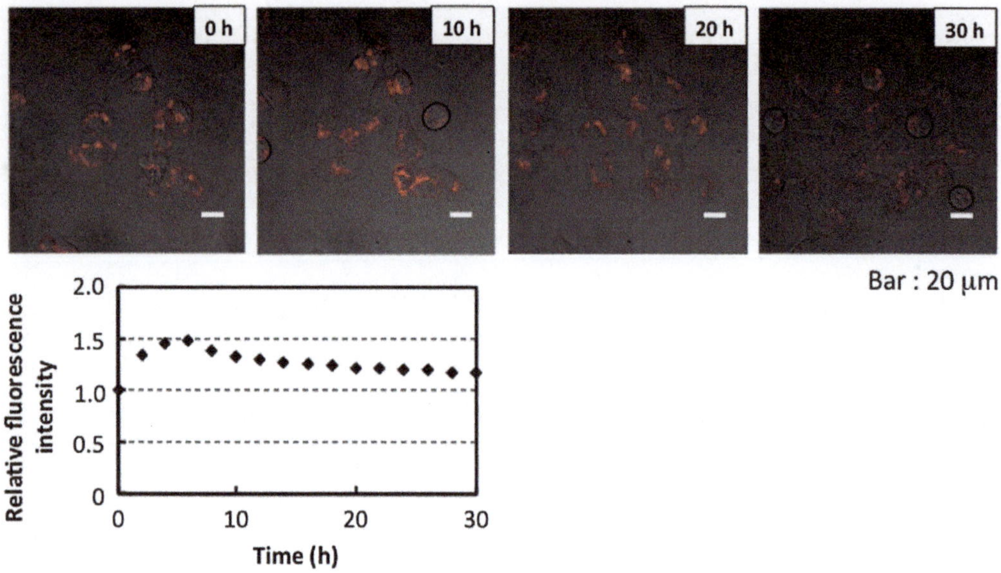

Figure 11. Distribution of R8-PMBN/PLA/QD nanoparticles in the cells during cell proliferation (upper panel) and relative fluorescence intensity of total microscope area (lower panel). (Reproduced from [102].)

nanostructures. Moreover, the interface between the nanostructure and biological environments becomes critical as researchers start to construct multifunctional nanostructures.

The design of the chemical structure of MPC polymers was strongly inspired by natural phospholipids at the cell membrane surface. This feedback from biological systems represents a rational strategy for obtaining new polymeric biomaterials. The development of MPC polymers did not have limitations in the induction of cell and tissue responses in biological systems. Thus, many kinds of MPC polymers have been designed and synthesized with flexibility for use in several applications. The PMBN described in this review is a useful polymer for constructing artificial cell membrane structures. By combining specific biomolecules on the PMBN surface, it can be used as a delivery system for biomolecules, as well as for the development of diagnostic and bioimaging probes. MPC units that are concentrated on the surface of nanodevices provide a high shielding effect against protein adsorption. The effect of MPC units does not interrupt the reactivity of immobilized biomolecules on the surfaces. The conjugation of biomolecules on the MPC polymer surface with a desirable arrangement is essential in achieving an excellent biointerface. The artificial cell membrane structure constructed by the MPC polymer and biomolecules can be applied for the preparation of in-cell nanodevices with the ability to completely control the interactions between nanodevices and cells. These facts underline the importance of biointerface technology in understanding the fundamentals of biomolecular behavior in cellular environments.

Acknowledgements

The authors thank Mr Yusuke Goto and Ms Kaori Hayashi, Department of Materials Engineering, The University of Tokyo for their assistance in preparation of this manuscript. This research was performed through a Grant-in-Aid for Scientific Research on Innovative Areas "Nanomedicine Molecular Science" [number 2306] from the Ministry of Education, Culture, Sports, Science, and Technology of Japan.

Disclosure statement

No potential conflict of interest was reported by the authors.

References

[1] Kim JW, Cho J, Cho J, et al. Synthesis of monodisperse bi-compartmentalized amphiphilic janus microparticles for tailored assembly at the oil–water interface Angew. Chem Int Ed. 2016;55:4509–4513.
[2] Abe H, Liu J, Ariga K. Catalytic nanoarchitectonics for environmentally compatible energy generation. Mater Today. 2016;19(1):12–18.
[3] Shin T-H, Choi Y, Kim S, et al. Recent advances in magnetic nanoparticle-based multi-modal imaging. Chem Soc Rev. 2015;44:4501–4516.
[4] Wang H, Liu S, Zhang Y-L, et al. Controllable assembly of silver nanoparticles induced by femtosecond laser direct writing. Sci Technol Adv Mater. 2015;16 (2015) 024805 (7pp)
[5] Sheng Y, De Liao LD, Thakor NV, et al. Nanoparticles for molecular imaging. J Biomed Nanotechnol. 2014;10(10):2641–2676.
[6] Tang Y, Cheng W. Key parameters governing metallic nanoparticle electrocatalysis. Nanoscale. 2015;7:16151–16164.

[7] Wu W, Wu Z, Yu T, et al. Recent progress on magnetic iron oxide nanoparticles: synthesis, surface functional strategies and biomedical applications. Sci Technol Adv Mater. 2015; 16. 023501(43pp)

[8] Kim S-J, Xu W, Ahmad MD, et al. Synthesis of nanoparticle CT contrast agents: *in vitro* and *in vivo* studies. Sci Technol Adv Mater. 2015;16. 055003 (9pp).

[9] Zeth K, Hoiczyk E, Okuda M. Ferroxidase-mediated iron oxide biomineralization: novel pathways to multifunctional nanoparticles. Trends Biochem. Sci. 2016;41:190–203.

[10] Wang Z, Yu J, Gui R, et al. Carbon nanomaterials-based electrochemical aptasensors. Biosens Bioelectron. 2016;79:136–149.

[11] Kandasamy G, Maity D. Recent advances in superparamagnetic iron oxide nanoparticles (SPIONs) for *in vitro* and *in vivo* cancer nanotheranostics. Int J Pharm. 2015;496(2):191–218.

[12] Samiei M, Farjami A, Dizaj SM, et al. Nanoparticles for antimicrobial purposes in Endodontics: A systematic review of *in vitro* studies. Mater Sci Eng C Mater Biol. 2016;58:1269–1278.

[13] Passos ML, Pinto PC, Santos JL, et al. Nanoparticle-based assays in automated flow systems: a review. Anal Chim Acta. 2015;889:22–34.

[14] Mou X, Ali Z, Li S, et al. Applications of magnetic nanoparticles in targeted drug delivery system. J Nanosci Nanotechnol. 2015;15(1):54–62.

[15] Mekaru H, Lu J, Tamanoi F. Development of mesoporous silica-based nanoparticles with controlled release capability for cancer therapy. Adv Drug Deliv Rev. 2015;95:40–49.

[16] Loos C, Syrovets T, Musyanovych A, et al. Functionalized polystyrene nanoparticles as a platform for studying bio-nano interactions. Beilstein J Nanotechnol. 2014;5:2403–2412.

[17] Pissuwan D, Niidome T. Polyelectrolyte-coated gold nanorods and their biomedical applications. Nanoscale. 2015;7(1):59–65.

[18] Bernsen MR, Guenoun J, van Tiel ST, et al. Nanoparticles and clinically applicable cell tracking. Br J Radiol. 2015;88(1054):20150375.

[19] Soenen SJ, Parak WJ, Rejman J, et al. (Intra)Cellular stability of inorganic nanoparticles: effects on cytotoxicity, particle functionality, and biomedical applications. Chem Rev. 2015;115(5):2109–2135.

[20] Doh KO, Yeo Y. Application of polysaccharides for surface modification of nanomedicines. Ther Deliv. 2012;3(12):1447–1456.

[21] Yang H-M, Park C-W, Ahn T, et al. A direct surface modification of iron oxide nanoparticles with various poly(amino acid)s for use as magnetic resonance probes. J Colloid Interface Sci. 2013;391:158–167.

[22] Huynh R, Chaubet F, Jozefonvicz J. Anticoagulant properties of dextranmethylcarboxylate benzylamide sulfate (DMCBSu); a new generation of bioactive functionalized dextran. Carbohydrate Res. 2001;332:75–83.

[23] Lemarchand C, Gref R, Couvreur P. Polycaccharide-decorated nanoparticles. Eur J Pharm Biopharm. 2004;58(2):327–341.

[24] Rabanel J-M, Hildgen P, Banquy X. Assessment of PEG on polymeric particles surface, a key step in drug carrier translation. J Controlled Release. 2014;185:71–87.

[25] Suk J-S, Xu Q, Kim N, et al. PEGylation as a strategy for improving nanoparticle-based drug and gene delivery. Adv Drug Delivery Rev. 2016;99 Part A:28–51.

[26] Shimizu T, Mima Y, Hashimoto Y, et al. Anti-PEG IgM and complement system are required for the association of second doses of PEGylated liposomes with splenic marginal zone B cells. Immunobiology. 2015;220(10):1151–1160.

[27] Mima Y, Hashimoto Y, Shimizu T, et al. Anti-PEG IgM is a major contributor to the accelerated blood clearance of polyethylene glycol-conjugated protein. Mol Pharm. 2015;12(7):2429–2435.

[28] Wang C, Cheng X, Su Y, et al. Accelerated blood clearance phenomenon upon cross-administration of PEGylated nanocarriers in beagle dogs. Int J Nanomedicine 10 3533–3545.

[29] Puri A, Blumenthal R. Polymeric lipid assemblies as novel theranostic tools. Acc Chem Res. 2011;44(10):1071–1079.

[30] Monge S, Canniccioni B, Graillot A, et al. Phosphorus-containing polymers: a great opportunity for the biomedical field. Biomacromolecules. 2011;12(6):1973–1982.

[31] Ruiz L, Hilborn JG, Léonard D, et al. Synthesis, structure and surface dynamics of phosphorylcholine functional biomimicking polymers. Biomaterials. 1998;19(11-12):987–998.

[32] Nagase Y, Nakajima S, Oku M, et al. Synthesis and properties of segmented poly(urethane-urea) s containing phosphorylcholine moiety in the side-chain. Polym J. 2008;40(12):1149–1156.

[33] Ishihara K, Ueda T, Nakabayashi N. Preparation of phospholipid polymers and properties as hydrogel membranes. Polym J. 1990;22:355–360.

[34] Ishihara K, Fukazawa K. 2-Methacryloyloxyethyl phosphorylcholine polymer. In Monge S and David G, editor. Phosphorus-based polymers: from synthesis to applications Chapter 5. Cambridge, UK: RSC publishing; 2014. p. 68–96.

[35] Ueda T, Oshida H, Kurita K, et al. Preparation of 2-methacryloyloxyethyl phosphorylcholine copolymers with alkyl methacrylates and their blood compatibility. Polym J. 1992;24(11):1259–1269.

[36] Ishihara K, Aragaki R, Ueda T, et al. Reduced thrombogenicity of polymers having phospholipid polar groups. J Biomed Mater Res. 1990;24:1069–1077.

[37] Ishihara K, Ziats NP, Tierney BP, et al. Protein adsorption from human plasma is reduced on phospholipid polymers. J Biomed Mater Res. 1991;25:1397–1407.

[38] Ishihara K, Nomura H, Mihara T, et al. Why do phospholipid polymer reduce protein adsorption ? J Biomed Mater Res. 1998;39:323–330.

[39] Moro T, Takatori Y, Ishihara K, et al. Surface grafting of artificial joints with a biocompatible polymer for preventing periprosthetic osteolysis. Nat Mater. 2004;3(11):829–836.

[40] Ishihara K. Bioinspired phospholipid polymer biomaterials for making high performance artificial organs. Sci Technol Adv Mater. 2000;1:131–138.

[41] Iwasaki Y, Ishihara K. Phosphorylcholine-containing polymers for biomedical applications. Ann Bioanal Chem. 2005;381:534–546.

[42] Lewis AL, Lloyd AW. Biomedical applications of biomimetic polymers: the phosphorylcholine-containing polymers in biomimetic, bioresponsive, and bioactive materials. In: Santin M and Phillips G, editors. An Introduction to Integrating Materials with Tissues. Hoboken (NJ): Wiley; 2012, p. 95–140.

[43] Iwasaki Y, Ishihara K. Cell membrane-inspired phospholipid polymers for developing medical devices with excellent biointerfaces. Sci Technol Adv Mater. 2012;13.064101 (14p).
[44] Ishihara K. Highly lubricated polymer interfaces for advanced artificial hip joints through biomimetic design. Polym J. 2015;47(9):585–597.
[45] Kihara S, Yamazaki K, Litwak KN, et al. In vivo evaluation of a mpc polymer coated continuous flow left ventricular assist system. Artif Organs. 2003;27(2):188–192.
[46] Lewis AL, Vick TA, Collias AC, et al. 2001. Phosphorylcholine-based polymer coatings for stent drug delivery. J Mater Sci Mater Med. 12(10/12) 865–870.
[47] Zhang Z, Cao X, Zhao X, et al. Controlled delivery of antisense oligodeoxynucleotide from cationically modified phosphorylcholine polymer films. Biomacromolecules. 2006;7(3):784–791.
[48] Zhong Q, Yan J, Qian X, et al. Atomic layer deposition enhanced grafting of phosphorylcholine on stainless steel for intravascular stents. Colloids Surf B Biointerfaces. 2014;121:238–247.
[49] Shimada T, Ueda M, Jinno H, et al. Development of targeted therapy with paclitaxel incorporated into EGF-conjugated nanoparticles. Anticancer Res. 2009;29:1009–1014.
[50] Miyata R, Ueda M, Jinno H, et al. Selective targeting by preS1 domain of hepatitis B surface antigen conjugated with phosphorylcholine-based amphiphilic block copolymer micelles as a biocompatible, drug delivery carrier for treatment of human hepatocellular carcinoma with paclitaxel. Int J Cancer. 2009;124(10):2460–2467.
[51] McRae Page SM, Henchey E, Chen X, et al. Efficacy of polyMPC–DOX prodrugs in 4t1 tumor-bearing mice. Mol Pharmaceutics. 2014;11:1715–1720.
[52] Kano T, Kakinuma C, Wada S, et al. Enhancement of drug solubility and absorption by copolymers of 2-Methacryloyloxyethyl phosphorylcholine and n-butyl methacrylate. Drug Metab Pharmacokinet. 2011;26(1):79–86.
[53] Onoue S, Kojo Y, Suzuki H, et al. Development of novel solid dispersion of tranilast using amphiphilic block copolymer for improved oral bioavailability. Int J Pharm. 2013;452(1-2):220–226.
[54] Onoue S, Suzuki H, Kojo Y, et al. Self-micellizing solid dispersion of cyclosporine A with improved dissolution and oral bioavailability. Eur J Pharm Sci. 2014;62:16–22.
[55] Ukawa M, Akita H, Masuda T, et al. 2-Methacryloyloxyethyl phosphorylcholine polymer (MPC)-coating improves the transfection activity of GALA-modified lipid nanoparticles by assisting the cellular uptake and intracellular dissociation of plasmid DNA in primary hepatocytes. Biomaterials. 2010;31(24):6355–6362.
[56] Wen Y, Zhang Z, Li J. Highly efficient multifunctional supermolecular gene carrier system self-assembled from redox-sensitive and zwitterionic polymer blocks. Adv Funct Mater. 2014;24:3874–3884.
[57] Lam JK, Armes SP, Lewis AL, et al. Folate conjugated phosphorylcholine-based polycations for specific targeting in nucleic acids delivery. J Drug Target. 2009;17(7):512–523.
[58] Qiang G, Lee I. Kinetics of swelling-shrinking rearrangement of a self-aggregated nanohydrogel at solid/liquid interfaces. J Exp Nanosci. 2011;6(5):509–520.
[59] Xu S, Ye Z, Wu P. Biomimetic controlling of $CaCO_3$ and $BaCO_3$ superstructures by zwitterionic polymer. ACS Sustainable Chem Eng. 2015;3:1810–1818.
[60] Jin S, Zhou N, Xu D, et al. Synthesis and characterization of poly(2-methacryloyloxyethyl pho sphorylcholine) onto graphene oxide. Polym Adv Technol. 2013;24:685–691.
[61] Dai F, Zhang M, Hu B, et al. Immunomagnetic nanoparticles based on a hydrophilic polymer coating for sensitive detection of Salmonella in raw milk by polymerase chain reaction. RSC Advence. 2015;5:3574–3580.
[62] Peacock AK, Cauët SI, Taylor A, et al. Poly[2-(methacryloyloxy)ethylphosphorylcholine]-coated iron oxide nanoparticles: synthesis, colloidal stability and evaluation for stem cell labelling. Chem Commun (Camb). 2012;48(75):9373–9375.
[63] Sun XY, Yu SS, Wan JQ, et al. Facile graft of poly(2-methacryloyloxyethyl phosphorylcholine) onto Fe(3) O(4) nanoparticles by ATRP: synthesis, properties, and biocompatibility. J Biomed Mater Res A. 2013;101A(2):607–612.
[64] Li X, Zhang B, Tian L, et al. Improvement of recognition specificity of surface protein-imprinted magnetic microspheres by reducing nonspecific adsorption of competitorsusing2-methacryloyloxyethylphosphorylcholine. Snes Actu B: Chemical. 2015;208:559–568.
[65] Yokoyama R, Suzuki S, Shirai K, et al. Preparation and properties of biocompatible polymer-grafted silica nanoparticle. Eur Polym J. 2006;42:3221–3229.
[66] Müllner M, Cui J, Noi KF, et al. Surface-initiated polymerization within mesoporous silica spheres for the modular design of charge-neutral polymer particles. Langmuir. 2014;30(21):6286–6293.
[67] Shao Z, Yang Y, Lee H, et al. Synthesis and suspension rheology of titania nanoparticles grafted with zwitterionic polymer brushes. J Colloid Interface Sci. 2012;386(1):135–140.
[68] Fuchs AV, Walter C, Landfester K, et al. Biomimetic silver-containing colloids of poly(2-methacryloyloxyethyl phosphorylcholine) and their film-Formation properties. Langmuir. 2012;28(11): 4974–4983.
[69] Fuchs AV, Ritz S, Putz S, et al 2-13. Bioinspired phosphorylcholine containing polymer films with silver nanoparticles combining antifouling and antibacterial properties. Biomater Sci. 1 470–477.
[70] Kitayama Y, Takeuchi T. Localized surface plasmon resonance nanosensing of C-reactive protein with poly(2-methacryloyloxyethyl phosphorylcholine)-grafted gold nanoparticles prepared by surface-initiated atom transfer radical polymerization. Anal Chem. 2014;86(11):5587–5594.
[71] Chen X, Lawrence J, Parelkar S, et al. Novel zwitterionic copolymers with dihydrolipoic acid: synthesis and preparation of nonfouling nanorods. Macromolecules. 2013;46:119–127.
[72] Bortolotto T, Facchinetto ST, Trindade SG, et al. Polymer-coated palladium nanoparticle catalysts for Suzuki coupling reactions. J Colloid Interface Sci. 2015;439:154–161.
[73] Matsuno R, Goto Y, Konno T, et al. Controllable nanostructured surface modification on quantum dot for biomedical application in aqueous medium. J Nanosci Nanotechnol. 2009;9(1):358–365.
[74] Liu Y, Inoue Y, Ishihara K. Surface functionalization of quantum dots with fine-structured pH-sensitive

phospholipid polymer chains. Colloids Surf B: Biointerfaces. 2015;135:490–496.
[75] Xu FM, Xu JP, Lv L-P, et al. Bowl- and porous sphere-shaped supramolecular assemblies and their application as templates for confirmed assembly of gold nanoparticles. Soft Matter. 2011;7:1114–1120.
[76] Goto Y, Matsuno R, Konno T, et al. Polymer nanoparticles covered with phosphorylcholine groups and immobilized with antibody for high-affinity separation of proteins. Biomacromolecules. 2008;9(3):828–833.
[77] Kim HI, Ishihara K. Phospholipid polymer can reduce cytocompatibility of poly(lactic acid) nanoparticles in a high-content screening assay. Biomater Biomed Engineer. 2014;1(2):85–94.
[78] Chen W, Inoue Y, Ishihara K. Preparation of photoreactive phospholipid polymer nanoparticles to immobilize and release protein by photoirradiation. Colloids Surf B: Biointerfaces. 2015;135:365–370.
[79] Osawa K, Imae T, Ujihara M, et al. Preparation of amphiphilic diblock copolymers with pendant hydrophilic phosphorylcholine and hydrophobic dendron groups and their self-association behavior in water. J Polym Sci Polym. Chem. 2013;51(22):4923–4931.
[80] Han H, Zhang S, Wang Y, et al. Biomimetic drug nanocarriers prepared by miniemulsion polymerization for near-infrared imaging and photothermal therapy. Polymer. 2016;82:255–261.
[81] Ishihara K, Iwasaki Y, Nakabayashi N. Polymeric lipid nanosphere constituted of poly(2-methacryloyloxyethyl phosphorylcholine-*co*-*n*-butyl methacrylate) Polym. J. 1999;31:1231–1236.
[82] Konno T, Ishihara K. Enhanced solubility of paclitaxel using water-soluble and biocompatible 2-methacryloyloxyethyl phosphorylcholine polymers. J Biomed Mater Res. 2003;65A(2):210–215.
[83] Lobb EJ, Ma I, Billingham NC, et al. Facile synthesis of well-defined, biocompatible phosphorylcholine-based methacrylate copolymers via atom transfer radical polymerization at 20 °C. J Am Chem Soc. 2001;123(32):7913–7914.
[84] Yusa S-I, Fukuda K, Yamamoto T, et al. Synthesis of well-defined amphiphilic block copolymers having phospholipid polymer sequences as a novel biocompatible polymer micelle reagent. Biomacromolecules. 2005;6(2):663–670.
[85] Shen L. Biocompatible polymer/quantum dots hybrid materials: current status and future developments. J Funct Biomater. 2011;2(4):355–372.
[86] Wegner KD, Hildebrandt N. Quantum dots: bright and versatile *in vitro* and *in vivo* fluorescence imaging biosensors. Chem Soc Rev. 2015;44(14):4792–4834.
[87] Wang Y, Hu R, Lin G, et al. Functionalized quantum dots for biosensing and bioimaging and concerns on toxicity. ACS Appl Mater Interfaces. 2013;5(8):2786–2799.
[88] Tripathi SK, Kaur G, Khurana RK, et al. Quantum dots and their potential role in cancer theranostics. Crit Rev Ther Drug Carrier Syst. 2015;32(6):461–502.
[89] Petryayeva E, Algar WR, Medintz IL. Quantum dots in bioanalysis: a review of applications across various platforms for fluorescence spectroscopy and imaging. Appl Spectrosc. 2013;67(3):215–252.
[90] Hu K, Wang H, Tang G, et al. In vivo cancer dual-targeting and dual-modality imaging with functionalized quantum dots. J Nucl Med. 2015;56(8):1278–1284.
[91] Liu H, Tang W, Li C, et al. CdSe/ZnS quantum dots-labeled mesenchymal stem cells for targeted fluorescence imaging of pancreas tissues and therapy of type 1 diabetic rats. Nanoscale Res Lett. 2015;10(1):959.
[92] Breus VV, Pietuch A, Tarantola M, et al. The effect of surface charge on nonspecific uptake and cytotoxicity of CdSe/ZnS core/shell quantum dots. Beilstein J Nanotechnol. 2015;6:281–292.
[93] Painuly D, Bhatt A, Krishnan VK. Physicochemical and *in vitro* biocompatibility evaluation of water-soluble CdSe/ZnS core/shell. J Biomater Appl. 2014;28(8):1125–1137.
[94] Jańczewski D, Tomczak N, Han M-Y, et al. Synthesis of functionalized amphiphilic polymers for coating quantum dots. Nat Protocols. 2011;6(10):1546–1553.
[95] Reissmann S. Cell penetration: scope and limitations by the application of cell-penetrating peptides. J Pept Sci. 2014;20(10):760–784.
[96] Brock R. The uptake of arginine-rich cell-penetrating peptides: putting the puzzle together. Bioconjug Chem. 2014;25(5):863–868.
[97] Goto Y, Matsuno R, Konno T, et al. Artificial cell membrane-covered nanoparticles embedding quantum dots as stable and highly sensitive fluorescence bioimaging probes. Biomacromolecules. 2008;9(11):3252–3257.
[98] Watanabe J, Ishihara K. Instantaneous determination via bimolecular recognition: usefulness of FRET in phosphorylcholine group enriched nanoparticles. Bioconjug Chem. 2007;18(6):1811–1817.
[99] Konno T, Watanabe J, Ishihara K. Conjugation of enzymes on polymer nanoparticles covered with phosphorylcholine groups. Biomacromolecules. 2004;5(2):342–347.
[100] Ito T, Watanabe J, Takai M, et al. Dual mode bioreactions on polymer nanoparticles covered with phosphorylcholine group. Colloids Surf B Biointerfaces. 2006;50(1):55–60.
[101] Ishihara K, Tsukamoto Y, Goto Y, et al. Enhanced and specific internalization of polymeric nanoparticles to cells. IFMBE Proceedings. 2013;49:262–265.
[102] Ishihara K. Novel bioinspired phospholipid polymer biomaterials for nanobioengineering. In: Ruiz-Molina D, Novio F, Roscini C, editors. Bio- and bioinspired nanomaterials. Weinhelm, Germany: Wiley-VCH; 2015. p. 369–389.

2

Electrical and photo-electrical properties of MoS_2 nanosheets with and without an Al_2O_3 capping layer under various environmental conditions

Muhammad Farooq Khan, Ghazanfar Nazir, Volodymyr M. Iermolenko and Jonghwa Eom

Department of Physics & Astronomy and Graphene Research Institute, Sejong University, Seoul 05006, Korea

ABSTRACT

The electrical and photo-electrical properties of exfoliated MoS_2 were investigated in the dark and in the presence of deep ultraviolet (DUV) light under various environmental conditions (vacuum, N_2 gas, air, and O_2 gas). We examined the effects of environmental gases on MoS_2 flakes in the dark and after DUV illumination through Raman spectroscopy and found that DUV light induced red and blue shifts of peaks (E^1_{2g} and A_{1g}) position in the presence of N_2 and O_2 gases, respectively. In the dark, the threshold voltage in the transfer characteristics of few-layer (FL) MoS_2 field-effect transistors (FETs) remained almost the same in vacuum and N_2 gas but shifted toward positive gate voltages in air or O_2 gas because of the adsorption of oxygen atoms/molecules on the MoS_2 surface. We analyzed light detection parameters such as responsivity, detectivity, external quantum efficiency, linear dynamic range, and relaxation time to characterize the photoresponse behavior of FL-MoS_2 FETs under various environmental conditions. All parameters were improved in their performances in N_2 gas, but deteriorated in O_2 gas environment. The photocurrent decayed with a large time constant in N_2 gas, but decayed with a small time constant in O_2 gas. We also investigated the characteristics of the devices after passivating by Al_2O_3 film on the MoS_2 surface. The devices became almost hysteresis-free in the transfer characteristics and stable with improved mobility. Given its outstanding performance under DUV light, the passivated device may be potentially used for applications in MoS_2-based integrated optoelectronic circuits, light sensing devices, and solar cells.

CLASSIFICATION
40 Optical; magnetic and electronic device materials; 105 Low-Dimension (1D/2D) materials; 100 Materials; 201 Electronics / Semiconductor / TCOs; 200 Applications; 212 Surface and interfaces; 200 Applications

KEYWORDS
MoS_2; transition metal dichalcogenides; electrical property; photoresponse parameters; Raman spectroscopy; DUV light

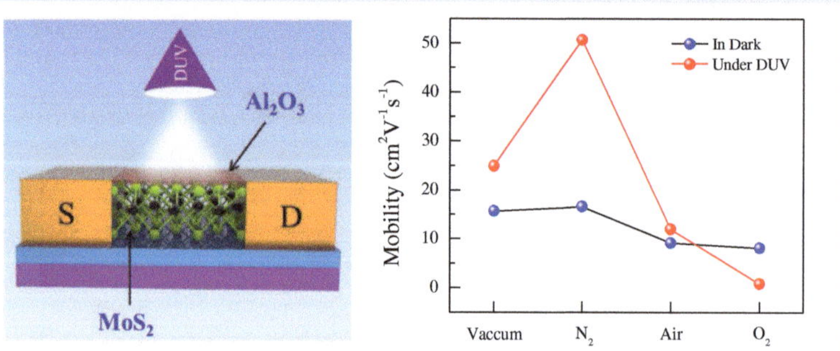

1. Introduction

Two-dimensional (2D) nanomaterials have sustained attention because of their unique properties and facile fabrication process despite the complexity of their structures. Graphene as a 2D material has become popular in recent years because of its outstanding linear dispersion relation, high charge carrier mobility, feasibility of chemical doping, and other unique physical properties conferred by its low dimensionality.[1–5] However, the zero band gap of graphene hinders its application in logic and optoelectronic devices. Graphene-based field–effect transistors (FETs) cannot be efficiently switched off; that is, off-current is comparable with on-current. However, 2D transition metal dichalcogenides (TMDCs) possess a suitable band gap of ~1–2 eV, making these materials potentially applicable in nanoelectronics, sensing, and photonics.[6–9] Several studies have analyzed 2D TMDC-based FETs, photodetectors, and gas sensors. The earliest TMDC, namely WSe_2 crystal, was used in FETs and exhibits a high mobility (>500 $cm^2 V^{-1}s^{-1}$) and ambipolar behavior with a ~10^4 on/off ratio at 60 K.[10] TMDCs are potential materials for molecular sensing

CONTACT Jonghwa Eom eom@sejong.ac.kr

application because of their high surface-to-volume ratio. MoS_2 sheets are sensitive detectors of NO, NO_2, NH_3, N_2, and triethylamine gases.[11–15] However, the large variation in the electrical transport properties of MoS_2 is generally attributed to extrinsic or environmental effects. These effects may significantly limit the exploration of the intrinsic properties of MoS_2. Investigating the environmental factors that influence the reliability and stability of MoS_2 FETs is important. Detailed and comparative investigations on environmental gases effects are needed to understand the overall performance of MoS_2 FETs and undesirable effects should be minimized or removed to enhance device reliability and stability.

In this study, we fabricated few-layer (FL) MoS_2 FETs and investigated electrical transport properties under various environmental conditions. We analyzed the change in on-state current under different environmental conditions. We observed the hysteresis behavior in transfer characteristics, where oxygen on the MoS_2 surface plays a vital role in hysteresis. Gas environment significantly affects the characteristics of MoS_2 FETs under deep ultraviolet (DUV) light. Thus, we investigated the time-dependent photoresponse of MoS_2 FETs in the presence of various environmental gases. We also studied the maximum photocurrent saturation, responsivity, detectivity, external quantum efficiency (EQE), linear dynamic range (LDR), and relaxation time after switching light ON and OFF under various environmental conditions. Finally, we investigated the characteristics of devices after passivating by Al_2O_3 film on the MoS_2 surface. The devices became almost hysteresis-free and stable FETs with improved mobility. The effect of Al_2O_3 on MoS_2 flakes was discussed to explore the basic and important parameters regarding light detection.

2. Experimental section

We fabricated FL-MoS_2 FETs using a conventional approach. FL-MoS_2 flakes were transferred to clean 300-nm-thick SiO_2 layer on p++ Si substrate with Scotch tape via mechanical cleavage method.[16] FL flakes were initially identified under an optical microscope (Figure 1(a)) and further confirmed through Raman spectroscopy. The thickness of the flakes was estimated as ~3.6 nm by atomic force microscopy, which indicated the presence of five layers (Supporting Information Figure S1). Large patterns formed through photo-lithography for each device, and fine electrodes for source and drain were completed after electron beam-lithography. The source–drain contacts of Cr/Au (10/80 nm) were deposited via thermal evaporation. Electrical transport measurements were performed using a Keithley 2400 source meter (Beaverton, OR, USA) and a Keithley 6485 picoammeter (Beaverton, OR, USA). The final optical device image is illustrated in Figure 1(b). All of the measurements were carried out at room temperature. Raman spectra were recorded with a Renishaw microspectrometer equipped with a 514 nm laser at room temperature. DUV light (λ = 220 nm and average intensity of 11 mW cm^{-2}) was used for illumination.

A 10-nm-thick Al_2O_3 passivation layer was deposited on the FL-MoS_2 at 200 °C through atomic layer deposition (ALD; Lucida D100 ALD by NCD Co., Ltd, Daejeon, Korea). During the growth process, 100 cycles with a growth rate of ~1 Å per cycle were completed to reach the desired thickness. Trimethylaluminum and deionized water were used as precursors. Ultra-pure N_2 (99.999%) was used as a carrier and purging gas. The pressure of the growth chamber was maintained at ~5.5×10^{-3} torr during deposition.

3. Results and discussion

Raman spectroscopy is a fast and reliable tool that provides information about structural and local perturbation of 2D materials. The Raman spectrum of MoS_2 is dominated by two vibrational modes: E^1_{2g}, which is attributed to in-plane vibrations of two S atoms with respect to the Mo atom, and A^1_g, which corresponds to the out-of-plane vibrations of S atoms in opposite directions. The E^1_{2g} and A^1_g modes are sensitive to the number of layers forming the structure of MoS_2.[17, 18] The Raman spectra of FL-MoS_2 layers were recorded at room temperature with a laser excitation of ~514 nm and a small laser power of ~1.0 mW to avoid the heating effect. Figure 1(c) shows the Raman spectra of pristine and gas-treated (in the dark and under DUV light) FL-MoS_2 (five layers; see Figure S1). The FL-MoS_2 exhibited strong bands at ~383.2 and ~407.7 cm^{-1}, which is ascribed to in-plane vibrational (E^1_{2g}) and out-of-plane vibrational (A_{1g}) modes, respectively. The vibrational modes of FL-MoS_2 did not significantly change after treating the samples with O_2 and N_2 gases in the dark. However, we observed a shift toward a lower wave number after DUV treatment of FL-MoS_2 in the presence of N_2 gas. Subsequently we observed a shift toward a higher wave number in the presence of O_2 gas under DUV light (Figure 1(c)). After the DUV treatment in N_2 gas, the red shift of A_{1g} and E^1_{2g} was ~2.2 and ~1.6 cm^{-1}, respectively. The subsequent DUV treatment in O_2 gas environment made the blue shift of A_{1g} and E^1_{2g} by ~2.1 and ~1.5 cm^{-1}, respectively. The red and blue shifts in peak (E^1_{2g} and A_{1g}) position are due to doping of electrons and holes on FL-MoS_2 by DUV treatment in N_2 and O_2 gas, respectively. The mechanism of Raman shifts by adsorption and desorption of these gas molecules is attributed to electron-phonon coupling of modes. The A_{1g} mode couples more strongly with electrons than the E^1_{2g} mode. The electron doping in MoS_2 causes softening specifically of its Raman-active A_{1g} phonon, accompanied by increase in line-width of its Raman peaks.[19] In contrast, the other Raman mode with E^1_{2g} symmetry is less sensitive to electron doping. The Raman spectrum

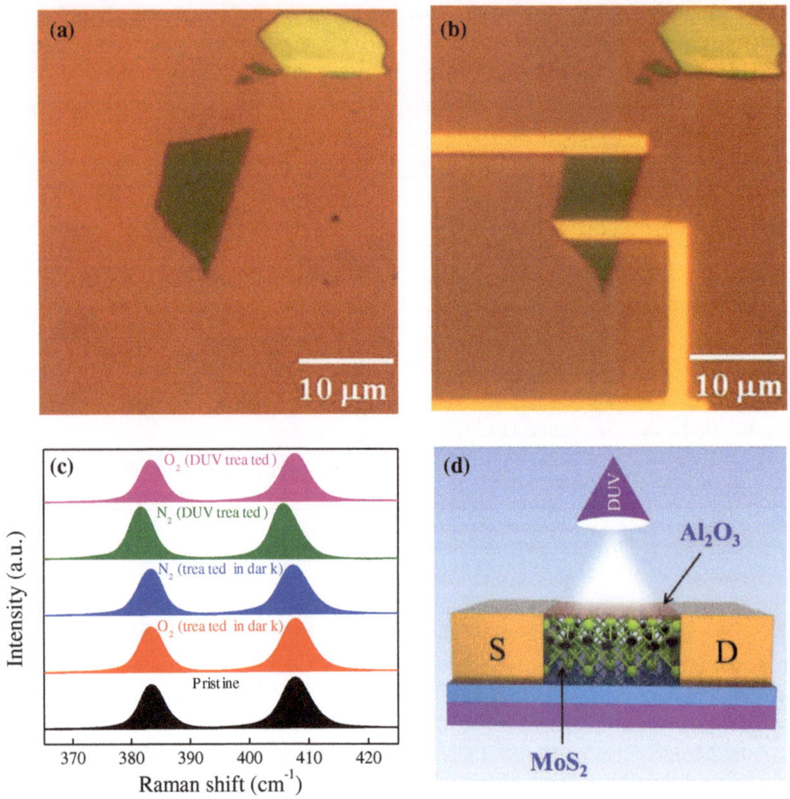

Figure 1. Few-layer MoS_2 FET. (a) Optical image of FL-MoS_2 flake serving as a conducting channel in the transistor. (b) Optical image of the final device based on the flake shown in (a). The two contacts were used for electrical transport measurements. (c) Raman spectra of the FL-MoS_2 pristine (black), O_2 treated in the dark (red), N_2 treated in the dark (blue), N_2 treated under DUV (green), and O_2 treated under DUV (pink). (d) Schematic view of the structure of the passivated (10 nm Al2O3 layer) FL-MoS_2 FET under DUV illumination. A FL-MoS_2 was transferred on a p-doped silicon substrate with a 300-nm-thick SiO_2 capping layer. The substrate served as a back gate.

for each case was obtained from five different points on the MoS_2 surface, and the results were consistent.

The electrical transport measurements of the FL-MoS_2 FETs (sample-1; L = 8.1 μm and W = 7.6 μm) were carried out under various environmental conditions. Transfer characteristics (drain–current I_D as a function of back-gate voltage V_g) were measured at room temperature with fixed source–drain voltage (V_{ds}) = 1 V. Figure 2(a) presents the transfer characteristics of the FL-MoS_2 FETs in the dark under various environmental conditions. The measurements were performed in vacuum, followed by N_2, air, and O_2 gas environment. The current level of the device was almost similar in the N_2 and vacuum environments but was substantially reduced in the air and O_2 environments. The I_{on}/I_{off} of our device was ~10^5 and remained almost same in vacuum, N_2, air, and O_2 gas flow (Supporting Information Figure S2a).

Figure 2(b) illustrates the hysteresis in the transfer characteristics of FL-MoS_2 in the dark under different environmental conditions. We swept the V_g from −40 to 80 V repeatedly and observed a hysteresis in vacuum. Later on, the same devices were exposed in N_2 gas and found a similar hysteresis trend to that in vacuum. However, further exposure of devices to air and oxygen significantly changed the hysteresis because of adsorption of oxygen atoms/molecules on top of MoS_2 surface.[20] This kind of oxygen adsorption resulted in formation of charge-trapping sites and made the hysteresis prominent in FL-MoS_2 FETs. In transfer characteristics in Figure S2a, we found that the threshold voltages (V_{th}) of FL-MoS_2 FETs was near −35, −37, −34, and −28 V in vacuum, N_2, air, and O_2, respectively. Furthermore, we observed that the V_{th} remained similar in vacuum and N_2 gas but shifted toward positive gate voltages when the devices were exposed to air and O_2 gas. The shift of V_{th} to positive back gate voltages indicates an electron deficiency in FL-MoS_2 due to exposure to air and oxygen. Such behavior was previously reported for MoS_2-related FET devices in O_2 environments, which was also attributed to the absorption of oxygen molecules into sulfur or defect states on the MoS_2 surface, which traps the charge carriers.[21, 22]

Figure 2(c) shows the transfer characteristics of the FL-MoS_2 FETs under various environmental conditions in the presence of DUV light. DUV light substantially

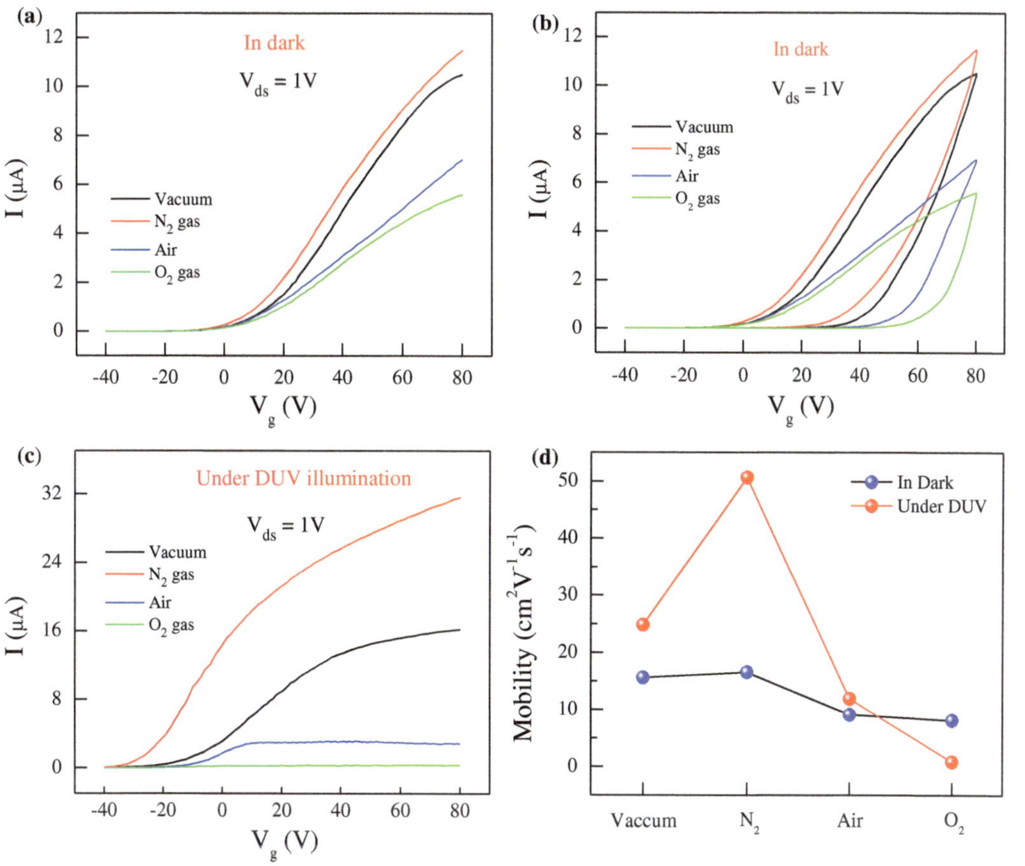

Figure 2. Electrical transport of FL-MoS_2 FET before passivation. (a) Transfer characteristics of the FL-MoS_2 FET device in the dark. Transfer characteristics were in vacuum, N_2, air, and O_2 environments. (b) Hysteresis in transfer characteristics of the FL-MoS_2 FET device in the dark. Transfer characteristics were obtained under various environmental conditions (vacuum, N_2 gas, air, and O_2 gas). (c) Transfer characteristics of the FL-MoS_2 FET under DUV illumination in vacuum, N_2, air, and O_2 environments. The power intensity and wavelength of light were 11 mW cm^{-2} and 220 nm, respectively. (d) Field-effect mobility of the FL-MoS_2 FET device in the dark and in the presence of DUV illumination under various environmental conditions.

improved the drain current of the FL-MoS_2 FETs in N_2 and in vacuum, but significantly decreased it in air and in O_2 gas environments. DUV light reduced the oxygen atom/molecules from the MoS_2 nanosheet surface in N_2 and vacuum environments. The adsorption of oxygen atom/molecules on the MoS_2 nanosheet reduced the number of charge carriers (electrons), causing a decline of drain current, which is consistent with previous reports.[23, 24] Figure 2(d) shows the field-effect mobility (μ) of the FL-MoS_2 device under various environmental conditions in the dark and in the presence of DUV light. The mobility (μ) of the FL-MoS_2 device was obtained using the following relation:

$$\mu = \frac{L}{WC_g V_{ds}} \frac{dI_{ds}}{dV_g} \qquad (1)$$

where dI_{ds}/dV_g is slope of the transfer curve in the linear region, C_g is the gate capacitance (~115 aF μm^{-2}) of Si/SiO_2 substrates (300 nm), V_{ds} is the source–drain current for each device, L is the channel length, and W is the channel width. In the dark, we estimated the μ_{vacuum}, μ_{N2}, μ_{air}, and μ_{O2} of the FL-MoS_2 FETs to be ~15.6, 16.5, 9.1, and 8.1 cm^2 V^{-1}s^{-1}, respectively. However, under DUV illumination, the μ_{vacuum}, μ_{N2}, μ_{air}, and μ_{O2} of the FL-MoS_2 FETs were ~24.8, 50.6, 11.9, and 0.8 cm^2 V^{-1}s^{-1}, respectively. The FL-MoS_2 FETs exhibited a high I_D and large mobility in N_2 but showed a small I_D in air and O_2 environment under DUV illumination.[25] Figure 3 presents the output characteristics of the FL-MoS_2 FETs in vacuum, N_2, air, and O_2 environments in the dark. The output characteristics (I_D versus V_{ds}) were measured at a fixed V_g ranging from −20 V to + 60 V with a step of 20 V. The nonlinear I-V characteristics supports that the Cr/Au metal electrodes make Schottky barrier due to the work function difference between metal and MoS_2.

To measure the photodetector response, the MoS_2 nanosheet device was irradiated with DUV light (sample-2; L = 1.3 μm and W = 3.15 μm). In Figure 4(a) we show the transient current response to a cycle of

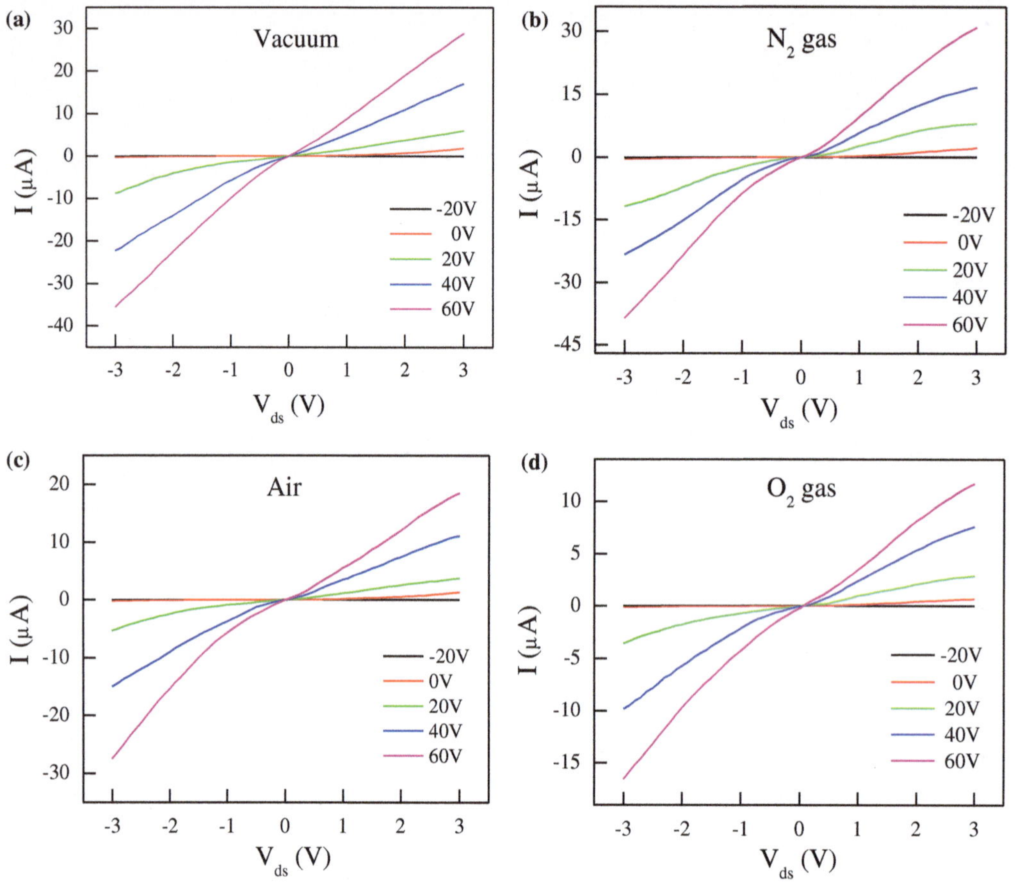

Figure 3. Drain-source characteristics of the device in the dark. The output characteristics (I_D-V_{ds}) of the FL-MoS_2 FET at fixed V_g in (a) vacuum, (b) N_2, (c) air, and (d) O_2 environments. V_g ranges from −20 V to + 60 V with a step of 20 V.

60 s with alternating ON and OFF of the light source (V_{ds} = 1 V and V_g = 0 V) under various environmental conditions. Figure 4(a) illustrates that the photocurrent response depended on various environments. The photocurrent response was weak in O_2 gas but strong in N_2 gas. The MoS_2 nanosheet exhibited a repeatable and reasonably stable response to incident DUV light under various environmental conditions. We also found that the photocurrent response increased with V_{ds}. The photocurrent ($\Delta I_{ph} = I_{ph} - I_{dark}$) after 60 s DUV illumination reached ~14.7, 39.7, 83.3, and 88.5 μA at V_{ds} = 1, 4, 6, 8 V, respectively (Supporting Information Figure S2b and c).

In Figure 4(b), the photocurrent saturation was examined under various environmental conditions. We illuminated the FL-MoS_2 FETs with DUV light until the photocurrent saturation was obtained and then turned off the DUV light to observe photocurrent decay behavior. The maximum photocurrent values ($\Delta I_{max} = I_{saturation} - I_{dark}$) of the FL-$MoS_2$ FETs were ~29, 12.3, 6.5, and 67.7 μA (V_{ds} = 1 V and V_g = 0 V) under various environmental conditions (Figure 4(c)). The largest and smallest photocurrent saturations were observed in N_2 and O_2 gas environments, respectively.

Two key figures of merit were measured to quantify detector performance. Current responsivity (R_λ) and detectivity (D^*) were defined as follows [26]:

$$R_\lambda = \frac{I_{ph}}{PA} \tag{2}$$

$$D^* = \frac{R_\lambda A^{\frac{1}{2}}}{\sqrt{(2eI_{dark})}} \tag{3}$$

where ΔI_{ph} is the photo-excited current, P is the light power intensity, and A is the effective area of the photodetector. Responsivity is defined as the photocurrent generated per unit power intensity of the incident light on the effective area of a photoconductor. The power intensity of our DUV incident light and the area of device were 11 mW cm^{-2} and 4.09 μm^2, respectively. In Equation (3), e is the absolute value of the electron charge and I_{dark} is the current density in dark. Figure 4(d) shows the responsivity and detectivity of our devices, where R_λ was estimated to be ~65.9, 27.9, 14.7, and 154

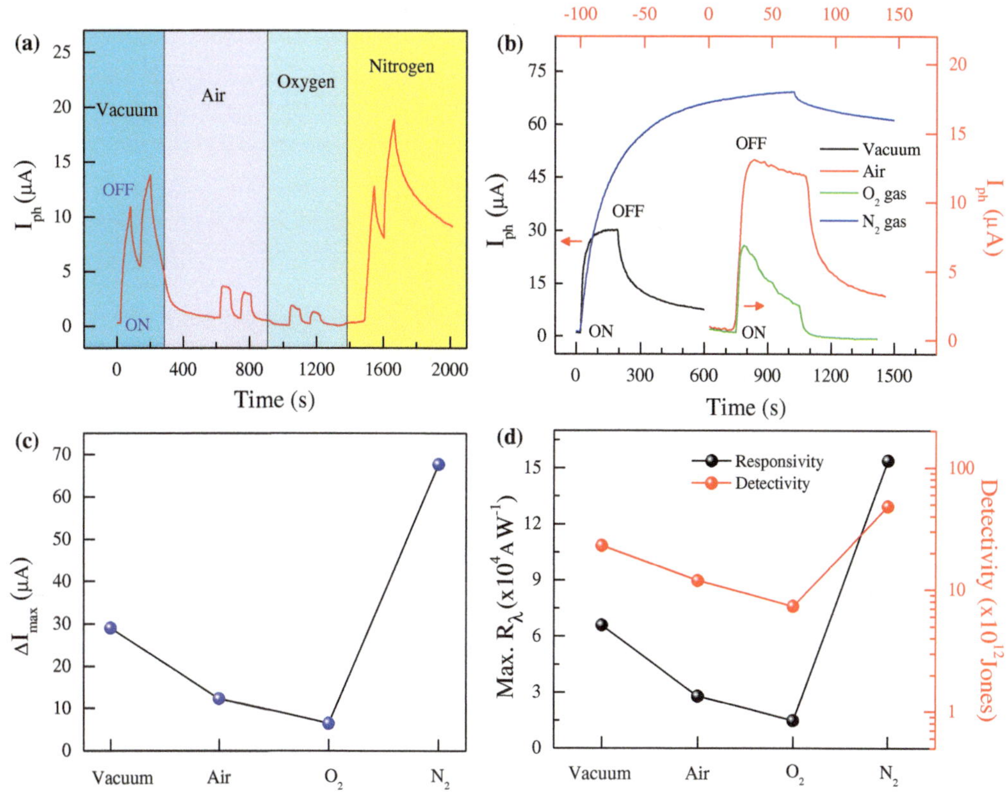

Figure 4. Photocurrent response in various environments. (a) Time-dependent photocurrent response of the device in different environments at V_{ds} = 1 V and V_g = 0 V. DUV light was first turned on for 60 s and then turned off for 60 s. (b) Photocurrent saturation of the FL-MoS_2 FET in vacuum and N_2 gas (left axis and bottom axis), and in air and O_2 gas (right axis and top axis) at (V_{ds} = 1 V and V_g = 0 V). (c) Saturated photocurrent values in vacuum, N_2 gas, air, and O_2 gas. (d) Modulation of photocurrent response parameters; maximum responsivity (left axis) and detectivity (right axis) in different environments.

kAW^{-1} in vacuum, air, O_2, and N_2 gas environments, respectively. D^*, which is measured in Jones units (1 Jones = 1 cm $Hz^{1/2}$ W^{-1}), ranged from ~10^{13}–10^{14}. These data indicated the best response in N_2 gas, which is more than ~10^4 times higher than that of the previously reported 2D, layered material devices on Si/SiO_2 substrates.[26–30]

To investigate photodetector performance further, two other critical parameters, namely EQE and LDR (typically presented in the unit of dB), were measured as seen in Figure 5(a) and (b), respectively. The EQE (= $hcR_\lambda/e\lambda$, where h is a plank constant, c is the speed of light, R_λ is the responsivity at a DUV wavelength of 220 nm, and e is the electron charge) is defined as the number of electron–hole pairs excited by one absorbed photon. The EQE was estimated to be ~3700, 1500, 830, and 8600% in vacuum, air, O_2, and N_2 environments, respectively. Compared with previously reported devices, our MoS_2 devices on the Si/SiO_2 substrate exhibited higher EQE in N_2 gas.[24,28,31,32] There are three factors which improve external quantum efficiency (EQE) of our devices; higher responsivity, light of lower wavelength and N_2 gas environment. The wavelength, λ, is an important factor which is the denominator in the formula (EQE = $hcR_\lambda/c\lambda$). We enhanced our EQE to 8600% using smaller wavelength together with N_2 gas environments. The N_2 gas provides supportive environment for photocurrent generation. The LRD (= 20 log (I_{ph}/I_{dark}), where I_{ph} is the photocurrent measured at a light intensity of 11 mW cm^{-2}) was measured under DUV illumination. The measured LDR values of the MoS_2 nanosheet device were ~29.5, 23.16, 20, and 34.4 dB in vacuum, air, O_2, and N_2 environments, respectively. The results demonstrate that MoS_2 nanosheet devices in N_2 gas can be potential photodetectors in the future.

Relaxation response time or decaying behavior is another key factor in photodetector performance. Decaying behavior is observed after photocurrent saturation under various environmental conditions. Relaxation data were extracted from Figure 4(b), when light was turned off after photocurrent saturation. The response time obtained from some 1D nanostructures or graphene oxide-based photodetectors ranges from seconds to several tens of minutes.[33–35] The wide range of response time is attributed to the difference in the materials or device structures. The dynamic response to

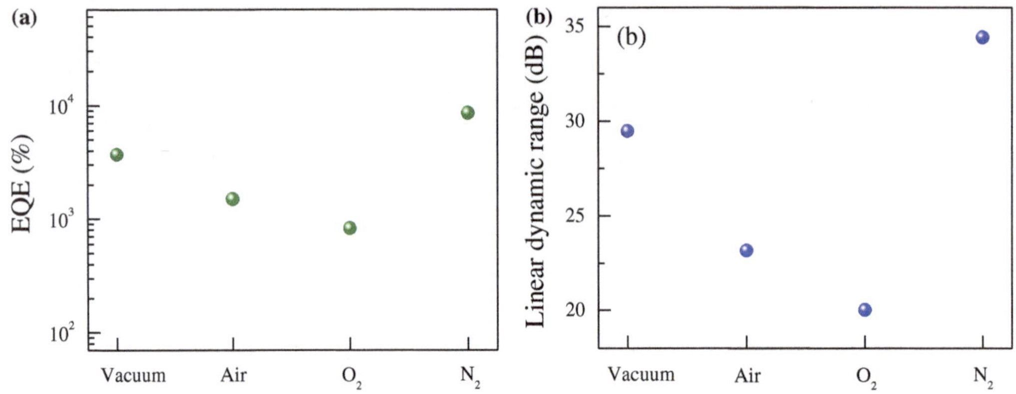

Figure 5. Light detection response. Crucial factors to be calculated in vacuum, N_2, air, and O_2 for (a) external quantum efficiency and (b) linear dynamic range. The data are obtained from Figure 4(b), (c), and (d).

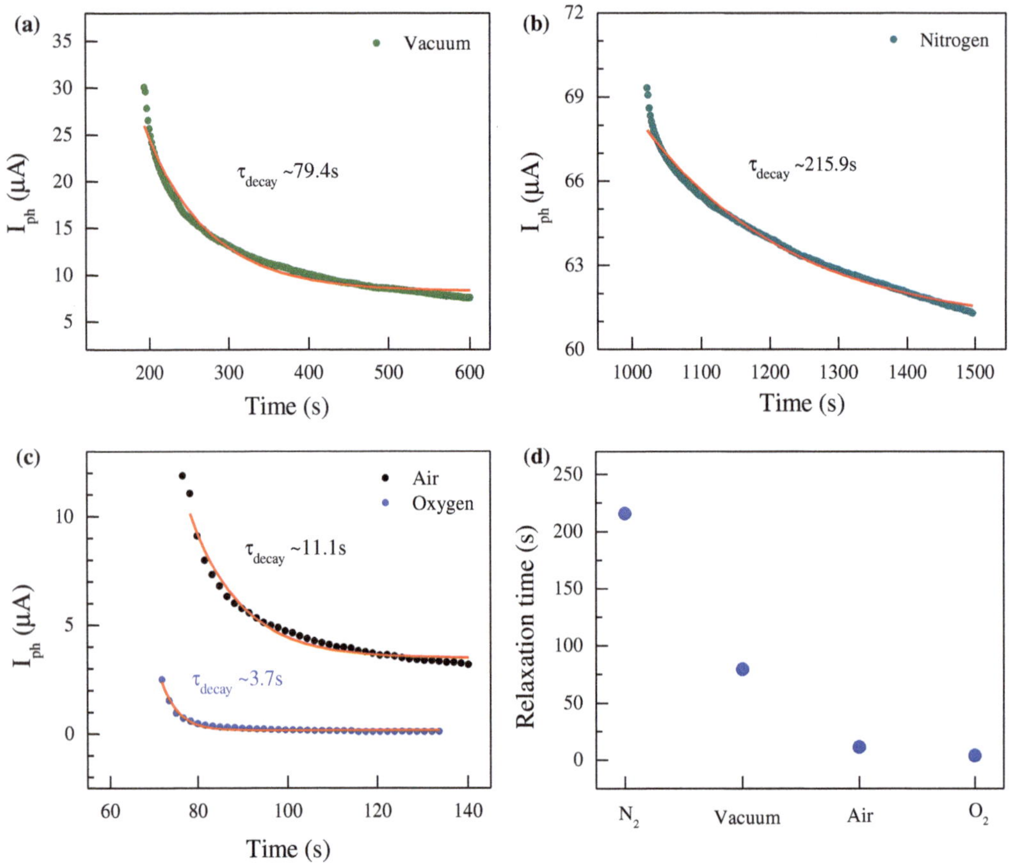

Figure 6. Photocurrent decaying behavior in different environments. Transient property of FL-MoS_2 FETs under different conditions and data fitting (red line) with a single exponential equation. (a) Decay of drain current in vacuum. (b) Decay of drain current in N_2. The device exhibits a slow decay with a large time constant. (c) Decay of drain current in air and O_2. The drain current decays with a small time constant. (d) Relaxation time constants of the device in vacuum, N_2, air, and O_2.

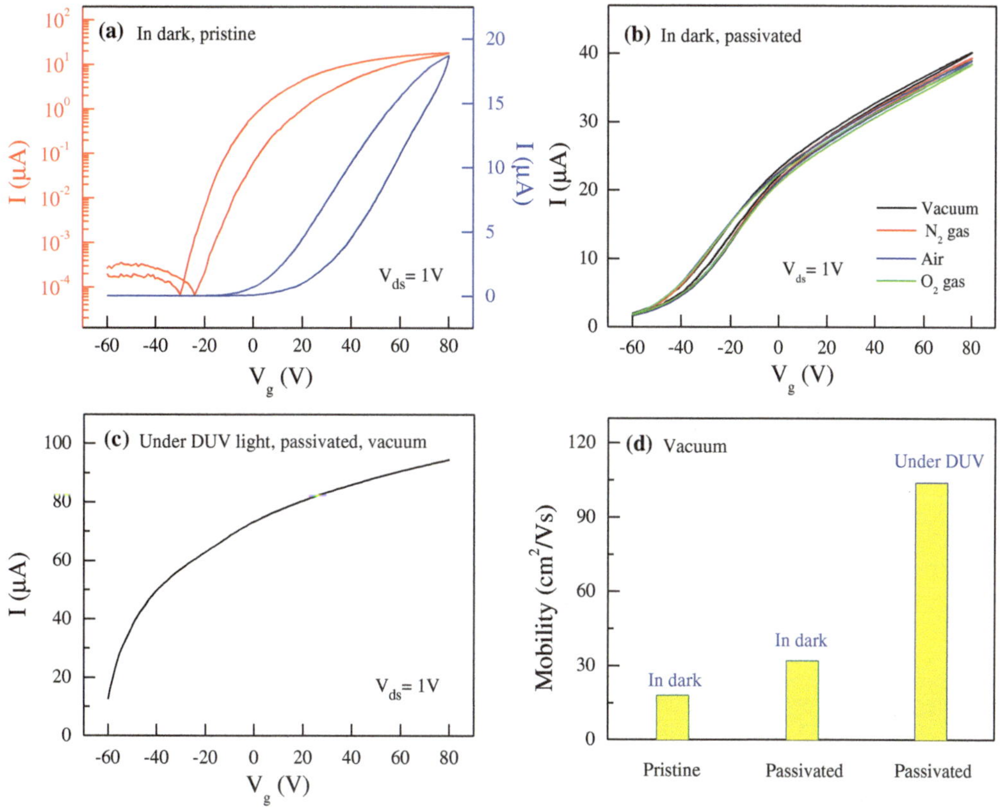

Figure 7. Electrical transport measurements of passivated FL-MoS_2 FET. (a) Transfer characteristics of the FL-MoS_2 FET device in the dark. Transfer characteristics are measured in vacuum. The linear scale is on the right axis, and the logarithm scale is on the left axis. (b) Hysteresis in transfer characteristics after passivation. The measurements were performed in the dark under different environmental conditions with V_g from −60 V to +80 V. (c) Transfer characteristics under DUV illumination. The measurements were carried out in vacuum at V_{ds} = 1 V. The power intensity and wavelength of light were 11 mW cm^{-2} and 220 nm, respectively. (d) Field–effect mobility of the passivated FL-MoS_2 FET device in the dark and under illumination. The measurement was conducted in vacuum.

DUV illumination for relaxation time can be expressed as [28, 36]:

$$I_{\text{ph}}(t) = I_{\text{dark}} + A\exp\left(-\frac{t}{\tau_{\text{decay}}}\right) \qquad (4)$$

where A is a scaling constant, τ_{decay} is a time constant for decaying, and t is the time after DUV light is switched on or off. The time constant (τ) can be calculated by fitting the experimental data. Figure 6(a–c) describes the decaying photocurrent behavior in N_2, vacuum, air, and O_2 environments, respectively, and the red lines indicate the fitting data. The relaxation time was calculated to be ~215.9, 79.4, 11.1, and 3.7 s in N_2, vacuum, air, and O_2, respectively (Figure 6(d)). These results demonstrated that the decay process was very slow in N_2 and significantly fast in O_2 gas. Thus, the adsorbates and defect states originating from ambient air, water, and oxygen atoms/molecules play an important role in photocurrent relaxation and electron-hole pair recombination dynamics. In general, the surface-adsorbed oxygen significantly affects the photoresponse of MoS_2 and ZnO films.[37,38] In the present study, electron-capturing impurity states were reduced under DUV light in N_2 gas environment but were enhanced under DUV light in O_2 gas environment. Thus, the decay time was significantly longer in N_2 environment and substantially shorter in O_2 environment. Evidently, decaying behavior was robustly dependent on the surrounding environment.

We measured the electrical transport of the FL-MoS_2 FETs (sample-3; L = 1.5 μm and W = 1.2 μm) in vacuum at a bias voltage V_{ds} = 1 V to study the passivation effect by high-k dielectric materials to circumvent the environmental effects. The transfer characteristic of the pristine device at room temperature in the dark is shown in Figure 7(a), where a pronounced hysteresis can be observed. A 10-nm-thick passivation layer of Al_2O_3 was deposited on the same device through atomic layer deposition (ALD). After depositing the passivation layer, the transfer characteristics of the device were examined in the dark under various environmental conditions (Figure 7(b)). The MoS_2 device was almost hysteresis-free with increased drain current and exhibited n-type. The Al_2O_3 layer on the MoS_2 surface served as a

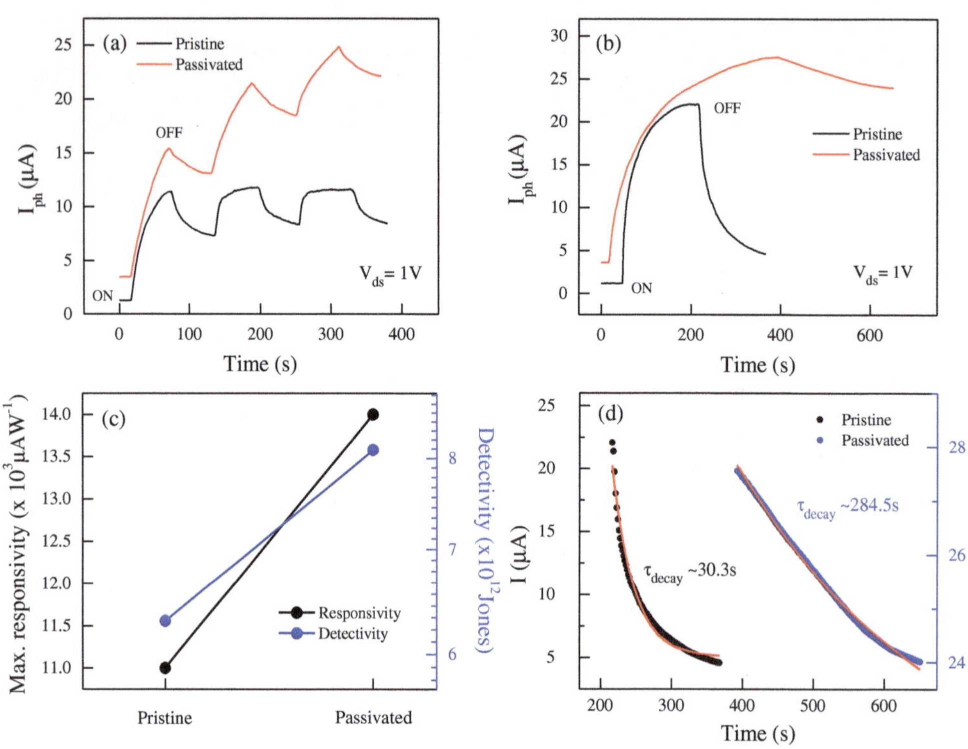

Figure 8. Photocurrent response. (a) Time-dependent photocurrent response of the pristine and passivated devices under vacuum at V_{ds} = 1 V and V_g = 0 V. DUV light was initially turned on for 60 s and then turned off for 60 s. (b) Photocurrent saturation of FL-MoS_2 FETs in vacuum for the pristine and passivated devices at V_{ds} = 1 V and V_g = 0 V. (c) Photocurrent response parameters, including maximum responsivity (left axis) and detectivity (right axis), of the pristine and passivated devices. The data were obtained from Figure 8(b). (d) Decaying behavior of the pristine and passivated devices when the light was turned off after photocurrent saturation. The red line presents the fitting by a single exponent equation.

protecting or capping layer. Hence, the effect of external environments was effectively disregarded.

In addition, field-effect mobility increased after passivation by Al_2O_3 film. The improvement of device performance after deposition of a high-k dielectric is associated with the suppression of Coulomb scattering, modification of phonon dispersion, the difference of Al_2O_3 and SiO_2 dielectric constants (3.9 for SiO_2 and 9.0 for Al_2O_3), and the removal of impurities during ALD growth at 200 °C.[39–41] Extensive theoretical work including the calculation of phonon dispersion relations in MoS_2, the calculation of scattering rates on phonons and charge impurities, is necessary to describe these phenomena completely. After the deposition of a 10-nm-thick Al_2O_3 layer, the electrical properties were also measured under DUV illumination in vacuum at V_{ds} = 1 V. A huge increment in drain current was observed (Figure 7(c)). The field-effect mobility of the pristine and passivated devices in the dark and in the presence of DUV illumination was measured in vacuum, and the mobility was ~18, 32, and 104 cm^2 V^{-1}s^{-1}, respectively (Figure 7(d)).

We explored the photoresponse of the FL-MoS_2 FETs (sample-4; L = 2 μm and W = 8.2 μm) passivated by the Al_2O_3 layer (10 nm). The transfer characteristics of the device are shown in Figure S2d (Supporting Information). The time-dependent response under DUV light is described in Figure 8(a) with switching light ON and OFF for a 60 s cycle at (V_{ds} = 1 V and V_g = 0 V) in vacuum. As shown in Figure 8(a), the passivated device exhibited a slow response because of the suppression of impurities and defect states by the Al_2O_3 layer. However, the photocurrent of the passivated device enhanced compared with that of the pristine device. In Figure 8(b), the maximum photocurrent saturation values (ΔI_{max} = $I_{saturation}$ - I_{dark}) of the pristine and passivated FL-MoS_2 FETs were found to be ~20.6 and 24.2 μA, respectively. R_λ and D^* were also investigated as shown in Figure 8(c). The maximum responsivity and detectivity of the pristine device were ~11 × 10^3 AW^{-1} and ~6.3 × 10^{12} Jones, respectively. However, those after deposition of the Al_2O_3 layer were ~14 × 10^3 AW^{-1} and ~8.2 × 10^{12} Jones, respectively. To investigate further the effect of passivation, the relaxation time (τ_{decay}) of the pristine and passivated devices was examined (Figure 8(d)). The data for decaying time were obtained from Figure 8(b) and exponentially fitted by the single exponential Equation (4). The carrier lifetimes for the pristine and passivated devices were ~30.3 and 284.5 s, respectively.

As depicted in Figure 8(d), the relaxation time of the passivated device was significantly longer than that of the pristine device. This result can be attributed to the fact that the intermediate states of defects and oxygen constitutes are reduced after Al_2O_3 deposition and those excited electrons take a long time to relax.

4. Conclusions

In conclusion, we fabricated FL-MoS_2 FETs and studied the effects of environmental gases in the dark and under DUV illumination. We determined that DUV light induced red and blue shifts of Raman peak (E^1_{2g} and A_{1g}) positions in the presence of N_2 and O_2 gases, respectively. We investigated the electrical transport measurements of FL-MoS_2 FETs under various environmental conditions and found that device performance did not degrade in N_2 gas in the dark. However, O_2 gas significantly reduced the ON-current state. In the dark, the environmental gases did not modify the I_{on}/I_{off} of the device and remained consistent at ~10^5. The threshold voltage and hysteresis were nearly similar in vacuum and N_2 gas but shifted toward positive gate voltage accompanied with a large hysteresis loop in air and O_2 gas. However, the environmental gases changed the characteristics of the FL-MoS_2 FETs under DUV illumination. The photocurrent response and saturated photocurrent of the FL-MoS_2 FETs under various environmental conditions were examined to investigate light detection parameters, such as responsivity, detectivity, external quantum efficiency, and linear dynamic range. The results demonstrated a strong photoresponse in N_2 gas but a poor photoresponse in O_2 gas. The photocurrent relaxation after turning off DUV light was also examined under various environmental conditions. The lifetime of a carrier was longer in N_2 gas but shorter in O_2 gas, indicating that N_2 gas helped the device recover from defects and impurities, while O_2 gas adversely affected device characteristics. Furthermore, we passivated the device by depositing the Al_2O_3 layer to protect MoS_2 from environmental effects. We observed almost hysteresis-free transfer characteristics with improved electrical and photoelectrical responses and field-effect mobility after deposition of the passivation layer. We also found that τ_{decay} increased after Al_2O_3 layer deposition. This phenomenon may be attributed to the screening of impurities and defects from the MoS_2 surface. Thus, we developed improved, stable, reliable, and hysteresis-free FL-MoS_2 FETs for various nanotechnology applications.

Acknowledgment

This research was supported by Nano-Material Technology Development Program (2012M3A7B4049888) through the National Research Foundation of Korea (NRF) funded by the Ministry of Science, ICT and Future Planning. This research was also supported by Priority Research Center Program (2010-0020207) and the Basic Science Research Program (2013R1A1A2061396) through NRF funded by the Ministry of Education.

Disclosure statement

No potential conflict of interest was reported by the authors.

References

[1] Geim AK. Graphene: status and prospects. Science. 2009;324:1530–1534.
[2] Bostwick A, Ohta T, Seyller T, et al. Quasiparticle dynamics in graphene. Nat Phys. 2007;3:36–40.
[3] Geim AK, Novoselov KS. The rise of graphene. Nat Mater. 2007;6:183–191.
[4] Khan MF, Iqbal MZ, Iqbal MW, et al. Improving the electrical properties of graphene layers by chemical doping. Sci Technol Adv Mater. 2014;15:055004.
[5] Khan MF, Iqbal MZ, Iqbal MW, et al. Stable and reversible doping of graphene by using KNO3 solution and photo-desorption current response. Rsc Adv. 2015;5:50040–50046.
[6] Han S, Kwon H, Kim SK, et al. Band-gap transition induced by interlayer van der Waals interaction in MoS 2. Phys Rev B. 2011;84:045409.
[7] Wilson J, Yoffe A. The transition metal dichalcogenides discussion and interpretation of the observed optical, electrical and structural properties. Adv Phys. 1969;18:193–335.
[8] Tsai D-S, Liu K-K, Lien D-H, et al. Few-layer MoS_2 with high broadband photogain and fast optical switching for use in harsh environments. ACS Nano. 2013;7:3905–3911.
[9] Tongay S, Zhou J, Ataca C, et al. Broad-range modulation of light emission in two-dimensional semiconductors by molecular physisorption gating. Nano Lett. 2013;13:2831–2836.
[10] Podzorov V, Gershenson ME, Kloc C, et al. High-mobility field-effect transistors based on transition metal dichalcogenides. Appl Phys Lett. 2004;84:3301–3303.
[11] Li H, Yin ZY, He QY, et al. Fabrication of single- and multilayer MoS_2 film-based field-effect transistors for sensing NO at room temperature. Small. 2012;8:63–67.
[12] He QY, Zeng ZY, Yin ZY, et al. Fabrication of flexible MoS_2 thin-film transistor arrays for practical gas-sensing applications. Small. 2012;8:2994–2999.
[13] Late DJ, Huang YK, Liu B, et al. Sensing behavior of atomically thin-layered MoS_2 transistors. ACS Nano. 2013;7:4879–4891.
[14] Perkins FK, Friedman AL, Cobas E, et al. Chemical vapor sensing with monolayer MoS_2. Nano Lett. 2013;13:668–673.
[15] Khan MF, Iqbal MW, Iqbal MZ, et al. Photocurrent response of MoS_2 field-effect transistor by deep ultraviolet light in atmospheric and N2 gas environments. Acs Appl Mater Inter. 2014;6:21645–21651.
[16] Novoselov K, Jiang D, Schedin F, et al. Two-dimensional atomic crystals. P Natl Acad Sci USA. 2005;102:10451–10453.
[17] Ferrari AC, Basko DM. Raman spectroscopy as a versatile tool for studying the properties of graphene. Nat Nanotechnol. 2013;8:235–246.
[18] Lee C, Yan H, Brus LE, et al. Anomalous lattice vibrations of single- and few-layer MoS_2. ACS Nano. 2010;4:2695–2700.

[19] Chakraborty B, Bera A, Muthu D V S, et al. Symmetry-dependent phonon renormalization in monolayer MoS_2 transistor. Physical Review B. 2012;85:161403.

[20] Late DJ, Liu B, Matte HS, et al. Hysteresis in single-layer MoS_2 field effect transistors. ACS Nano. 2012;6:5635–5641.

[21] Qiu H, Pan LJ, Yao ZN, et al. Electrical characterization of back-gated bi-layer MoS_2 field-effect transistors and the effect of ambient on their performances. Appl Phys Lett. 2012;100:123104.

[22] Park W, Park J, Jang J, et al. Oxygen environmental and passivation effects on molybdenum disulfide field effect transistors. Nanotechnology. 2013;24:095202.

[23] Singh AK, Andleeb S, Singh J, et al. Ultraviolet-light-induced reversible and stable carrier modulation in MoS_2 field-effect transistors. Adv Funct Mater. 2014;24:7125–7132.

[24] Jariwala D, Sangwan VK, Wu CC, et al. Gate-tunable carbon nanotube-MoS_2 heterojunction p-n diode. Proc Natl Acad Sci USA. 2013;110:18076–18080.

[25] Furchi MM, Polyushkin DK, Pospischil A, et al. Mechanisms of photoconductivity in atomically thin MoS_2. Nano Lett. 2014;14:6165–6170.

[26] Jacobs-Gedrim RB, Shanmugam M, Jain N, et al. Extraordinary photoresponse in two-dimensional In2Se3 nanosheets. ACS Nano. 2014;8:514–521.

[27] Hu PA, Wang LF, Yoon M, et al. Highly responsive ultrathin gas nanosheet photodetectors on rigid and flexible substrates. Nano Lett. 2013;13:1649–1654.

[28] Hu PA, Wen ZZ, Wang LF, et al. Synthesis of few-layer GaSe Nanosheets for high performance photodetectors. Acs Nano. 2012;6:5988–5994.

[29] Zhang YZ, Liu T, Meng B, et al. Broadband high photoresponse from pure monolayer graphene photodetector. Nat Commun. 2013;4:1811.

[30] Yin ZY, Li H, Li H, et al. Single-layer MoS_2 phototransistors. ACS Nano. 2012;6:74–80.

[31] Xia FN, Mueller T, Lin YM, et al. Ultrafast graphene photodetector. Nat Nanotechnol. 2009;4:839–843.

[32] Chitara B, Panchakarla LS, Krupanidhi SB, et al. Infrared photodetectors based on reduced graphene oxide and graphene nanoribbons. Adv Mater. 2011;23:5419.

[33] Konstantatos G, Howard I, Fischer A, et al. Ultra sensitive solution-cast quantum dot photodetectors. Nature. 2006;442:180–183.

[34] Fang X, Bando Y, Liao M, et al. An efficient way to assemble zns nanobelts as ultraviolet-light sensors with enhanced photocurrent and stability. Adv Funct Mater. 2010;20:500–508.

[35] Zhang A, Kim H, Cheng J, et al. Ultrahigh responsivity visible and infrared detection using silicon nanowire phototransistors. Nano Lett. 2010;10:2117–2120.

[36] Hu P, Zhang J, Yoon M, et al. Highly sensitive phototransistors based on two-dimensional GaTe nanosheets with direct bandgap. Nano Research. 2014;7:694–703.

[37] Zhang W, Huang JK, Chen CH, et al. High-Gain phototransistors based on a CVD MoS_2 monolayer. Adv Mater. 2013;25:3456–3461.

[38] Soci C, Zhang A, Xiang B, et al. ZnO nanowire UV photodetectors with high internal gain. Nano Lett. 2007;7:1003–1009.

[39] Cho A-J, Yang S, Park K, et al 2014 Multi-layer MoS_2 FET with small hysteresis by using atomic layer deposition Al2O3 as gate insulator. ECS Solid State Letters 3 Q67–Q9

[40] Na J, Joo M-K, Shin M, et al. Low-frequency noise in multilayer MoS 2 field-effect transistors: the effect of high-k passivation. Nanoscale. 2014;6:433–441.

[41] Fonoberov VA, Balandin AA. Giant enhancement of the carrier mobility in silicon nanowires with diamond coating. Nano Lett. 2006;6:2442–2446.

3

Preparation of magnetic mesoporous silica nanoparticles as a multifunctional platform for potential drug delivery and hyperthermia

Xia Yu[a] and Yufang Zhu[b]

[a]School of Medical Instrument and Food Engineering, University of Shanghai for Science and Technology, Shanghai, P.R. China
[b]School of Materials Science and Engineering, University of Shanghai for Science and Technology, Shanghai, P.R. China

ABSTRACT
We report the preparation of magnetic mesoporous silica (MMS) nanoparticles with the potential multifunctionality of drug delivery and magnetic hyperthermia. Carbon-encapsulated magnetic colloidal nanoparticles (MCN@C) were used to coat mesoporous silica shells for the formation of the core-shell structured MMS nanoparticles (MCN@C/mSiO_2), and the rattle-type structured MMS nanoparticles (MCN/mSiO_2) were obtained after the removal of the carbon layers from MCN@C/mSiO_2 nanoparticles. The morphology, structure, magnetic hyperthermia ability, drug release behavior, *in vitro* cytotoxicity and cellular uptake of MMS nanoparticles were investigated. The results revealed that the MCN@C/mSiO_2 and MCN/mSiO_2 nanoparticles had spherical morphology and average particle sizes of 390 and 320 nm, respectively. The MCN@C/mSiO_2 nanoparticles exhibited higher magnetic hyperthermia ability compared to the MCN/mSiO_2 nanoparticles, but the MCN/mSiO_2 nanoparticles had higher drug loading capacity. Both MCN@C/mSiO_2 and MCN/mSiO_2 nanoparticles had similar drug release behavior with pH-controlled release and temperature-accelerated release. Furthermore, the MCN@C/mSiO_2 and MCN/mSiO_2 nanoparticles showed low cytotoxicity and could be internalized into HeLa cells. Therefore, the MCN@C/mSiO_2 and MCN/mSiO_2 nanoparticles would be promising for the combination of drug delivery and magnetic hyperthermia treatment in cancer therapy.

KEYWORDS
Mesoporous nanoparticles; drug delivery; magnetic hyperthermia; multifunctionality

CLASSIFICATION
30 Bio-inspired and biomedical materials; 102 Porous/Nanoporous/Nanostructured materials; 211 Scaffold/Tissue engineering/Drug delivery

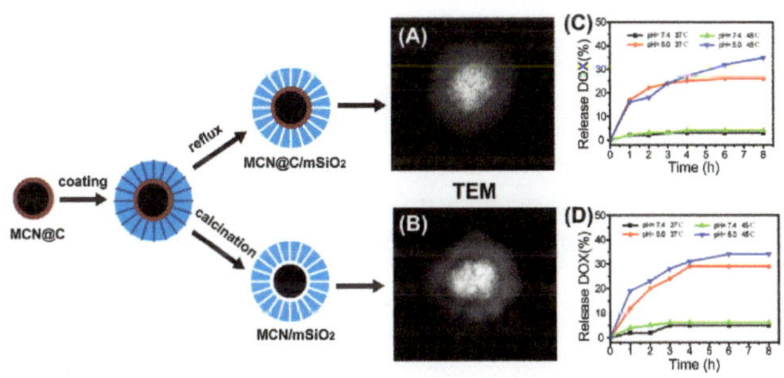

1. Introduction

Chemotherapy is often used for cancer therapy, but the side-effect of toxic free anticancer drugs is very serious to the body.[1] Therefore, the use of carriers for drug delivery is of great important and promising in cancer therapy.

To date, a variety of carriers including organic and inorganic materials have been studied for anticancer drug delivery.[2] Recently, mesoporous silica nanoparticles (MSNs) have been intensively studied as nanocarriers for drug delivery owing to their high surface area, tunable particle size and mesopore size, good biocompatibility and facial surface functionalization.[3–5] Compared to organic carriers such as liposomes, micelles and biodegradable polymers, MSNs can be free from various biochemical attacks. Furthermore, MSNs can load a high amount of drugs and control drug release behavior. For example, Niu et al. [6] developed a kind of core-shell MSN with smaller mesopore in the shell and larger mesopore in the core, which could

CONTACT Yufang Zhu ✉ zjf2412@163.com

achieve a three-step release behavior controlled by tuning the shell thickness from 5 nm to 60 nm. Li et al. [7] designed PEGylated mesoporous silica nanorattles as nanocarriers for docetaxel delivery to liver cancer, and the docetaxel-loaded mesoporous silica nanorattles showed increased tumor inhibition ability compared to the clinical formulation of docetaxel.

However, cancer recurrence often occurs when only chemotherapy is used to treat the tumors, although the use of carriers can allow targeted drug delivery and controlled drug release. Therefore, much effort has been made to enhance cancer therapeutic efficacy by combining other efficient approaches, such as gene therapy, immunotherapy, radiation therapy, and magnetic hyperthermia, to achieve a synergistic effect.[8–10] Among them, magnetic hyperthermia is considered an effective treatment option for cancer without adverse side effects. This method involves raising the temperature to 43–48°C to deactivate cancer cells,[11,12] where the temperature increase (or heat generation) is caused by magnetic nanoparticles (MNPs) due to the hysteresis loss and/or Néel and Brownian relaxations of MNPs under an alternating magnetic field.[10] Heat can increase the efficacy of different chemotherapeutic drugs in the hyperthermia temperature range.[13] On the other hand, studies demonstrated that MNPs-based hyperthermia treatment may induce anticancer immunity, and can also be used for controlled drug delivery.[14] The synergistic therapeutic effects may be achieved by the combination of chemotherapy with magnetic hyperthermia.[15–17] However, the critical issue is to prepare a multifunctional platform with the simultaneous controlled anticancer drug release and magnetic hyperthermia ability.

To solve this critical issue, magnetic mesoporous silica (MMS) nanoparticles are proposed to be a multifunctional platform for drug delivery and magnetic hyperthermia. Mesoporous silica can deliver drugs owing to the high drug loading in mesoporous channels.[3–5] MNPs can be embedded in mesoporous silica, which is useful for the enhancement of the stability of MNPs and the decrease of potential toxicity effect of MNPs on cells.[18] Therefore, preparation of MMS nanoparticles for potential drug delivery or magnetic hyperthermia has gained more attention.[19–30] Gai et al. [19] prepared fibrous-structured magnetic mesoporous particles (Fe_3O_4/FMSMs), which showed a sustained drug release behavior and excellent magnetic responsivity to external magnetic field. Furthermore, the Fe_3O_4/FMSMs carriers could deliver DOX into the cell cytoplasm and cell nucleus. Shi and coworkers [20–22] reported a series of Fe_3O_4/Fe_2O_3@mSiO_2 nanocapsules with a single or double mesoporous silica shell, and these nanocapsules had sufficient magnetic responsivity and high anticancer drug loading capacity. Also, the DOX-loaded nanocapsules showed greater cytotoxicity to induce MCF-7 cell death than free DOX. Julian-Lopez et al. [23] for the first time reported mesoporous maghemite-organosilica microspheres with the multifunctionality of MR imaging and hyperthermia therapy. Vallet-Regí and coworkers [24,25] synthesized magnetic γ-Fe_2O_3-encapsulated mesoporous silica microspheres, and these magnetic microspheres showed the controlled release of therapeutic drugs and good magnetic heating capacity under a low-frequency alternating magnetic field. Lu et al. [26] prepared magnetic iron oxide-loaded hollow mesoporous silica nanocapsules with a particle size of 100 nm, and they found that magnetic nanocapsules induced heating upon exposure to magnetic field and remotely triggered drug release. However, there are limited studies on investigating MMS nanoparticles for drug delivery with synergistic magnetic hyperthermia until now.

Recently, we prepared Fe_3O_4/SiO_2 MMS nanoparticles by encapsulating Fe_3O_4 nanoparticles in mesoporous silica using a sol-gel method, which had pH-controlled DOX release behavior and magnetic heating capacity under an alternating magnetic field.[27] Using DNA to cap the mesopore openings of MMS nanoparticles, the DOX-loaded nanosystem showed potential temperature-controlled DOX release induced by magnetic heating.[28] However, the Fe_3O_4 amount was limited in the prepared Fe_3O_4/SiO_2 MMS nanoparticles, which results in the lower magnetic heating capacity.

In this study, we report a kind of MMS nanoparticle as a multifunctional platform for drug delivery with synergistic magnetic hyperthermia ability. Preparation of MMS nanoparticles involves the synthesis of carbon-encapsulated magnetic colloidal nanoparticles (MCN@C) by solvothermal treatment [31] and subsequently coating mesoporous silica on MCN@C to form the core-shell structured MCN@C/mSiO_2 nanoparticles. Here, the carbon layer could enhance the ability to protect the magnetic colloidal nanoparticles from oxidation or dissolution in an acidic environment. After the treatment of the MCN@C/mSiO_2 nanoparticles by calcination to remove the carbon layers, rattle-type MCN/mSiO_2 nanoparticles were obtained. Furthermore, magnetic hyperthermia ability, drug release behavior, biocompatibility and cellular uptake of the MCN@C/mSiO_2 and MCN/mSiO_2 nanoparticles were investigated.

2. Experimental details

2.1. Materials

Ferrocene ($Fe(C_5H_5)_2$, ≥98%), hydrogen peroxide (H_2O_2, 30%), acetone (C_3H_6O, ≥99%), cetyl trimethyl ammonium bromide (CTAB), triethanolamine (TEA) and tetraethyl orthosilicate (TEOS) were of analytic grade from the Sinopharm Chemical Reagent (Shanghai) Co., Ltd, Shanghai, China. Doxorubicin hydrochloride (DOX) was purchased from Sangon Biotech (Shanghai) Co., Ltd (Shanghai, China). Dodecane was obtained from TCI, Shanghai, China. All chemicals were used as received without further purification.

2.2. Preparation of carbon-encapsulated magnetic colloidal (MCN@C) nanoparticles

MCN@C nanoparticles were synthesized according to the previous report after minor modification.[31] Typically, 0.30 g of ferrocene was completely dissolved in 30 ml of acetone, and 1.50 ml of H_2O_2 was slowly added into the above solution to form the precursor solution. After the vigorous magnetic stirring for 30 min, the precursor solution was transferred to a Teflon-lined stainless autoclave, and then heated to 230°C for 48 h. After cooling to room temperature, the precipitates were collected by a magnet and washed with ethanol several times. Finally, the black products were dried at room temperature in a vacuum oven to obtain MCN@C nanoparticles.

2.3. Preparation of MMS nanoparticles

MMS nanoparticles were prepared via a one-pot biphase stratification approach after some modification.[32] Typically, 0.2 g of MCN@C nanoparticles was added into 60 ml of water, and followed the ultrasonication for 30 min to disperse MCN@C nanoparticles. Subsequently, the MCN@C suspension was transferred into a 100-ml round bottom flask including 0.5 g of CTAB and 0.36 g of TEA. After stirring gently at 60°C for 1 h, 20 ml of TEOS solution (5% v/v in dodecane) was carefully added to the flask. The reaction was kept at 60°C with continuous stirring for 8 h. Then, the colloidal nanoparticles were separated by centrifugation, and washed with ethanol several times. Finally, the obtained colloidal nanoparticles were treated in ammonium nitrate (NH_4NO_3) ethanol solution (0.6 wt%) at 60°C for 6 h thrice to remove the surfactant, which was denoted as MCN@C/mSiO_2 nanoparticles. Rattle-type MMS nanoparticles were obtained by the calcination of the MCN@C/mSiO_2 nanoparticles at 540°C for 7 h, which was denoted as MCN/mSiO_2 nanoparticles.

2.4. Characterization

The wide-angle X-ray diffraction (WAXRD) patterns were performed on a D8 ADVANCE powder diffractometer (Bruker, Germany) using Cu Kα1 radiation (1.5405 Å). Scanning electron microscope (SEM) observations were carried out with a FEI Quanta 450 field emission setup. Scanning transmission electron microscopy (STEM) images were obtained with a Tecnai G2 F30 electron microscope (FEI, Netherlands) operated at an acceleration voltage of 300 kV. UV-vis analysis was performed on a NanoDrop2000C spectrophotometer (Thermo Fisher Scientific, USA). N_2 sorption measurements were carried out on a Micromeritics Tristar 3020 automated surface area and pore size analyzer (Micromeritics, USA) at 77 K under continuous adsorption conditions. Brunauer–Emmett–Teller (BET) and Barrett–Joyner–Halenda (BJH) methods were used to determine the surface area and pore size distribution. Magnetization curves were carried out using a LakeShore 7407 vibrating sample magnetometer (VSM, Lake Shore, USA) at 298 K.

2.5. Drug loading and release from MMS nanoparticles

In this study, anticancer drug (doxorubicin hydrochloride, DOX) was introduced into the MCN@C/mSiO_2 and MCN/mSiO_2 nanoparticles to verify drug release behavior. For drug loading, DOX was dissolved into PBS to obtain DOX solution with a concentration of 0.5 mg ml^{-1}. Subsequently, 60 mg of MCN@C/mSiO_2 or MCN/mSiO_2 nanoparticles was dispersed into 12 ml of DOX solution. The mixture solution was agitated for 24 h under dark conditions, and then collected by centrifugation to obtain the DOX-loaded MCN@C/mSiO_2 or MCN/mSiO_2 nanoparticles. The DOX-loaded nanoparticles were washed with PBS twice to remove DOX adsorbed on the surfaces. The DOX loading amount was determined by measuring the initial and residual DOX solutions using UV-vis analysis at the wavelength of 481 nm.

The DOX release behaviors from the MCN@C/mSiO_2 or MCN/mSiO_2 nanoparticles were performed as follows: 60 mg of the DOX-loaded MCN@C/mSiO_2 or MCN/mSiO_2 nanoparticles was dispersed in 10 ml of the release medium (pH 5.0 or pH 7.4 at 37°C and pH 5.0 or pH 7.4 at 45°C). At the predetermined time intervals, 20 μl of the released solution was withdrew and centrifuged for UV-vis analysis. After the UV-vis analysis, the residual solution was returned to the release system.

2.6. Magnetic hyperthermia ability of MMS nanoparticles

The evaluation of magnetic hyperthermia ability of the MCN@C, MCN@C/mSiO_2 and MCN/mSiO_2 nanoparticles was carried out on a DM100 System (NanoScale Biomagnetics, Spain). The MCN@C, MCN@C/mSiO_2 and MCN/mSiO_2 nanoparticles were dispersed in H_2O, and the concentration was 30 mg ml^{-1}, respectively. Subsequently, 1.0 ml of the nanoparticles suspension was put in the test vessel and then fixed under an alternating magnetic field. The magnetic field strength and frequency were set to be 180 Gauss and 409 kHz, respectively. The specific absorption rate (SAR) was calculated to evaluate the magnetic hyperthermia ability of the MCN@C, MCN@C/mSiO_2 and MCN/mSiO_2 nanoparticles.

2.7. Cell culture

In this study, HeLa cells lines were used to test *in vitro* cytotoxicity and cellular uptake of MMS nanoparticles. HeLa cell lines were maintained in MEM medium containing 10% fetal bovine serum, 100 units ml^{-1} penicillin, and 100 mg ml^{-1} streptomycin. Cells were cultured with the complete medium in 5% CO_2 at 37°C.

2.8. *In vitro* cytotoxicity of MMS nanoparticles

In vitro cytotoxicities of the MCN@C, MCN@C/mSiO_2 and MCN/mSiO_2 nanoparticles were evaluated using Cell Counting Kit-8 (CCK-8) assay (Dojindo, Japan). Before cell seeding, the MCN@C, MCN@C/mSiO_2 and MCN/mSiO_2 nanoparticles were dispersed in MEM medium with a centration of 1 mg ml^{-1}. HeLa cells were seeded into a 96-well plate at a density of 5000 cells in each well. Subsequently, the nanoparticles suspension were added in each well, and the final concentrations of the MCN@C, MCN@C/mSiO_2 and MCN/mSiO_2 nanoparticles were 0, 25, 50, 75 and 100 μg ml^{-1} with the medium volume of 100 μl in each well. After 24 h incubation of cells, 10 μl of CCK-8 solution was added into each well, and the cells were incubated for another 3 h. The optical density (OD) was then measured by the absorbance at 450 nm using a microplate reader (Bio-Rad 680, California, USA).

2.9. Cellular uptake of MMS nanoparticles

For observation of cellular uptake of the MCN@C, MCN@C/mSiO_2 and MCN/mSiO_2 nanoparticles, three kinds of nanoparticles were labeled with rhodamine B isothiocyanate (RBITC) to form RBITC-labeled nanoparticles. Typically, 1.0×10^5 cells were seeded in a 35-mm Petri dish. After the cells were attached to the bottom of Petri dish they were washed twice with PBS. Subsequently, RBITC-labeled nanoparticles were added to the Petri dishes (100 μg ml^{-1}). After another 4 h incubation, the cells were washed with PBS three times to remove the remaining nanoparticles and dead cells. Then, the nuclei of cells were stained for 25 min using Hoechst 33342 solution (0.5 μg ml^{-1}). After washing with PBS to remove the residual dye molecules, the cells were fixed with 4% of paraformaldehyde for 20 min. Finally, the cells were washed with PBS for three times, and observed under a confocal laser scanning microscopy (SP5, Leica, Hamburg, Germany).

3. Results and discussion

3.1. Characterization of MCN@C, MCN@C/mSiO_2 and MCN/mSiO_2 nanoparticles

The wide-angle XRD patterns of MCN@C, MCN@C/mSiO_2 and MCN/mSiO_2 nanoparticles are shown in Figure 1(A). MCN@C nanoparticles were synthesized by the previously reported method,[31] and the XRD pattern indicated that the MCN@C nanoparticles mainly had magnetite (JCPDS file 19-0629) or maghemite (JCPDS file 39-1346). For MCN@C/mSiO_2

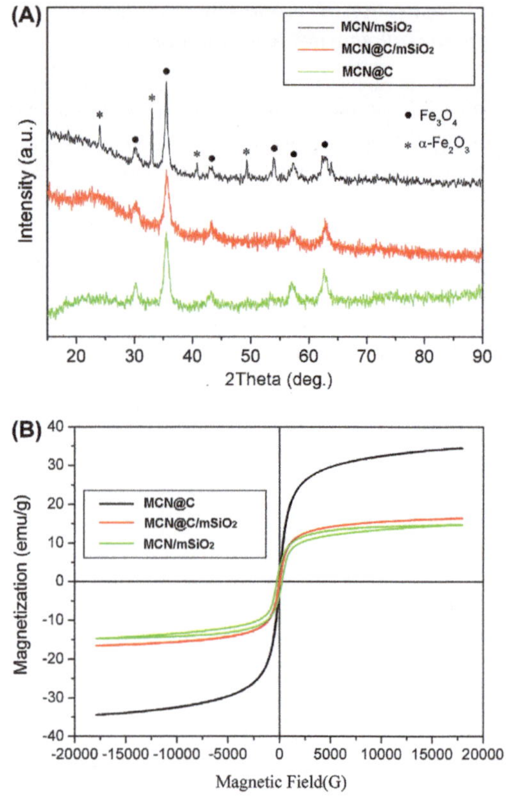

Figure 1. The wide-angle XRD patterns (A) and magnetization curves measured at room temperature (B) of the MCN@C, MCN@C/mSiO_2 and MCN/mSiO_2 nanoparticles.

nanoparticles, the XRD pattern showed a broad reflection at $2\theta = 20–25°$, suggesting an amorphous structure of the mesoporous silica coating; but other diffraction peaks were the same as those of MCN@C nanoparticles, indicating that the crystalline structure of the MCN@C nanoparticles did not change after the coating of mesoporous silica and the removal of surfactant. But for MCN/mSiO_2 nanoparticles, several diffraction peaks indexed to hematite (JCPDS file 33-664) were observed on the XRD pattern, except for the peaks for magnetite or maghemite, which suggests that some magnetite or maghemite in MCN@C nanoparticles have converted to hematite in MCN/mSiO_2 nanoparticles due to the calcination treatment for the removal of carbon and surfactant. On the other hand, compared to the MCN@C and MCN@C/mSiO_2 nanoparticles, the width of the diffraction peaks of the MCN/mSiO_2 nanoparticles became much narrower, which suggested that the calcination treatment of the MCN@C/mSiO_2 nanoparticles induced the formation of larger crystals of magnetite or maghemite. According to Scherrer's formula, the magnetic nanocrystal size was increased from 16.1 nm in the MCN@C/mSiO_2 nanoparticles to 18.2 nm in the MCN/mSiO_2 nanoparticles.

Figure 1(B) shows the magnetization curves of the MCN@C, MCN@C/mSiO_2 and MCN/mSiO_2 nanoparticles measured at 298 K with a vibrating sample magnetometer (VSM). Similar to the previous reports,[31] the magnetic property of the MCN@C nanoparticles demonstrated a very small remanence and reveals its superparamagnetic nature at 298 K, and the magnetization saturation value (M_s) of the MCN@C nanoparticles was estimated to be 34.4 emu g^{-1}. It can also be observed that a very small hysteresis loop is for the MCN@C/mSiO_2 nanoparticles and the M_s decreased to be 16.5 emu g^{-1}. This indicated that the MCN@C nanoparticles still maintained superparamagnetic behavior. The decrease in M_s was attributed to the non-magnetic mesoporous silica coating on the MCN@C nanoparticles. However, a relatively obvious hysteresis loop was observed in the magnetization curve of the MCN/mSiO_2 nanoparticles. It might be that the larger crystals of magnetite or maghemite and hematite formation contributed the ferromagnetic behavior of the MCN/mSiO_2 nanoparticles. On the other hand, the M_s of the MCN/mSiO_2 nanoparticles decreased to be 14.7 emu g^{-1} owing to the formation of some hematite crystals from magnetite or maghemite during the calcination treatment. Compared to our previously reported MMS nanoparticles (M_s of 4.2–8.1 emu g^{-1}),[27] the MCN@C/mSiO_2 and MCN/mSiO_2 nanoparticles exhibited much higher magnetization saturation, which might enhance the magnetic heating capacity.

Figure 2 shows SEM and STEM images of MCN@C, MCN@C/mSiO_2 and MCN/mSiO_2 nanoparticles. The MCN@C nanoparticles had a narrow particle size distribution, and the average particle size was about 170 nm (Figure 2(A)). STEM images revealed that the MCN@C nanoparticles had an obvious core-shell structure, similar to the previous reports,[31] and magnetic cores were composed of small magnetite or maghemite nanocrystals and thin carbon shells were coated on the surfaces of magnetic cores. After coating with mesoporous silica on the MCN@C nanoparticles, the MCN@C/mSiO_2 and MCN/mSiO_2 nanoparticles still had spherical morphology, but the average particle size increased to about 390 and 320 nm, respectively. STEM images showed different structures for the MCN@C/mSiO_2 and MCN/mSiO_2 nanoparticles. The MCN@C/mSiO_2 nanoparticles had the core/shell structure, while the MCN/mSiO_2 nanoparticles had a rattle-type structure, e.g. iron oxide nanocrystals were encapsulated in hollow cores of the MCN/mSiO_2 nanoparticles. On the other hand, dendritic mesoporous channels can be clearly observed on both MCN@C/mSiO_2 and MCN/mSiO_2 nanoparticles, indicating the possibility of drug loading in the MCN@C/mSiO_2 and MCN/mSiO_2 nanoparticles for drug delivery.

Figure 3 shows N_2 adsorption–desorption isotherms of the MCN@C, MCN@C/mSiO_2 and MCN/mSiO_2 nanoparticles and their corresponding pore size distribution curves. The MCN@C nanoparticles had much lower N_2 adsorption and no mesopore size distribution, the BET surface area (S_{BET}) was about 70 m^2 g^{-1}. However, both isotherms of the MCN@C/mSiO_2 and MCN/mSiO_2 nanoparticles belong to type IV isotherm, indicating the mesoporous structure of the MCN@C/mSiO_2 and MCN/mSiO_2 nanoparticles. The S_{BET} of the MCN@C/mSiO_2 and MCN/mSiO_2 nanoparticles were 606 and 709 m^2 g^{-1}, respectively. The pore volumes (V_p) at P/P_0 = 0.97 for the MCN@C/mSiO_2 and MCN/mSiO_2 nanoparticles were 0.95 and 1.08 cm^3 g^{-1}, respectively (Table 1). However, the MCN@C/mSiO_2 and MCN/

Figure 2. SEM and STEM images of the MCN@C (A, D); MCN@C/mSiO_2 (B, E); and MCN/mSiO_2 (C, F) nanoparticles.

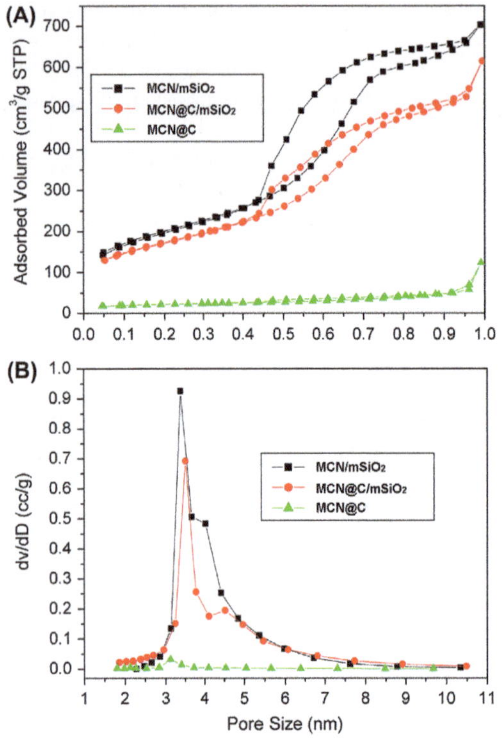

Figure 3. N_2 adsorption-desorption isotherms (A) of the MCN@C, MCN@C/$mSiO_2$ and MCN/$mSiO_2$ nanoparticles and their corresponding pore size distribution curves (B).

$mSiO_2$ nanoparticles exhibited the similar pore size distributions peaked at 3.4 and 3.2 nm. Therefore, the coating of mesoporous silica on the MCN@C nanoparticles provides high surface area and mesoporous channels for loading a high amount of drugs. Furthermore, the MCN/$mSiO_2$ nanoparticles created hollow cores and might contribute much higher drug loading capacity.

3.2. Magnetic hyperthermia ability of the MCN@C, MCN@C/mSiO$_2$ and MCN/mSiO$_2$ nanoparticles

Figure 4 shows the magnetic heating curves of the MCN@C, MCN@C/$mSiO_2$ and MCN/$mSiO_2$ nanoparticles with a concentration of 30 mg ml^{-1} evaluated under the same alternating magnetic field. From the temperature kinetic curve of the MCN@C nanoparticles, the temperature increased from 37 to 59°C within 5 min, suggesting that alternating magnetic field can induce the MCN@C nanoparticles to generate heat and reach the hyperthermia temperature within a very short time. For the MCN@C/$mSiO_2$ and MCN/$mSiO_2$ nanoparticles, the temperature increase with time was a little slower than the MCN@C nanoparticles. But the absolute temperature increases in 5 min for the MCN@C/$mSiO_2$ and MCN/$mSiO_2$ nanoparticles were estimated to be 10.4 and 13.5°C, respectively. That is to say, the MCN@C/$mSiO_2$ and MCN/$mSiO_2$ nanoparticles can also generate heat for hyperthermia treatment. Correspondingly, the specific absorption rate (SAR) of the MCN@C, MCN@C/$mSiO_2$ and MCN/$mSiO_2$ nanoparticles, which was quantified for the magnetic heating capacity of magnetic materials, was estimated to be 54.7, 34.8 and 25.7 W g^{-1}, respectively. The MCN@C/$mSiO_2$ and MCN/$mSiO_2$ nanoparticles also showed good magnetic hyperthermia ability, although the magnetic hyperthermia ability of the MCN@C/$mSiO_2$ and MCN/$mSiO_2$ nanoparticles was lower than that of the MCN@C nanoparticles, owing to the coating of non-magnetic mesoporous silica on the MCN@C nanoparticles. On the other hand, the MCN@C/$mSiO_2$ nanoparticles showed a little higher magnetic heating capacity than the MCN/$mSiO_2$ nanoparticles, which is attributed to the maintaining of iron oxide crystals in the MCN@C/$mSiO_2$ nanoparticles during extracting treatment, but a small amount of hematite crystals were formed in the MCN/$mSiO_2$ nanoparticles during the calcination treatment.

3.3. Drug release behavior from the MCN@C/mSiO$_2$ and MCN/mSiO$_2$ nanoparticles

Doxorubicin hydrochloride (DOX) was used as a model anticancer drug to investigate the drug release behavior from the MCN@C/$mSiO_2$ and MCN/$mSiO_2$ nanoparticles in this study. The DOX loading amounts in the MCN@C/$mSiO_2$ and MCN/$mSiO_2$ nanoparticles were 58.5 and 96.2 μg DOX per 1 mg of the MCN@C/$mSiO_2$ or MCN/$mSiO_2$ nanoparticles, respectively. Here, the higher DOX loading capacity of the MCN/$mSiO_2$ nanoparticles was attributed to its higher surface area and hollow cavity.

Figure 5 shows the DOX release profiles from the MCN@C/$mSiO_2$ and MCN/$mSiO_2$ nanoparticles in the release media with pH 7.4 and 5.0 at temperatures of 37 and 45°C. Here, the temperature of 45°C was used to evaluate the possibility of drug release in the range of hyperthermia temperature. Both MCN@C/$mSiO_2$ and MCN/$mSiO_2$ nanoparticles exhibited pH dependent DOX release behavior. The DOX releases from the

Table 1. The structure parameters, DOX loading capacities and SAR values of the MCN@C, MCN@C/$mSiO_2$ and MCN/$mSiO_2$ nanoparticles.

Samples	S_{BET} (m^2 g^{-1})	V_p (cm^3 g^{-1})	D_p (nm)	DOX loading (μg/mg)	SAR (W g^{-1})
MCN@C	70	0.16	/	/	54.7
MCN@C/$mSiO_2$	606	0.95	3.4	58.5	34.8
MCN/$mSiO_2$	709	1.08	3.5	96.2	25.7

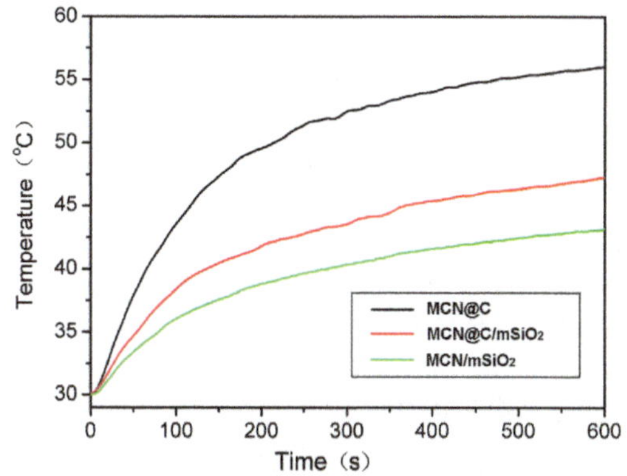

Figure 4. The magnetic heating curves of the MCN@C, MCN@C/mSiO$_2$ and MCN/mSiO$_2$ nanoparticles with a concentration of 30 mg ml^{-1} in H_2O evaluated under an alternating magnetic field. The magnetic field strength and frequency are 180 Gauss and 409 kHz, respectively.

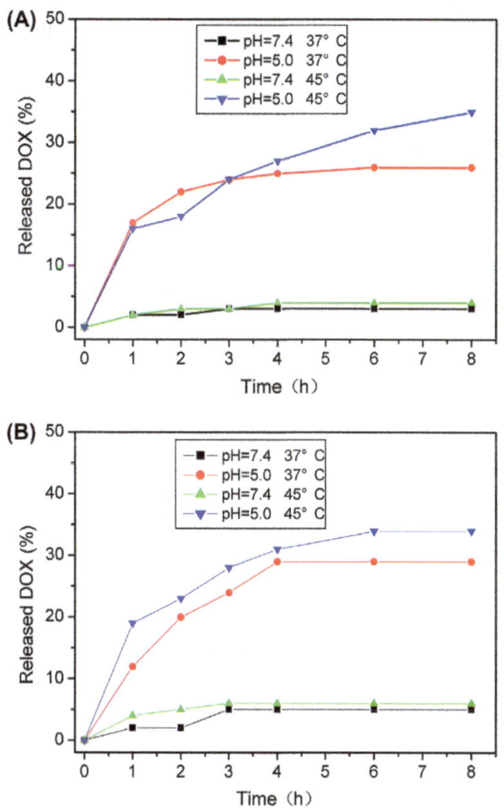

Figure 5. The DOX release behaviors from the MCN@C/mSiO$_2$ (A) and MCN/mSiO$_2$ (B) nanoparticles in the release media of pH 7.4 and 5.0 and at temperatures of 37 and 45°C.

MCN@C/mSiO$_2$ and MCN/mSiO$_2$ nanoparticles were very slow in pH 7.4 medium (only *c.*2–5% of DOX release in 8 h), but the DOX release became faster in pH 5.0 medium (*c.*25–30% of DOX release in 8 h). The DOX release behaviors from the MCN@C/mSiO$_2$ and MCN/mSiO$_2$ nanoparticles are similar to those in previous reports on the DOX release from mesoporous silica nanoparticles.[27] It has been accepted that the electrostatic interaction and hydrogen bonding between DOX molecules and the surfaces of silica nanoparticles could become weaker with lowering pH value, owing to the protonation of surface silanols. Thus, DOX molecules are easy to detach from the surfaces of silica nanoparticles in acidic environment, and thereby fasten DOX release.

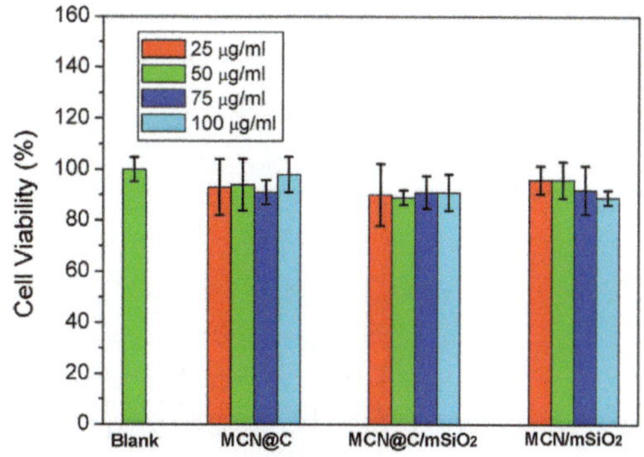

Figure 6. Viabilities of the MCN@C, MCN@C/mSiO$_2$ and MCN/mSiO$_2$ nanoparticles to HeLa cells after incubation for 24 h evaluated using CCK-8 assay.

On the other hand, for both MCN@C/mSiO$_2$ and MCN/mSiO$_2$ nanoparticles, the DOX release at 45°C was slightly faster compared to that at 37°C, which might be because the increased temperature can accelerate the diffusion of DOX molecules in the mesoporous channels. In this study, the MCN@C/mSiO$_2$ and MCN/mSiO$_2$ nanoparticles can generate heat under an alternating magnetic field to raise temperature, and thereby stimulate drug diffusion.

Therefore, such drug release behavior could decrease the side-effect of toxic anticancer drug delivery, because anticancer drugs mostly do not release from the carriers during the circulation of the drug delivery system in the blood stream (pH 7.4). When the drug delivery system is internalized in the targeted cancer cells, faster DOX release could enhance the therapeutic efficiency owing to the acidic environment in the endosome/lysosome and cytosol of cancer cells (pH 5.0–5.5). The combination with magnetic hyperthermia could further accelerate drug release after the endocytosis of drug delivery system by cancer cells.

3.4. In vitro cytotoxicity and cellular uptake of MMS nanoparticles

Figure 6 shows *in vitro* cytotoxicities of the MCN@C, MCN@C/mSiO$_2$ and MCN/mSiO$_2$ nanoparticles against HeLa cells evaluated using CCK-8 assay. Cell viabilities of the MCN@C, MCN@C/mSiO$_2$ and MCN/mSiO$_2$ nanoparticles were more than 90% even up to a

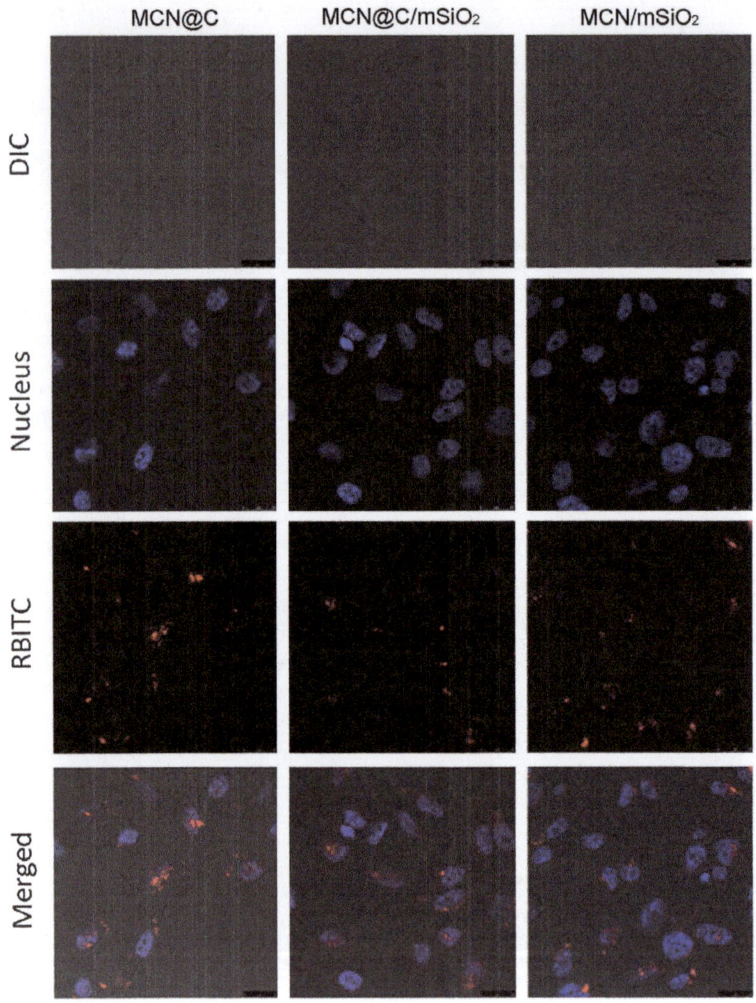

Figure 7. Cellular uptake of the MCN@C, MCN@C/mSiO_2 and MCN/mSiO_2 nanoparticles in HeLa cells after 4 h of incubation with a concentration of 100 μg ml^{-1}: DIC shows the optical photograph of HeLa cells; Nucleus shows the nuclei of HeLa cells stained with Hoechst 33342; RBITC shows the location of RBITC-labeled nanoparticles; Merged shows the overlap DIC, Nucleus and RBITC images, indicating that RBITC-labeled MCN@C, MCN@C/mSiO_2 and MCN/mSiO_2 nanoparticles were taken up by HeLa cells. All scale bars are 25 μm.

concentration of 100 μg ml^{-1}, and they had no significant difference compared to those of the control group. The results revealed that the MCN@C, MCN@C/mSiO_2 and MCN/mSiO_2 nanoparticles are safe for drug delivery.

Cellular uptake of the drug carriers is very important for enhancing the drug delivery efficiency. On the one hand, cellular uptake of drug carriers could be beneficial for magnetic hyperthermia therapy owing to magnetic local heating in cancer cells. On the other hand, cellular uptake of drug carriers could also enhance the chemotherapeutic efficiency due to the intracellular delivery of drugs. In this study, RBITC-labeled MCN@C, MCN@C/mSiO_2 and MCN/mSiO_2 nanoparticles were incubated with HeLa cells for 4 h to observe the cellular uptake. As shown in Figure 7, red fluorescence could be observed in the cells for each type of magnetic mesoporous nanoparticles, which suggests that the MCN@C, MCN@C/mSiO_2 and MCN/mSiO_2 nanoparticles were taken up into HeLa cells after endocytosis. However, the MCN@C, MCN@C/mSiO_2 and MCN/mSiO_2 nanoparticles have not been internalized into the nuclei. Furthermore, the distributions of red fluorescence had no significant difference among the MCN@C, MCN@C/mSiO_2 and MCN/mSiO_2 nanoparticles. Therefore, the MCN@C/mSiO_2 and MCN/mSiO_2 nanoparticles as nanocarriers for anticancer drug delivery could achieve local drug delivery and magnetic hyperthermia, and thereby enhance the cancer therapeutic efficiency.

4. Conclusions

In this study, we have synthesized MMS nanoparticles ($MCN@C/mSiO_2$ and $MCN/mSiO_2$) as a multifunctional platform for potential drug delivery and magnetic hyperthermia. The $MCN@C/mSiO_2$ nanoparticles had core-shell structure with higher magnetic heating capacity compared to the $MCN/mSiO_2$ nanoparticles, but the $MCN/mSiO_2$ nanoparticles had higher drug loading capacity. Both $MCN@C/mSiO_2$ and $MCN/mSiO_2$ nanoparticles had similar drug release behavior with pH-controlled release and temperature-accelerated release. Furthermore, the $MCN@C/mSiO_2$ and $MCN/mSiO_2$ nanoparticles showed low cytotoxicity and could be internalized into cells within a short period. Therefore, the $MCN@C/mSiO_2$ and $MCN/mSiO_2$ nanoparticles would be promising for chemotherapy with synergistic magnetic hyperthermia in cancer therapy.

Disclosure statement

No potential conflict of interest was reported by the authors.

Funding

The authors gratefully acknowledge the support by National Natural Science Foundation of China [number 51572172, 51102166], Shanghai Shuguang Project [number 12SG39], Program for New Century Excellent Talent in University [number NCET-12-1053] and the Hujiang Foundation of China [number B14006].

References

[1] Jelovac D, Armstrong DK. Recent progress in the diagnosis and treatment of ovarian cancer *CA Cancer.* J Clin. 2011;61:183–203.

[2] Cho K, Wang X, Nie S. Therapeutic nanoparticles for drug delivery in Cancer Clin. Cancer Res. 2008;14:1310–1316.

[3] He Q, Shi JL. Mesoporous silica nanoparticle based nano drug delivery systems: synthesis, controlled drug release and delivery, pharmacokinetics and biocompatibility. J Mater Chem. 2011;21:5845–5855.

[4] Wang BX, Li CH, Peng XF. Research on stability of nano-particle suspension. J. Univ. Shanghai Sci. Technol. 2003;3:209–212.

[5] Tang F, Li L, Chen D. Mesoporous silica nanoparticles: synthesis, biocompatibility and drug delivery. Adv Mater. 2012;24:1504–1534.

[6] Niu DC, Ma Z, Li YS, et al. Synthesis of core–shell structured dual-mesoporous silica spheres with tunable pore size and controllable shell thickness. J Am Chem Soc. 2010;132:15144–15147.

[7] Li L, Tang F, Liu H, et al. In Vivo delivery of silica nanorattle encapsulated docetaxel for liver cancer therapy with low toxicity and high efficacy. ACS Nano. 2010;4:6874–6882.

[8] Jia HZ, Zhang W, Zhu JY, et al. Hyperbranched–hyperbranched polymeric nanoassembly to mediate controllable co-delivery of siRNA and drug for synergistic tumor therapy. J. Controlled Release. 2015;216:9–17.

[9] Tagliamonte M, Petrizzo A, Napolitano M, et al. Novel metronomic chemotherapy and cancer vaccine combinatorial strategy for hepatocellular carcinoma in a mouse model. Cancer Immunol Immunother. 2015;64:1305–1314.

[10] Kumar CSSR, Mohammad F. Magnetic nanomaterials for hyperthermia-based therapy and controlled drug delivery. Adv Drug Del. Rev. 2011;63:789–808.

[11] Falk MH, Issels RD. Hyperthermia in oncology. Int J Hyperthermia. 2001;17:1–18.

[12] Kobayashi T. Cancer hyperthermia using magnetic nanoparticles. Biothechnol J. 2011;6:1342–1347.

[13] Lee H, Kim S, Choi BH, et al. Hyperthermia improves therapeutic efficacy of doxorubicin carried by mesoporous silica nanocontainers in human lung cancer cells. Int J Hyperthermia. 2011;27:698–707.

[14] Ito A, Shinkai M, Honda H, et al. Medical application of functionalized magnetic nanoparticles. J Biosci Bioeng. 2005;100:1–11.

[15] Taratula O, Dani RK, Schumann C, et al. Multifunctional nanomedicine platform for concurrent delivery of chemotherapeutic drugs and mild hyperthermia to ovarian cancer cells. Int J Pharm. 2013;458:169–180.

[16] Ren Y, Zhang H, Chen B, et al. Multifunctional magnetic Fe3O4 nanoparticles combined with chemotherapy and hyperthermia to overcome multidrug resistance. Int J Nanomed. 2012;7:2261–2269.

[17] Pradhan P, Giri J, Rieken F, et al. Targeted temperature sensitive magnetic liposomes for thermo-chemotherapy. J Controlled Release. 2010;142:108–121.

[18] Gupta AK, Gupta M. Synthesis and surface engineering of iron oxide nanoparticles for biomedical applications. Biomaterials. 2005;26:3995–4021.

[19] Gai S, Yang P, Ma PA, et al. Fibrous-structured magnetic and mesoporous Fe3O4/silica microspheres: synthesis and intracellular doxorubicin delivery. J Mater Chem. 2011;21:16420–16426.

[20] Wu H, Liu G, Zhang S, et al. Biocompatibility, MR imaging and targeted drug delivery of a rattle-type magnetic mesoporous silica nanosphere system conjugated with PEG and cancer-cell-specific ligands. J Mater Chem. 2011;21:3037–3045.

[21] Chen Y, Chen H, Zeng D, et al. Core/Shell structured hollow mesoporous nanocapsules: A potential platform for simultaneous cell imaging and anticancer drug delivery. ACS Nano. 2010;4:6001–6013.

[22] Zhao W, Chen H, Li Y, et al. Uniform rattle-type hollow magnetic mesoporous spheres as drug delivery carriers and their sustained-release property. Adv Funct Mater. 2008;18:2780–2788.

[23] Julian-Lopez B, Boissiere C, Chaneac C, et al. Mesoporous maghemite–organosilica microspheres: a promising route towards multifunctional platforms for smart diagnosis and therapy. J Mater Chem. 2007;17:1563–1569.

[24] Ruiz-Hernandez E, Lopez-Noriega A, Arcos D, et al. Aerosol-assisted synthesis of magnetic mesoporous silica spheres for drug targeting. Chem Mater. 2007;19:3455–3463.

[25] Martin-Saavedra FM, Ruiz-Hernandez E, Bore A, et al. Magnetic mesoporous silica spheres for hyperthermia therapy. Acta Biomater. 2010;6:4522–4531.

[26] Lu F, Popa A, Zhou S, et al. Iron oxide-loaded hollow mesoporous silica nanocapsules for controlled drug release and hyperthermia. Chem Commun. 2013;49:11436–11438.

[27] Tao C, Zhu Y. Magnetic mesoporous silica nano particles for potential delivery of chemotherapeutic drugs and hyperthermia. Dalton Trans. 2014;43: 15482–15490.
[28] Zhu Y and Tao C. DNA-capped Fe3O4/SiO2 magnetic mesoporous silica nanoparticles for potential controlled drug release and hyperthermia. RSC Adv. 2015;5:22365–22372.
[29] Liu J, Qiao SZ, Hartono SB, et al. Monodisperse yolk-shell nanoparticles with a hierarchical porous structure for delivery vehicles and nanoreactors. Angew Chem Int Ed. 2010;49:4981–4985.
[30] Souza KC, Mohallem NDS, Sousa EMB. Mesoporous silica-magnetite nanocomposite: facile synthesis route for application in hyperthermia. J Sol-Gel Sci Technol. 2010;53:418–427.
[31] Wang H, Chen QW, Yu YF, et al. Size- and solvent-dependent magnetically responsive optical diffraction of carbon-encapsulated superparamagnetic colloidal photonic crystals. J Phys Chem C. 2011;115:11427–11434.
[32] Shen D, Yang J, Li X, et al. Biphase stratification approach to three-dimensional dendritic biodegradable mesoporous silica nanospheres. Nano Lett. 2014;14: 923–932.

4

NEXAFS study of electronic and atomic structure of active layer in Al/indium tin oxide/TiO_2 stack during resistive switching

Elena Filatova[a], Aleksei Konashuk[a], Yuri Petrov[a], Evgeny Ubyivovk[a], Andrey Sokolov[b], Andrei Selivanov[a] and Victor Drozd[a]

[a]Institute of Physics, St Petersburg State University, Ul'yanovskaya Str. 1, Peterhof, 198504, St Petersburg, Russia
[b]Institute Nanometre Optics and Technology (FG-INT), Helmholtz-Zentrum Berlin für Materialien und Energie GmbH, Albert Einstein Str. 15, 12489, Berlin, Germany

ABSTRACT
We have studied the stability of the resistive switching process in the $Al/(In_2O_3)_{0.9}(SnO_2)_{0.1}/TiO_2$ assembly grown by atomic layer deposition. Besides electrical characterization the effect of electric field on the atomic electronic structure of the TiO_2 layer was studied using near edge X-ray absorption fine structure (NEXAFS) spectroscopy. The region of the current instability in the I-V characteristics was revealed. Presumably this current instability is supported by the amorphous structure of the TiO_2 film but is initiated by the surface morphology of the Al substrate. A formation of the O_2 molecules was established which occurs specifically in the region of the current instability that is a result of electrical Joule heating manifestation.

KEYWORDS
ReRAM; NEXAFS; current instability; local heating; molecular oxygen

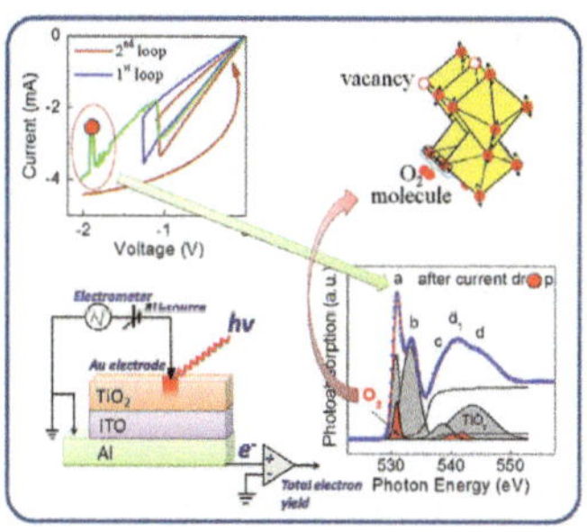

Electrophysical and NEXAFS studies were implemented in the same point of $Al/ITO/TiO_2$ assembly (left bottom corner). The region of the current instability in the I-V characteristics was revealed. The current instability resulted in transition to a new hysteresis loop (left upper corner). A formation of the O_2 molecules in the volume of TiO_2 active layer has been established specifically in the region of current instability (right upper and bottom corners).

1. Introduction

Resistance-change random access memory (ReRAM) has attracted extensive attention as a promising candidate for non-volatile memory due to its high switching speed, high scalability, multi-bit storage potential and simple structure.[1–14] The principle of operation of ReRAMs involves a functional dependence of resistance of the material on the charge passing through it [15] that is realized in devices as resistive switching.[16] There is consensus now that resistive switching is performed by the conductive filaments formation/rupture process in many systems, and the conductive filament model is now accepted extensively.[3,4,17–21] Note that conductive filaments need to be initially formed in the insulator.[22,23] However, there still exist some issues in ReRAM devices based on the filament formation/rupture process that are critical to guarantee the stability and reproducibility of the operations in the ReRAM. Currently, [16,24–26]the

CONTACT Elena Filatova feo@ef14131.spb.edu

possible causes for the lack of stability of resistance values in the low resistive state (R_L) and high resistive state (R_H) are being actively discussed.[16,24–26] It is noted [24,25,27] that the low resistance state is more stable than the high resistance state. According to the model proposed in [28], the distribution of filaments in the switching matrix is generally random, allowing a large variation of switching voltages for different cells from time to time. One of the factors affecting the stability of the resistive switching is a microstructure of the active layer.[29–31] In these references it was found that with increase of crystallinity (decrease of grain boundary density), the resistances and switching voltages become much more stable. According to [32], the location and growth direction of the filaments are confined by the grain boundaries.

The goal of the current paper is a study of the stability of the resistive switching process in the Al/ITO/TiO_2 (ITO is $(In_2O_3)_{0.9}(SnO_2)_{0.1}$) assembly. Earlier [33] we revealed that metal/TCO/TiO_2 (TCO is a transparent conducting oxide) assemblies demonstrate the memristor effect after synthesis (without additional annealing) that provides progress towards understanding the nature of the memristor effect. Also it was established that after the switching process these structures can be retained for a long time. The production-grade Al wafer has been chosen as the substrate because of the specific morphology of its surface (grainy morphology). This allows us to analyze the effect of the surface morphology on the microstructure of the film and as a consequence on the stability of the resistive switching in the system. As a reference system the deg-Si/ITO/TiO_2 assembly was used, where deg-Si is degenerate silicon substrate with atomically smooth surface and electrical properties nearly the same as those of a metal substrate. In order to gain insight into the electronic and atomic structure of the TiO_2 film in addition to electrophysical investigations we have carried out near edge X-ray absorption fine structure (NEXAFS) spectroscopic studies of the assemblies after applying the electric field. We have studied the effect of the electric field at different stages, characterized by different values of current and voltage, on the atomic and electronic structure of assemblies. The NEXAFS arises from the excitations into unoccupied molecular orbitals. NEXAFS is dominated by multiple scattering of a low-energy photoelectron in the valence potential set up by the nearest surroundings. Spectral "fingerprint" techniques can be used to identify the local bonding environment. This fact defines the highest sensitivity of NEXAFS, to distinguish chemical bonds and nearest neighbors. Thus NEXAFS provides information about local (associated with a hole localization in the core shell) and partial (allowing for certain angular momentum symmetry) electronic density of states of the conduction band.

2. Experimental methods

TiO_2 film of 20 nm thickness was grown by the atomic layer deposition (ALD) technique on production-grade Al and deg-Si wafers. ALD provides thickness control and aggressive conformality of TiO_2 thin films at a reasonably low processing temperature. The TiO_2 film was deposited in a cross-flow "Nanoserf" reactor at a temperature of 200°C using $TiCl_4$, $Ti[OCH(CH_3)_2]_4$, $[(CH_3)_2N]_4Ti$ and H_2O precursors. A homogeneous ITO buffer layer of thickness 100 nm was grown on the Al and deg-Si substrates prior to the ALD deposition of TiO_2 film.

The surface topography of Al/ITO/TiO_2 and deg-Si/ITO/TiO_2 assemblies was examined with the helium ion microscope (HIM) at a chamber base pressure of 10^{-7} Torr. A focused beam of single-charged helium ions accelerated to 35 keV scanned over selected area of the sample. A beam current from 0.1 to 1 pA was used for imaging to avoid ion-induced sputtering of the sample. Images were obtained in the secondary electron detection mode with a conventional Everhart-Thornley detector. The high resolution transmission electron microscopy (HRTEM) images were obtained with a Carl Zeiss Libra 200 FE 200 keV microscope.

The I-V characteristics of the synthesized structures were measured at room temperature at ambient atmosphere using an original setup in the I-V sweep mode. The measurements were performed by applying a linearly varying voltage in the range of ±2 V to the top electrode with the bottom electrode grounded (Figure 1); current was changed within ± 4 mA. The applied voltage allowed creation of an electric field of the order of 1 MV/cm in the TiO_2 layer. The Au circular-shaped electrodes were used as external point contacts. The contact area was about 1×10^{-4} cm^2. The electrode holder allowed contact with a constant clamp force. Measurement was performed in a few steps (Figure 2): positioning of the electrode over measuring point, gentle landing of the electrode, carrying out desired sweeps/switching, lifting the electrode.

To study the atomic and electronic structure of the active layer of TiO_2 at the point of impact of a high electric field (at the point of resistive switching) a special

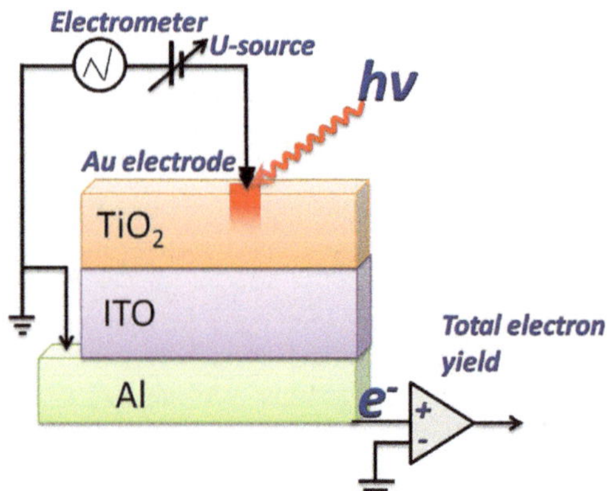

Figure 1. Schematic representation of electrophysical and NEXAFS measurements implemented at the same point of the sample.

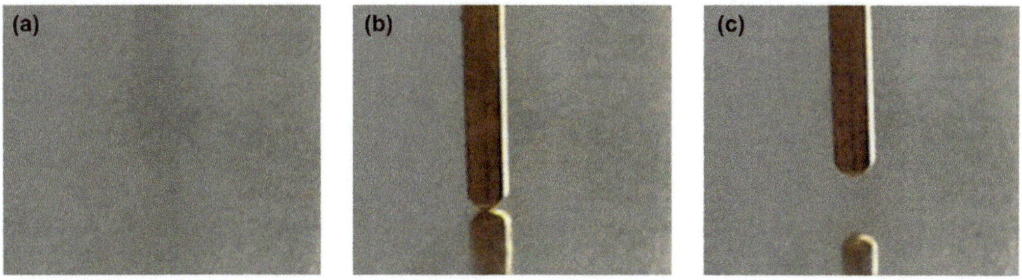

Figure 2. Photograph of the sample surface before and after electrode landing and retraction: sample placed on measuring table (a), electrode landed (b), electrode retracted (c). The diameter of electrode was about 1 mm, the contact area was about 1×10^{-4} cm^2.

fiducial grid was applied on the surface of the sample, enabling measurement of the current-voltage characteristics and NEXAFS spectra in the same point of the film: the light spot on the sample was combined with contact area between the Au electrode and film TiO_2. For this purpose the aluminum fiducial marks were deposited on the surface of pristine TiO_2 by magnetron sputtering. All the marks were located on the same line.

The resistive switching in all the systems was performed outside the vacuum chamber of the spectrometer, in the air. Note that earlier [33] the possibility of switching in the air in all the studied samples was confirmed by implementing the resistive switching in a specially designed box under a nitrogen atmosphere. The resistive switching processes were shown to be almost identical in ambient and nitrogen atmosphere. A specially designed table with a vernier was used. The system allowed the position of the Au electrode to be controlled: it could be mechanically moved along a straight line. The resistive switching was carried out at the points located midway between the Al marks.

After that the sample was placed in the vacuum chamber of the spectrometer so that a straight line passing through the centers of marks were coincided with the line at which the light spot could move during the translation the sample holder. Note that it was possible to move the sample perpendicularly to the beam during the experiment without upsetting the vacuum; this has permitted different points on its surface to be studied. In order to align the scale of the Au electrode movement in air with movement of the sample in the vacuum chamber, the scanning all over the length of the sample was conducted at fixed photon energy. We used an energy of 465 eV, which corresponds to Ti $L_{2,3}$ absorption edge where a strong absorption by titanium atoms occurs. An example of such a scan is shown in Figure 3. The minima on the curve correspond to the aluminum marks, and the maxima correspond to the TiO_2 film located between labels.

As a result the points at which the implementation of high electric field (resistive switching) was carried were established. Further, the Ti $L_{2,3}$- and O K absorption spectra were measured at these points. Schematic representation of the idea of the experiment is shown in Figure 1.

The NEXAFS measurements were performed at the reflectometer set-up mounted on the optics beamline (D-08-1B2) of the Berlin Synchrotron Radiation facility BESSY-II of the Helmholtz Zentrum Berlin (HZB). A Hamamatsu GaAsP diode (4 x 4 mm window), together with a Keithley 617 electrometer, was used as a detector. NEXAFS spectra were measured at the incident angle of 45° in the vicinity of Ti $L_{2,3}$ and O K absorption edges with energy resolution better than $E/\Delta E = 3000$. The spectra were obtained by monitoring the total electron yield from the samples in a current mode. Apertures were used to narrow the area of the light spot on the film. Taking into account the size of X-ray beam which was 150×200 μm^2 and the grazing incidence angle 45°, the size of the studied area was about 200×200 μm^2. Thus the size of the studied area was approximately equal to the active area in which resistive switching occurs. It is important to emphasize that after applying the voltage the electrode was retracted from the film TiO_2, and further NEXAFS measurements were carried out directly at the contact area at the TiO_2 film without electrode. As a consequence, the obtained NEXAFS spectra provided information on changes in the electronic and atomic structure itself from the electric field impact area. The schematic representation of electrophysical and NEXAFS measurements is shown in Figure 1. The paper is focused on the Al/ITO/TiO_2 assembly, which was studied before (as pristine) and after electric field effect. Additionally the deg-Si/ITO/TiO_2 assembly was studied as a reference system.

3. Results and discussion

The stable reversible resistive switchings in the Al/ITO/TiO_2 assembly were achieved by applying a linearly varying voltage. The stable hysteresis loop is shown in Figure 4(a) (blue curve). This state will be termed the first stable hysteresis loop. Only anti-clockwise bipolar resistive switching was realized. Since no instabilities were observed in the range of the stable hysteresis loop the sweep up to −2 V was performed beyond the voltage (−1.3 V) needed for the $R_L \rightarrow R_H$ transition. The instabilities of current started from the −1.7 V. It is important that a current drop had occurred at −1.9 V. The voltage was switched off at −2 V after the reverse current jump.

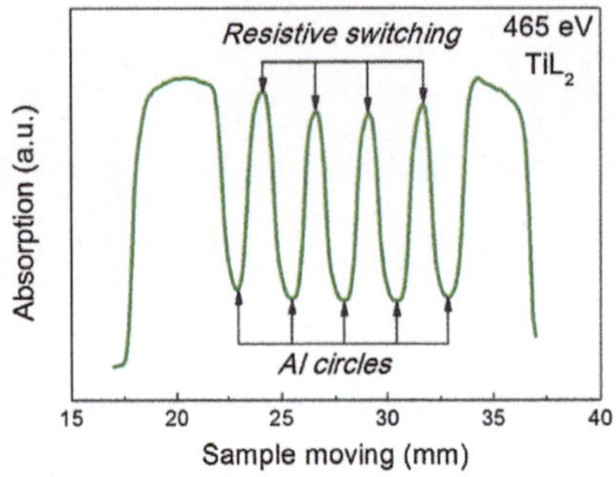

Figure 3. The example of scanning over the length of the sample carried out at 465 eV. The energy 465 eV corresponds to the Ti $L_{2,3}$ absorption edge where a strong absorption by titanium atoms occurs. The minima on the curve correspond to the aluminum marks, maxima correspond to the TiO_2 film located between aluminum marks.

The current instability resulted in transition to a new hysteresis loop that is shown in the inset in Figure 4(a) (brown curve). An initial state in the new loop was R_L. The values of R_L and R_H had decreased in comparison with the values for the initial hysteresis loop. It should be noted that the values of R_L and R_H established for first and second stable hysteresis loops were repeatable many times (at least 15 times) with an uncertainty of ±0.5%.

It was plausible to assume that the process of the current instability was accompanied by irreversible changes in electronic and atomic structure of the TiO_2 because the initial hysteresis loop was no longer obtainable. In order to gain insight into the electronic and atomic structure of the active TiO_2 layer the NEXAFS study was conducted at different steps of the electric field effect, which are shown in Figure 4(a) by red circles. Note that the 20 nm thick film was studied. Taking into account the value of a probing depth (of the order of 7 nm) [34,35] in the NEXAFS technique, it is obvious that we studied the processes occurring in the volume (closer to the surface) of the film but not at the interface between the TiO_2 film and ITO. Actually we studied the response of the electronic and atomic structure of the film to the effect of electric field.

In the first step of the NEXAFS experiment the film (pristine structure) was studied at different points. The absorption spectra in the vicinity of Ti $L_{2,3}$(2p) and O K(1s) absorption edges were measured (Figure 5(a) and (b)). It is important to emphasize that within the statistical error (in the order of 3%) the spectra measured at different points on the surface of the sample coincided.

According to the classical conception, the NEXAFS excitation at the Ti 2p threshold in TiO_2 occurs mainly within the octahedral environment of oxygen atoms and the Ti 2p absorption spectrum should reflect the energies of the free Ti 3d states, because it is dominated by the 2p → 3d dipole transitions in the Ti atoms.[36] One can see that the Ti 2p absorption spectrum reflects clearly the spin-orbit splitting of the Ti 2p level. The Ti $2p_{1/2}$ structures are marked by asterisks in Figure 5(a). The peaks ***a*** and ***b*** in Ti $2p_{3/2}$ spectrum stem from dipole allowed transitions of Ti $2p_{3/2}$ electrons to unoccupied 3d states, which are split into $3dt_{2g}$ (peak ***a***) and 3 de_g (peak ***b***) components by the octahedral crystal field created mainly by the O ions. The line shape analysis of the core-to-e_g transitions attracts the highest attention in order to reveal the crystal modification of TiO_2 film. The absence of the splitting the peak ***b*** (doubly degenerated e_g component) indicates an amorphous state of the studied TiO_2 film (Figure 5(a)). It should be noted that

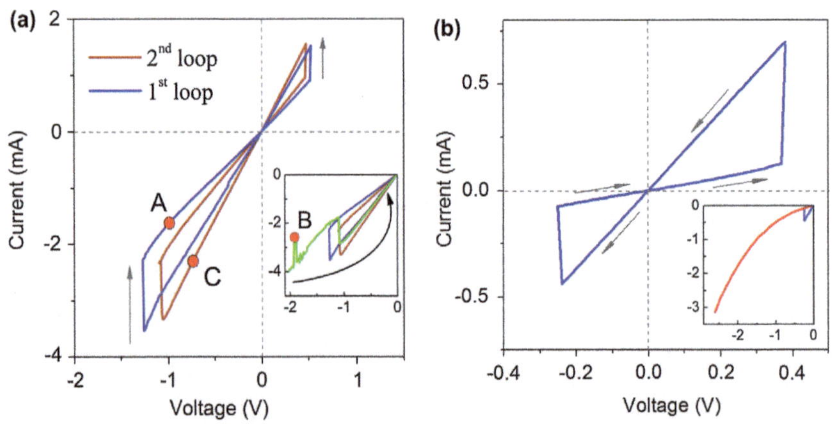

Figure 4. I-V characteristics for the substrate/ITO/TiO_2 assemblies (a) Al substrate; (b) deg-Si substrate. In panel (a) the blue curve is a first stable hysteresis loop. The brown curve is a second stable hysteresis loop obtained right after the current instabilities in the green curve (inset). The inset zooms into the region where the current instability occurs. The green curve is one sweep from 0 to −2 V after obtaining the first stable hysteresis loop, current instabilities were observed starting from −1.7 V. Both hysteresis loops were repeatable at least 15 times with uncertainty of 0.5%. The red circles in the curves denote the steps of I-V sweep on which the NEXAFS measurements were performed. A, R_H state in the first loop; B, current drop in the region of instability; C, R_L state in the second loop. Inset in the panel (b) shows the sweep beyond the stable hysteresis loop, no instabilities were observed.

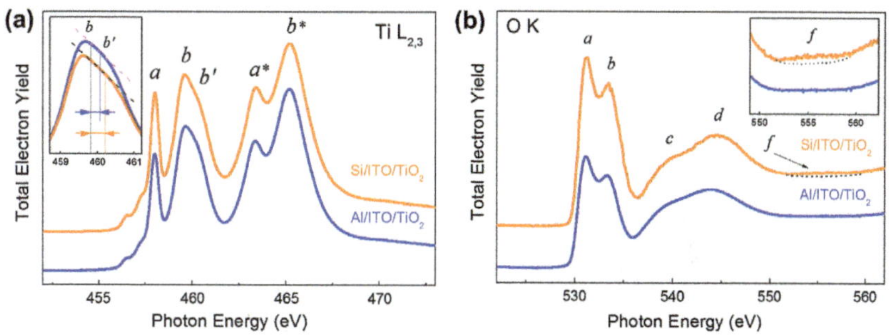

Figure 5. (a) Ti $L_{2,3}$ (2p)- and (b) O K (1s)-absorption spectra of the Al/ITO/TiO_2 and deg-Si/ITO/TiO_2 assemblies (pristine structure). The insets in panels (a) and (b) are zooms of features *b-b'* and *f* respectively.

the Ti 2p- absorption spectra measured before and after the electric field effect were almost indistinguishable, which means the main changes of the structure occur in the sublattice of oxygen. Due to this reason only the O 1s-absorption spectra will be discussed further.

The relative intensities of all the O 1s-absorption spectra have been normalized to the continuum jump (at the photon energy of 560 eV) after subtraction of a sloping background, which was extrapolated from the linear region below O 1s-absorption onset. Such normalization provides about the same total oscillator strength for all the O 1s-absorption spectra over the photon energy range of 520–560 eV in accordance with a general idea of oscillator strength distribution for the atomic X-ray absorption.[36]

Figure 5(b) shows the O 1s-absorption spectrum of the pristine structure of the TiO_2 film. According to [37–40], the molecular orbitals of TiO_2 derived from a linear combination of atomic orbitals (LCAO) are characterized by four unoccupied orbitals: $2t_{2g}$, $3e_g$, $3a_{1g}$ and $4t_{1u}$. In TiO_2, which has zero d-electrons, all four molecular orbitals are completely empty. In this classification the O 1s-edge features (labeled as ***a***, ***b***, ***c*** and ***d***) can be assigned to one electron transition from the O 1s orbital to the $2t_{2g}$, $3e_g$, $3a_{1g}$ and $4t_{1u}$ orbitals of TiO_2, respectively. Thus ***a*** and ***b*** peaks reflect the core-electron transitions in the oxygen atoms to the lowest unoccupied Ti $3dt_{2g}$ and 3 de_g electronic states that are mixed with the 2p states of the ligand (oxygen) atoms. The analysis of the energy position of the peaks shows that within the experimental accuracy (10 meV) the energy separation between ***a*** and ***b*** peaks is equal to 2.3 eV, which correlates well with the value for amorphous TiO_2.[41,42] It is important to stress that according to [40] the feature ***c*** originates from the states, which are dominated by oxygen 2p character and they are best assigned as antibonding combinations of direct oxygen–oxygen interactions. The feature ***d*** can be assigned to titanium 4 sp band mixing with O 2p states. The absence of the feature at the 553 eV in the O 1s-absorption spectrum is additional evidence of the amorphous structure of the film.[42–44]

As it has been mentioned in the introduction, according to the model proposed in [28], the distribution of filaments in the switching matrix is generally random, which allows a large variation of switching voltages for different cells from time to time. One of the factors affecting the stability of the resistive switching is a microstructure of the active layer.[29–31] These authors found that with increase of crystallinity (decrease of grain boundary density), the resistances and switching voltages become much more stable. According to [32] the location and growth direction of the filaments are confined by the grain boundaries. Thus it is plausible to assume that amorphous structure of TiO_2 film may be a reason for current instabilities observed.

In order to gain further insight into a microstructure of TiO_2 layer we have studied a surface topography of TiO_2 film and Al substrate by helium ion microscopy (HIM) and high resolution transmission electron microscopy (HRTEM). The TEM image of Al/ITO/TiO_2 assembly is shown in Figure 6(a). One can see that Al substrate has a pronounced surface relief with characteristic size no less than 20 nm. The ITO layer and TiO_2 film replicate this relief. It is important that TiO_2 film uniformly coats the surface without any discontinuity. The scaled HRTEM image in the area of TiO_2 film is shown in Figure 6(b). No crystal order can be seen in the structure of the TiO_2 film, which confirms its amorphous structure.

The HIM image of the surface topography of the TiO_2 film deposited on Al substrate is shown in Figure 7(a). The surface of the TiO_2 film formed on the Al substrate reveals a grainy morphology with size of grains in the range 20–30 nm. It is important that a clearly defined relief of the surface in the form of rolled strips is traced. The depth of the groove is no less than 10–15 nm and occasionally reaches the value of 20 nm. The surface of the Al substrate has the same morphology features, particularly rolled strips (Figure 7(b)). One can presume that such substrate relief causes the grainy structure of the TiO_2 film, leading to the current instabilities.

To support our idea that amorphous structure of the TiO_2 film may be a reason for the observed current

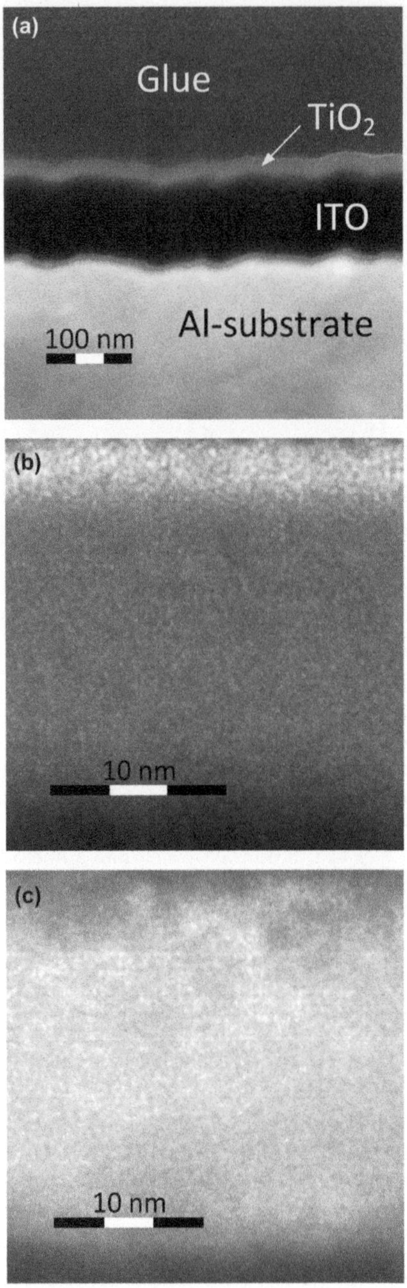

Figure 6. TEM image of Al/ITO/TiO_2 assembly (a). Panels (b) and (c) show the HRTEM images of TiO_2 film in Al/ITO/TiO_2 and deg-Si/ITO/TiO_2, respectively.

instabilities we have additionally studied the deg-Si/ITO/TiO_2 assembly. The I-V characteristic for the deg-Si/ITO/TiO_2 assembly is shown in Figure 4(b). No instability was observed in this system, either inside the hysteresis loop or beyond that, as is shown in the inset of Figure 4(b). As follows from Figure 7(c) the film grown on the deg-Si substrate shows a rather smooth uniform surface. Let us look at the Ti $L_{2,3}$ absorption spectrum of this system in pristine state, shown in Figure 5(a). The spectrum of the film synthesized on deg-Si differs from the spectrum of film grown on the Al substrate in the contrast and the shape of the feature ***b*** (appearance of a shoulder ***b'***). This doubling of the Ti 2p→3 de_g transitions in TiO_2 is well known [40,41,43–45] and is a result of symmetry breaking on Ti atom, which is typical for different crystal modifications (anatase, rutile, and brookite) of TiO_2. The manifestation of doubling of the Ti 2p→3 de_g transitions depends on the crystalline structure of TiO_2 and is expressed in the relative intensities of features ***b*** and ***b'***. In anatase the intensity of feature ***b*** is substantially higher than intensity of the feature ***b'***, while in rutile the opposite situation is observed.[43,45] Analysis of Ti $L_{2,3}$ absorption spectra allows us to assume the presence of signs of an ordered structure of the anatase type in the film synthesized on deg-Si, as the shoulder ***b'*** is more pronounced for deg-Si/ITO/TiO_2 (Figure 5(a) inset; the spectra shown here are normalized to the intensity of first peak ***a***). One can see a decreasing intensity of features ***b-b'*** relative to peak ***a*** for deg-Si/ITO/TiO_2 assembly. According to [46,47], the intensity of t_{2g} state (feature ***a***) dominates over e_g state (feature ***b-b'***) in the Ti $L_{2,3}$ absorption spectra of crystalline TiO_2, while the opposite situation is observed for amorphous TiO_2. Thus the higher intensity ratio of t_{2g} - e_g components (features ***a*** and ***b-b'***) may be considered as additional evidence of better order in TiO_2 film in deg-Si/ITO/TiO_2 assembly. The analysis of the shape of the O K absorption spectra of deg-Si/ITO/TiO_2 and Al/ITO/TiO_2 (Figure 5(b)) can further support our assumption. According to [46,47], the higher intensity of peaks ***a*** and ***b*** in Figure 5(b), together with their narrower width and shift of feature ***d*** toward the higher energy side in comparison with Al/ITO/TiO_2, also points to a better structural order of TiO_2 film in deg-Si/ITO/TiO_2 assembly. Finally, the presence of the broad and weak feature ***f*** in the O K absorption spectrum of the TiO_2 film grown on the deg-Si is an evidence of signs of ordered structure of the anatase type in this film.[40] In view of the electrophysical studies mentioned above, one can conclude the correlation exists between the crystallinity of the film structure and resistances and switching voltages stability that agrees well with [29–32].

Further, the film was studied after the electric field effect. The red circles in the I-V curves in Figure 4(a) denote the following steps of I-V sweep on which the NEXAFS measurements were performed: A denotes the R_H state in the first loop, B was chosen in the range of current instability, specifically after a current drop, and C was chosen after a reverse current jump, which had caused R_L state at new hysteresis loop.

Figure 8 shows the O 1s-absorption spectra of the TiO_2 film at the different steps of the electric field effect. In order to enable a comparison of the spectral features, the relative intensities of the spectra have been normalized to the same continuum jump at the photon energy of 560 eV, after subtraction of a sloping background, which was extrapolated from the linear region

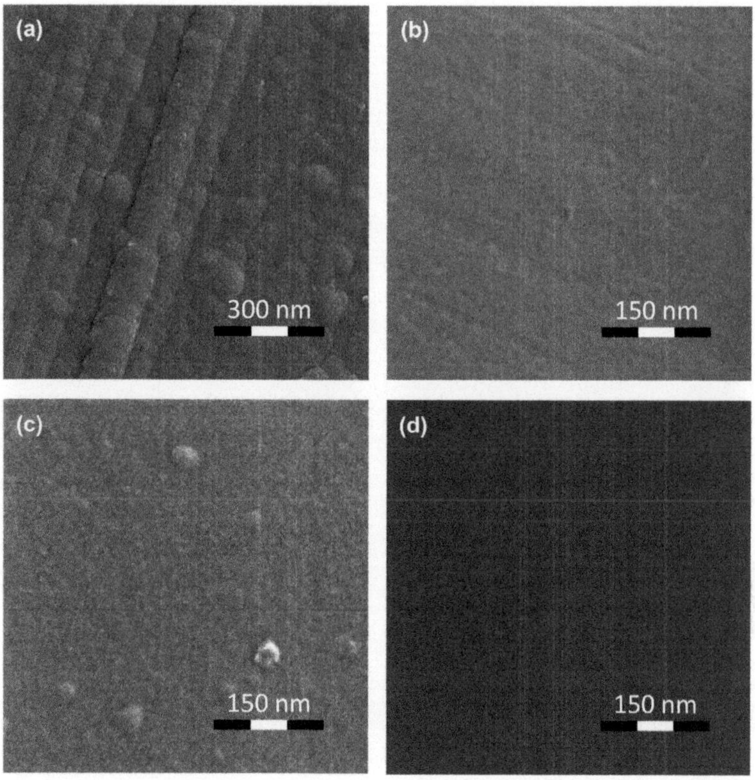

Figure 7. HIM image of TiO_2 film deposited on (a) Al/ITO and (c) deg-Si/ITO substrates and of an Al substrate (b) and a deg-Si substrate (d).

below the O 1s absorption onset. Such normalization provides about the same total oscillator strength for all the O 1s-absorption spectra over the photon energy range of 525–560 eV, in accordance with a general idea of oscillator strength distribution for the atomic X-ray absorption.[36] As it follows from Figure 8 the spectra correlate well in number and energy position with the main features of the fine structure. At the same time a clear change of the integrated intensity of peak ***a*** before and after inducing the electrical filed can be traced. Two different tendencies can be noticed. A significant intensity decrease is observed for peaks ***a*** and ***b*** depending on the step in orderly sequence: pristine structure, R_H state at first hysteresis loop (after electroforming) (A), and in the R_L state at second hysteresis loop (C). Contrary to these spectra the peak intensity ***a*** increases considerably and peak ***d1*** arises in the spectrum corresponding to the current drop (B) in the region of the current instability at the I-V characteristic.

Let us consider jointly the measured O 1s-absorption spectra and I-V characteristics. When a high electric field is applied to pristine structure (electroforming process) the oxygen atoms are knocked out from the lattice and move to the electrode with the positive voltage creating defects that causes formation of the conducting levels in the band gap. Note that in octahedral complexes the metal e_g orbitals (peak ***b***) are directed toward the oxygen atoms and have a strong overlap with the oxygen 2p orbitals. As a result, the O 1s-near edge structure is very sensitive to changes in the local environment of Ti atom.[43,48,49] The above-mentioned change of the integrated intensity of peak ***a*** before and after inducing the electrical field (decrease of the integrated intensity of peak ***a*** after electroforming process) together with the results of studies [48,49], according to which the peak ***a*** of the O 1s-absorption spectrum strongly depends on the concentration of oxygen vacancies, support the physical model of the electroforming process.

As follows from Figure 8 (4th curve), a subsequent decrease of the resistance (transition to R_L state at second hysteresis loop) is also accompanied by some decrease of the peaks intensity ***a*** and ***b***. It is plausible to assume that this transition is connected with further the oxygen atoms leaving the lattice and formation of the oxygen vacancies under the action of the electric field. The electric field stimulates the directed oxygen ions drift from the region of the formed vacancy localization, providing a spatial separation of the vacancy and oxygen ion preventing the possibility of reverse capture. The shift of the oxygen ions O^{2-} can occur also due to diffusion that can be promoted by the local heating.

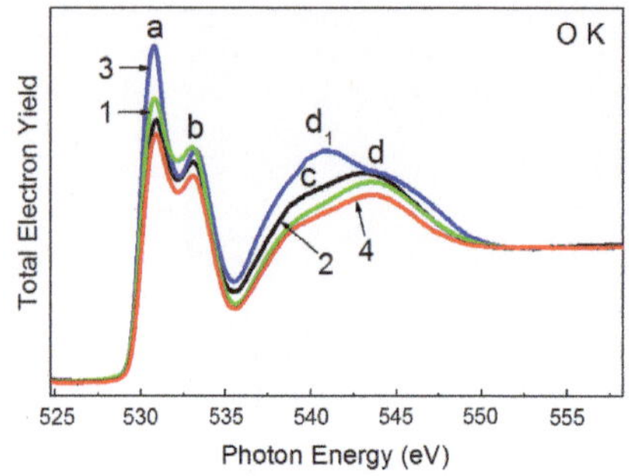

Figure 8. O K (1s)-absorption spectra of the Al/ITO/TiO_2 assembly measured at different stages of the electric field effect, characterized by different values of current and voltage: 1, pristine structure; 2, R_H state at first hysteresis loop (after electroforming); 3, after current drop in the region of current instability; 4, R_L state at second hysteresis loop. The negative polarity of bias was used.

Figure 9(a) shows the O 1s-absorption spectrum of the TiO_2 film measured after the current drop in the region of the instability at the I-V characteristic. The deconvolution of the spectrum performed using free software "Athena" [50] is also shown. The symmetrical Gaussian shape of peaks was used. The continuous edge jump at ionization potential has been represented by the arctangent function. The position of edge jump 536 eV was estimated from sum of O1s binding energy [52] and work function [53] in TiO_2. It can be seen from the presented deconvolution that the discussed spectrum is a sum of two spectra. The first one correlates with the spectrum of amorphous TiO_2.[41,42] The second spectrum demonstrates the presence of a localized narrow peak and four rather wide peaks. The first narrow peak correlates well in width and energy position with the first sharp resonance in absorption spectra of molecular O_2 presented in [51,55] which is due to$1\sigma_u \rightarrow 1\pi_g^*$ transition. The spectra of molecular O_2 contain the second relatively broad doublet band which is related to the $1\sigma_g \rightarrow 3\sigma_u^*$ excitation. Doubling of this band results from the spin-up and spin-down configurations in the final state. The doubling of this band and its relative intensity depend on the state of O_2 but it is essential that the energy position of the center of the σ^* band is not changed for gas-phase O_2, condensed or physisorbed O_2 on the metal surface. A resulting curve for sum of π^* and σ^* peaks in deconvolution of spectrum measured in the region of current instability (Figure 9(a)) is shown in Figure 9(b) in comparison with the spectra of gas-phase O_2 and O_2 physisorbed on Pt(111) from [51]. It is important that the center of σ^* band in the resulting envelope curve is almost the same as that in the presented spectra of gas-phase O_2 and physisorbed O_2. This allows us to affirm that molecular O_2 is being formed in the TiO_2 bulk during current instabilities. The absence of actual doubling of the σ^* band in the resulting envelope curve may indicate that not exactly gaseous O_2 is being formed in TiO_2 at the moment of current instability but some intermediate form of O_2. The bond length is almost the same for the intermediate and gaseous O_2 forms and these intermediate O_2 molecules are not chemically bonded to TiO_2 matrix, as the energy separation between π^* and σ^* bands and the shape of σ^* band are significantly different for lengthened O_2 molecules as it takes place in case of chemisorbed oxygen. [51] A formation of titanium dioxide molecule dimers $(TiO_2)_2$ seems also not to be the case. It is shown in [56] that titanium dioxide molecule dimers have symmetry significantly lower than TiO_6 octahedra. Changing the symmetry of Ti surroundings would surely provoke changing the shape of Ti $L_{2,3}$ absorption spectra but actually the shape of those was almost unchanged at all studied stages of electric field effect, as mentioned above.

The formation of O_2 molecules was indirectly indicated in a number of studies [57–60] by means of observation of bubbles on the surface of the top electrode of some assemblies after resistive switching. All knowledge of the occurrence of oxygen gas has till now been obtained from scanning electron microscopy (SEM) or atomic force microscopy (AFM) images and does not give a clear idea

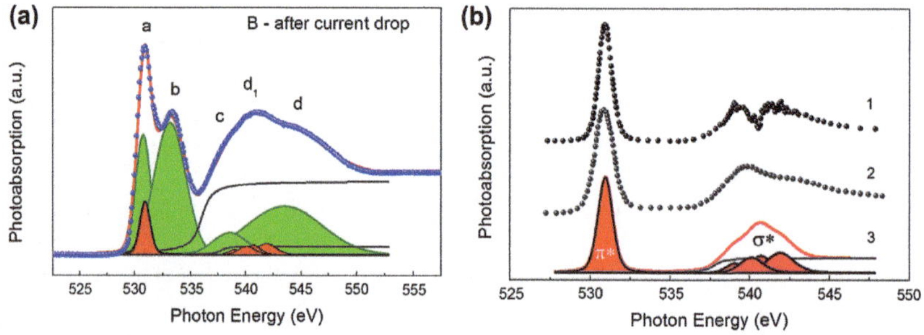

Figure 9. (a) O 1s-absorption spectrum of the Al/ITO/TiO_2 assembly measured in the region B after current drop in the region of the current instability. The deconvolution of the spectrum was performed using the free software "Athena" [50]. Green peaks correspond to t_{2g}, e_g, a_{1g}, and t_{1u} states of TiO_2; red peaks correspond to π^* and σ^* states of molecular oxygen. (b) 1 and 2, O 1s-absorption spectra of molecular oxygen in gas-phase and physisorbed on Pt(111) respectively from [51]; 3, a resulting curve for sum of π^* and σ^* peaks of molecular oxygen in deconvolution of O 1s-absorption spectrum shown in (a).

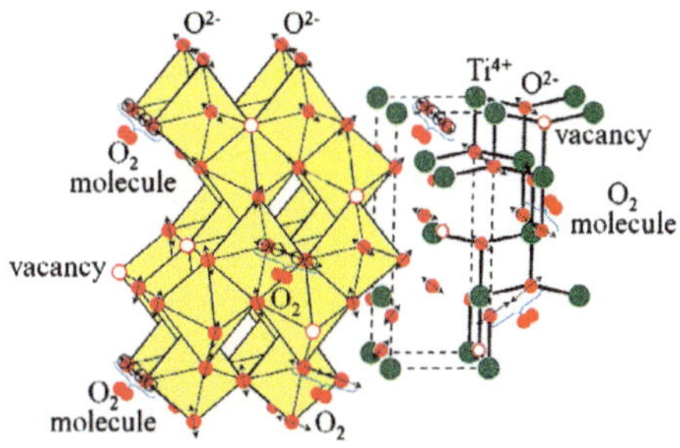

Figure 10. Schematic presentation of the oxygen atoms leaving the lattice, formation of the oxygen vacancies under the action of electric field and local heating and formation of the O_2 molecules in the Al/ITO/TiO_2 assembly.

at which specific stage of the switching process the formation of molecular oxygen occurs. One can see from our result that the O_2 molecules are formed exactly in the region of instability of current because neither the O 1s-absorption spectrum after electroforming nor that for R_L state in the second loop contains the contribution of O_2 molecules (Figure 8). One can see that the influence of electric field on the film in this region leads to formation of the molecular oxygen that is accompanied by the momentary decrease of the current (Figure 4(a)).

In our opinion the observed phenomenon is a manifestation of the joint action of electric field and local heating (self-heating), where the dominant mechanism is heating, wrenching the oxygen atoms off the regular bonds with formation of the oxygen vacancies. In reality the further increase of the voltage up to the region of instability leads to the increase of the amplitude of the oxygen atoms vibrations due to electrical Joule heating and as a result to the creation of the O_2 molecules. We have established that the observed current instability does not depend on the direction of the electric field, which confirms our supposition. The instability occurs at a voltage specific to the system. Exactly at these voltages the creation of the O_2 molecules occurs. Figure 10 summarizes the described processes and gives a schematic presentation of the formation of the O_2 molecules in the Al/ITO/TiO_2 assembly.

4. Conclusions

We have characterized an Al/ITO/TiO_2 assembly, where a TiO_2 layer of 20 nm thickness was grown by ALD. The electric field effect on the TiO_2 structure was studied by NEXAFS in addition to electrophysical measurements. We have observed current instability in the I-V characteristics and have shown its independence on the direction of the electric field. It has been revealed that this current instability is supported by the amorphous structure of the TiO_2 film but is initiated by the surface morphology of the Al substrate. A formation of the O_2 molecules in the TiO_2 active layer has been established specifically in the region of current instability. The formation of the oxygen bubbles on the surface of the top electrode of some assemblies during resistive switching was reported in [57–60]. All the knowledge of occurrence of oxygen gas has until now been obtained from SEM or AFM images and does not give a clear idea at which specific stage of the switching process the formation of molecular oxygen occurs. It has been only noted that the occurrence of the bubbles impedes the stable reversible resistance switching.

In the present studies we have directly observed formation of the O_2 molecules using NEXAFS spectroscopy and have shown that formation of the O_2 molecules occurs specifically in the region of the current instability as a result of electrical Joule heating manifestation. The obtained results are very important for a deeper understanding of the mechanism of resistance switching and charge transport in TiO_2 films and its future application as switching material in resistance change memory based devices.

Acknowledgments

The work was supported by St Petersburg State University [grant 11.37.656.2013]. The authors gratefully acknowledge the financial support of Helmholtz Zentrum Berlin (HZB) and also thank HZB for the allocation of synchrotron radiation beamtime. The work was performed using equipment of interdisciplinary resource center for "Nanotechnology" and resource center "Innovative technologies of composite nanomaterials" of St Petersburg State University.

Disclosure statement

No potential conflict of interest was reported by the authors.

Funding

This work was supported by the Helmholtz-Zentrum Berlin.

References

[1] Szot K, Speier W, Bihlmayer G, et al. Switching the electrical resistance of individual dislocations in single-crystalline $SrTiO_3$. Nat Mater. 2006;5:312–320.
[2] Terabe K, Hasegawa T, Nakayama T, et al. Quantized conductance atomic switch. Nature. 2004;433:47–50.
[3] Waser R, Aono M. Nanoionics-based resistive switching memories. Nat Mater. 2007;6:833–840.
[4] Sawa A. Resistive switching in transition metal oxides. Mater Today. 2008;11:28–36.
[5] Yi HT, Choi T, Cheong S-W. Reversible colossal resistance switching in (La,Pr,Ca)MnO_3: Cryogenic nonvolatile memories. Appl Phys Lett. 2009;95:063509–63511.

[6] Watanabe Y, Bednorz JG, Bietsch A, et al. Current-driven insulator–conductor transition and nonvolatile memory in chromium-doped $SrTiO_3$ single crystals. Appl Phys Lett. 2001;78:3738–3740.
[7] Beck A, Bednorz JG, Gerber CH, et al. Reproducible switching effect in thin oxide films for memory applications. Appl Phys Lett. 2000;77:139–141.
[8] Liu SQ, Wu NJ, Ignatiev A. Electric-pulse-induced reversible resistance change effect in magnetoresistive films. Appl Phys Lett. 2000;76:2749–2751.
[9] Choi BJ, Jeong DS, Kim SK. Resistive switching mechanism of TiO_2 thin films grown by atomic-layer deposition. J Appl Phys. 2005;98:033715–33724.
[10] Wu X, Zhou P, Li J, et al. Reproducible unipolar resistance switching in stoichiometric ZrO_2 films. Appl Phys Lett. 2007;90:183507–183509.
[11] Xu N, Liu LF, Liu XY, et al. Characteristics and mechanism of conduction/set process in TiN/ZnO/Pt resistance switching random-access memories. Appl Phys Lett. 2008;92:232112–232114.
[12] Rohde C, Choi BJ, Jeong DS, et al. Identification of a determining parameter for resistive switching of thin films. Appl Phys Lett. 2005;86:262907–262909.
[13] Choi BJ, Choi S, Kim KM, et al. Study on the resistive switching time of TiO_2 thin films. Appl Phys Lett. 2006;89:012906–12908.
[14] Rossel C, Meijer GI, Bremaud D, et al. Electrical current distribution across a metal–insulator–metal structure during bistable switching. J Appl Phys. 2001;90:2892–2898.
[15] Strukov DB, et al. The missing memristor found. Nature. 2008;453:80–83.
[16] Meijer GI. Who Wins the Nonvolatile Memory Race. Science. 2008;319:1625–1626.
[17] Kwon D-H, et al. Atomic structure of conducting nanofilaments in TiO_2 resistive switching memory. Nat Nanotechnol. 2010;5:148–153.
[18] Soni R, Meuffels P, Kohlstedt H, et al. Reliability analysis of the low resistance state stability of $Ge_{0.3}Se_{0.7}$ based solid electrolyte nonvolatile memory cells. Appl Phys Lett. 2009;94:123503–123505.
[19] Schindler C, Staikov G, Waser R. Electrode kinetics of Cu-SiO_2-based resistive switching cells: Overcoming the voltage-time dilemma of electrochemical metallization memories. Appl Phys Lett. 2009;94:072109–72111.
[20] Russo U, Kamalanathan D, Ielmini D, et al. Study of Multilevel Programming in Programmable Metallization Cell (PMC) Memory. IEEE Trans Electron Devices. 2009;56:1040–1047.
[21] Chang SH, et al. Effects of heat dissipation on unipolar resistance switching in capacitors. Appl Phys Lett. 2008;92:183507–183509.
[22] Choi SJ, Park GS, Kim KH, et al. In Situ Observation of Voltage-Induced Multilevel Resistive Switching in Solid Electrolyte Memory. Adv Mater. 2011;23:3272–3277.
[23] Waser R, Dittmann R, Staikov G, et al. Redox-Based Resistive Switching Memories – Nanoionic Mechanisms, Prospects, and Challenges. Adv Mater. 2009;21:2632–2663.
[24] Yang YC, Pan F, Zeng F. Bipolar resistance switching in high-performance Cu/ZnO:Mn/Pt nonvolatile memories: active region and influence of Joule heating. New J Phys. 2010;12:023008–23018.
[25] Yang M, Pei-Jian Z, Zi-Yu L, et al. Enhanced resistance switching stability of transparent ITO/TiO_2/ITO sandwiches. Chin Phys B. 2010;19:037304–37308.
[26] Wang L-G, Qian X, Cao Y-Q, et al. Excellent resistive switching properties of atomic layer-deposited Al_2O_3/HfO_2/Al_2O_3 trilayer structures for non-volatile memory applications. Nanoscale Res Lett. 2015;10:254028-142.
[27] Xu J, Yang Z, Zhang Y, et al. Bipolar resistive switching behaviours in ZnMnO film deposited on p-Si substrate by chemical solution deposition. Bull Mater Sci. 2014;37:1657–1661.
[28] Chae SC. Random Circuit Breaker Network Model for Unipolar Resistance Switching. Adv Mater. 2008;20:1154–1159.
[29] Zhu X-J, Shang J, Li R-W. An overview of materials issues in resistive random access memory. Front Mater Sci. 2012;6:183–206.
[30] Lanza M, Zhang K, Porti M, et al. Grain boundaries as preferential sites for resistive switching in the HfO_2 resistive random access memory structures. Appl Phys Lett. 2012;100:123508–123511.
[31] Lanza M, Bersuker G, Porti M, et al. Resistive switching in hafnium dioxide layers: Local phenomenon at grain boundaries. Appl Phys Lett. 2012;101:193502–193506.
[32] Zhuge F, Peng S, He C, et al. Improvement of resistive switching in Cu/ZnO/Pt sandwiches by weakening the randomicity of the formation/rupture of Cu filaments. Nanotechnol. 2011;22:275204–275210.
[33] Filatova EO, Baraban AP, Konashuk A, et al. Transparent-conductive-oxide (TCO) buffer layer effect on the resistive switching process in metal/ TiO_2/ TCO/metal assemblies. New J Phys. 2014;16:113014–113029.
[34] Stöhr J. NEXAFS Spectroscopy. Berlin: Springer; 1992.
[35] Hähner G. Near edge X-ray absorption fine structure spectroscopy as a tool to probe electronic and structural properties of thin organic films and liquids. Chem. Soc. Rev. 2006;35:1244–1255.
[36] Fano U, Cooper JW. Spectral Distribution of Atomic Oscillator Strengths. Rev Mod Phys. 1968;40:441–507.
[37] Fischer DW. X-Ray Band Spectra and Molecular-Orbital Structure of Rutile TiO_2. Phys Rev B. 1972;5:4219–4226.
[38] Stoyanov E. The effect of valence state and site geometry on Ti $L_{3,2}$ and O K electron energy-loss spectra of Ti_xO_y phases. Am. Mineral. 2007;92:557–586.
[39] Groot FMF, Figueiredo MO, Basto MJ, et al. 2p X-ray Absorption of Titanium in Minerals. Phys Chem Miner. 1992;19:140–147.
[40] Groot FMF, Faber J, Michiels JJM, et al. Oxygen 1s x-ray absorption of tetravalent titanium oxides: A comparison with single-particle calculations. Phys Rev B. 1993;48:2074–2080.
[41] Rath S, Gracia F, Yubero F, et al. Angle dependence of the O K edge absorption spectraof TiO_2 thin films with preferential texture. Nucl Instrum Methods Phys Res B. 2003;200:248–254.
[42] Filatova E, Taracheva E, Shevchenko G, et al. Atomic ordering in TiO_2 thin films studied by X-ray reflection spectroscopy. Phys Status Solidi B. 2009;246:1454–1458.
[43] Brydson R, Sauer H, Engel W, et al. Electron energy loss and X-ray absorption spectroscopy of rutile and anatase: a test of structural sensitivity. J Phys: Condens Matter. 1989;1:797–812.
[44] Groot FMF, Fuggle JC, Thole BT, et al. $L_{2,3}$ x-ray-absorption edges of d_0 compounds: K^+, Ca_2^+, Sc_3^+, and Ti_4^+ in Oh (octahedral) symmetry. Phys Rev B. 1990;41:928–937.

[45] Ruus R, Kikas A, Saar A, et al. Ti 2p and O1s X-ray absorption of TiO_2 polymorphs. Solid State Commun. 1997;104:199–203.
[46] Kucheyev SO, Buuren T, Baumann TF, et al. Electronic structure of titania aerogels from soft x-ray absorption spectroscopy. Phys Rev B. 2004;69:245102–245108.
[47] Casu MB, Braun W, Bauchspieß K R, et al. A multi-technique investigation of TiO_2 films prepared by magnetron sputtering. Surf Sci. 2008;602:1599–1606.
[48] Filatova EO, Sokolov A A, Egorova Yu V, et al. X-ray spectroscopic study of $SrTiO_x$ films with different interlayers. J Appl Phys. 2013;113:224301–224309.
[49] Muller DA, Nakagawa N, Ohtomo A, et al. Atomic-scale imaging of nanoengineered oxygen vacancy profiles in $SrTiO_3$. Nature. 2004;430:657–661.
[50] Ravel B, Newville, ATHENA, ARTEMIS, HEPHAESTUS: data analysis for X-ray absorption spectroscopy using IFEFFIT. J Synchrotron Radiat. 2005;12:537–541.
[51] Chen JG. NEXAFS investigations of transition metal oxides, nitrides, carbides, sulfides and other interstitial compounds. Surf Sci Rep. 1997;30:1–152.
[52] Gouttebaron R, Cornelissen D, Snyders R, et al. XPS study of TiO_x thin films prepared by d.c. magnetron sputtering in Ar–O_2 gas mixtures. Surf Interface Anal. 2000;30:527–530.
[53] Diebold U, Madey TE. TiO_2 by XPS. Surf Sci Spectra. 1998;4:227–231.
[54] Scanlon DO, Dunnill C W, Buckeridge J, et al. Band alignment of rutile and anatase TiO_2. Nat Mater. 2013;12:798–801.
[55] Kuiper P, Dunlap B. The σ^* absorption peak at the oxygen 1s edge of O_2: Exchange splitting, ultrafast dissociation, and atomiclike Auger spectra. J Chem Phys. 1994;100:4087–4092.
[56] Dargouthi S, Boughdiri S, Tangour B. Stabilizing of the Transitory Species $(TiO_2)_2$ by Encapsulation Into Carbon Nanotubes. Acta Chim Slov. 2015;62:445–451.
[57] Kwon D-H, et al. Atomic structure of conducting nanofilaments in TiO_2 resistive switching memory. Nat Nanotechnol. 2010;5:148–153.
[58] Yang JJ, Miao F, Pickett M D, et al. The mechanism of electroforming of metal oxide memristive switches. Nanotechnol. 2009;20:215201–215210.
[59] Huang J-J, Kuo C-W, Chang W-C, et al. Transition of stable rectification to resistive-switching in Ti/TiO_2/Pt oxide. Appl Phys Lett. 2010;96:262901–262904.
[60] Mehonic A, Buckwell M, Montesi L, et al. Structural changes and conductance thresholds in metal-free intrinsic SiO_x resistive random access memory. J Appl Phys. 2015;117:124505–124513.

Fabrication of photoactive heterostructures based on quantum dots decorated with Au nanoparticles

Elisabetta Fanizza[a,b], Carmine Urso[a], R. Maria Iacobazzi[c,d], Nicoletta Depalo[b], Michela Corricelli[a,b], Annamaria Panniello[b], Angela Agostiano[a,b], Nunzio Denora[c], Valentino Laquintana[c], Marinella Striccoli[b] and M. Lucia Curri[b]

[a]Dipartimento di Chimica, Università degli Studi di Bari, Via Orabona 4, 70126 Bari, Italy
[b]Istituto per i Processi Chimico Fisici IPCF Consiglio Nazionale delle Ricerche CNR, Via Orabona 4, 70126 Bari, Italy
[c]Dipartimento di Farmacia – Scienze del Farmaco, Università degli Studi di Bari, Via Orabona 4, 70126 Bari, Italy
[d]Istituto tumori IRCCS Giovanni Paolo II, Bari, Italy

ABSTRACT
Silica based multifunctional heterostructures, exhibiting near infrared (NIR) absorption (650–1200 nm) and luminescence in the visible region, represent innovative nanosystems useful for diagnostic or theranostic applications. Herein, colloidal synthetic procedures are applied to design a photoactive multifunctional nanosystem. Luminescent silica (SiO_2) coated quantum dots (QDs) have been used as versatile nanoplatforms to assemble on their surface gold (Au) seeds, further grown into Au spackled structures. The synthesized nanostructures combine the QD emission in the visible region, and, concomitantly, the distinctive NIR absorption of Au nanodomains. The possibility of having multiple QDs in a single heterostructure, the SiO_2 shell thickness, and the extent of Au deposition onto SiO_2 surface have been carefully controlled. The work shows that a single QD entrapped in 16 nm thick SiO_2 shell, coated with Au speckles, represents the most suitable geometry to preserve the QD emission in the visible region and to generate NIR absorption from metal NPs. The resulting architectures present a biomedical potential as an effective optical multimodal probes and as promising therapeutic agents due to the Au NP mediated photothermal effect.

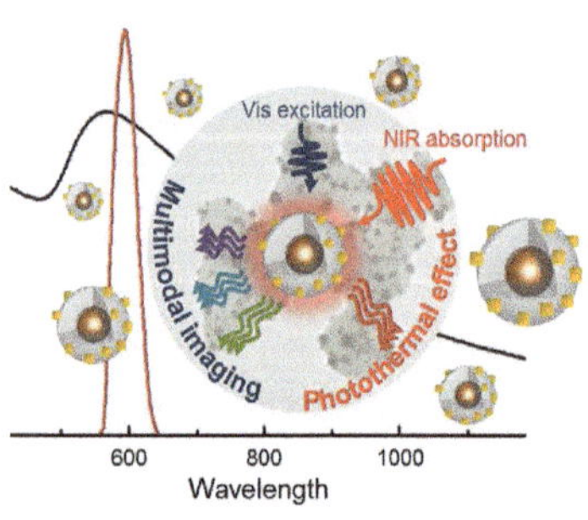

KEYWORDS
Quantum dots; metal nanoparticles; plasmonic luminescent nanostructures; multimodal bioimaging

CLASSIFICATION
10 Engineering and Structural materials

1. Introduction

Engineered photoactive nanomaterials represent a great promise for the advance of the next generation of nanomedicine tools, such as tags for imaging and early-stage diagnostic and therapeutic monitoring of diseases [1–3]. Current research efforts are focused on the design, development and application of multifunctional nanoparticles (NPs) that simultaneously provide cancer diagnostics and therapy, such as nanostructures able to specifically locate and selectively destroy cancer cells.

Efforts to create these multicomponent nanostructures have largely been driven by their specific chemical and physical properties and by their expected functionality, which make them advantageous in applications otherwise inaccessible by their single components.

Plasmonic NPs, exhibiting absorption in the visible and near infrared (vis-NIR) window (700–1200 nm), are

CONTACT Elisabetta Fanizza elisabetta.fanizza@uniba.it

clinically important owing to the low extinction coefficients of water, tissue, and blood in this region of the electromagnetic spectrum. Several gold (Au) based NIR absorbing materials have recently been developed, such as Au NPs of different shapes (nanorods, nanocages, agglomerates, hollow, pentagons, and large prisms) and Au composites (SiO_2@Au nanoshells and superparamagnetic iron oxide nanoshells) [4–8], which can be investigated by different imaging modalities, such as confocal microscopy, photoacoustic tomography, computed tomography, and dark-field light scattering imaging, and can be used to deliver localized photothermal therapy [9–11].

Heterostructures, that combine the NIR absorption of plasmonic NPs with supplementary functionalities, such as photoluminescence or magnetism, appear attractive for application in nanomedicine [1–3, 6, 8, 10–13]. Such nanostructures have potential as versatile multimodal bioimaging agents and efficient therapeutic tools by photothermal ablation, and hence for a simultaneous early stage detection and non-invasive treatment of various diseases, especially of cancer [1–3].

Here, the integration of NIR absorbing and luminescent materials into a single architecture with a size of a few tens of nanometers is presented. The all-inorganic heterostructure proposed in this work is based on plasmonic Au NPs and fluorescent CdSe@ZnS quantum dots (QDs), more stable with respect to organic dyes. Only a few reports have so far dealt with this type of heterostructure [15–18], applied as contrast agents for plasmon and fluorescence-based imaging and detection, with potential for image-guided photoablation therapy.

However the fabrication of this class of heterostructures imposes the control of the interaction between the plasmonic and luminescent domains, which depends on the mutual distance between plasmonic and fluorescent particles down to the nanometer level. It is known that QD emission intensity can be quenched or, alternatively, enhanced by carefully tailoring the distance from the plasmonic moiety [15–18]. In addition the plasmonic component must be properly designed in order to have absorption in the NIR region.

Here, a colloidal silica shell (SiO_2) has been selected as a robust scaffold and a versatile nanoplatform for combining the distinct spectral features of the fluorescent and plasmonic component [8]. By means of a multistep synthetic approach, QDs have been coated with a silica shell that has been further functionalized with Au nanodomains. The tunable thickness of the SiO_2 represents an adjustable spacer to properly modify the distance between metal NPs and luminescent QDs, and, hence, to control their interaction.

Recently, it has been reported [19] that SiO_2 NPs, coated with an Au shell, resulted in nanostructures with strong absorption and scattering ranging from the visible to near-IR region, depending on the core/shell size ratio. However most works concern Au nanoshells grown onto SiO_2 NPs of large diameter (between 80 and 150 nm), that show restricted application *in vivo* due to the limited diffusion in tissue of such large particles [19]. Few studies of NIR absorbing plasmonic NPs with sizes below this range have been reported [19]. The nanostructure proposed here, purposely designed with a size of a few tens of nanometers, can be envisaged to provide a significant contribution to this perspective.

A comprehensive optical (UV-vis absorbance, photoluminescence and infrared spectroscopy) and structural (transmission electron microscopy) investigation of the prepared structures has been performed in order to define the experimental conditions to achieve the most suited architecture able to combine NIR absorption and luminescence in the visible range. The cytotoxicity of nanostructures has been carefully evaluated after each fabrication step by means of *in vitro* studies on rat C6 glioma cells line, in order to assess their impact in a biological environment and thus makes such a multifunctional structure, potential for simultaneous multimodal imaging and therapy [15]. In addition, both SiO_2 and Au, generally regarded as safe biomaterials [20–25], provide surface anchoring points for efficient biomolecular conjugation for targeting applications [26–31].

2. Materials and methods

2.1. Chemicals

The CdSe@ZnS QD synthesis and surface processing were achieved through the following reagents: cadmium oxide (99.5% powder, CdO), trioctylphosphine oxide (99%, TOPO), t-butylphosphonic acid (98%, tBuPOH), hexadecylamine (technical grade 90%, HDA, Fluka, Milan, Italy), tributylphosphine (97%, TBP), selenium (99.99% powder, Se), sulfur (99.9999% powder, S), trioctylphosphine (90%, TOP), diethylzinc solution (1 M in heptane, Et_2Zn), hexamethyldisilathiane (HMST). The SiO_2 shell growth was carried out by: tetraethyl orthosilicate (98% d = 0.934 g $^{ml-1}$, TEOS), aqueous ammonia (d = 0.900 g $^{ml-1}$, NH_4OH), 5 polyoxyethylene nonylphenylether (igepal CO-520, Mn = 441), and surface functionalization with (3-aminopropyl)triethoxysilane (97%, APS), sodium hydroxide (NaOH), tetrakis(hydroxymethyl) phosphonium chloride (THPC, 80% in H_2O), gold(III) chloride hydrate ($HAuCl_4$•H_2O, 99.9%), potassium carbonate (K_2CO_3 99.8%), formaldehyde. For the ninhydrin test, ninhydrin and 2,6-lutidine (98+%, Alfa Aesar, Milan, Italy) were used. All chemicals, purchased from Sigma Aldrich (Milan, Italy) if not specified otherwise, were used without purification. All solvents, namely methanol, ethanol, chloroform, cyclohexane were of the highest purity available.

2.2. Synthesis of SiO_2 coated CdSe@ZnS QDs (QD@SiO_2)

The silica shell growth onto the 'as-synthesized' CdSe@ZnS QDs (see Supplementary Data for experimental details) was carried out by using a water-in-oil microemulsion approach [32–36]. Sequentially, 350 μl of IGEPAL CO-520, 200 μl of NH_4OH and a varied volume of TEOS (in the range of 20 to 50 μl) were injected to 6 ml of a cyclohexane solution of QDs ($1.5{\cdot}10^{-6}$ M). The solution was then kept under vigorous stirring and controlled temperature 28°C for 18 h.

Methanol was added to disrupt the micelle and the QD@SiO_2 NPs were collected by centrifugation at 7800 g for 20 min (Beckman J2-21, Rodano (MI), Italy). Repeated cycles of NP dispersion in ethanol and centrifugation were carried out to completely remove surfactant residuals. The prepared NPs were finally dispersed in 4 ml of ethanol. The QD@SiO_2 concentration in solution was nearly 10^{15} NPs in 4 ml of ethanol, and was calculated by using the method described in the Supplementary Material.

2.3. Surface functionalization with amine groups and characterization of QD@SiO_2 nanoparticles

Surface functionalization reactions of QD@SiO_2 NPs with amine groups were carried out adding equal volume of ammonia solution and APS, 20, 80 or 120 μl, respectively, to 2 ml of NP suspension prepared by diluting 400 μl of QD@SiO_2 samples with ethanol. The final suspension was stirred overnight at 28°C and the functionalized NPs were collected by centrifugation and cleaned up from unreacted reagents by repeated cycles of centrifugation/redispersion in ethanol.

The ninhydrin test, which allows an effective characterization and quantitative analysis of primary amine (NH_2) groups in different types of system, including biomolecules [37], was here modified to investigate the functionalization step of QD@SiO_2 and to quantify the yield of surface functionalization. A ninhydrin solution was preliminary prepared by dissolving 110 mg of ninhydrin in 16 ml of ethanol (0.68% w/v) and adding 4 ml of 2,6-lutidine.

Preliminary calibration experiments were performed by adding an excess of ninhydrin solution in ethanol to APS standard solutions ranging from 50 μM to 15 mM. Upon heating in a thermostatic bath at 90°C for 5 min, in the presence of NH_2 groups, the solution color changes from pale yellow (free ninhydrin) to blue due to the formation of Ruhemann's Blue byproduct, thus providing an evidence of the presence of the amine groups. The absorption intensity at 570 nm, characteristic of the Ruhemann's Blue byproduct, monitored by UV-vis absorbance spectroscopy, allows us to plot a calibration line and to determine the extinction coefficient ($\varepsilon_{570\ nm}$ = 49 l mol^{-1} cm).

The addition of an excess of ninhydrin solution to the NH_2-functionalized NP suspension and the spectrophotometric characterization were carried out. In particular, 1.9 ml of ninhydrin solution ($3{\cdot}10^{-2}$ M) were added to 0.1 ml of NPs collected after the functionalization reaction. The suspension was then warmed up at 90°C for 5 min. The change in the QD@SiO_2 suspension color from orange to blue (color of the Ruhemann's Blue assay byproduct) accounts for the presence of NH_2 groups, and the absorbance intensity measured at 570 nm, characteristic of the Ruhemann's byproduct, was used to titrate the amount of amino groups in each sample. The absorbance intensity at 570 nm of the NH_2-functionalized QD@SiO_2 samples was corrected by considering the contribution of the bare QD@SiO_2 to the absorption.

2.4. Growth of Au nanoparticles onto QD@SiO_2

The growth of Au NPs onto the silica surface was carried out following a reported method [38, 39] based on the three following steps: synthesis of 2 nm Au seeds, their assembly onto NH_2-functionalized QD@SiO_2 and finally the Au growth [38]. Au NPs were synthesized as reported elsewhere: 2 ml of 30 mM $HAuCl_4$ solution in milliQ H_2O was added in a three neck flask containing the reaction mixture based on 0.1 M NH_4OH and 0.17 M THPC solution, in a total volume of 48 ml of milliQ H_2O [38]. The solution was stirred at room temperature for about 30 min, meanwhile it turned from a pale yellow color of the Au salt, to a brown tint, corresponding to formation of small Au NPs.

The decoration of the NH_2-functionalized QD@SiO_2 with Au seeds was carried out by simply mixing 5 ml of the Au seed solution to 0.5 ml of NH_2-functionalized NPs under stirring for about 2 h. The beads were recovered by centrifugation at 7800 g, which allowed removal of excessive seeds, as well. The final precipitate was dissolved in 0.5 ml of milliQ H_2O.

The Au seeds decorating the silica surface were then used to grow Au nanoislands by addition of 4 ml of the Au salt solution in milliQ H_2O, containing also K_2CO_3 and $HAuCl_4$. Different K_2CO_3/$HAuCl_4$ (concentration of K_2CO_3 and $HAuCl_4$ equal to 1.8 mM/0.4 mM, 0.9 mM/0.2 mM and 0.2 mM/0.05 mM in milliQ H_2O) were tested to study the influence of the Au salt concentration on the final Au nanostructure size. After about 30 min stirring at room temperature, 10 μl of formaldehyde solution 0.3 M in milliQ H_2O was swiftly added. The reaction typically completed in 30 min.

The Au seed decoration process and subsequent growth of Au nanodomains onto the luminescent QD@SiO_2 were monitored by transmission electron microscopy (TEM) and optical absorbance and photoluminescence (PL).

2.5. Optical characterization

UV-vis absorption spectra were recorded with a Cary 5000 (Varian, Leini, (TO), Italy) UV/Vis/NIR spectrophotometer. PL spectra were recorded by using a Fluorolog 3 spectrofluorimeter (HORIBA Jobin-Yvon, Edison, New Jersey, USA), equipped with double grating excitation and emission monochromators. All optical measurements were performed at room temperature on samples obtained directly from synthesis without any size sorting treatment. Transient-PL measurements were performed using the time-correlated single photon counting (TCSPC) technique, with a FluoroHub (HORIBA Jobin-Yvon). The samples were excited at 375 nm by a picosecond laser diode (NanoLED 375L) emitting $\tau\approx80$ ps pulses at a 1 MHz repetition rate. The PL signals were dispersed by a double grating monochromator and detected by a picosecond photon counter (TBX ps Photon Detection Module, HORIBA Jobin-Yvon, Edison, New Jersey, USA). The temporal resolution of the experimental setup was ≈200 ps.

Relative PL quantum yield (QY) was measured by dispersing the samples in dimethylsulfoxide (DMSO) and using cumarine 153 in DMSO as a standard (absolute QY in DMSO 75%) due to its spectroscopic properties (emission centered at 530 nm and excitation spectra in the range of 370–450 nm, thus suitably matching the emission spectrum range of QDs). The excitation wavelength used for the PL QY measurements was 395 nm.

The absorbance of the Ruhemann's byproduct at 570 nm was determined using a Perkin Elmer (Monza, MI, Italy) 2030 multilabel reader Victor TM X3.

2.6. Transmission electron microscope

A JEOL JEM 1011 electron microscope operating at 100 kV and equipped with a high resolution CCD camera was used for TEM analysis. The sample was prepared by casting a drop of suitably diluted solution of NPs and QDs on the surface of a carbon deposited copper grid. Statistical analysis of the size (NP average size and size distribution) of the samples has been performed by use of a freeware image analysis program. In particular the average NP size and the percentage relative standard deviation (σ%) have been calculated for each sample. Such final statistical determination provides information on the NP size distribution. Its value is based on the spread of size compared to the average value and is expressed as a percentage.

2.7. Particle size, size distribution and surface charge

Hydrodynamic diameter (size), size distribution and colloidal stability of the nanostructures, before and after the growth of Au nanodomains on luminescent silica coated NPs, were detected using a Zetasizer Nano ZS, Malvern Instruments Ltd, Worcestershire, UK (DTS 5.00). In particular, size and size distribution were determined by means of dynamic light scattering (DLS) after sample dilution in demineralized water. Hydrodynamic diameter is reported by number and size distribution is described in terms of polydispersity index (PDI). The ζ-potential measurements, i.e. the surface charges, were carried out by using a laser Doppler velocimetry (LDV) after sample dilution in KCl aqueous solution (1 mM). All reported data are presented as mean values ± standard deviation of three replicates.

2.8. In vitro cytotoxicity of NPs and QDs

Cytotoxicity assays were carried out against rat C6 glioma cells. This cell line was maintained at 37°C in a humidified incubator containing 5% CO_2 in Dulbecco's modified Eagle's medium (Lonza, Caravaggio (BG), Italy) nutrient supplemented with 10% heat inactivated fetal bovine serum, 2 mM L-glutamine, 100 U ml^{-1} penicillin and 100 μg ml^{-1} streptomycin. The cell viability was determined using the 3-(4,5-dimethylthiazol-2-yl)-2,5-diphenyl-tetrazolium bromide (MTT) assay. Cells were dispensed into 96-well microtiter plates at a density of 5000 cells/well. Following overnight incubation, cells were treated with the different compounds, QD@SiO_2, APS-functionalized QD@SiO_2 and Au speckled QD@SiO_2, at different concentrations (ranging from 0.1 to 0.001 mg ml^{-1}). Then the plates were incubated at 37°C for 72 h. 10 μl of 0.5% w/v MTT were further added to each well and the plates were incubated for an additional 4 h at 37°C. Finally the cells were lysed by addition of 100 μl of 1:1 v/v DMSO:EtOH solution. The absorbance at 570 nm was measured by using a Perkin Elmer 2030 multilabel reader Victor TM X3. Each test was performed in triplicate in three separate experiments. The result is expressed as the IC_{50} value that is the concentration necessary for 50% inhibition.

3. Results and discussion

The work aims at preparing nanostructures, with size of a few tens of nanometers, having visible light emission and NIR absorption, as potential colloidal systems for multimodal imaging and photothermal therapy.

Scheme 1 reports the fabrication routes designed to accomplish the proposed multifunctional nanostructures, which, in particular, combines CdSe@ZnS QDs as inorganic fluorophore [29–31], with emission in the visible light region, and Au metal NPs, whose plasmon absorption can be tuned up to the NIR region of the electromagnetic spectrum [19, 38–41].

Scheme 1. General scheme for the QD functionalization process leading to SiO_2 shell growth (step 1), functionalization with amine groups upon reaction with (3-aminopropyl)triethoxysilane (step 2) and subsequent assembly of Au seeds and progressive formation of Au NPs upon the addition of K_2CO_3 and $HAuCl_4$ (steps 3, 4, 5).

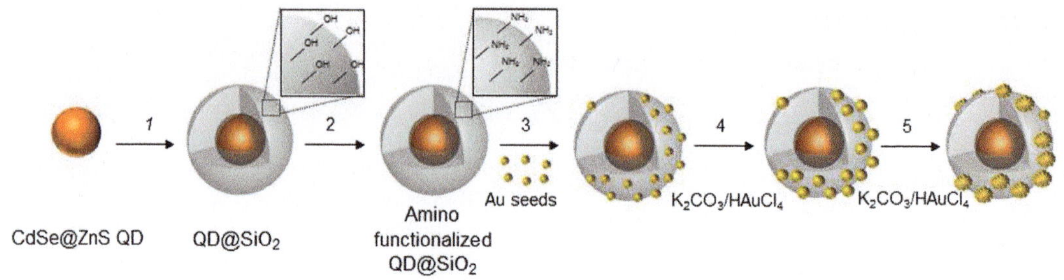

Scheme 1. General scheme for the QD functionalization process leading to SiO2 shell growth (step 1), functionalization with amine groups upon reaction with (3-aminopropyl)triethoxysilane (step 2) and subsequent assembly of Au seeds and progressive formation of Au NPs upon the addition of K2CO3 and HAuCl4 (steps 3, 4, 5).

The presence of an SiO_2 shell allows the combination of the distinct spectral features of fluorescent quantum dots and the metal NPs, by encapsulating one component inside and binding the other ones on its surface. Thanks to the control on its thickness, it represents an adjustable spacer to modulate the distance between metal NPs and luminescent QDs, circumventing significant QD quenching phenomena and thus providing a detectable fluorescence. In addition the dielectric–metal interface enables the geometric 'tunability' of the plasmon resonances of Au NP, inducing size dependent broad absorption, ranging from visible to NIR region.

Furthermore, growth of a silica shell onto Cd based QDs is an efficient method to reduce their toxicity and impart them dispersibility in aqueous media, thus enabling development of safe systems for biomedical applications [23–25]. Such a surface also offers a versatile platform not only for further functionalization with NIR absorbent domain, but also for further bioconjugation with specific biomolecules [32–36].

The overall multistep synthetic process has been designed to provide a single multifunctional heterostructure dispersible and stable in aqueous media with a rational control of the optical properties and interaction between the photoactive components.

3.1. Spectroscopic and morphological characterization of QD@SiO_2

As-synthesized luminescent organic-capped CdSe@ZnS QDs (see Supplementary Material Figure SI1) have been coated with a silica shell by following a well-established water-in-oil microemulsion synthetic route, which has been demonstrated as a suitable and reliable strategy for hydrophobic NPs in the range of a few nanometers [32–41]. This synthetic strategy limits the size of the final core/shell structure below 100 nm in diameter, and hence produces NPs that may more efficiently penetrate the cell membrane [19].

In a typical experiment, QDs are dispersed in cyclohexane and Igepal CO-520 as surfactants (which are non-toxic, cheap and easy to handle), and ammonia solution and TEOS are added to the reaction flask and stirred at room temperature. The TEOS concentration is varied, keeping constant all the other parameters and the final core/shell heterostructure morphology and optical properties are investigated.

Figure 1 shows TEM images (A–C), UV-vis absorbance (D) and PL (E) spectra of the QD@SiO_2 sample prepared at 10^{-6} M QD concentration (in cyclohexane) with TEOS amounts of 20, 30, and 50 µl (resulting in the samples QD@SiO_2_20, QD@SiO_2_30 and QD@SiO_2_50 respectively).

The TEM characterization highlights that nanostructure morphology and silica shell thickness can be tuned by means of a careful control of the amount of TEOS in the reaction. At 20 µl of TEOS, the resulting nanostructures (Figure 1(A)) show a large fraction of multiple QD in the cores (nearly 92% multicores versus 8% of single core NPs), a size of nearly 16 nm, a broad size distribution ($\sigma\%$ = 20%), and a poorly uniform shell (thickness is 6 nm, $\sigma\%$ = 17%). Multiple QD cores and single QD core silica coated core-shell nanostructures will be hereafter called QD_n@SiO_2 and QD@SiO_2, respectively.

By increasing the TEOS volume from 20 to 50 µl, a thicker and more uniform silica shell can be obtained (Figure 1(B) and (C)). The addition of 30 µl of TEOS in the reaction resulted in a decreased core multiplicity, with a higher incidence (58%) of single core NPs (QD@SiO_2 NPs, Figure 1(B)). Meanwhile an increase in the size of the NPs (23 nm, $\sigma_{\%}$ = 20%) and shell thickness (nearly 10 nm, $\sigma_{\%}$ = 17%) occurs.

Use of a larger volume (50 µl) of TEOS (Figure 1(C)) has produced a highly monodisperse single core QD@SiO_2 NPs (size 38 nm, $\sigma_{\%}$ = 8%), very uniform in shape, with a silica shell thickness of 16 nm ($\sigma_{\%}$ = 13%). The very thin silica shell and the concomitant occurrence of NPs with multiple QD cores (Figure 1(A)), observed at low TEOS content, can be explained considering that the limited amount of hydrolyzed TEOS is not sufficient to uniformly coat the QD surface. In fact, at low TEOS content, assemblies of QDs rather than single QDs are incorporated in the aqueous domains of the micelle, finally resulting in multiple core structures. By increasing up to 50 µl the amount of TEOS, the amount of silica precursor

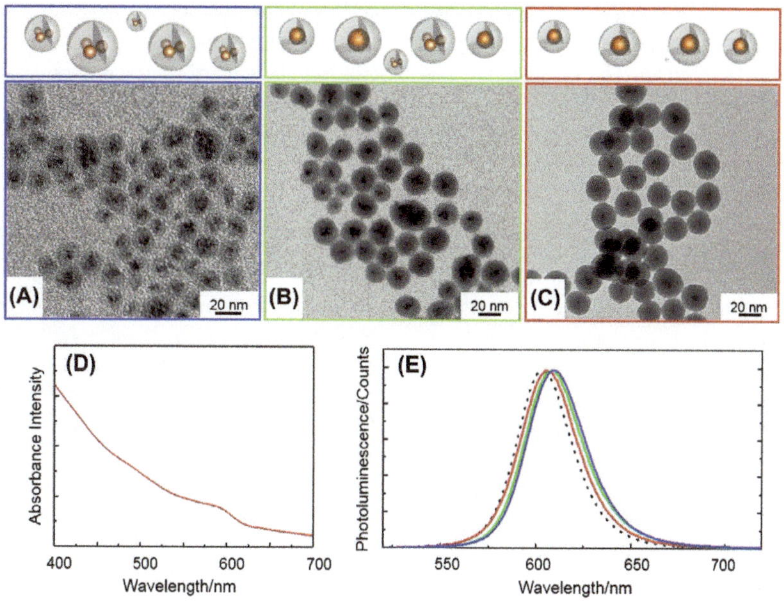

Figure 1. TEM images (A–C), UV-vis absorbance (D) and photoluminescence spectra (E) of QD@SiO_2 prepared at a QD concentration of 10^{-6} M in cyclohexane, IGEPAL CO520 (350 µl), NH_4OH (200 µl), by adding 20 µl (A, E blue line), 30 µl (B, E green line) and 50 µl (C, D, E red line) of TEOS. PL spectrum of bare QDs in chloroform (E, dashed line).

becomes sufficient to coat single QDs, thus resulting in a higher yield of single core NPs. In addition, such an amount of silica precursor is abundant enough to allow the growth of a thicker shell, as shown in Figure 1(B) and (C). Nevertheless such a TEOS content is still below the amount that results in homogeneous nucleation of empty SiO_2 NPs [42, 43].

As potential optical probes, the emission properties of the nanostructures have also been investigated. First, a slight red shift of the PL peak is observed in the spectra of the QD@SiO_2 samples (Figure 1(E) blue, green and red line) compared to that of the bare QDs (Figure 1(E) dashed line), the extent of which is a function of the volume of TEOS precursor used, as the lower the amount of TEOS, the larger the red shift of the emission peak. The shift with respect to native QDs can be definitely associated to the presence of the silica shell while the shift moving from QD@SiO_2_20 to QD@SiO_2_50 samples (Figure 1(A)–(C)) can be ascribed to collective energy transfer phenomena among the QDs trapped in the same SiO_2 matrix.

The changes in the QD@SiO_2 nanostructure geometry, induced by TEOS concentration, only slightly affect the PL quantum yield (QY). Compared to the bare QDs, whose PL QY is 18%, a partial decrease in PL QY is observed in the silica coated samples, although a higher relative PL QY has been calculated for QD@SiO_2 obtained with larger volume of TEOS (11%) with respect to QD_n@SiO_2, with PL QY of 10%. To better rationalize these results, PL recombination dynamics of as-synthesized QDs (Figure 2 black trace), QDn@SiO_2 (Figure 2

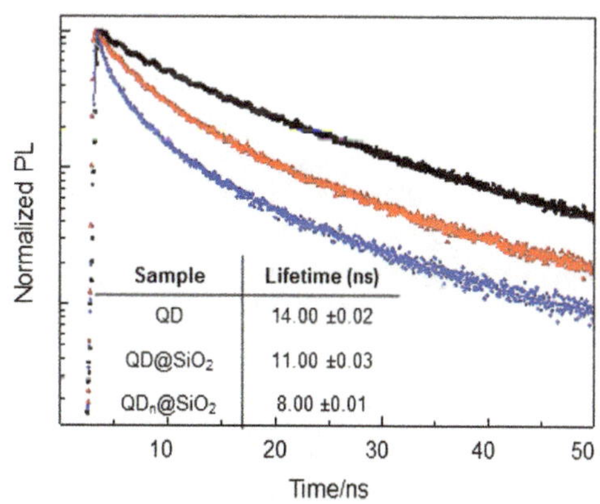

Figure 2. Time-resolved fluorescence intensity decays (A) and average lifetime (B) of QD (black line, QD@SiO_2 (red trace) and QD_n@SiO_2 (blue trace).

blue trace, corresponding to QD@SiO_2_20 sample) and QD@SiO_2 (Figure 2 red trace, corresponding to QD@ SiO_2_50 sample), have been measured. The decays can be well fitted with a typical three exponential function and the average lifetimes have been calculated (table in Figure 2). The data show that the PL decay of the SiO_2 coated QDs is faster than that observed for the as-synthesized QDs. Such a more rapid recombination can be ascribed to ligand exchange reactions caused by the SiO_2 shell growth [12], that lowers QD passivation, increasing defect surface states, which can act as recombination centers for charge carriers, decreasing the average lifetimes. In addition, the presence of the silica shell brings a modification in the external dielectric constant experienced by the QDs and a corresponding redistribution of the energetic levels (as confirmed by the slight red shift of PL in Figure 1(E)).

The faster charge carrier recombination measured in presence of multicores $QD_n@SiO_2$ (Figure 2, blue line) can be accounted by energy transfer mechanisms among luminescent QDs embedded in the same silica shell, which are possible due to the very short inter-particle distance.

The spectroscopic and morphologic investigation clearly indicates that the nanostructures with a single QD in the core, presenting a high monodispersity, a very uniform silica shell and a reasonable PL QY, are the best candidate for luminescent nanoprobes to be used as scaffold for further decoration with the NIR absorbing Au NPs.

Indeed the silica shell with a thickness of 16 nm is expected to be large enough to avoid energy transfer processes from the photoexcited QD to the metal NPs and hence quenching phenomena.

3.2. Functionalization of QD@SiO2 NPs with amine groups

Primary amine groups (NH_2) have been grafted on the silanized surface of $QD@SiO_2$ (NH_2-functionalized $QD@SiO_2$ NPs) by reaction with 20, 80 and 120 μl APS in alcoholic solution under basic pH. Titration of the NH_2 groups has been performed by means of the assay reported in [37], based on the selective reaction of ninhydrin with NH_2 groups in the presence of 2,6-lutidine. The formation of Ruhemann's Blue as a byproduct (see Supplementary Material Figure SI2), with its characteristic intense absorption band in the visible (λ_{max} = 570 nm) allows us to assess the presence and concentration of NH_2 groups in the sample. From the ninhydrin test using the Lambert–Beer equation the concentration of NH_2 groups in the final volume of 1.5 ml of suspension (containing approximately $0.4 \cdot 10^{15}$ NPs) has been determined as nearly 7 and 10 mM for 20 and 80 μl APS, respectively. Further increase in the APS volume used for the functionalization reaction did not induce a remarkable increase in the amine group concentration, highlighting that an almost complete surface saturation has been accomplished by using 80 μl APS [20], with a density of nearly 5 NH_2 groups nm^{-2}.

ζ-potential measurements have been also carried out to investigate the change in the surface charge of the NPs upon functionalization with APS [16]. The negative ζ-potential value of the $QD@SiO_2$ NPs (−37.3 ± 0.3 mV), ascribable to partially ionized silanol groups on the surface, becomes positive (+45 ± 0.1 mV) upon surface saturation with NH_2 groups. This high ζ-potential value also indicates a very good colloidal stability of amine grafted luminescent silica coated NPs.

DLS analysis has clearly pointed out, for each step, a monomodal size distribution, thus suggesting that no NP aggregation occurs in aqueous medium and no significant increase of the average diameter passing from the $QD@SiO_2$ NPs to the amine grafted $QD@SiO_2$ NPs (see Supplementary Material Figure SI4). In particular, average diameter of 41.9 ± 0.9 (PDI 0.18 ± 0.02), 42.7 ± 0.6 (PDI 0.20 ± 0.02) has been recorded for the two nanostructures, respectively.

3.3. Decoration of amino-functionalized $QD@SiO_2$ with Au nanoparticles

The positively charged NH_2-functionalized $QD@SiO_2$ NPs have been mixed with the negatively charged hydrophilic THPC-stabilized Au NPs [44, 45].

Figure 3(A) and (B) shows the TEM micrographs of the NH_2-functionalized $QD@SiO_2$ NPs before (A) and after (B) functionalization with the Au seeds. The dark spots at the surface of SiO_2 in Figure 3(B) can be clearly ascribed to the Au NPs assembled onto the surface of NH_2-functionalized $QD@SiO_2$ NPs (Figure 3(A)), and a surface coverage of 20–30% has been calculated on the base of TEM investigation, in agreement with previously reported studies [46, 47]. Non-specific adsorption was demonstrated to be absent (see Supplementary Material Figure SI5).

In Figure 3(F), the UV-vis absorbance spectra of the Au-decorated $QD@SiO_2$ (black line) show a featureless band, that can be ascribed to very small Au NPs [46]. PL spectrum and time-resolved fluorescence intensity decays (see Supplementary Data Figure SI6) show that decoration of the $QD@SiO_2$ surface with Au NPs does not significantly affect the luminescence signal. The average lifetime values (see Supplementary Data Figure SI6) of the bare $QD@SiO_2$ (11.0 ns ± 0.3) remain almost unchanged after functionalization with APS and subsequent decoration with Au NPs (10.0 ns ± 0.2), thus allowing us to exclude any energy transfer phenomena between the metal and the QDs. This evidence is consistent with the fact that 16 nm thick silica shell of the QD@ SiO_2 sample is thick enough to prevent any significant energy transfer phenomena [16–18].

In order to induce isotropic Au growth on the single Au seed particle, the functionalized $QD@SiO_2$ NPs have been successively mixed with suitable quantities of Au precursor, $HAuCl_4$ and of K_2CO_3 and formaldehyde as reducing agents, in order to evaluate the influence of the molar content of the Au precursor on the final nanostructure. In Figure 3, the vis-NIR absorption spectra and the morphology of these heterostructures at increasing $K_2CO_3/HAuCl_4$ are reported. At $K_2CO_3/HAuCl_4$ concentrations of 0.2 mM/0.05 mM, the localized surface plasmon resonance (LSPR) peak is broad and centered at 526 nm (Figure 3(F) blue line).

The LSPR broadening can be ascribed to the large size distribution of Au islands, as confirmed by the corresponding TEM micrograph.

By increasing the Au precursor concentration the size of the Au islands progressively increases, resulting in

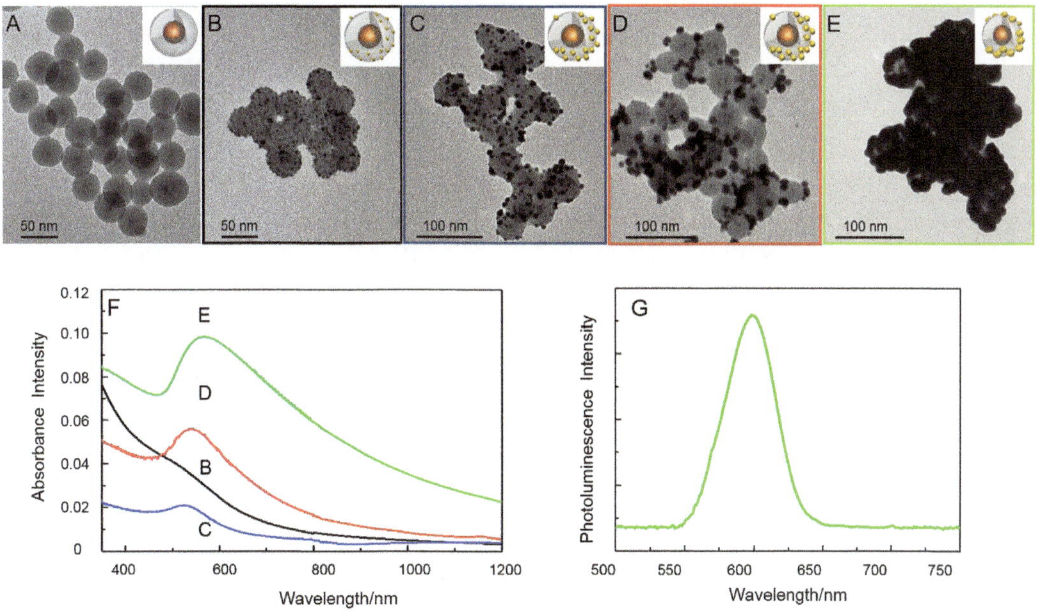

Figure 3. TEM micrographs of amino-functionalized QD@SiO2 before (A), and after (B) assembly of Au seeds, and (C–E) further Au deposition; each frame shows the nanoshell at different concentrations of Au precursor solutions: 0.5 M (C), 2 M (D) and 4 M (E) The UV-vis-NIR absorbance spectra of samples C–E are reported in panel F (each suspension has been diluted 1:5 in order to reduce scattering contribution to the absorption). Emission spectra of sample in Figure 3(E) (λ_{ex}=400 nm).

larger and larger particles, formed by partially merging neighbor Au nanodomains onto the silica surface (Figure 3(C)). At the precursor concentrations of 0.9 mM/0.2 mM ($K_2CO_3/HAuCl_4$), the LSPR presents a red-shift (Figure 3(F) red line), as a consequence of the further growth of the Au domains. Indeed, the corresponding TEM micrograph shows a larger Au NP coverage of silica surface (Figure 3(D)). Finally, for a K_2CO_3/$HAuCl_4$ concentration ratio of 1.8 mM/0.4 mM, the LSPR appeared as a broaden peak at 569 nm (Figure 3(F) green line). The modifications of the LSPR band are clear indications of Au nanostructure growth [46], that results in the formation of first an incomplete (Figure 3(E)), and then a complete (see Figure SI6 Supplementary Material) Au shell, deriving from the coalescence between neighbor Au domains, as demonstrated by the corresponding TEM micrographs. Interestingly, the whole range of the investigated precursor concentrations results in NPs with a partial Au domain coverage, hence incomplete shells. Such Au domains growth with increasing concentrations of K_2CO_3 and $HAuCl_4$. The LSPR positions measured for the prepared samples are in good agreement with previous findings on analogous systems, based on Au NPs growth on 30 nm SiO_2 particles [46].

The sample reported in Figure 3(E) clearly shows a NIR absorbance (green line spectrum in Figure 3(F)) and a retained luminescence of the QDs. Although a partial quenching has been observed and a final PL QY of 5% has been achieved (Figure 3(G)), the obtained nanostructures can still be used as effective luminescent probes for bioimaging. The not negligible absorbance in the biologically invisible region typically used for photothermal cancer therapy (i.e. laser excitation at 808 nm) makes these nanostructures appealing to tackle the treatment of cancer cells by converting the excited state photon energy into heat.

Further growth of a complete and thick Au nanoshell brings the absorption of the nanostructure to change into a well-defined plasmon band, with no absorption tail in the NIR region, as reported [19] (Supplementary Data Figure SI7), and the emission completely disappears.

A hybrid system based on metallic and fluorescent structures has been extensively studied and PL quenching or enhancement has been observed depending on the distance and the orientation of the fluorophore with respect to the metal surface as well as on the topology of the metal surface. The Purcell effect has mainly accounted for the fluorescence enhancement arising from the plasmon and exciton coupling in nanostructures based on a continuous shell or ring of metallic materials close to a fluorophore, behaving as a metallic resonant cavity [18]. Conversely, quenching can result from charge transfer to the metal or to energy transfer and dissipation. The partial PL quenching of the here-described Au speckled nanostructures can be ascribed to the screening effect of the Au nanodomains, that absorb the excitation radiation and induce a reduction of the number of photons that reaches the QD cores, rather than being caused by interaction between metal and luminescent NPs. Due to the thickness of the silica shell, energy transfer phenomena between the metal and the QDs can be excluded in the

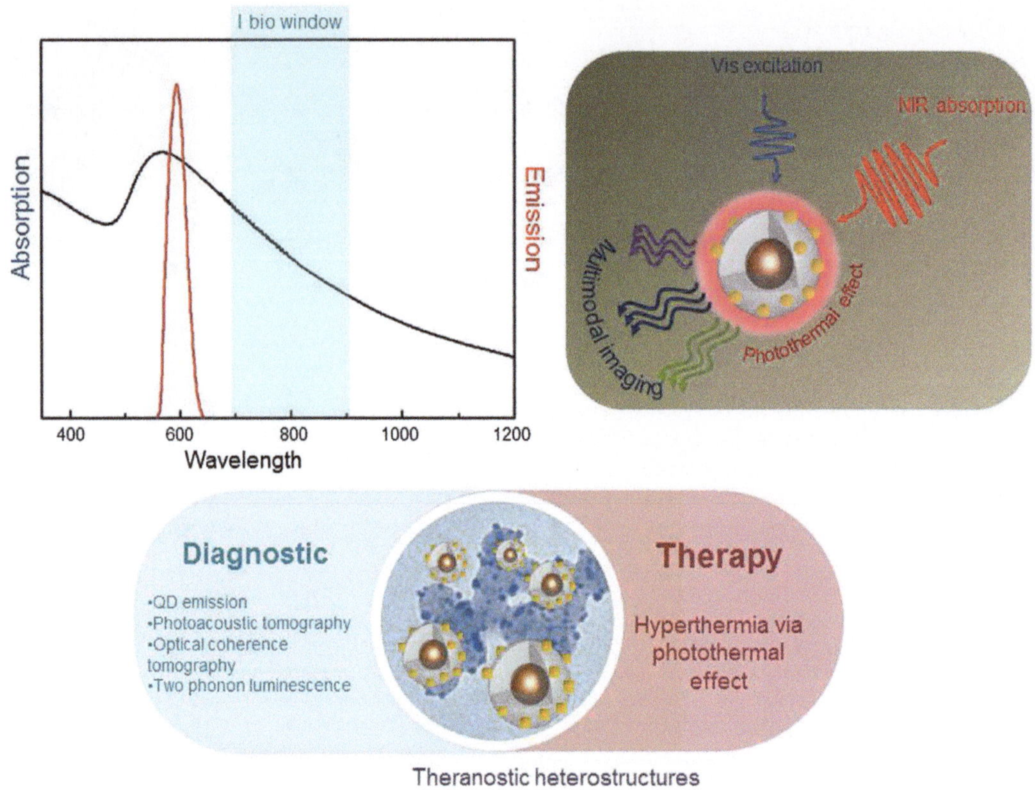

Figure 4. Summary of the spectral features of the obtained Au speckled QDs, which make the multifunctional architectures ideal candidate for theranostic application.

final nanostructures, as demonstrated by the PL time-resolved characterization that shows PL decay (and average lifetime) of the NPs to be almost unchanged after decoration with Au NPs (see Supplementary Data Figure S6). On the other hand, PL enhancement due to the Purcell effect is not expected due to the size regime, geometry and morphology of the Au nanodomain decorated silica coated QD nanostructures, that, for the discontinue nature of the gold shell, is far to behave as a plasmonic fluorescent resonator.

The reported data highlight that a careful control of the fabrication route of water dispersible Au-decorated QD@SiO_2 NPs is strictly required when absorption in the NIR 'biological tissue transparency window' and luminescence in the visible range need to be concomitantly conveyed in the same nanostructured object (Figure 4) having a size of a few tens of nanometers. DLS investigation performed on sample E has revealed a monomodal size distribution and an average diameter of 50.6 ± 0.5 nm (PDI 0.21 ± 0.03), highlighting an increase of size passing from the amine grafted QD@SiO_2 NPs to Au decorated nanostructures, as expected (see Supplementary Material Figure SI4). Finally, ζ-potential measurements have provided a slightly less positive value (+20 ± 0.2 mV) of Au decorated NPs with respect to that of the amino functionalized nanostructures, which can be ascribed to the decrease of the free amino groups present at the surface of Au-decorated QD@SiO_2 NPs, due to their partial neutralization by the growth of metal nanodomains.

In order to test the viability of the prepared Au-decorated QD@SiO_2 nanostructures in biologically relevant *in vitro* studies, their cytotoxicity response has been investigated, by testing the nanomaterials at the different stages of their processing.

For this purpose *in vitro* growth inhibition of rat C6 rat glioma cells and MTT assay have been used for the cytotoxicity assessment, as C6 rat glioma cells represent an experimental model system for glioblastoma multiforme, having been also widely studied for elucidating mechanism of tumor growth, angiogenesis and invasion, and in design and evaluation of anti-cancer therapies [47].

In particular, cytotoxicity has been tested for a long (72 h) incubation period, exploring NP concentrations ranging from 0.1 mg ml^{-1} to 0.001 mg ml^{-1}. In Table 1, IC_{50} values for QD@SiO_2, APS-functionalized QD@SiO_2 and Au decorated QD@SiO_2 have been reported.

The obtained results indicate that functionalization of QD@SiO_2 NPs with APS causes a decrease in cell vitality (IC_{50} = 9 μg ml^{-1}), reasonably ascribed to the positive charge expressed by the NP surface.

Table 1. *In vitro* growth inhibition of rat C6 glioma cells treated with the different types of nanostructure at increasing concentrations.

Nanomaterial sample	IC_{50} (μg ml^{-1})[a]
QD@SiO_2	30.1 ± 0.9
NH_2-functionalized QD@SiO_2	9.2 ± 0.6
Au NPs-decorated QD@SiO_2	13.5 ± 0.8

[a]Data are presented as mean ± standard error of three separate experiments performed in triplicate.

These results are in good agreement with those previously observed for other positively charged NPs [48], that show a higher cytotoxicity as long as the amount of amine groups on the NP surface increases. Interestingly upon functionalization with Au NPs, the IC_{50} value slightly increases with respect to that observed for NH_2-functionalized QD@SiO_2 (13.5 μg ml^{-1} and 9.2 μg ml^{-1}, respectively), thus suggesting that the presence of Au NPs at the QD@SiO_2 surface enhances their biological viability and demonstrating that a suitably surface engineering procedure can control their biocompatibility.

4. Conclusions

A visible-NIR photoactive heterostructure based on Au nanodomain decorated luminescent QD@SiO_2 NPs with size in the submicrometer range have been successfully fabricated by suitably tailored colloidal strategies.

Different types of heterostructure, differing in QD multiplicity inside the SiO_2 NPs, the SiO_2 shell thickness and extent of Au nanodomain deposition onto SiO_2 surface, have been prepared by tuning the experimental parameters, in order to accomplish a nanostructure whose optical properties fit with the aims of the work. The deep morphological and optical characterization shows that a multifunctional heterostructure able to preserve the QD emission in the visible region and to generate NIR absorption from metal nanodomains has been achieved.

The synthesized nanostructure represents one of the few examples of all inorganic colloidal NPs which combines in a submicrometer sized heterostructure the QD luminescence and Au NIR absorption.

The good photostability of QDs, the high stability in aqueous solution and their biological viability make these multifunctional heterostructures a promising nanoplatform highly relevant for potential applications in nanobiotechnology, as multimodal bioimaging agents and as mediators of photothermal therapy.

Disclosure statement

No potential conflict of interest was reported by the authors.

Funding

The work has been supported by PON R&C 2007–2013 MAAT-Molecular Nanotechnology for Health and Environment [project number: PON02_00563_3316357], National Sens&Micro LAB [POFESR 2007–2013], FIRB Futuro in Ricerca [RBFR122HFZ] and PRIN 2010–2011 [2010C4R8M8], and Nanomax-integrable sensors for pathological biomarkers diagnosis [N-CHEM], and PRIN 2012 Italian National Project [Prot. 2012T9XHH7] projects, Laboratorio Regionale di Sintesi e Caratterizzazione di Nuovi Materiali Organici e Nanostrutturati per Elettronica, Fotonica e Tecnologie Avanzate.

References

[1] Singh AK, Hahn MA, Gutwein LG, et al. Theranostic nanoparticle platform for bioimaging and cancer therapy. Inter J Nanomedicine. 2012;7:2739–2750.

[2] Menon JU, Jadeja P, Tambe P, et al. Nanomaterials for photo-based diagnostic and therapeutic applications. Theranostics. 2013;3:152–166.

[3] Lee D-E, Koo H, Sun IC, et al. Multifunctional nanoparticles for multimodal imaging and theragnosis. Chem Soc Rev. 2012;41:2656–2672.

[4] Kim JW, Galanzha EI, Shashkov EV, et al. Golden carbon nanotubes as multimodal photoacoustic and photothermal high-contrast molecular agents. Nat Nanotechnol. 2009;4:688–694.

[5] Maltzahn GVC, Park AJ, Ramanathan R, et al. SERS-coded gold nanorods as a multifunctional platform for densely multiplexed near-infrared imaging and photothermal heating. Adv Mater. 2009;21: 3175–3180.

[6] La S, Clare SE, Halas NJ. Nanoshell-enabled photothermal cancer therapy: impending clinical impact. Acc Chem Res. 2008;41:1842–1851.

[7] Saha A, Basiruddin SK, Sarkar R, et al. Functionalized plasmonic–fluorescent nanoparticles for imaging and detection. J Phys Chem C. 2009;113:18492–18498.

[8] Salguerino-Maceira V, Correa-Duarte MA, Farle M, et al. Bifunctional gold-coated magnetic silica spheres. Chem Mater. 2006;18:2701–2706.

[9] Sharma P, Brown SC, Bengtsson N, et al. Gold-speckled multimodal nanoparticles for noninvasive bioimaging. Chem Mater. 2008;20:6087–6094.

[10] Vankayala R, Lin CC, Kalluru P, et al. Gold nanoshells-mediated bimodal photodynamic and photothermal cancer treatment using ultra-low doses of near infra-red light. Biomaterials. 2014;35:5527–5538.

[11] Cheng FY, Chen CT, Yeh CS. Comparative efficiencies of photothermal destruction of malignant cells using antibody-coated silica@au nanoshells, hollow au/ag nanospheres and au nanorods. Nanotechnology. 2009;20:421504.

[12] Melancon MP, Lu W, Yang Z, et al. In vitro and in vivo targeting of hollow gold nanoshells directed at epidermal growth factor receptors for photothermal ablation therapy. Mol Cancer Ther. 2008;7:1730–1739.

[13] Michalet X, Pinaud FF, Bentolila LA, et al. Quantum dots for live cells, in vivo imaging, and diagnostics. Science. 2005;307:538–544.

[14] Sharma P, Brown SC, Singh A, et al. Near-infrared absorbing and luminescent gold speckled silica nanoparticles for photothermal therapy. J. Mater. Chem. 2010;20:5182–85(b)Jin Y, Gao X. Plasmonic

fluorescent quantum dots. Nat Nanotechnol. 2009;4:571–576.
[15] Jin Y, Gao X. Plasmonic fluorescent quantum dots. Nat Nanotechnol. 2009;4:571–576.
[16] Liu N, Prall BS, Klimov VI. Hybrid gold/silica/nanocrystal-quantum-dot superstructures: synthesis and analysis of semiconductor–metal interactions. J Am Chem Soc. 2006;128:15362–15363.
[17] Castello Serrano I, Vazquez-Vazquez C, Matas Adams A, et al The effect of the silica thickness on the enhanced emission in single particle quantum dots coated with gold nanoparticles. RSC Adv. 2013;3:10691–10695
[18] Ji B, Giovanelli E, Harbert B, et al. Non-blinking quantum dot with a plasmonic nanoshell resonator. Nat Nanotechnol. 2015;10:170–175.
[19] Rasch MR, Konstantin VS, Korgel BA. Limitations on the optical tunability of small diameter gold nanoshells. Langmuir. 2009;25:11777–11785.
[20] Perrault SD, Walkey C, Jennings T, et al. Mediating tumor targeting efficiency of nanoparticles through design. Nano Lett. 2009;9:1909–1915.
[21] Piao Y, Burns A, Kim J, et al. Designed fabrication of silica-based nanostructured particle systems for nanomedicine applications. Adv Fun Mater. 2008;18:3745–3758.
[22] Wang L, Wang K, Santra S, et al. Watching silica nanoparticles glow in the biological world. Anal Chem. 2006;78:646–654.
[23] Smith JE, Wang L, Tan W. Bioconjugated silica-coated nanoparticles for bioseparation and bioanalysis. Trend Anal Chem. 2006;25:848–855.
[24] Wang L, Zhao W, Tan W. Bioconjugated silica nanoparticles: development and applications. Nano Res. 2008;1:99–115.
[25] Lee JE, Lee N, Kim T, et al. Multifunctional mesoporous silica nanocomposite nanoparticles for theranostic applications. Acc Chem Res. 2011;44:893–902.
[26] Santra S. Fluorescent silica nanoparticles for cancer imaging. Methods Mol Biol. 2010;624:151–162.
[27] Larson DR, Ow H, Vishqasro D, et al. Silica nanoparticle architecture determines radiative properties of encapsulated fluorophores. Chem Mater. 2008;20:2677–2684.
[28] Wang K, He X, Yang XH, et al. Functionalized silica nanoparticles: a platform for fluorescence imaging at the cell and small animal levels. Acc Chem. Res. 2013;46:1367–1376.
[29] Alivisatos P, Gu W, Larabell C. Quantum dots as cellular probes. Annu Rev Biomed Eng. 2005;7: 55–76.
[30] Dembski S, Graf C, Krüger T, et al. Photoactivation of CdSe/ZnS quantum dots embedded in silica colloids. Small. 2008;4:1516–1526.
[31] Guerrero Marinez A, Perez-Juste J, Liz-Marzan LM. Silica-coated nanomaterials: recent progress on silica coating of nanoparticles and related nanomaterials. Adv Fun Mater. 2010;22:1182–1195.
[32] Fanizza E, Urso C, Pinto V, et al. Single white light emitting hybrid nanoarchitectures based on functionalized quantum dots. J. Mater. Chem. C. 2014;2:5286–5291.
[33] Fanizza E, Depalo N, Clary L, et al. A combined size sorting strategy for monodisperse plasmonic nanostructures. Nanoscale. 2013;5:3272–3282.
[34] Selvan ST, Tan TT, Ying JY. Robust, non-cytotoxic, silica-coated cdse quantum dots with efficient photoluminescence. Adv Mater. 2005;17:1620–1625.
[35] Koole R, van Schooneveld MM, Hilhorst J, et al. On the incorporation mechanism of hydrophobic quantum dots in silica spheres by a reverse microemulsion method. Chem Mater. 2008;20:2503–2512.
[36] Yang P, Murase N, Suzuki M, et al. Bright, non-blinking, and less-cytotoxic SiO_2 beads with multiple CdSe/ZnS nanocrystals. Chem Comm. 2010;46:4595–4597.
[37] Depalo N, Catucci L, Mallardi A, et al. Enrichment of cardiolipin content throughout the purification procedure of photosystem II. Bioelectrochemistry. 2004;63:103–106.
[38] Pham T, Jackson JB, Halas NJ, et al. Preparation and characterization of gold nanoshells coated with self-assembled monolayers. Langmuir. 2002;18:4915–4920.
[39] Huschka R, Barhoumi A, Liu Q, et al. Gene silencing by gold nanoshell-mediated delivery and laser-triggered release of antisense oligonucleotide and siRNA. ACS Nano. 2012;6:7681–7691.
[40] Graf C, Gao Q, Schütz I, et al. Surface functionalization of silica nanoparticles supports colloidal stability in physiological media and facilitates internalization in cells. Langmuir. 2012;28:7598–7613.
[41] Westcott SL, Oldenburg V, Lee TR, et al. Formation and adsorption of clusters of gold nanoparticles onto functionalized silica nanoparticle surfaces. Langmuir. 1998;14:5396–5401.
[42] Vogt C, Toprak MS, Muhammed M, et al. High quality and tuneable silica shell-magnetic core nanoparticles. Nanopart Res. 2010;12:1137–1147.
[43] Ding HL, Zhang YX, Wang S, et al. $Fe3O_4$@SiO_2 core/shell nanoparticles: the silica coating regulations with a single core for different core sizes and shell thicknesses. Chem Mater. 2012;24:4572–4580.
[44] Meijerink A. 2008. Exciton dynamics and energy transfer processes in semiconductor nanocrystals. In Rogach, A., editor. Semiconductor Nanocrystal Quantum Dots. Vienna: Springer; p. 277
[45] Jones M, Scholes GD. On the use of time-resolved photoluminescence as a probe of nanocrystal photoexcitation dynamics. J. Mater. Chem. 2010;20:3533–3538.
[46] Jain PK, Lee KS, El-Sayed IH, et al. Calculated absorption and scattering properties of gold nanoparticles of different size, shape, and composition: applications in biological imaging and biomedicine. J Phys Chem B. 2006;110:7238–7248.
[47] Grobben B, De Deyn PP, Slegers H. Rat C6 glioma as experimental model system for the study of glioblastoma growth and invasion. Cell Tissue Res. 2002;310:257–270.
[48] Denora N, Laquintana V, Lopalco A, et al. In vitro targeting and imaging the translocator protein TSPO 18-kDa through G(4)-PAMAM–FITC labeled dendrimer. J Control Release. 2013;172:1111–1125.

6

Experimental study and modeling of atomic-scale friction in zigzag and armchair lattice orientations of MoS_2

Meng Li[a,b], Jialin Shi[a,b], Lianqing Liu[a], Peng Yu[a], Ning Xi[a,c] and Yuechao Wang[a]

[a]State Key Laboratory of Robotics, Shenyang Institute of Automation, Chinese Academy of Sciences, Shenyang 110016, PR China
[b]University of the Chinese Academy of Sciences, Beijing 100049, PR China
[c]Department of Electrical and Computer Engineering, Michigan State University, East Lansing, Michigan 48824, PR USA

ABSTRACT

Physical properties of two-dimensional materials, such as graphene, black phosphorus, molybdenum disulfide (MoS_2) and tungsten disulfide, exhibit significant dependence on their lattice orientations, especially for zigzag and armchair lattice orientations. Understanding of the atomic probe motion on surfaces with different orientations helps in the study of anisotropic materials. Unfortunately, there is no comprehensive model that can describe the probe motion mechanism. In this paper, we report a tribological study of MoS_2 in zigzag and armchair orientations. We observed a characteristic power spectrum and friction force values. To explain our results, we developed a modified, two-dimensional, stick-slip Tomlinson model that allows simulation of the probe motion on MoS_2 surfaces by combining the motion in the Mo layer and S layer. Our model fits well with the experimental data and provides a theoretical basis for tribological studies of two-dimensional materials.

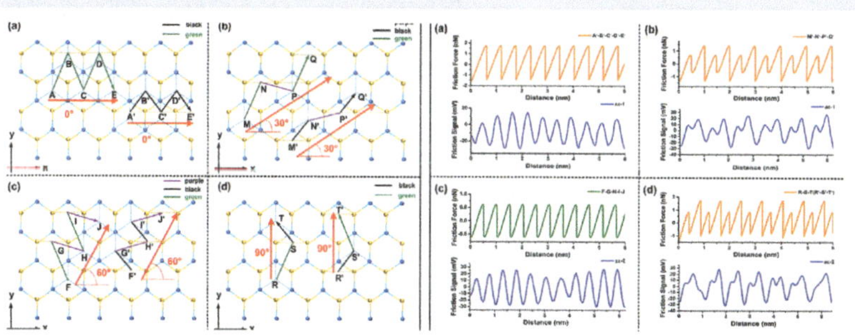

KEYWORDS
MoS_2; atomic-scale friction; two-dimensional model; lattice orientation

CLASSIFICATION
40 Optical, magnetic and electronic device materials; 105 Low-Dimension (1D/2D) materials; 100 Materials; 212 Surface and interfaces; 200 Applications; 303 Mechanical / Physical processing; 300 Processing / Synthesis and Recycling

1. Introduction

Two-dimensional materials with strong in-plane covalent bonding and weak van der Waals interaction are the most common class of crystalline structures that can be easily exfoliated into atomically thin layers. Achievements in fabrication have led to a new research field devoted to the study of their novel physical, chemical and mechanical properties,[1–6] which have fascinated the scientific community. Among the family of two-dimensional materials, the form that attracts the most interest is graphene, which has shown superior carrier mobility of up to 200,000 cm^2V^{-1} and record thermal conductivity and stiffness.[7–9] The discovery of graphene has also unveiled other layered-structure materials, which include the graphene analogs hexagonal boron nitride (h-BN), black phosphorus, and transition metal dichalcogenides (TMDC). For instance, the perfectly flat and electrically insulating properties allow the use of h-BN as an alternative substrate and excellent gate dielectric.[10,11] A predicted direct bandgap of 2 eV and its anisotropic nature make black phosphorus an ideal material for the infrared regime and new types of plasmonic devices.[12,13]

Among the two-dimensional materials, graphene, with inherent zero-band gap,[14] usually requires complex bandgap engineering processes to fabricate field-effect transistors (FET).[15,16] The wide bandgap of h-BN makes it unsuitable for electronic applications.[17] Black phosphorus, however, is chemically inert in atmosphere, and its monolayer form is expected to be destroyed upon

CONTACT Meng Li limeng@sia.cn

exposure to air,[18] which remains an important challenge restricting its future applications. By comparison, several TMDCs have sizable bandgaps of 1–2 eV and possess novel properties that are complementary to, yet distinctive from, those of graphene, ranging from electronic, optical, mechanical, chemical and thermal properties.[19,20] Furthermore, TMDCs are generally air-stable and naturally abundant,[21] rendering them promising channel materials to replace silicon.

In particular, MoS_2, a TMDC, is useful in device applications due to its satisfactory bandgaps, good flexibility, thermal stability, excellent carrier mobility, and compatibility with silicon CMOS (Complimentary Metal Oxide Semiconductor),[22,23] and is being heavily investigated for wide-ranging application in electronics and optoelectronics.[24–27] For example, the high on-off current ratio, up to 200 $cm^2 V^{-1}$ carrier mobility and good sub-threshold swing have been utilized to fabricate more complex, two-dimensional, hetero-CMOS inverters with low power consumption.[28] An indirect-to-direct bandgap transition occurs when bulk MoS_2 is thinned to a monolayer, which induces strong photoluminescence (PL) with sensitive photoresponse, thus rendering it attractive for optoelectronic devices.[29,30] In sum, the intriguing prospect of the potential nanoelectronic applications, which may take advantage of the novel physical properties, has inspired researchers to explore the unknown characteristics of MoS_2.

Recently, intensive theoretical research on MoS_2 nanoribbons has shown that armchair nanoribbons are nonmagnetic and semiconducting,[31] whereas zigzag nanoribbons are ferromagnetic metallic.[32] Thus, most of the aforementioned electrical and optical applications largely depend on the edge state because metallic edges may result in current leakage and thereby immensely hinder normal device function.[33] However, for piezotronics, zigzag edge is preferable because it induces and accumulates piezoelectric polarization charges to produce piezoelectric current and voltage responses due to its ferromagnetic property.[34] In view of the fact that different chirality directly leads to distinct properties and practical applications, it is essential to extensively investigate these two lattice orientations, towards finding a convenient method to effectively differentiate them in the future.

The specific chirality of a material can be determined by the lattice orientation detection technique,[35,36] which presents the atomic resolution of the corresponding material and enables us to investigate the unknown microscopic domains. Current lattice orientation detection techniques include transmission electron microscopy (TEM), scanning tunneling microscopy (STM) and atomic force microscopy (AFM).[37–40] TEM and STM techniques require either extremely expensive equipment, ultra-high vacuum (UHV) environment or additional gold coating procedures to produce a conductive substrate, which greatly increases the difficulty of detecting multiple lattice orientation. AFM is also able to achieve a high-resolution atomic image, and has the advantage over TEM and STM that it can simultaneously acquire the lattice orientation and friction information through the atomic image of the material surface in lateral force microscopy (LFM) mode. Meanwhile, as a lubricant material, the tribological properties of MoS_2 have a significant impact on many fields of science and technology, and the relevant research mainly focuses on the friction force of interlayer sliding [41–43] or the friction motion on various substrates.[44] Atomic-scale friction has been studied but knowledge is not sufficient. Research on atomic-scale friction between the probe and surface is scarce and does not address the relationship between friction and lattice orientation. In view of the shortcomings in TEM and STM techniques and the enormous significance of atomic-scale tribological research, AFM is more suitable for our study and is employed to obtain the friction information.

This study aims to carry out research on atomic-scale friction of MoS_2 and to work towards identifying lattice orientation, and is the first study of the friction variation and friction force in zigzag and armchair lattice orientations of MoS_2 using AFM and LFM. A modified model is established to explain thoroughly the motion mechanism of the probe on the MoS_2 surface in zigzag and armchair orientations. For a better comparison of the friction signal variation, fast Fourier transform (FFT) is performed. Finally, measurements of the friction force in the experiment and simulation are conducted to explore uncharted territories in the atomic-scale friction of MoS_2.

2. Modeling of the lateral friction in zigzag and armchair orientations

In the lateral friction experiment, stick-slip movement occurs between the probe and MoS_2 surface.[45] To derive the relationship between the friction force and the lattice orientation and to further establish a comprehensive model of probe motion, a two-dimension Tomlinson model,[46] which describes the movement in idealized LFM constant-force mode, is adopted for the simulation modeling:

$$\begin{aligned} m_x\ddot{x}_t &= k_x(x_M - x_t) - \frac{\partial V(x_t, y_t)}{\partial x_t} - \gamma_x \dot{x}_t, \\ m_y\ddot{y}_t &= k_y(y_M - y_t) - \frac{\partial V(x_t, y_t)}{\partial y_t} - \gamma_y \dot{y}_t, \end{aligned} \tag{1}$$

where m_x and m_y are the effective mass of the system, x_t and y_t are the actual positions of probe at particular time, x_M and y_M are the equilibrium positions of the probe, k_x and k_y are the elasticity of the probe, and γ_x and γ_y are the damping constants of the system. In theory, the degree of freedom for the motion of atoms is high in two-dimensional systems in which non-adiabatic motion is avoided.[47] In our two-dimensional Tomlinson model, the sweeping of the probe can be regarded as adiabatic.

To obtain the solutions of Equation (1), an appropriate and simple interaction potential $V(x_t, y_t)$ has to be selected. The original interaction potential is described as:

$$V(x_t, y_t) = V_0 \cos\left(\frac{2\pi}{a_x} x_t\right) \cos\left(\frac{2\pi}{a_y} y_t\right) \tag{2}$$

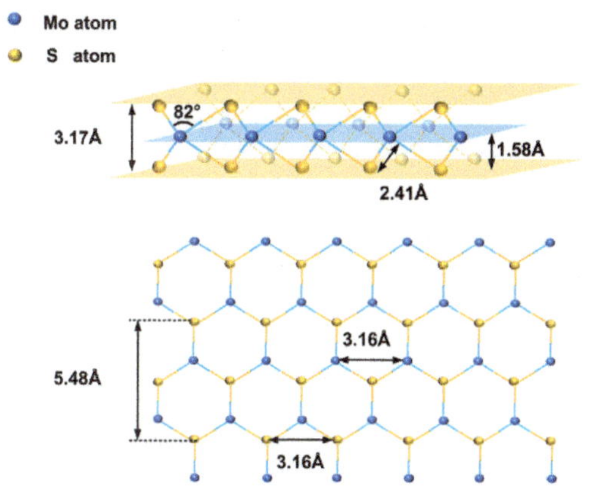

Figure 1. Lattice constants of MoS_2. The top part shows a lateral view of MoS_2, while the bottom part depicts a top view. The yellow and blue spheres indicate S and Mo atoms of MoS_2, respectively.

where a_x and a_y are the lattice constants of S layer in a MoS_2 film. Typical LFM parameters are:[46] $V_0 = 1.0ev$, $k_x = k_y = 10\text{Nm}^{-1}$, V_m=400 Å S^{-1}, V_0 is the initial value of tip-sample interaction, and V_m is the scan speed. Then, the damping constant of the system can be obtained $\gamma_x = \gamma_y = 2\sqrt{k_x m_x} \approx 10^{-3}$Ns/m. MoS_2 film can be envisioned as a sandwich-like slab that consists of two individual atomic layers, a Mo layer and an S layer. As shown in the inset of Figure 1, Mo and S atoms are covalently linked in a sandwich structure. According to the literature, the characteristic value of MoS_2 can be approximated as follows: in the same layer, the distances between Mo atom and S atom and between the two S atoms are 3.16 Å and 2.41 Å, respectively; however, in different layers, the perpendicular distance between two S atoms and between S and Mo atoms is 3.17 Å and 1.58 Å, respectively. The bond angle between two S atoms and a Mo atom is 82°.

Given the stick-slip movement, the probe shuttles between the upper layer (S layer) and lower layer (Mo layer) in the actual LFM experiment. Although the interaction between the probe and lower layer might be weaker compared to that of the upper layer, it should not be ignored to establish a comprehensive motion model. In the following two sections, the probe motion in zigzag and armchair orientations combining two layers will be discussed in detail.

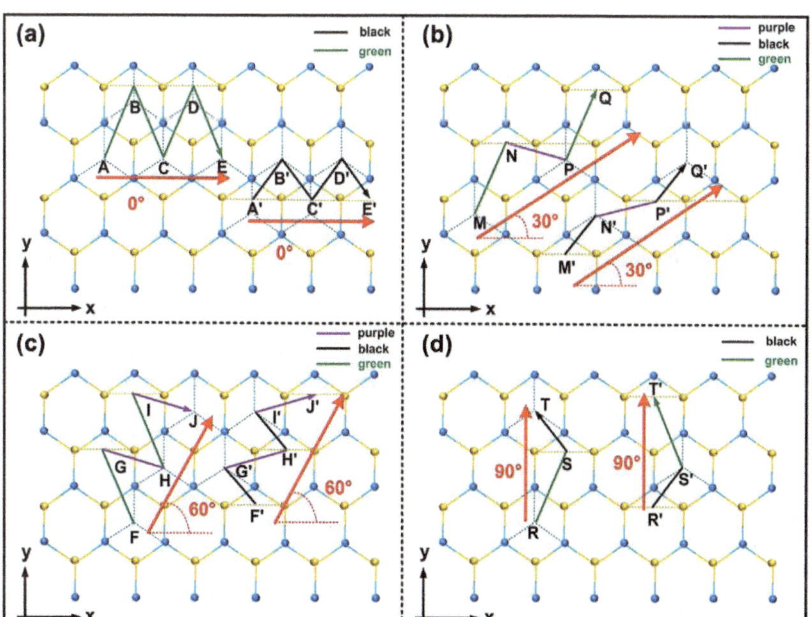

Figure 2. Moving trajectories in various lattice orientations: (a) zz-1 orientation; (b) ac-1 orientation; (c) zz-2 orientation; (d) ac-2 orientation. The black, green and purple arrow lines indicate the moving trajectories of the probe along different lattice orientations. The red arrow lines indicate the scan direction of the probe. The dashed yellow lines connect two adjacent S atoms via the central points of S crystal cell. The dashed blue lines connect two adjacent Mo atoms via the central points of Mo crystal cell.

2.1. Motion in zigzag orientation

For zigzag orientation, the probe moves along two distinct paths due to the disparity in the relative position of the MoS_2 lattice structure, which are the first type zigzag (zz-1) and the second type zigzag (zz-2), as depicted in Figures 2(a) and (c), respectively. The scan direction of zz-1 is defined as 0°. For zz-2, the scan direction is 60°. In view of the characteristic of stick-slip movement, we assume the probe moves via the central points in each layer. Then, considering the two different types of atoms in MoS_2, the probe motion in each path of zigzag orientation can be further divided into two distinct cases: one case starts from the central points of the Mo crystal cell and terminates at the central points of the S crystal cell; the other case is the opposite. For instance, in the case of zz-1, as shown in Figure 2(a), two moving trajectories are possible, A-B-C-D-E or A′-B′-C′-D′-E′. For A-B-C-D-E, the trajectory consists of various lines connecting the central points of the Mo crystal cell and that of the adjacent S crystal cell. In contrast, the component lines of A′-B′-C′-D′-E′ connect the central points of the S crystal cell to that of the adjacent Mo crystal cell. The moving trajectories of zz-2, F-G-H-I-J, and F′-G′-H′-I′-J′, displayed in Figure 2(c), are constructed in the same way. In both paths, these lines are marked with different colors (black, green and purple) according to their specific lengths. The calculation is discussed in the third part of this chapter.

2.2. Motion in armchair orientation

In a similar way, the moving path of armchair orientation can be divided into two distinct paths: the first type armchair (ac-1) and the second type armchair (ac-2), as depicted in Figures 2(b) and (d). The scan directions of these two paths are 30° and 90°, respectively. Each path contains two different trajectories. As shown in Figures 2(b) and (d), in armchair orientation, the probe moves along M-N-P-Q or M′-N′-P′-Q′ and R-S-T or R′-S′-T′. The component of each trajectory is the same as in zigzag orientation discussed above: the line segments between the central points of S or Mo crystal cell and that of their adjacent crystal cell. These lines segments are also marked in different colors in terms of length. Notably, as shown in Ffigure 2(d), the length of the moving trajectories R-S-T and R′-S′-T′ in the second type of armchair orientation are equal due to the specific lattice symmetry.

2.3. Simulation results

According to the analysis above, the movements of the probe in both S layer and Mo layer must be considered together. The lattice constants in the original interaction potential described in Equation (2) need to be modified because they simply reflect the movement in S layers. To obtain the actual lattice constants combining the movement in two layers, the accurate length of the various trajectories should be calculated first. Based on the triangle cosine theorem, the length of the line segments in different colors can be obtained by taking these parameters into the calculation. Furthermore, the length of various trajectories, R, both in zigzag and armchair orientations, can be obtained by simple addition. The results are summarized in Table 1.

Based on the data, the original interaction potential can be modified as:

$$V(x_t, y_t) = V_0 \cos\left(\frac{2\pi}{R^*} x_t\right) \cos\left(\frac{2\pi}{\frac{R^*}{a_x} a_y} y_t\right) \quad (3)$$

Where R^* is the actual lattice constant considering the moving trajectory R in various lattice orientations. The expression of R^* is as follows:

$$R^* = R\cos\theta \quad (4)$$

where θ is the corresponding lattice orientation. For zz-1, zz-2, and ac-1, the values of θ are 0°, 60°, and 30°, respectively. For ac-2, considering the particularity of the cosine function (cos90° = 0), 90° is treated as 30°.

Then, simulations of the lateral force friction under zigzag and armchair orientations are performed using Matlab software (MathWorks, USA, version 7.0). The simulation results are displayed in Figure 3.

3. Experimental method

The substrate material used for room-temperature tests was silicon covered with a 285-nm-thick oxide. To improve the measurement accuracy, two types of multilayer MoS_2 film samples were fabricated. The first type, labeled sample-1, was grown by chemical vapor deposition at the Shanghai Institute of Optics and Fine Mechanics. The original solution was diluted three times and then sonicated for 12 minutes at 59 Hz in an ultrasonic oscillator (SK5210 LHC). One 2 μl drop was then cast on a freshly cleaned silicon substrate and dried at

Table 1. Results of the trajectory length in various lattice orientations.

Color	Computation result (Å)	Lattice orientation	Expression of motion period	R (Å)
Black (b)	2.85	zz-1	2 x b / 2 x g	5.7 / 8.12
Green (g)	4.06	zz-2	b + p / g + p	6.51 / 7.72
Purple (p)	3.66	ac-1	b + p / g + p	6.51 / 7.72
		ac-2	b + g	6.91

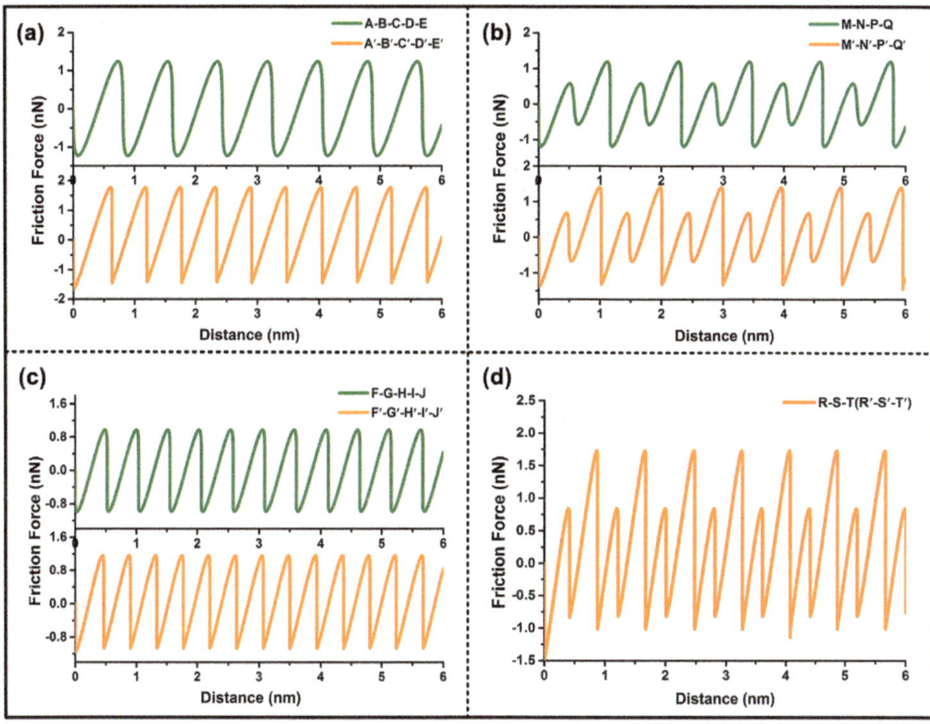

Figure 3. Simulation results of the lateral friction for various lattice orientations of MoS_2: (a) zz-1 orientation; (b) ac-1 orientation; (c) zz-2 orientation; (d) ac-2 orientation. The simulated waves marked in green indicate the first-case trajectory in each path of zigzag orientation or armchair orientation. The simulated waves marked in orange indicate the second-case trajectory in each path of zigzag orientation or armchair orientation.

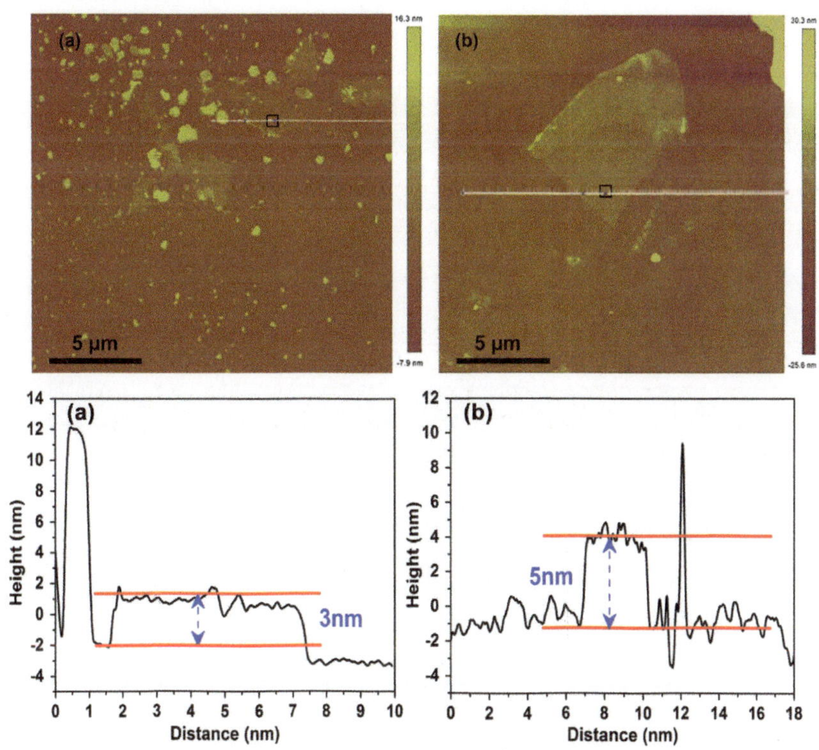

Figure 4. (a, b) Surface morphology of the multilayer MoS_2 films: (a) sample-1; (b) sample-2. (c, d) Height profile along the white line in (c) panel a; (d) panel b. The black rectangles indicate the scan areas.

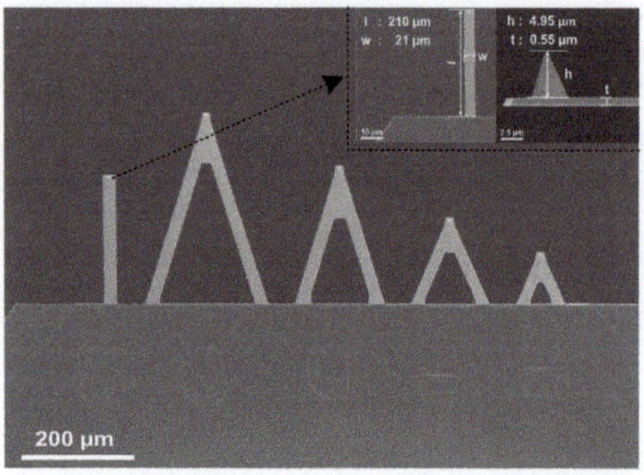

Figure 5. Scanning electron microscopy image of the probe used for the LFM experiment. *l*, *w*, *h*, and *t* define the length, width, height and thickness of the probe, respectively.

room temperature. The second film type, labeled sample-2, was mechanically peeled from bulk MoS_2 (provided by XFNANO, Nanjing, China).

AFM was utilized to observe and analyze the surface morphology. A flat and clean surface of the sample is required to satisfy the adiabatic condition and to ensure the experiment proceeds smoothly. For this purpose, we prepared flat and smooth pieces of film before the experiment. For the first type of MoS_2 film, shown in Figure 4(a), we use a relatively flatter and larger sample, whose size is approximately 7.5 × 5.3 μm and average height is approximately 3 nm, indicating that it is a five-layer MoS_2. For the second type of MoS_2 film, shown in Figure 4(b), the size is approximately 9 × 6 μm and the average height is approximately 5 nm, suggesting it is an eight-layer MoS_2.

Subsequently, to acquire the atomic image, an LFM experiment was performed in the contact mode of a multimode AFM in air under ambient conditions (25–60% relative humidity, 20–50 °C). To guarantee a rigorous contrast between the two samples, LFM experiments for two cases were performed under the same conditions. We adopted the same type of probe, i.e. a normal contact silicon nitride AFM probe (MLC-B, Bruker, Inc.[48]) The length, width, height, radius, and thickness of the probe were 210 μm, 21 μm, 4.95 μm, 10 nm and 0.5 m, respectively, as shown in the inset of Figure 5. The spring constant of the probe was 0.02 N m^{-1}. During the experiment for two cases, the scan size and rate were 10 nm and 29.5 Hz, respectively. The LFM experiment typically achieved the best atomic-scale imaging under 90° scanning. The scan angle in two cases remained constant at 90°. In addition, the proportional gain, integral gain and setpoint value were kept at 0.2, 0.3 and 0.5, respectively. Finally, the total number of the lines for each atomic image, i.e. the pixel resolution of the image, was 256.

4. Results and discussion

4.1. Atomic images of two types of MoS_2 multilayers

After FFT filtering, atomic images for the two types of MoS_2 films were acquired. The surface morphology and atomic images of the central parts in MoS_2 films are displayed in Figure 6(a) and (b). Because the scanning angle in two cases was constant at 90°, the lattice structure was achieved by rotating the original lattice by 90° either clockwise or counter-clockwise (the obtained results were the same in both cases). In addition, for a single crystal, the lattice distribution of the entire sample is continuous and unbroken. Therefore, the edge state is consistent with the inner structure. The right sharp edge of the mechanical exfoliated single crystal MoS_2 film can be safely deduced through the central atomic distribution image because it is parallel to zz-2′ orientation, as indicated by the red arrow in Figure 6(b). After acquiring these atomic images, the useful information they contain is further extracted and discussed in detail in the following section.

4.2. Characteristics of actual lateral friction signals

After discerning the different lattice orientations, the lateral friction signal waves of the corresponding film can be obtained by pulling a line along each angle in the atomic image, as displayed in Figures 7(a) and (b).

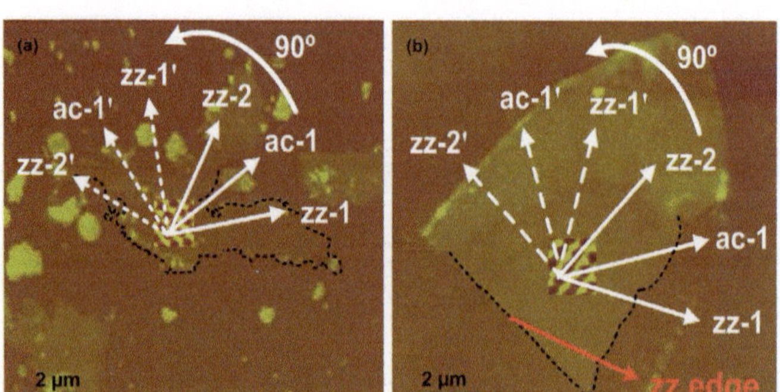

Figure 6. Lattice orientation identification of the multilayer MoS_2 films: (a) sample-1; (b) sample-2. The white arrow lines indicate the zigzag and armchair orientations in the atomic image. The dotted white arrow lines indicate the actual zigzag and armchair orientations of MoS_2 films after rotating 90°. The black dotted lines depict the partial edges of MoS_2 films.

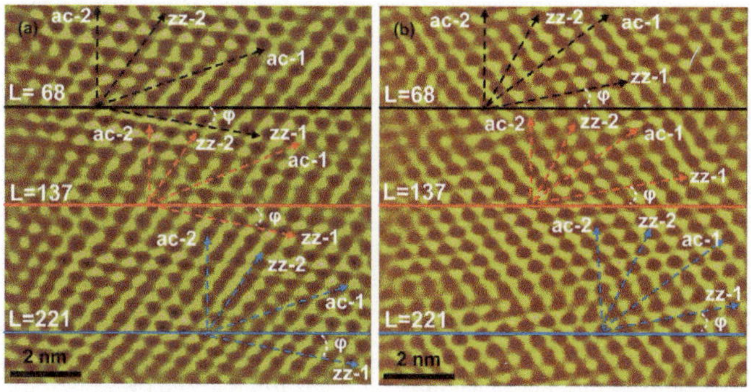

Figure 7. Atomic images of multilayer MoS_2 films. (a) Sample-1 and (b) Sample-2. The black, red and blue lines indicate the 68th, 137th and 221st rows of the atomic image, respectively. The four dashed lines marked different colors indicate zz-1, zz-2, ac-1 and ac-2 orientations at different locations, respectively.

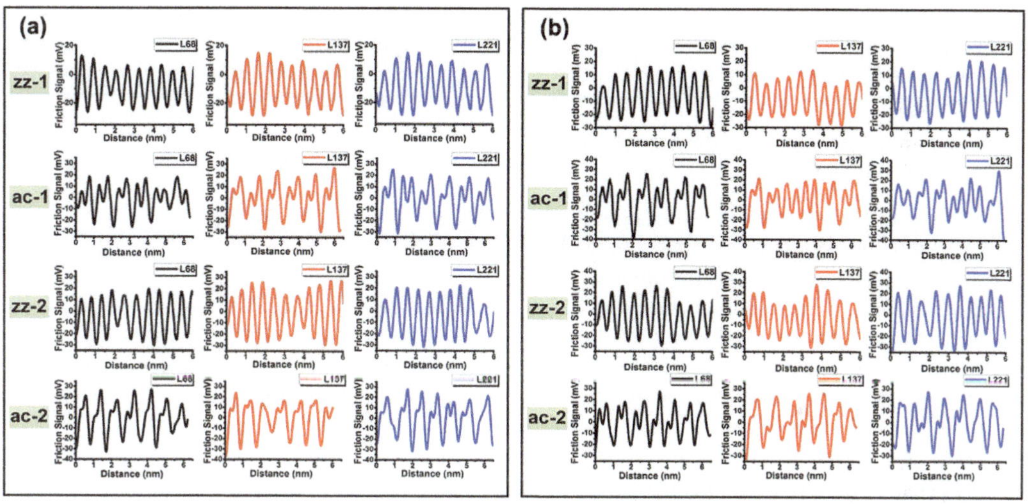

Figure 8. Lateral friction signal of multilayer MoS_2 films. (a) Lateral friction signal of sample-1. (b) Lateral friction signal of sample-2. The waves in black, red and blue indicate the friction signal obtained from 68th, 137th and 221st rows of the atomic image, respectively.

Table 2. Relevant data of the simulation zigzag and armchair orientation parameters using in the simulation.

Lattice orientation type	Number of wave peaks in simulation result	Number of wave peaks in actual friction signal	Corresponding trajectory of actual friction signal	Shape of the wave in simulation result	Shape of the wave in actual friction signal
zz-1	10/7	10	A′-B′-C′-D′-E′	Saw-tooth wave	Saw-tooth wave
zz-2	13//11	11	F-G-H-I-J	Saw-tooth wave	Saw-tooth wave
ac-1	6/5	6	M′-N′-P′-Q′	Combination of two types of saw-tooth wave varying in height	Combination of two types of saw-tooth wave varying in height
ac-2	7	7	R-S-T (R′-S′-T′)	Combination of two types of saw-tooth wave varying in height	Combination of two types of saw-tooth wave varying in height

For the sake of accuracy, each angle was processed at three different locations of the friction atomic image. Because the total number of the rows per image is 256, the three locations can be differentiated by the number of rows they locate. Three distinct locations are colored in black, red and blue. In addition, φ, representing the actual location of the films, is the angle between the 0° scanning direction and 0° lattice orientation, which remains constant in the same friction atomic image. The values of φ are approximately 10° and 7°. By pulling the line along different angles of 0° and 60° or 30° and 90°, lateral friction signal waves of zigzag and armchair lattice orientations can be obtained, respectively. The lateral friction signal waves in two cases are plotted in Figure 8(a) and (b).

Based on the actual lateral friction signal waves obtained from the two samples, we compare the simulation results and the actual friction signal; the results are summarized in Table 2.

According to Table 2, most of the data in the real signal profile exhibit consistent correlation with the simulation results with respect to the shape and the numbers of wave peaks. However, inconsistency between the simulation and the actual signal in the number of wave peaks is observed. The simulation is based on the assumption that each lattice direction may contain two trajectories. But in realistic friction experiments, only one of the postulated motion curves exists. According to the number of wave peaks in the actual friction signal, the possible paths for zz-1, zz-2, ac-1 and ac-2, are A′-B′-C′-D′-E′, F-G-H-I-J, M′-N′-P′-Q′, and R-S-T, respectively. The theoretical result is simulated under ideal conditions, whereas the experimental data with uncertainties leads to bias in various levels.

4.3. Power spectrum analysis of the friction signal variation in zigzag and armchair orientations

The direct comparison between the simulation result and the actual friction signal demonstrates the validity of the established motion model for the probe motion on the MoS_2 surface. We conducted an extensive analysis of the actual friction signals between the two samples. As the AFM and LFM experiments are performed under ambient conditions, several external factors, such as alteration of the experimental parameters, environmental noise and condition of the apparatus, affect the experimental results to varying degrees. Hence, the actual friction signals include disturbances, and direct comparison is not suitable. To address this problem, an appropriate technique should be selected.

FFT is a computationally effective technique in signal processing.[48] With the aid of FFT, the irrelevant external influences we mentioned above could be reduced and even eliminated. For these reasons, we employ this approach to process all of the waves in the various lattice orientations obtained from the friction atomic image. After processing the original friction signal waves in Figure 8(a) and (b), power spectra of each lattice orientation can be obtained, as shown in Figure 9(a) and (b).

Figures 9(a) and (b) show the distinct variation of the actual friction signals. For zz-1 and zz-2 lattice orientations, only one major peak appears in the corresponding power spectrum diagrams, whereas for ac-1 and ac-2 lattice orientations, the number of major peaks is generally two. This phenomenon is reasonable considering the shapes of the waves listed in Table 2. However, to thoroughly explain this phenomenon, we need to return to the modified interaction potential. In view of the physical structure of MoS_2 and the relationship between ay and axis $a_y = \sqrt{3}a_x$, Equation 3 can be re-written as:

$$V(x_t, y_t) = V_0 \cos\left(\frac{2\pi}{R^*}x_t\right)\cos\left(\frac{2\pi}{\sqrt{3}R^*}y_t\right) \quad (5)$$

In addition, x_t and y_t can be replaced by $x_t = r\cos\theta, y_t = r\sin\theta$. By substituting these two relations into Equation (5), the inherent frequency depending on the angle can be obtained:

$$w_1 = \left|(\cos\theta + \sin\theta/\sqrt{3})/R^*\right| \quad (6)$$

$$w_2 = \left|(\cos\theta - \sin\theta/\sqrt{3})/R^*\right| \quad (7)$$

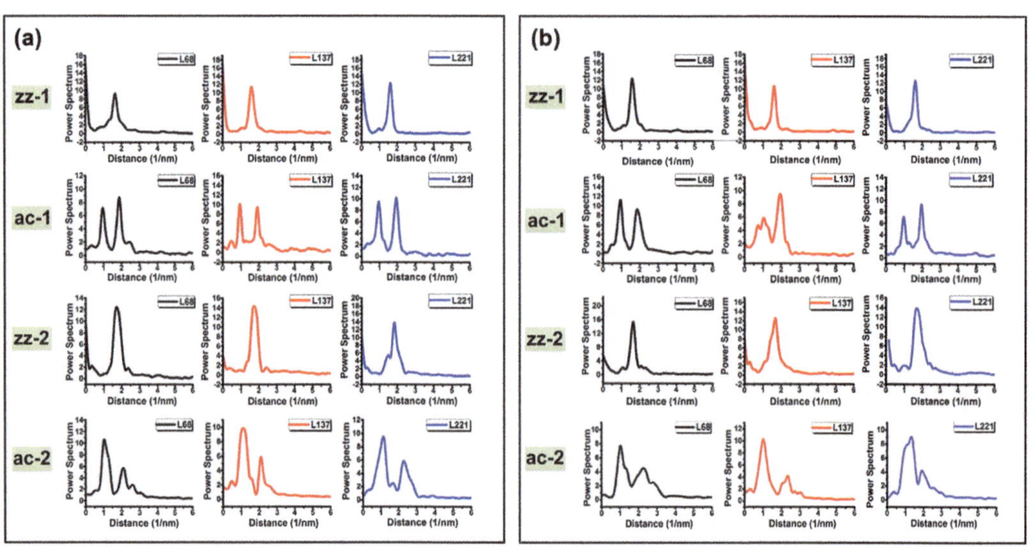

Figure 9. Power spectrum diagrams of the lateral fiction signal for (a) Sample-1 and (b) Sample-2. The waves in black, red and blue indicate the power spectra obtained from friction signal at the 68th, 137th and 221st rows of the atomic image, respectively.

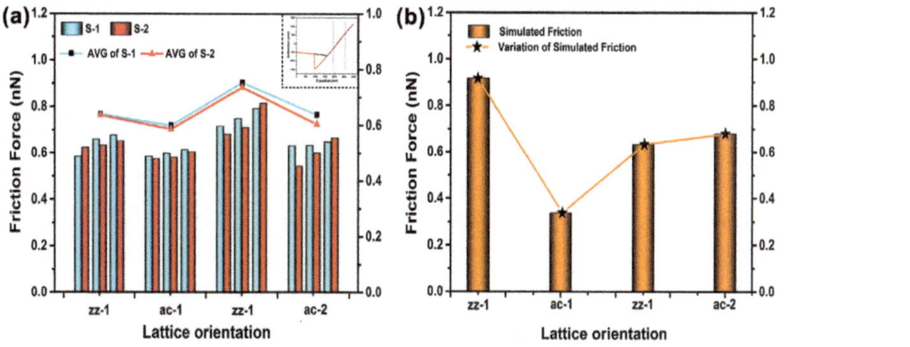

Figure 10. Friction force in multilayer MoS_2 films. (a) Measured (inset: force curve of the probe) and (b) Simulated.

By substituting the corresponding angle and the value of R^*, the frequency in various lattice orientations can be obtained. For zz-1 and zz-2, $w_1 = w_2 = \frac{1}{R^*} = \frac{1}{R\cos\theta}$. The angle of ac-2 is replaced by 30°, as discussed previously. Then, for ac-1 and ac-2, $w_1 = \frac{2}{\sqrt{3}R}$, $w_2 = \frac{1}{\sqrt{3}R}$. Only one frequency is obtained in zz-1 and zz-2, and two different frequencies are obtained in ac-1 and ac-2, which explains the one or two major peaks in the corresponding power spectrum diagrams.

4.4. *Friction force in armchair and zigzag lattice orientations of two types of MoS_2 multilayers*

Finally, we discuss the atomic-scale friction force of multilayer MoS_2 films. When the probe slides on the surface of MoS_2 films, the lateral force F_1 applied to the rectangular cantilever can be calculated by the following equation:[49]

$$F_l = \frac{Gwt^3}{3l(h + t/2)} S_l V_l \tag{8}$$

where G is the shear modulus of the silicon nitride cantilever, and w, l and t are the effective width, length and thickness of the cantilever, respectively. h is the height of the probe. MLCT-B probes have the following dimensions and material constant: $w = 2.1 \times 10^{-5}$ m, $l = 2.1 \times 10^{-4}$ m, $t = 0.55 \times 10^{-6}$ m, $h = 4.95 \times 10^{-6}$ m, and $G \approx 0.53 \times 10^{11}$ N m^{-2}. S_l is the lateral sensitivity of the photodetector, and V_l is the corresponding photodetector lateral voltage output. According to the laser beam mechanism, the displacement of the laser spot on the photodetector is determined by the distance between the reflecting point and spot position on the photodetector and the reflecting cantilever's angular deflection. The distance can usually be regarded as a constant; then, the lateral sensitivity S_l can be assumed to be equal to the normal sensitivity S_n based on the hypothesis that the photodetector is rotationally symmetric. Then, S_l can be calculated by:

$$S_l = S_n = \frac{3}{2l}\frac{dx_n}{dV_n} \tag{9}$$

where $\frac{dx_n}{dV_n}$ can be derived from the slope of the force curve, which is presented in the inset of Figure 10(a). According to the calculation, $\frac{dx_n}{dV_n} = 130.51\text{nm}V^1$. Then, the friction force of various lattice orientations can be calculated from the following equation:

$$F_l \approx 0.052n\text{Nm} - 1\text{V}^{-1} \times V_l \tag{10}$$

The actual friction forces of two MoS_2 films in various lattice orientations are displayed in Figures 10(a) and (b). As shown in Figure 10(a), the average actual friction force in the thinner film, S-1, is higher than that of the thicker film, S-2, which is consistent with published reports.[50]

For more proof, we calculated the friction force of the simulation results. Based on the data displayed in Ffigure 8, the simulated friction force can be directly obtained by averaging the absolute values for various lattice orientations. The simulated friction force is displayed in Figure 10(b).

As shown in Figure 10(a) and (b), the friction force varies with the lattice orientations in MoS_2. The discovery of this phenomenon will play a critical role in building a closed-loop fabrication strategy with real-time force sensor feedback to detect and control the lattice orientation. We also observed that the friction force in the actual situation does not perfectly match with the simulated results in terms of the friction force distribution in zigzag and armchair lattice orientations. The inconsistency is due to unavoidable experimental error. The exact reason remains unclear.

5. Conclusions

In summary, highly distinct, atomic-scale friction in zigzag and armchair lattice orientations of MoS_2 is revealed through simulation and experiment. In particular, the friction signals of the two orientations are thoroughly compared. First, to reflect the relationship between the lateral friction and the lattice orientation, we propose the motion trajectories of the probe in both zigzag and armchair orientations and establish a comprehensive motion model. Then, the simulation of the friction force is completed using Matlab software. Furthermore, an AFM and LFM based experiment is conducted to obtain the actual friction signal. The consistency between the simulation results and the experimental data confirms the validity of the friction modeling for MoS_2. By employing the FFT technique, the power spectra of the lateral friction in zigzag and armchair orientations are analyzed, which provides a more convenient and intuitive comparison. Finally, the actual and simulated friction forces are calculated. Although some discrepancies exist between the two results, the force variation dependence on the lattice orientation may hint a corresponding orientation detecting strategy in the future.

Our results may help to develop a reliable and effective method to experimentally differentiate these two important lattice orientations. The theoretical and experimental demonstration provides solid evidence of regular atomic-scale friction variation dependence on zigzag and armchair lattice orientations. Moreover, the distinct characteristics in both lattice orientations provide the possibility of establishing an effective lattice chirality detecting method or developing a real-time force sensing and feedback mechanism, which would be potentially useful for MoS_2 and other two-dimensional materials.

Acknowledgments

We thank the Shanghai Institute of Optics and Fine Mechanics for the preparation of MoS_2 films.

Disclosure statement

No potential conflict of interest was reported by the authors.

Funding

This work was supported by the National Natural Science Foundation of China [grant number 61522312], [grant number 61375107] and the CAS FEA International Partnership Program for Creative Research Teams.

Notes on Contributors

Meng Li is currently working toward the Ph.D. degree from the State Key Laboratory of Robotics, Shenyang Institute of Automation, Chinese Academy of Sciences, Shenyang, China. Her current research interests include the atomic-scale research on the nano materials MoS_2 and nano electrical devices.

Jialin Shi is currently working toward the Ph.D. degree from the State Key Laboratory of Robotics, Shenyang Institute of Automation, Chinese Academy of Sciences, Shenyang, China. His current research interests include ultrasonic vibration-assisted AFM technique and and nanomanipulation based on nanorobots.

Lianqing Liu is currently a Professor at the Shenyang Institute of Automation. His current research interests include nanorobotics, intelligent control, and biosensors.

References

[1] Butler SZ, Hollen, SM, Cao, L, et al. Progress, challenges, and opportunities in two-dimensional materials beyond graphene. ACS Nano. 2013;7:2898–2926.

[2] Chhowalla M, Shin HS, Eda G, et al. The chemistry of two-dimensional layered transition metal dichalcogenide nanosheets. Nat Chem. 2013;5:263–275.

[3] Johari P, Shenoy VB. Tuning the electronic properties of semiconducting transition metal dichalcogenides by applying mechanical strains. ACS nano. 2012;6:5449–5456.

[4] Kumar A, Low T, Fung KH, et al. Tunable light-matter interaction and the role of hyperbolicity in graphene-hBN system. Nano Lett. 2015;15:3172–3180.

[5] Zhao W, Ghorannevis Z, Chu L, et al. Evolution of electronic structure in atomically thin sheets of WS2 and WSe2. ACS Nano. 2013;7:791–797.

[6] Andleeb S, Kumar Singh A, Eom J. Chemical doping of MoS 2 multilayer by p-toluene sulfonic acid. Sci Technol Adv Mater. 2015;16:035009.

[7] Bolotin KI, Sikes KJ, Jiang Z, et al. Ultrahigh electron mobility in suspended graphene. Solid State Comm. 2008;146:351–355.

[8] Balandin AA, Ghosh S, Bao W, et al. Superior thermal conductivity of single-layer graphene. Nano Lett. 2008;8:902–907.

[9] Wei Y, Wang B, Wu J, et al. Bending rigidity and gaussian bending stiffness of single-layered grapheme. Nano Lett. 2013;13:26–30.

[10] Liu Z, Song L, Zhao S, et al. Direct growth of graphene/hexagonal boron nitride stacked layers. Nano Lett. 2011;11:2032–2037.

[11] Dean CR, Young, AF, Meric, I, et al. Boron nitride substrates for high-quality graphene electronics. Nat Nanotechnol. 2010;5:722–726..

[12] Xia F, Wang H, Jia Y. Rediscovering black phosphorus as an anisotropic layered material for optoelectronics and electronics. Nature Comm. 2014;5:4458.

[13] Li L, Yu Y, Ye GJ, et al. Black phosphorus field-effect transistors. Nat Nanotechnol. 2014;9:372–377.

[14] Barone V, Hod O, Scuseria GE. Electronic structure and stability of semiconducting graphene nanoribbons. Nano Lett. 2006;6:2748–2754.

[15] Balog R, Jørgensen Bjarke, Nilsson Louis, et al. Bandgap opening in graphene induced by patterned hydrogen adsorption Nat Mater. 2010;9:315–319.

[16] Iqbal MW, Iqbal MZ, Khan MF, et al. Modification of the structural and electrical properties of graphene layers by Pt adsorbates Sci. Technol Adv Mater. 2014;15:055002.

[17] Ma S. Bandgap modulation and hydrogen storage with Cr-doped BN sheets. Int J Mod Phys B. 2015;29:1550160.

[18] Castellanos-Gomez A, Vicarelli, L, Prada, E, et al. Isolation and characterization of few-layer black phosphorus. 2D Mater. 2014;1 025001.

[19] Fang H, Tosun M, Seol G, et al. Degenerate n-Doping of few-layer transition metal dichalcogenides by potassium. Nano Lett. 2013;13:1991–1995.

[20] Wang QH, Kalantar-Zadeh K, Kis A, et al. Electronics and optoelectronics of two-dimensional transition metal dichalcogenides. Nature Nanotech. 2012;7:699–712.

[21] Rai A, Valsaraj, A, Movva, Hema CP, et al. Air stable doping and intrinsic mobility enhancement in monolayer molybdenum disulfide by amorphous titanium suboxide encapsulation. Nano lett. 2015;15:4329–4336.

[22] Liu H, Neal AT, Ye PD. Channel length scaling of MoS2 MOSFETs. ACS Nano. 2012;6:8563–8569.

[23] Yoon Y, Ganapathi K, Salahuddin S. How good can monolayer MoS_2 transistors be? Nano Lett. 2011;11:3768–3773.

[24] Wang QH, Kalantar-Zadeh K, Kis A, et al. Electronics and optoelectronics of two-dimensional transition metal dichalcogenides. Nat Nanotechnol. 2012;7:699–712.

[25] Buscema M, Barkelid M, Zwiller V, et al. Large and tunable photothermoelectric effect in single-layer MoS2. Nano Lett. 2013;13:358–363.

[26] Das S, Chen H-Y, Penumatcha AV, et al. High performance multilayer MoS2 transistors with scandium contacts. Nano Lett. 2013;13:100–105.

[27] Sundaram RS, Engel M, Lombardo A, et al. Electroluminescence in single layer MoS2. Nano Lett. 2013;13:1416–1421.

[28] Jeon PJ, Kim JS, Lim JY, et al. Low power consumption complementary inverters with n-MoS2 and p-WSe2 dichalcogenide nanosheets on glass for logic and light-emitting diode circuits. ACS Appl Mater Interfaces. 2015;7:22333–22340.

[29] Ji Q, Zhang, Y, Gao, T, et al. Epitaxial monolayer MoS2 on mica with novel photoluminescence. Nano Lett. 2013;13:3870–3877.

[30] Yin Z, Li H, Li H, et al. Single-layer MoS2 phototransistors. ACS Nano. 2012;6:74–80.

[31] Dolui K, Pemmaraju CD, Sanvito S. Electric field effects on armchair MoS2 nanoribbons. ACS Nano. 2012;6:4823–4834.

[32] Botello-Méndez AR, López-Urías F, Terrones M, et al. Metallic and ferromagnetic edges in molybdenum disulfide nanoribbons. Nanotechnology. 2009;20:325703.

[33] Lucking MC, Bang J, Terrones H, et al. Multivalency-induced band gap opening at MoS 2 edges. Chem Mater. 2015;27:3326–3331.

[34] Wu W, Wang, L, Li, Y, et al. Piezoelectricity of single-atomic-layer MoS2 for energy conversion and piezotronics. Nature. 2014;514:470–474.

[35] Zhou W, Zou X, Najmaei S, et al. Intrinsic structural defects in monolayer molybdenum disulfide. Nano Lett. 2013;13:2615–2622.

[36] Helveg, S, Lauritsen, JV, Laegsgaard, E, et al. Atomic-scale structure of single-layer MoS2 nanoclusters. Phys Rev Lett. 2000;84:951–954.

[37] Dumcenco D, Ovchinnikov D, Lopez Sanchez O, et al. Large-area MoS_2 grown using H_2 S as the sulphur source. 2D Mater. 2015;2:044005.

[38] Prezzi D, Eom D, Rim KT, et al. Edge structures for nanoscale graphene islands on co(0001) surfaces. ACS Nano. 2014;8:5765–5773.

[39] Machida R, Toda R, Fujikawa S, et al. Effects of Ga-induced reconstructed surfaces and atomic steps on the morphology of GaSb islands on Si(100). Appl Surf Sci. 2015;351:686–692.

[40] Levita G, Cavaleiro A, Molinari E, et al. Sliding properties of MoS 2 layers: load and interlayer orientation effects. J Phys Chem C. 2014;118:13809–13816.

[41] Onodera T, et al. A computational chemistry study on friction of h-MoS_2. Part II. Friction anisotropy. J Phys Chem. B. 2010;114:15832–15838.

[42] Wang L-F, Ma T-B, Hu Y-Z, et al. Superlubricity of two-dimensional fluorographene/MoS2 heterostructure: a first-principles study. Nanotechnology. 2014;25: 385701.

[43] Quereda J, Castellanos-Gomez A, Agraït N, et al. Single-layer MoS2 roughness and sliding friction quenching by interaction with atomically flat substrates. Appl Phys Lett. 2014;105:53111.

[44] Hölscher H. Friction at atomic-scale surface steps: experiment and theory. Phys Rev. Lett. 2008;101:246105.

[45] Hölscher H, Schwarz UD, Wiesendanger R. Modelling of the scan process in lateral force microscopy. Surf Sci. 1997;375:395–402.

[46] Hirano M. Atomistics of friction. Surf Sci Rep. 2006;60:159–201.

[47] Takeda M, Ina H, Kobayashi S. Fourier-transform method of fringe-pattern analysis for computer-based topography and interferometry. J Opt Soc Am. 1982;72:156.

[48] Xie H, Vitard J, Haliyo S, et al.Calibration of lateral force measurements in atomic force microscopy with a piezoresistive force sensor. Rev Sci Instrum. (2014);79:33708[Online]. Available: http://www.brukerafmprobes.com/Product.aspx?ProductID=3394

[49] Xie H, Vitard J, Haliyo S, et al. Calibration of lateral force measurements in atomic force microscopy with a piezoresistive force sensor. Rev Sci Instrum. 2008;79:33708.

[50] Lee C, Li Q, Kalb W, et al. Frictional characteristics of atomically thin sheets. Sci. 2010;328:76–80.

7

Plastic accommodation at homophase interfaces between nanotwinned and recrystallized grains in an austenitic duplex-microstructured steel

Iván Gutierrez-Urrutia[a,b], Fady Archie[a], Dierk Raabe[a], Feng-Kai Yan[c], Nai-Rong Tao[c] and Ke Lu[c,d]

[a]Microstructure Physics and Alloy Design, Max-Planck-Institut für Eisenforschung, Max-Planck Str. 1, D-40237 Düsseldorf, Germany; [b]Structural Materials Unit, Research Center for Strategic Materials, National Institute for Materials Science, Tsukuba, Ibaraki 305-0047, Japan; [c]Institute of Metal Research, Chinese Academy of Sciences, Shenyang National Laboratory for Materials Science, Shenyang 110016, PR China; [d]Herbert Gleiter Institute of Nanoscience, Nanjing University of Science & Technology, Nanjing 210094, PR China

ABSTRACT

The plastic co-deformation behavior at the homophase interfaces between the hard nanotwinned grain inclusions and the soft recrystallized matrix grains in a duplex-microstructured AISI 316L austenitic stainless steel is examined through the analysis of long-range orientation gradients within the matrix grains by electron backscatter diffraction and transmission electron microcopy. Our analysis reveals that the mechanical accommodation of homophase interfaces until a macroscopic strain of 22% is realized within a small area of soft grains (about four grains) adjacent to the homophase interface. The activation of deformation twinning in the first two grain layers results in the occurrence of a 'hump' in the orientation gradient profile. We ascribe this effect to the role of deformation twinning on the generation of geometrically necessary dislocations. The smooth profile of the orientation gradient amplitude within the first 10 grain layers indicates a gradual plastic accommodation of the homophase interfaces upon straining. As a consequence, damage nucleation at such interfaces is impeded, resulting in an enhanced ductility of the single phase duplex-microstructured steel.

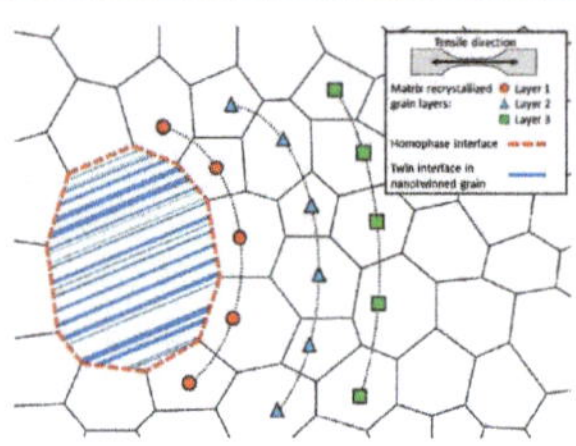

1. Introduction

Duplex-microstructured steels have been intensively investigated in the past years as a high performance class of advanced structural materials. These microstructures are commonly based on composite-type morphologies containing a hard phase, typically 20–30 vol.% martensite, and a soft phase, namely, ferrite [1–3]. They are very attractive to the automotive industry because of their favorable combination of high strength and good formability [1,3,4]. The austenite to martensite transformation in these steel grades is accompanied by a 2–4% volume expansion, generating residual stresses and complex strain gradients in the matrix ferrite adjacent to the ferrite/martensite interface [5,6]. The residual stresses tend to enhance the plastic flow in ferrite and decrease the elastic limit while strain gradients accommodated by geometrically necessary dislocations (GNDs) contribute to the continuous yielding behavior that characterizes the strain-hardening behavior [5–9].

The large number of relevant microstructure parameters, the complexity of the underlying deformation mechanisms and the limited ductility when exposed to complex strain path changes make the design of these steel grades a challenging task [10–13]. Hence, recently, different approaches have been proposed to design single-phase heterogeneous structural steels containing homophase interfaces [14–21]. Among them, the processing strategies typically involve severe plastic deformation to high strain levels and subsequent annealing treatments to create fully or partially recrystallized microstructures containing different density/types of crystal defects or grain size distributions [14–19,22].

CONTACT Iván Gutierrez-Urrutia gutierrezurrutia.Ivan@nims.go.jp

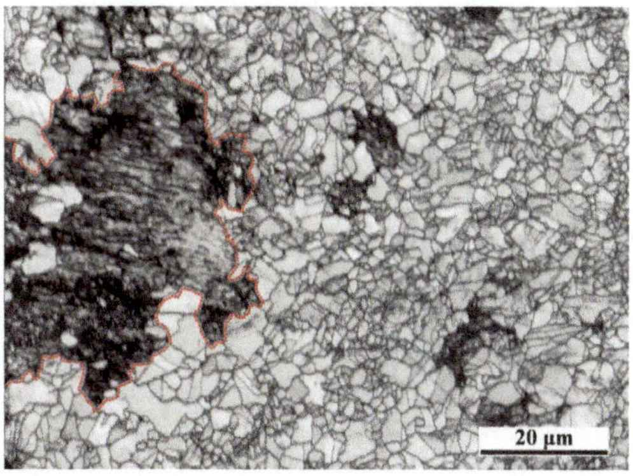

Figure 1. EBSD map of the duplex microstructured AISI 316L stainless steel consisting of coarse nanotwinned grains and fine recrystallized grains. The homophase between the two grain types is indicated by a red dotted line.

One of these approaches has successfully produced austenitic single phase duplex-microstructured steels consisting of coarse nanotwinned grains embedded into fine recrystallized matrix grains by means of dynamic plastic deformation (DPD) [17–19,23]. From a micromechanical standpoint, such microstructures can be considered as a single-phase composite consisting of hard inclusions, namely the nanotwinned grains, surrounded by softer recrystallized grains. Specifically, the tensile strength of nanotwinned grains can be as high as ~1.5–2.0 GPa, i.e. higher than martensite, but can yet sustain ~5% uniform tensile strain [18,19].

Homophase interfaces such as dislocation boundaries and twin interfaces play a significant role on the strain-hardening behavior, and hence on the mechanical properties of polycrystalline materials. The most evident interface parameter that contributes to the strain-hardening behavior of polycrystalline materials is the interface spacing. According to dislocation-mean free path theories of strain-hardening [24,25], homophase interface spacing contributes to the macroscopic flow stress by scaling laws such as the Hall-Petch relation and the similitude relation [26–29]. However, other homophase interface parameters have a significant contribution to the strain-hardening behavior as well: misorientation of the dislocation boundary and twin thickness determine the critical stress required to transfer plasticity across dislocation boundaries and twin interfaces, respectively [30,31]; elastic and plastic mismatch between the iso-phase control the plastic accommodation of the homophase interface, and hence the formation of strain gradients accommodated by geometrically necessary dislocations (GNDs) around such interfaces [32–34].

In a previous work [23], we have investigated the plastic deformation mechanisms of a novel austenitic duplex microstructured steel fabricated by dynamic plastic deformation (DPD). The duplex microstructure consists of strong nanotwinned grains embedded in soft recrystallized grains. We observed that at low strain levels (below 5%), the material deforms homogeneously by gradual co-deformation between the hard and soft grains without producing noticeable strain localization at the homophase interfaces between the two types of grains. With further straining (over 10%), a strain gradient is developed within the softer grains as a function of the distance from the homophase interfaces. This effect, together with the activation of localized deformation in the form of shear banding within the coarse nanotwinned grains, results in an inhomogeneous deformation behavior of the duplex steel. It is thus clear that the mechanical behavior of the homophase interfaces between the hard and soft grains plays a significant role on the deformation behavior, and hence, on the mechanical behavior of this advanced steel.

The present work investigates the details of the plastic accommodation of homophase interfaces through the analysis of the evolving long-range orientation gradients within the recrystallized matrix grains by electron backscatter diffraction (EBSD) and transmission electron microscopy (TEM). Our analysis of in-grain orientation gradients reveals that the mechanical accommodation of homophase interfaces until a macroscopic strain of 22% is realized within a small area of soft grains (about four grains) adjacent to such interface. The activation of deformation twinning in the first two soft grain layers close to the homophase interface results in the occurrence of a 'hump' in the orientation gradient profile. We ascribe this effect to the role of deformation twinning on the generation of geometrically necessary dislocations. The smooth profile of the orientation gradient amplitude within the first 10 grain layers indicates a gradual plastic accommodation of such interface upon straining. We associate this finding to the good ductility exhibited by the present steel (total elongation of about 46%).

2. Experimental methodology

The material used in this work was an AISI 316L austenitic single phase-duplex microstructured steel. The duplex microstructure consists of coarse inclusion grains (average grain size of 20 μm) with an area fraction of ~23%, and fine recrystallized matrix grains with an average grain size of ~2 μm (Figure 1). Most of the coarse grains contain a lamellar twin structure of nano-scale twins (thickness ~23 nm) arranged into bundles. The coarse twinned grains are considered as a hard inclusion phase and the recrystallized grains act as a soft matrix phase [19,23]. The duplex microstructure was produced via a two-step process, namely, imposing first a dynamic plastic deformation (DPD) processing through multiple impacts to a total strain of 1.6, and a subsequent annealing step at 750°C for 45 min [17,19,35]. Interrupted tensile tests to engineering strain levels of ε=0.05, 0.12 and 0.22 were performed to investigate the evolution of the

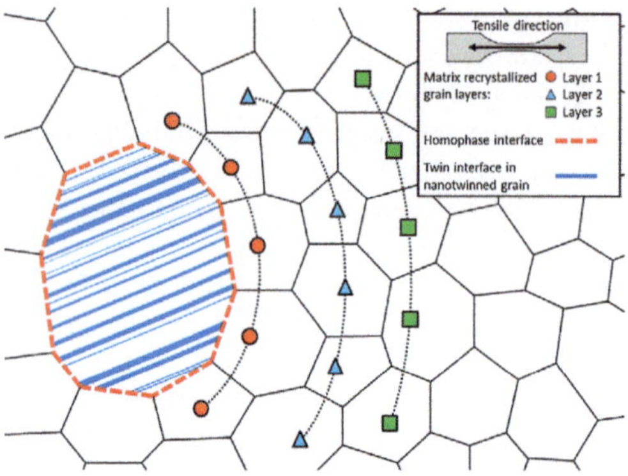

Figure 2. Sketch of the duplex-microstructure at the homophase interface region. The classification of matrix grains surrounding the hard grain inclusion into layers is depicted.

deformation behavior of the duplex microstructured steel. The tensile samples had 5 mm gage length, 2 mm gage width and 1 mm gage thickness. The monotonic tensile deformation experiments were carried out on a tensile Kammrath & Weiss GmbH test instrument (44141 Dortmund, Germany) equipped with a digital image correlation (DIC) system (ARAMIS system, GOM-Gesellschaft für Optische Messtechnik mbH, 38106 Braunschweig, Germany) to measure the local and the macroscopic strain distribution [36]. The surface pattern required for DIC was obtained by applying two different color sprays on the sample surface. First, a white spray was used to obtain a homogeneous background, and then a black spray was applied to obtain a spotted pattern [37]. Averaged engineering strain values were retrieved from the corresponding strain maps. At each strain level, microstructure was characterized by EBSD after DIC surface pattern removal. The observation direction was parallel to the tensile axis. EBSD maps were acquired with a 6500 F JEOL field emission gun-scanning electron microscope (FEGSEM) equipped with a TSL OIM EBSD system at 15 kV acceleration voltage, working distance of 15 mm and step size of 50 nm. The details of TEM characterization are described in [23].

3. Results and discussion

3.1. Analysis method of orientation gradients

We investigated the deformation behavior in the nanotwinned grain/recrystallized grain (nt-grain/Rx-grain) interface regions by means of an EBSD approach based upon grain reference orientation deviation (GROD) maps. The method is outlined below in more detail. The deformation behavior in the homophase interface regions was evaluated along several matrix Rx-grain layers surrounding the hard inclusion grains, as depicted schematically in Figure 2. We have only considered areas containing similar grain sizes of the recrystallized matrix grains of about 2–5 μm to avoid blurring of the results due to grain size effects. The approach chosen here allows obtaining a sound microstructure–orientation gradient correlation for such heterostructure homophase morphologies in the present alloys. The rationale behind this method is to quantify the degree of strain localization near the interface of the nanotwinned grain inclusions. The evaluation method is as follows. First, the matrix Rx-grains surrounding a hard inclusion grain are classified into several layers according to their distance to the nearest homophase interface (Figure 2). Second, grain reference orientation deviation (GROD) maps are calculated from the EBSD data as a function of the angular deviation from a reference orientation within a given grain. Basically, GROD maps display in-grain misorientations with respect to the selected reference orientation. In the present case, we have set the reference orientation as the one containing the lowest Kernel average misorientation (KAM) value. This parameter is calculated as the average misorientation, Δg_K, of a given point relative to its neighbors, with the exclusion of misorientation values Δg_{Ai} that exceed a maximum tolerance value of 2° [38,39]:

$$\Delta g_K = \frac{1}{4}\left(\Delta g_{A1} + \Delta g_{A2} + \Delta g_{A3} + \Delta g_{A4}\right) \qquad [1]$$

where Δg_{Ai} refers to the misorientation between a given point A and the neighbor i. Figure 3(a) shows an example of a GROD map corresponding to a matrix Rx-grain close to a homophase interface. This figure reveals that plastic deformation is concentrated at the grain boundaries rather than at the grain interiors, as reflected by the high local misorientation values. This effect can be ascribed to elastic and plastic incompatibility effects between neighboring grains which promote the activation of a higher number of slip systems compared to the grain interiors [40,41]. Figure 3(a) also reveals the development of several regions with high localized in-grain orientation gradients, namely H1 and H2, as expected in polycrystal plasticity owing to different boundary conditions and, hence, different accommodation gradients on opposite sides of the same grain [42,43]. As a general measure to quantify such in-grain orientation gradients in the different matrix Rx-grain layers around the homophase interface, we assign to each grain the orientation gradient amplitude with the highest GROD value, for instance the amplitude H2 in Figure 3(a), as depicted in Figure 3(b). We then calculate the average orientation gradient amplitude for each grain layer (about 10 grains per layer) and analyze them as a function of progressing sample deformation (the same grains are analyzed in each layer at evolving strain level).

3.2. Microstructure characterization

As Figure 1 reveals, the microstructure of the AISI 316L austenitic single phase-duplex microstructured steel can be considered as a partially recrystallized

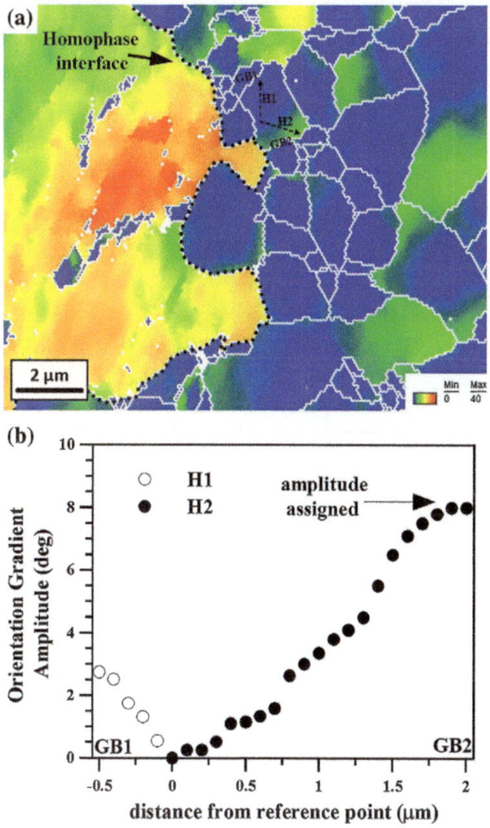

Figure 3. Sketch of the analysis of orientation gradients. (a) Example of grain reference orientation deviation (GROD) map. H1 and H2 refer to in-grain orientation gradients around grain boundaries labeled as GB1 and GB2, respectively. (b) Schematic plot of the amplitude of the orientation gradients shown in (a).

composite-type structure formed by micron-sized Rx austenitic grains containing hard austenitic inclusions (outlined, non-recrystallized grains, i.e. mainly nt-grains). Figure 4 shows the corresponding GROD maps of the tracked area at several engineering strain levels [23]. These maps show the formation of local orientation gradients within the matrix Rx-grains, i.e. the soft phase, close to the homophase interface (indicated by a red dotted line in Figure 4) upon macroscopic straining. In the as-processed state, the coarse nt-grain (outlined by a red dotted line) contains orientation gradients which are ascribed to the high density of crystal defects whereas the Rx-grains are initially free of orientation gradients; see Figure 4(a). Orientation gradients within Rx-grains adjacent to the homophase interface are visible at a strain of 0.12; see Figure 4(c). Further straining to 0.22 engineering strain results in a similar distribution of orientation gradients within all the Rx-grains; see Figure 4(d).

The evolution of the average orientation gradient amplitude as a function of the imposed macroscopic strain in the 10 matrix Rx-grain layers adjacent to the homophase interface of Figure 4 is shown in Figure 5. The main characteristics of this analysis are: first, we observe the development of a gradual transition from strong orientation gradients that are located within the first four grain layers to small orientation gradients occurring in grains further away from the immediate homophase interface. The intensity of the strong orientation gradients is about 1.5 times higher than the grains farther away from the interface. Second, we observe the occurrence of a 'hump' in the orientation gradient profile located at about the third grain layer. These observations indicate that the homophase interface plays a significant role on the deformation behavior of the nearest four grain layers to such interface. The development of an area of soft matrix Rx-grains with enhanced plastic activity, i.e. high density of geometrically necessary dislocations (GNDs), around a homophase interface can be ascribed to the mechanical incompatibility between the hard nt-grain and the soft Rx-grains resulting in a load transfer effect similar to that occurring in metal matrix composites [44] and second phase-particle containing materials [32–34]. The present results indicate that such mechanical incompatibility is mainly relaxed within the soft Rx-grains by the occurrence of strain gradients extending over about four grains.

The occurrence of a 'hump' in the profile of the average orientation gradient amplitude deviates from the typical trend $1/\lambda$, where λ is the distance from a given interface, reported in several prior studies [5,6,34]. Figure 5 also shows that the height of the 'hump' scales with strain. This observation suggests the occurrence of a governing relation between the evolution of the deformation substructure and the development of orientation gradients within the first four matrix Rx-grain layers around the homophase interface. TEM observations revealed a pronounced activation of deformation twins within the Rx-grains. As Figure 6 shows, at a macroscopic strain of 0.12, Rx-grains adjacent to the homophase interface, namely grains labeled as (a) and (b), develop a lamellar twin-type structure, as shown in the corresponding diffraction patterns. In contrast, Rx-grains located away from such homophase interfaces contain dislocation substructures free of deformation twins. As a previous work has recently shown [23, figures 1(b), 5(a) and 8], in the present single phase-duplex microstructured steel strained to low strain levels (below 5%), the homophase interfaces between the nanotwinned grains and the recrystallized grains are accommodated by dislocation plasticity. The nt-grains deform in a homogeneous fashion in conjunction with the surrounding Rx-grains without generating significant strain localization near their interfaces, as revealed by the homogeneous dislocation density distribution within the Rx-grains adjacent to such interfaces [23, figure 5(b)].

Deformation twinning in fcc steels is a stress-assisted mechanism which is dependent on the crystallographic grain orientation through the Schmid factor, and the grain size [31,37,45]. In the present composite-structure,

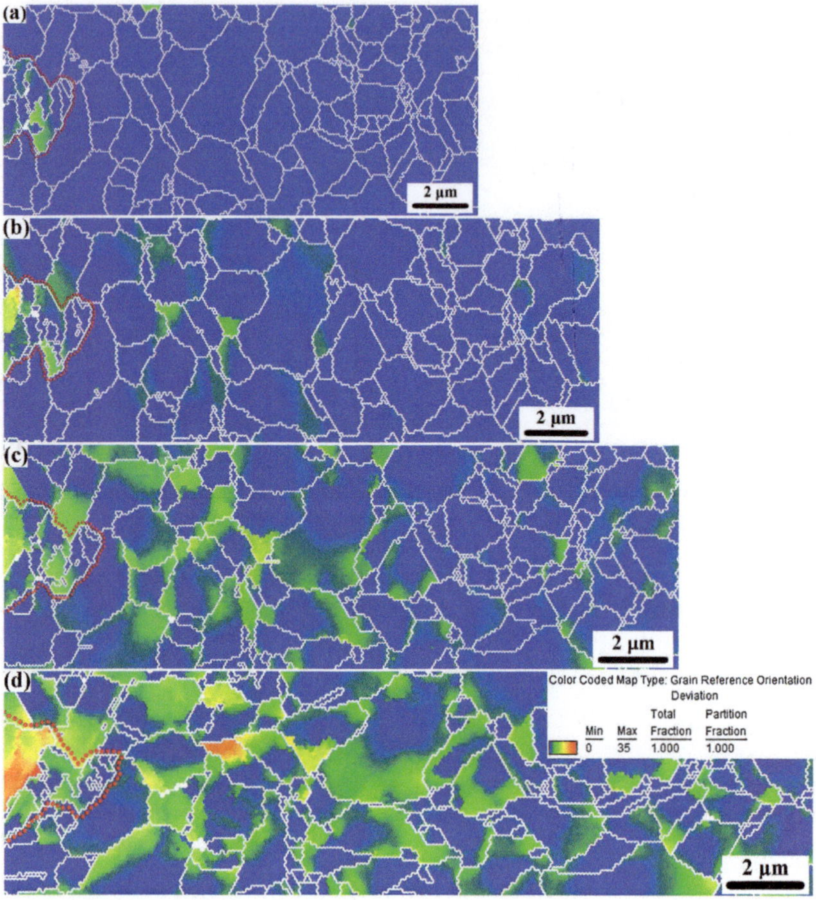

Figure 4. Grain reference orientation deviation (GROD) maps of the tracked duplex microstructure at several strain levels [23]. The homophase interface is indicated by a red dotted line. (a) undeformed; (b) engineering strain of 0.05; (c) 0.12; (d) 0.22.

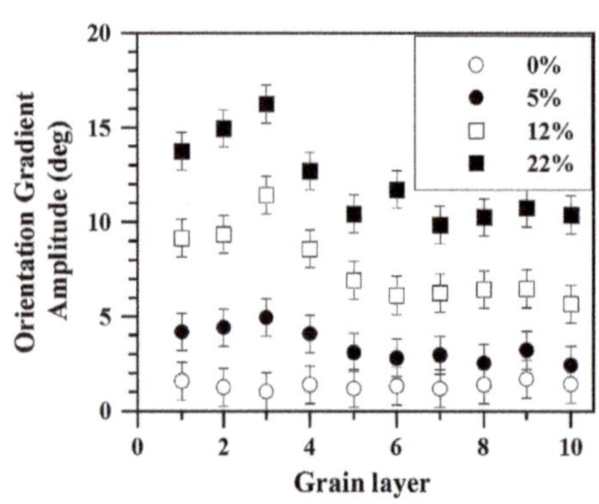

Figure 5. Evolution of the average orientation gradient amplitude with the imposed engineering strain along 10 matrix grain layers adjacent to the homophase interface shown in Figure 4. Error bars indicate statistical and experimental angular error (± 1°).

we do not observe any preferential crystallographic orientation for twinning of the matrix Rx-grain layers. EBSD mapping reveals that these grains develop a typical α-fiber with texture components oriented along the line between the <001>//TA and <111>//TA crystallographic directions (TA: tensile axis). As these grains have similar average grain sizes, this lack of preferential crystallographic orientation suggests that the activation of deformation twinning in these grains is controlled by the local stress state, which can strongly vary from the macroscopic stress state [37,45]. Such local high stress concentrations result from the plastic strain mismatch between the stiff nt-grain and the soft Rx-grain.

The occurrence of a 'hump' in the profile of the average orientation gradient amplitude can be therefore explained as follows. Orientation gradients in second-phase containing materials are determined by the generation of geometrically necessary dislocations, ρ_{GND}, required to accommodate the plastic gradient ascribed to the mechanical incompatibility between the hard inclusion and the soft matrix. This effect can be roughly estimated as $\rho_{GND} \sim \varepsilon/\lambda$, where λ is the distance from a homophase interface and ε is the difference of plastic strain between the soft phase and hard phase [33,34]. However, the activation of deformation

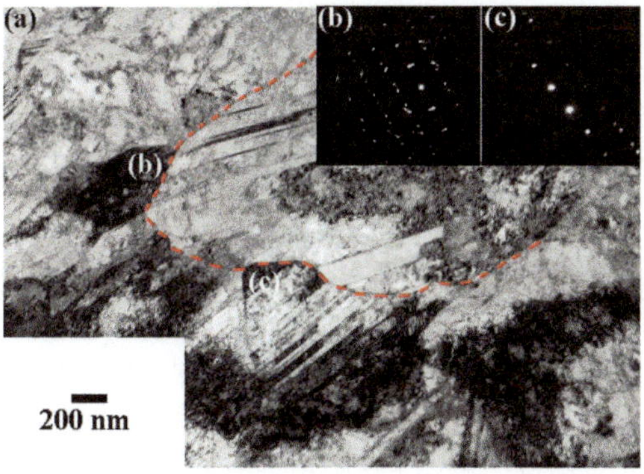

Figure 6. (a) Bright-field TEM image of the deformation structure of an area containing a single coarse nanotwinned grain (indicated by a red dotted line) and several adjacent recrystallized matrix grains. The sample was deformed in tension to a strain of 0.12. (b, c) Corresponding diffraction patterns of recrystallized matrix grains labeled as b and c in (a).

twinning modifies this picture. TEM observations in the present material reveal that at a macroscopic strain of 0.12–0.22 twins are mainly visible within the first and second Rx-grains adjacent to the homophase interface [23]. These observations indicate that grain deformation can be carried out in these grains by slip and twinning. We suggest that the competition between these two deformation modes to accommodate the plastic gradient associated to the mechanical incompatibility between the hard inclusion and the soft matrix, as well as slip hardening due to twin-slip interaction result in smaller ρ_{GND} compared to the case where only slip is available, such in Rx-grains away from the homophase interface. As a consequence, the orientation gradient profile deviates from a simple trend ~ $1/\lambda$ but exhibits a more complicated behavior.

At this point, it is also relevant to discuss the role of crystallographic grain orientation on slip transfer, and hence, on the occurrence of local orientation gradients. The geometry of slip transfer between two slip systems on either side of a boundary is usually defined by three angles, namely, the angle between slip vectors (κ), the angle between slip plane normals (ψ), and the angle between the two slip plane intersections with the grain boundary plane (φ) [46]. So far, two criteria have been proposed, which are based on the maximizing of a parameter that is a product of the cosine of some of these angles. Following these criteria, grain boundaries can be classified as impenetrable, penetrable and transparent according to their ability to transfer an incoming slip system. In the present work, we have investigated the occurrence of in-grain orientation gradients with respect to the distance of a specific homophase interface by EBSD. We define such distance as grain layers (up to 10), which are defined as grain neighbors to the interface. We then calculate the orientation gradient amplitude of each grain layer as the average amplitude of the highest GROD amplitude of 10 grains. In other words, the orientation gradient amplitude plotted in Figure 5 corresponds to the average of the orientation gradient amplitude of 10 grains per layer. Taking into account that the crystallographic orientations of these grains are contained within an α-fiber, i.e. they contain discrete grain orientations, and the intrinsic EBSD resolution (~1–2° [47]), we consider that grain orientation effects on slip transfer, and hence on the amplitude of in-grain orientation gradients, are smoothed out in the present analysis.

It is important to recognize the excellent mechanical compatibility exhibited by the present homophase interface between the hard nanotwinned grain inclusion and soft Rx-grains. The present results reveal that the accommodation of the mechanical incompatibility between the two grain types is carried out in a gradual fashion both by slip and twinning along the four adjacent matrix Rx-grains to the interface, as reflected by the smooth profile of the orientation gradient amplitude. This result has a pronounced effect in attaining a high ductility (total elongation of the present single phase duplex-microstructured steel is about 46%). Both experimental and computational studies report that damage in particle-free materials commonly nucleates at locations of large strain incompatibilities such as grain or phase boundaries where high heterogeneous strain gradients can be developed [41,43,46]. Specifically, damage nucleation is dependent on several aspects such as the boundary orientation and structure, the grain boundary slip transfer geometry, i.e. slip/twin planes and directions of active deformation systems on either side of the boundary, and the stress–strain gradient history in the grains on either side of an interface [41,43,46,48,49]. The development of long-scale heterogeneous strain gradients at such homophase interfaces may cause large tensile tractions in the boundary resulting in damage nucleation [48,50]. The present analysis suggests that the gradual accommodation of the plastic deformation within the homophase interface region mitigates damage nucleation at such interfaces, and accordingly enhances the ductility of the duplex-microstructured steel.

5. Conclusions

We have investigated the plastic co-deformation behavior of an austenitic duplex-microstructured AISI 316L stainless steel at the homophase interfaces between hard nanotwinned grain inclusions and soft recrystallized matrix grains. The evolution of the underlying deformation structure as a function of strain was investigated by using EBSD and TEM. In-grain orientation gradients along the recrystallized matrix grains are analyzed by means of a grain reference orientation deviation-type approach based on EBSD data. The following conclusions can be drawn:

- Our analysis of in-grain orientation gradients reveals that the mechanical accommodation of homophase interfaces until a macroscopic strain of 22% is realized within a small area of soft grains (about four grains) adjacent to the homophase interface. The activation of deformation twinning in the first two soft grain layers close to the homophase interfaces results in the occurrence of a 'hump' in the orientation gradient profile. We ascribe this effect to the role of deformation twinning on the generation of geometrically necessary dislocations.

- The smooth profile of the orientation gradient amplitude within the first 10 Rx-grain layers indicates that the mechanical accommodation of the interface is realized in a gradual fashion along the adjacent matrix grains surrounding the hard inclusion grains.
- We ascribe the good ductility exhibited by the present austenitic duplex-microstructured steel (total elongation of about 46%) to the gradual accommodation of plastic deformation by multiple slip system activation within the homophase interface region that leads to GNDs, which hinders the occurrence of high local orientation gradients that can nucleate damage at such interfaces, and accordingly enhances the total elongation.

Disclosure statement

No potential conflict of interest was reported by the authors.

Funding

The authors are grateful for financial support of the Ministry of Science and Technology of China [grant 2012CB932201]; and the National Natural Science Foundation [grants 51371172, 51231006]. The German Research Foundation (Deutsche Forschungsgemeinschaft DFG) is gratefully acknowledged for financial support through the project [SFB-761 – steel *ab initio*].

References

[1] Bhadeshia HKDH, Honeycombe R. Steels. Microstructure and properties. Oxford: Elsevier Ltd; 2006.
[2] Speich GR, Miller RL. Mechanical properties of ferrite-martensite steels, structure and properties of dual-phase steels. In: TMS-AIME, Kot RA, Morris JW, editor. Structure and Properties of Dual-Phase Steels. New York (NY); 1979. p. 145–182.
[3] Rashid MS. Dual-phase steels. Annu Rev Mater Sci. 1981;11:245.
[4] Calcagnotto M, Ponge D, Raabe D. Effect of grain refinement to 1mm on strength and toughness of dual-phase steels. Mater Sci Eng A. 2010;527:7832.
[5] Kadkhodapour J, Schmauder S, Raabe D, et al. Experimental and numerical study on geometrically necessary dislocations and non-homogenous mechanical properties of the ferrite phase in dual phase steels. Acta Mater. 2011;59:4387.
[6] Calcagnotto M, Ponge D, Demir E, et al. Orientation gradients and geometrically necessary dislocations in ultrafine grained dual-phase steels studied by 2D and 3D EBSD. Mater Sci Eng A. 2010;527:2738.
[7] Jiang Z, Guan Z, Lian J. Effects of microstructural variables on the deformation behaviour of dual-phase steel. Mater Sci Eng A. 1995;190:55.
[8] Ghassemi-Armaki H, Maas R, Bhat SP, et al. Deformation response of ferrite and martensite in a dual-phase steel. Acta Mater. 2014;62:197.
[9] Delinc- M, Jacques PJ, Pardoen T. Separation of size-dependent strengthening contributions in fine-grained Dual Phase steels by nanoindentation. Acta Mater. 2006;54:3395.
[10] Delincé M, Bréchet Y, Embury JD, et al. Structure–property optimization of ultrafine-grained dual-phase steels using a microstructure-based strain hardening model. Acta Mater. 2007;55:2337.
[11] Woo W, Em VT, Kim E-Y, et al. Stress–strain relationship between ferrite and martensite in a dual-phase steel studied by in situ neutron diffraction and crystal plasticity theories. Acta Mater. 2012;60:6972.
[12] Ramazani A, Mukherjee K, Schwedt A, et al. Quantification of the effect of transformation-induced geometrically necessary dislocations on the flow-curve modelling of dual-phase steels. Int J Plast. 2013;43:128.
[13] Chen P, Ghassemi-Armaki H, Kumar S, et al. Microscale-calibrated modeling of the deformation response of dual-phase steels. Acta Mater. 2014;65:133.
[14] Tsuji N, Ueji R, Minamino Y. Nanoscale crystallographic analysis of ultrafine grained IF steel fabricated by ARB process. Scripta Mater. 2002;47:69.
[15] Chen XH, Lu J, Lu L, et al. Tensile properties of a nanocrystalline 316L austenitic stainless steel. Scripta Mater. 2005;52:1039.
[16] Weng Y. Ultra-fine grained steels. Berlin: Springer-Verlag; 2009.
[17] Lu K, Yan FK, Wang HT, et al. Strengthening austenitic steels by using nanotwinned austenitic grains. Scripta Mater. 2012;66:878.
[18] Wang HT, Tao NR, Lu K. Strengthening an austenitic Fe–Mn steel using nanotwinned austenitic grains. Acta Mater. 2012;60:4027.
[19] Yan FK, Liu GZ, Tao NR, et al. Strength and ductility of 316L austenitic stainless steel strengthened by nano-scale twin bundles. Acta Mater. 2012;60:1059.
[20] Gutierrez-Urrutia I, Raabe D. Multistage strain hardening through dislocation substructure and twinning in a high strength and ductile weight-reduced Fe-Mn-Al-C steel. Acta Mater. 2012;60:5791.
[21] Gutierrez-Urrutia I, Raabe D. Microbanding mechanism in a Fe-22Mn-0.6C (wt.%) high-Mn twinning induced plasticity steel. Scripta Mater. 2013;69:53.
[22] Li S, Gazder AA, Beyerlein IJ, et al. Microstructure and texture evolution during equal channel angular extrusion of interstitial-free steel: Effects of die angle and processing route. Acta Mater. 2007;55:1017.
[23] Yan FK, Tao NR, Archie F, et al. Deformation mechanisms in an austenitic single-phase duplex microstructured steel with nanotwinned grains. Acta Mater. 2014;81:487.

[24] Nes E. Modelling of work hardening and stress saturation in FCC metals. Prog Mater Sci. 1998;41:129.
[25] Kocks UF, Mecking H. Physics and phenomenology of strain hardening: the FCC case. Prog Mater Sci. 2003;48:171.
[26] Sauzay M, Kubin LP. Scaling laws for dislocation microstructures in monotonic and cyclic deformation of fcc metals. Prog Mater Sci. 2011;56:725.
[27] Kuhlmann-Wilsdorf D. Theory of plastic deformation:-properties of low energy dislocation structures. Mater Sci Eng A. 1989; 113:1.
[28] Kuhlmann-Wilsdorf D. "Regular" deformation bands (DBs) and the LEDS hypothesis. Acta Mater. 1999;47:1697.
[29] Asgari S, El-Danaf E, Kalidindi SR, et al. Strain hardening regimes and microstructural evolution during large strain compression of low stacking fault energy fcc alloys that form deformation twins. Metall Mater Trans A. 1997;28:1781.
[30] Hughes DA, Hansen N. Microstructure and strength of nickel at large strains. Acta Mater. 2000;48:2985.
[31] Christian JW, Mahajan S. Deformation twinning. Prog Mater Sci. 1995;39:1.
[32] Humphreys FJ, Kalu PN. The plasticity of particle-containing polycrystals. Acta Metall Mater. 1990;38:917.
[33] Ashby MF. The deformation of plastically non-homogeneous materials. Phil Mag. 1970;21:399.
[34] Humphreys FJ. Local lattice rotations at second phase particles in deformed metals. Acta Metall. 1979;27:1801.
[35] Li YS, Tao NR, Lu K. Microstructural evolution and nanostructure formation in copper during dynamic plastic deformation at cryogenic temperatures. Acta Mater. 2008;56:230.
[36] Zhao Z, Ramesh M, D Raabe, et al. Investigation of three-dimensional aspects of grain-scale plastic surface deformation of an aluminum oligocrystal. Int. J. Plast. 2008;24:2278.
[37] Gutierrez-Urrutia I, Zaefferer S, Raabe D. The effect of grain size and grain orientation on deformation twinning in a Fe–22 wt.% Mn–0.6 wt.% C TWIP steel. Mater Sci Eng A. 2010;527:3552.
[38] Wright SI, Nowell MM, Field DP. A Review of Strain Analysis Using Electron Backscatter Diffraction. Microsc Microanal. 2011;17:316.
[39] Schwartz AJ, Kumar M, Adams BL, Field DP. Electron backscatter diffraction in materials science. New York (NY): Springer; 2009.
[40] Hirth JP. The Influence of Grain Boundaries on Mechanical Properties. Metall Trans. 1972;3:3047.
[41] Roters F, Eisenlohr P, Hantcherli L, et al. Overview of constitutive laws, kinematics, homogenization and multiscale methods in crystal plasticity finite-element modeling: Theory, experiments, applications. Acta Mater. 2010;58:1152.
[42] Raabe D, Sachtleber M, Zhao Z, et al. Micromechanical and macromechanical effects in grain scale polycrystal plasticity. Experimantion and simulation. Acta Mater. 2001;49:3433.
[43] Bieler TR, Eisenlohr P, Roters F, et al. The role of heterogeneous deformation on damage nucleation at grain boundaries in single phase metals. Int J Plast. 2009;25:1655.
[44] Arsenault RJ, Fishman S, Taya M. Deformation and fracture behavior of metal-ceramic matrix composite materials. Prog Mater Sci. 1994;38:1.
[45] Steinmetz DR, J ein T, Wietbrock B, et al. Revealing the strain hardening behavior of twinning induced plasticity steels through a dislocation density- and twin evolution-based constitutive model: theory, simulations, experiments. Acta Mater. 2013;61:494.
[46] Bieler TR, Eisenlohr P, Zhang C, et al. Grain boundaries and interfaces in slip transfer. Curr Opin Solid State Mater Sci. 2014;18:212.
[47] Zaefferer S. On the formation mechanisms, spatial resolution and intensity of backscatter Kikuchi patterns. Ultramicroscopy. 2007;107:254.
[48] Querin JA, Schneider JA, Horstemeyer MF. Analysis of micro void formation at grain boundary triple points in monotonically strained AA6022-T43 sheet metal. Mater Sci Eng A. 2007;463:101.
[49] Livingston JD, Chalmers B. Multiple slip in bicrystal deformation. Acta Metall. 1957;5:322.
[50] Bieler TR, Fallahi A, Ng BC, et al. Fracture initiation/propagation parameters for duplex TiAl grain boundaries based on twinning, slip, crystal orientation, and boundary misorientation. Intermetallics. 2005; 13:979.

8

Properties of epitaxial, (001)- and (110)-oriented $(PbMg_{1/3}Nb_{2/3}O_3)_{2/3}$-$(PbTiO_3)_{1/3}$ films on silicon described by polarization rotation

Muhammad Boota[a,b], Evert P. Houwman[a], Matthijn Dekkers[c], Minh D. Nguyen[a,c], Kurt H. Vergeer[a], Giulia Lanzara[b], Gertjan Koster[a] and Guus Rijnders[a]

[a]Faculty of Science and Technology, MESA+ Institute for Nanotechnology, University of Twente, P.O. Box 217, 7500AE Enschede, the Netherlands
[b]Engineering Department, University of Rome "ROMA TRE", Via della Vasca Navale 79, 00146 Rome, Italy
[c]SolMates BV, Drienerlolaan 5, Building 6, 7522NB Enschede, the Netherlands

ABSTRACT
Epitaxial $(PbMg_{1/3}Nb_{2/3}O_3)_{2/3}$-$(PbTiO_3)_{1/3}$ (PMN-PT) films with different out-of-plane orientations were prepared using a CeO_2/yttria stabilized ZrO_2 bilayer buffer and symmetric $SrRuO_3$ electrodes on silicon substrates by pulsed laser deposition. The orientation of the $SrRuO_3$ bottom electrode, either (110) or (001), was controlled by the deposition conditions and the subsequent PMN-PT layer followed the orientation of the bottom electrode. The ferroelectric, dielectric and piezo electric properties of the ($SrRuO_3$/PMN-PT/$SrRuO_3$) ferroelectric capacitors exhibit orientation dependence. The properties of the films are explained in terms of a model based on polarization rotation. At low applied fields domain switching dominates the polarization change. The model indicates that polarization rotation is easier in the (110) film, which is ascribed to a smaller effect of the clamping on the shearing of the pseudo-cubic unit cell compared to the (001) case.

KEYWORDS
PMN-PT; pulsed laser deposition; orientation control; ferroelectricity; piezoelectricity; thin film; epitaxy

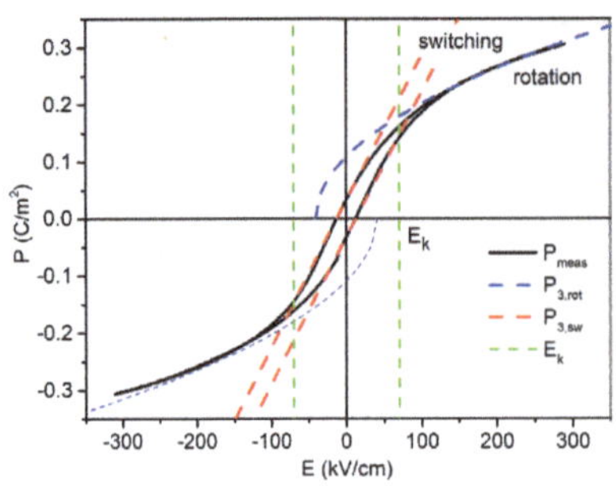

1. Introduction

Piezoelectric ceramic materials like $PbZr_xTi_{1-x}O_3$ (PZT) and $(PbMg_{1/3}Nb_{2/3}O_3)_{1-x}$-$(PbTiO_3)_x$ (PMN-PT) are used in a wide range of devices in both sensing and actuation applications, employing their piezoelectric property to directly convert mechanical energy into electrical energy or vice versa. There is a strong effort to fabricate epitaxial ferroelectric PZT and relaxor PMN-PT thin film based devices on silicon substrates for ferroelectric oxide thin film based electronic, photonic and MEMS (microelectromechanical systems) devices [1,2]. Bulk relaxor ferroelectrics like PMN-PT ($x \approx 0.33$) show superior piezo-response, exhibiting 5–10 times larger piezoelectric coefficients than bulk PZT ceramics. This material has also a very large electromechanical coupling coefficient, $k_{33} \approx 0.9$ [3,4]. Therefore it may also be an important material for thin film applications, necessitating the development of reliable thin film deposition techniques and the understanding of the properties of such thin films.

The growth of epitaxial PMN-PT films using appropriate buffers is challenging because of the large lattice mismatch between the film and the Si substrate. Another challenge is to control the crystallographic orientation of the PMN-PT films. The ferro- and piezoelectric properties of the most noticeable ferroelectric oxides (specifically the pure ferroelectric (PZT) and the relaxor

CONTACT Evert P. Houwman e.p.houwman@utwente.nl

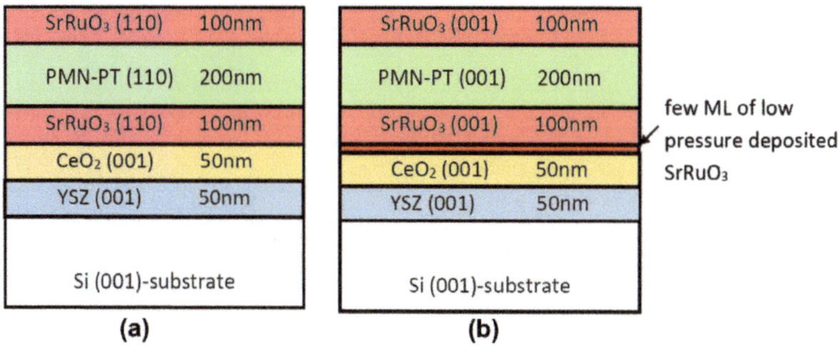

Figure 1. Schematic layer stack of (a) (110)-oriented and (b) (001)-oriented SRO/PMN-PT/SRO ferroelectric capacitors. ML stands for monolayer.

ferroelectric PMN-PT) are strongly correlated with the crystallographic orientation of the thin films.

A number of heterostructures has been developed to achieve epitaxial growth of PMN-PT films on silicon. Baek et al. [4] reported the deposition of high crystalline quality, epitaxial PMN-PT films on a $SrTiO_3$ buffered silicon substrate. Wang et al. [5] developed a $SrTiO_3$/TiN/Si heterostructure, whereas Tsang et al. [6] demonstrated that the MgO/TiN/Si bilayer system can be successfully used to obtain epitaxial growth of PMN-PT on silicon. The yttria-stabilized zirconia/ceria (YSZ/CeO_2) heterostructure is one of the most extensively investigated layer stacks for epitaxial growth of ferroelectrics oxide films on silicon [7,8]. There are reports on the deposition of (001)- and (110)-oriented epitaxial PMN-PT films using appropriate buffer layer(s) and/or electrode material.[4–10]. However, there is no literature about the control of the crystallographic orientation of epitaxial PMN-PT films on a silicon substrate with the same buffer layer(s) and/or electrode material. If one can ensure that film growth is the same in both orientations, while the intrinsic material properties in the film are kept the same, such a model system would allow one to compare the measured properties of these films and relate possible differences to the different crystal growth orientations only.

In this work, epitaxial PMN-PT films with (001) or (110) orientation are controllably fabricated on silicon substrates by pulsed laser deposition (PLD) using the *same* buffer layer/base electrode materials stack. The top monolayer of the base electrode is modified by choosing different deposition conditions (see details below) to realize a certain growth orientation. In this way stable PMN-PT films in the same substrate-induced in-plane strain state are realized and a clean model system is obtained to investigate and compare the effect of structure and crystallographic orientation on the ferro- and piezoelectric properties of epitaxial PMN-PT films on a silicon substrate.

We describe the ferroelectric, piezoelectric and dielectric properties of the films in terms of a model based on polarization rotation. At lower fields polarization switching dominates the film properties. Due to the different growth orientations (but equal in-plane strain state) we hypothesize that the polarization vector in our PMN-PT films rotates along different paths of the intrinsic energy landscape under the influence of the applied field, when scanning through polarization loops. The measurements indeed show that within the framework of the model the field dependence of the polarization angle depends on the growth orientation and thus the rotation path.

2. Experimental methods

Two types of PMN-PT based capacitor structures were grown on a Si substrate, with either (001) or (110) growth orientation. The layer stack of the capacitor is shown schematically in Figure 1. The 200 nm thick epitaxial PMN-PT films were fabricated using 100 nm thick $SrRuO_3$ (SRO) base and top electrodes. The SRO/PMN-PT/SRO structures are deposited on a (001)-oriented silicon substrate buffered with a CeO_2(001)/YSZ(001) bilayer stack (each layer is 50 nm thick). All layers are deposited in series, without breaking the vacuum, by means of pulsed laser deposition (PLD) using a KrF Excimer laser (248 nm wavelength). The substrate was placed at a distance of 6 cm from the target. The PMN-PT films deposition was performed at a laser fluence of 2.25 J cm^{-2} and 4 Hz repetition rate.

In order to achieve epitaxial growth of the oxides on Si a buffer layer of YSZ is used, which was grown at a substrate temperature of 800°C. YSZ can grow heteroepitaxially on silicon as it scavenges the native Si oxide layer under specific deposition conditions, allowing the reproducible coherent growth of the oxides on a silicon substrate [11]. A following layer of ceria was grown at 800°C on the YSZ buffer layer with a cube-on-cube epitaxial relationship which can therefore be used as a second buffer layer to reduce the lattice mismatch between YSZ and the base electrode SRO layer. The in-plane mismatch between the fluorite structure of CeO_2 and the perovskite pseudocubic SRO crystal is only 2.2% if the perovskite cubic unit cell of SRO is rotated by 45° with respect to the CeO_2 unit cell [12,13]. To achieve the

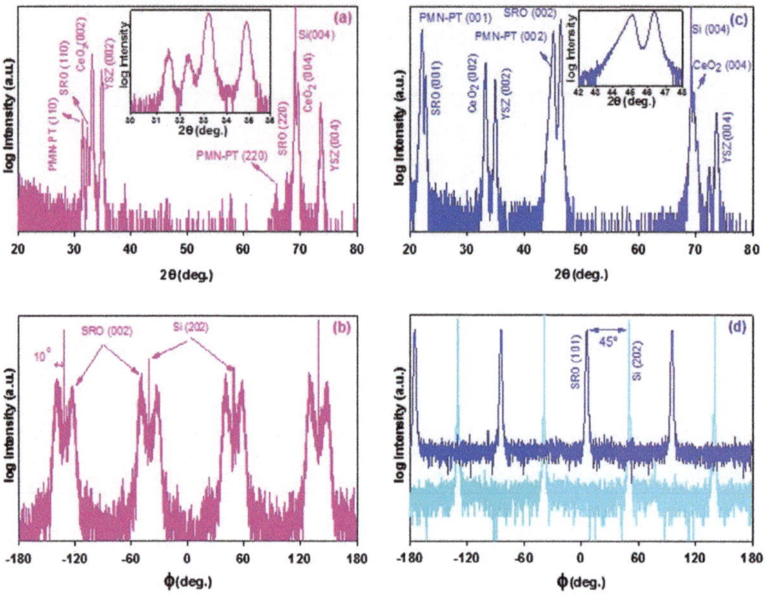

Figure 2. (a) XRD-2 diffractogram of 200 nm thick (110)-oriented PMN-PT film deposited on SRO(110)/CeO_2(001)/YSZ(001) trilayer system on a silicon substrate. The inset shows an expansion of the (110) reflections. (b) Corresponding φ-scan of the Si substrate and SRO bottom electrode. (c) XRD pattern of 200 nm thick (001)-oriented PMN-PT film deposited on a SRO(001/CeO_2(001)/YSZ(001) buffered Si substrate. The inset shows the (002) reflections. (d) Corresponding φ-scan of the Si substrate and SRO bottom electrode.

two desired orientations the first SRO monolayers can be deposited under either reduced oxygen pressure conditions ($<10^{-5}$ mbar) and a high substrate temperature of 800°C leading to (001)-oriented growth followed by deposition of the main part of the SRO bottom electrode at 600°C, which leads to (001) growth of the SRO or skipping the low pressure/high temperature step resulting in (110)-oriented SRO [7,14,15]. The subsequent PMN-PT layer deposited at a substrate temperature of 600°C and an oxygen pressure of 0.28 mbar copies the orientation of the underlying SRO layer. The SRO top electrode was again deposited at 600°C. After deposition the samples were cooled down *in situ* from deposition temperature to room temperature under a 1 bar oxygen pressure.

The structural properties and epitaxial relationships were examined by X-ray diffraction (XRD; PANalytical X'pert PRO MRD, Almelo, the Netherlands). Ferroelectric capacitor devices, 200 × 200 μm² in size, were patterned by photolithography and structured by argon ion beam milling. Ferroelectric hysteresis ($P - E$) loops were measured with a ferroelectric tester (AixACCT TF Analyser 3000, Aachen, Germany), using a bipolar triangular pulse at a frequency of 1 kHz and varying amplitude. Fatigue measurements were performed using a rectangular bipolar pulse train at a frequency of 10 kHz and with 150 kV cm^{-1} amplitude. The effective longitudinal piezoelectric coefficient ($d_{33} - E$) loops were measured with a laser Doppler vibrometer (Polytec MSA-400, Irvine, CA, USA), with a sinusoidal excitation of 8 kHz on top of a stepwise changing *DC* bias voltage using a lock-in technique. The relative dielectric constant versus electric field ($C - E$) measurements were performed with a Keithley 4200 (Tektronix, Beaverton, OR, USA) instrument at 10 kHz.

3. Experimental results

3.1. Structural analysis

Figure 2(a) shows the $\theta - 2\theta$ diffraction pattern of a PMN-PT film grown on the SRO/CeO_2/YSZ heterostructure on silicon. The reflection peaks indicate that the PMN-PT film is in the perovskite phase and has a (110) orientation. No extra peaks that correspond to impurity phases or other orientations are observed. The FWHM of the rocking curve of the PMN-PT (110) peak is 1.6°. φ-scans of the CeO_2/YSZ bilayer stack show four identical sets of peaks located at the same angles as the Si-substrate, separated by 90°, proving that CeO_2(001)/YSZ(001) bilayer is grown with a relaxed epitaxial relationship with the Si(001) substrate [8,9]. The φ-scan in Figure 2(b) shows a peak doubling of the pseudocubic (222) direction of the 45° tilted SRO pseudocube peaks positioned at + 10° and −10° with respect to the position of the silicon (202) peak. This indicates the presence of twin domains in the film as was previously also observed by Hou et al. [16] for SRO deposited directly on YSZ/Si(001) and by Dekkers et al. [7] for SRO on CeO_2/YSZ/Si(001). Because of the very small lattice mismatch between the SRO bottom electrode and the successive PMN-PT layer, the growth of PMN-PT follows the orientation of the bottom electrode. Therefore the twin

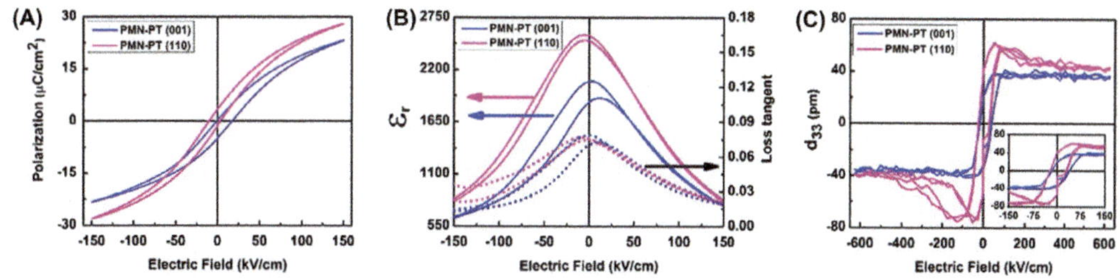

Figure 3. (a) Polarization hysteresis loop before fatigue treatment; (b) relative dielectric constant ε_r and dielectric loss tangent loops; (c) piezoelectric coefficient d_{33} of 200 nm thick (001)- and (110)-oriented PMN-PT films on CeO_2/YSZ buffered Si(001) substrate, using SRO electrodes.

domain structure also exists in the PMN-PT film, as was determined by ϕ-scan XRD analysis (not shown). The fairly large rocking curve width of the PMN-PT (110) peak is ascribed to the twin domain structure. The out-of-plane lattice constant of the PMN-PT (110) is obtained as 0.567 nm, which is slightly (−0.4%) less than that of an unstrained 45° tilted pseudocube (0.569 nm = $\sqrt{2}a_{pc}$, with a_{pc} = 0.4022 nm), hence the PMN-PT (110) film is slightly tensile strained in the film plane. Figure 2(c) shows the $\theta - 2\theta$ scan of the (001)-oriented PMN-PT films. Also here no additional peaks corresponding to impurity phases or other orientations were detected. The FWHM of the rocking curve of the PMN-PT (002) peak is 0.7°, which is half that of the (110)-oriented PMN-PT films. From the inset of Figure 2(c) we deduce that part of the PMN-PT (002) reflection is slightly shifted to lower angles from the angle expected for rhombohedral, unstrained PMN-PT. The broad shoulder is attributed to be due to a *compressive* strain of the in-plane lattice constant of the part of the layer in contact with the bottom SRO layer, relaxing to the bulk lattice parameter [17]. The maximum of the reflection corresponds to an out-of-plane lattice parameter of 0.4018 nm, indicating that the bulk of the film is slightly *tensile* strained, as is also expected from the differences in thermal expansion coefficients of film and substrate. Thus the compressive strain in the interface layer is expected to arise from cube-on-cube epitaxial growth of PMN-PT on the SRO with a much smaller lattice constant, while with increasing PMN-PT thickness the epitaxial strain is relaxed by defect incorporation.

3.2. Ferroelectric, dielectric and piezoelectric properties

First, we note that the observed film properties are reproducible for different films (for each orientation several films were grown and measured), indicating that the observed differences are related to the growth orientation and not to variations in growth conditions. Typical ferroelectric hysteresis loops ($P - E$) of the PMN-PT films with different crystallographic orientation, measured before any fatigue treatment, are shown in Figure 3(a). The loops exhibit strong slanting and narrow hysteresis typical for bulk relaxors. The $P - E$ loop of the (001) film shows a positive voltage shift, whereas that of the (110) films is slightly negatively shifted. Apart from the voltage shifts the loops appear to be symmetric. The polarization of the (110)-oriented film at high fields is significantly larger than that of the (001)-oriented film.

Somewhat naively one can interpret the difference in polarization values to be due to the different orientations of the pseudocubic $[111]_{pc}$ direction, which is the preferential polarization direction in bulk PMN-PT. The length of the projection of this vector on the film normal direction is a factor $\sqrt{2/3}$ for a (110)-oriented film and $\sqrt{1/3}$ for a (001)-oriented film, respectively, of the length of the $[111]_{pc}$ vector (see Figure 6). Thus one would expect in a first approximation that the ratio of the measured polarizations for the films with different growth orientations to be $P_{(110)}/P_{(001)} = \sqrt{2}$. Experimentally the ratio is about 1.2 at high fields. However if one assumes that the polarization can rotate freely in the relaxor material, this argument is not valid, since the polarization angle then depends on the strain state and applied field, as we will see below within the polarization rotation model.

Another difference between the loops is the relatively large shift of the loop of the (001) film towards positive field direction, indicating the existence of a positive self-bias E_{sb} ($E_{sb} = (E_c^+ + E_c^-)/2$) of magnitude + 8.1 kV cm^{-1} (E_c^+ and E_c^- are the coercive fields of the rising and falling branches of the $P - E$ loop). The $P - E$ loop of the (110)-oriented PMN-PT films shows a comparatively small, negative self-bias of −3.6 kV cm^{-1}. Typically, in perovskite ferroelectrics a compressive in-plane stress results into a positive imprint, whereas in-plane tensile stress acts in the opposite way [18]. Note that the difference of thermal expansion coefficients between PMN-PT and the silicon substrate would suggest the presence of in-plane tensile strain in both films [19], which cannot therefore explain the difference in field shift for both cases. The XRD analysis suggests that the near bottom electrode part of the (001)-oriented PMN-PT film shows a strain gradient, with maximum strain at the base electrode/PMN-PT interface and relaxing into the film by the incorporation of lattice defects.

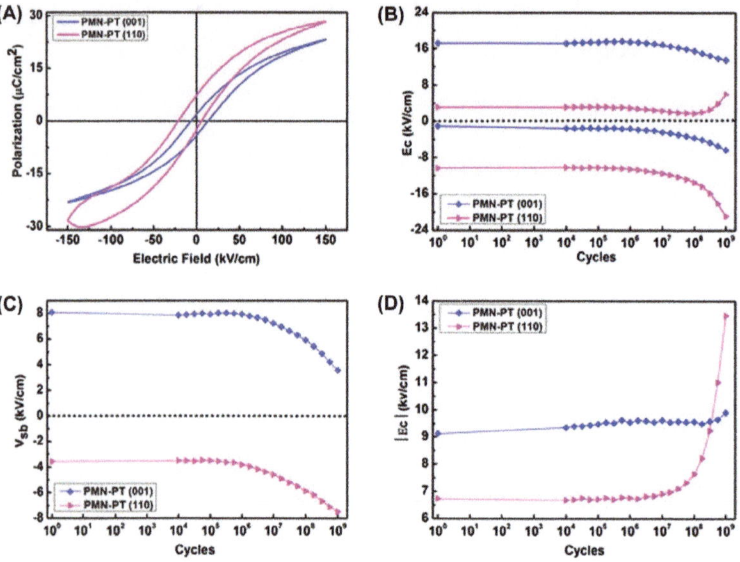

Figure 4. (a) $P-E$ loops after fatigue treatment (10^9 cycles); (b) coercive fields; (c) self-bias field; and (d) average coercive field versus number of ageing cycles performance of (001)- and (110)-oriented PMN-PT films on CeO_2/YSZ buffered Si(001)-substrates, using SRO electrodes.

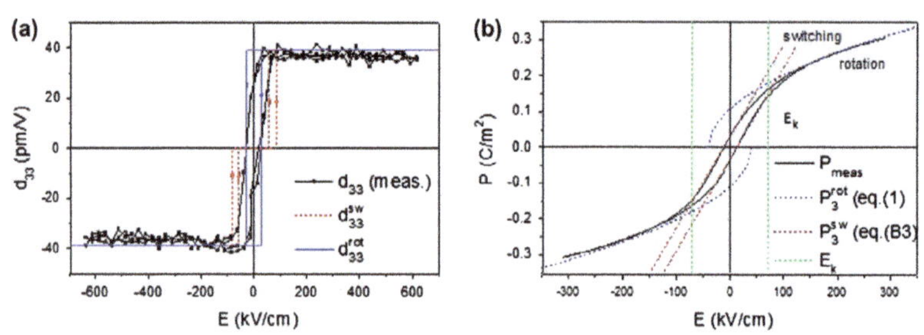

Figure 5. (a) $d_{33}-E$ hysteresis loop of (001)-oriented PMN-PT film, fitted with the rotation and switching model curves. (b) Polarization hysteresis loop, fitted with polarization rotation at high fields and switching model at low fields. (c) Polarization angle and shear angle versus applied field.

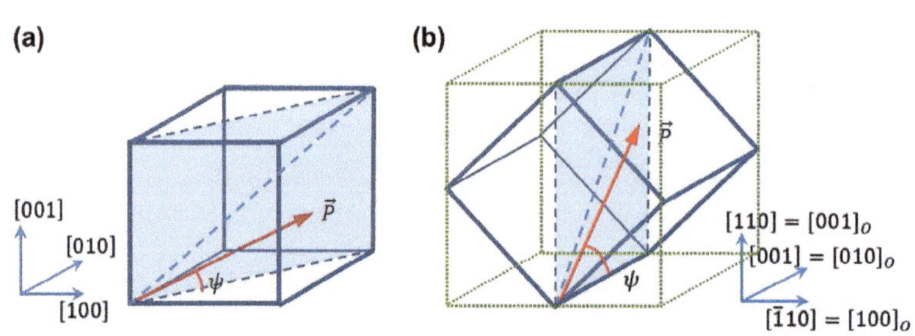

Figure 6. Rotation of the polarization vector in the ($\bar{1}$ 10) plane of (a) the (001)-oriented pseudocubic unit cell under the influence of the applied field E_3; (b) the (110)-oriented unit cell. The green frame gives the orthorhombic unit cell.

It was previously shown that such a strain gradient can give rise to a large build-in electrical field, which is a possible cause for the self-bias voltage of this film [17]. The (110)-oriented film shows a smaller self-bias with opposite sign. In this film strain relaxation is much easier because of the high density of structural defects, due to the twinning of the crystal structure of the bottom electrode and PMN-PT film. This is consistent with the XRD rocking curve measurement showing more peak broadening, hence a larger tilt angle range of the crystallites within the film and therefore a higher density of structural defects as compared with the (001) films. The increased strain relaxation causes the strain gradient layer to be very thin, as is evidenced by the observation that the (110) reflection does not show an obvious asymmetry (shoulder). Also the self-bias voltage is low and even negative. The latter observation suggests that any build-in field – if present at all – is (over-) compensated by negatively charged defects in the film–substrate interface layer [17].

It is instructive to compare the $P-E$ loops with those of (001)-poled single-crystal PMN-PT as presented in [27]. There the loops show no hysteresis, no self-bias, a linear field dependence in a very narrow field range between about −5 and 5 kV cm^{-1}, associated with domain wall motion, and a gradual increase up to saturation outside this field range due to polarization rotation. For the films discussed here there is significant hysteresis, hence domain wall pinning; a finite field-bias; a 20-fold wider linear section of the loop, and a much slower increase of the polarization by rotation with increasing field. These effects are attributed to the clamping to the substrate.

The relative dielectric constant ε_r and dielectric loss tangent curves of the PMN-PT films are shown in Figure 3(b). A very small distance between the peaks of the two branches of the loop, much smaller than the width of the opening in the hysteresis loop, is observed. The (110)-oriented film has a larger dielectric constant than the (001)-oriented film over the full field range. The losses are approximately equal.

The loop of the measured effective piezoelectric coefficient $d_{33} = \partial z / \partial V$ of the (001)-oriented film (Figure 3(c)) is square, symmetric and saturates at about 39 pm V^{-1}. The loop of the (110) film is asymmetric and shows strong maxima. These maxima occur at different negative applied fields, depending on the direction of the field change. The field range of the sloped part of both hysteresis loops is approximately as wide as that of the $P-E$ loops, while the coercive fields of the $d_{33}-E$ loops are significantly larger than of the $P-E$ loops. The nearly constant d_{33} of the (001) film implies that the (average) unit cell height increases linearly with applied field, whereas the dimension of the unit cell in the [110] direction of the (110) film is more sensitive to the change in field at low fields than at higher field strengths. However d_{33} saturates at nearly the same value as for the (001) film. The maximum d_{33} values attained in these films are significantly smaller than the bulk values reported, up to 2800 pm V^{-1} [20]. The large effective d_{33} values reported for PMN-PT single domain, single crystals, poled in the (001) or (110) direction are ascribed to the large contribution of the large shear component d_{15}^R of the rhombohedral phase to d_{33}. However, in the films considered here, shearing is largely obstructed by the clamping to the substrate and the counteracting shear of adjacent grains with the polar axis in different in-plane directions.

3.3. *Cycling stability of the ferroelectric response*

To analyse the effect of orientation and structure on the stability of the ferroelectric response, both films were subjected to a large number (10^9) of switching cycles. The main fatigue feature of the (001)-oriented films is a small field shift of the $P-E$ loop upon cycling, reflected in shifting coercive fields (Figure 4(b)) and thus a changing self-bias field (Figure 4(c)). On cycling the self-bias field, generally ascribed to an inhomogeneous fixed charge distribution in the piezoelectric, asymmetric electrodes or a thin insulating layer separating one of the electrodes and the piezoelectric, is compensated by a redistribution or compensation of the fixed charges through a slow field driven process. The average coercive field, hence the opening of the $P-E$ loop, remains nearly constant upon cycling (Figure 4(d)). The $P-E$ loop of this film (Figure 4(a)) after fatigue treatment is very similar to the initial loops, indicating that there is no increase of leakage currents. The (110)-oriented film also shows a slight shift along the field axis of the $P-E$ loop (negative field direction) upon cycling up to about 10^8 cycles, thus slight shifting of the coercive fields and a small change in self-bias field. However, above 10^8 cycles the loop opening $|E_c|$ strongly increases and the remanent polarization doubles. Simultaneously the $P-E$ loop of the (110) film shows strongly increased leakage currents. We interpret these changes as being due to an increased conductivity, especially of the grain boundaries of this film, due to the voltage cycling and the associated mechanical cycling via the piezo-electric effect. The (110) film is more sensitive to this fatigue mechanism, because the as-grown film has already more defective grain boundaries due to the crystallographic twinning.

4. Theory and applications

In relaxors the change in polarization is considered to be due to continuous rotation of the polarization vector (without significant extension of the polarization vector) and domain switching under influence of the applied electrical field [20,21]. In the following we develop a simple analytic approach based on these assumptions. This allows us to derive equations from the experimental data that capture the essential field dependence of the material properties and connects quantitatively the measured polarization, dielectric and piezo-electric hysteresis loops. This makes it possible to separate intrinsic effects from extrinsic effects, such as domain wall motion.

In an unstrained (001)-oriented film of PMN-PT material with rhombohedral symmetry the polarization can be oriented in the eight equivalent <111> directions, thus one expects the material to be divided in polarization domains with these eight polarization directions. At high enough field in the out-of-plane direction only the four domains with a polarization component in the direction of the applied field remain (in bulk this correspond to the '4R' domain structure and the '4M$_C$'

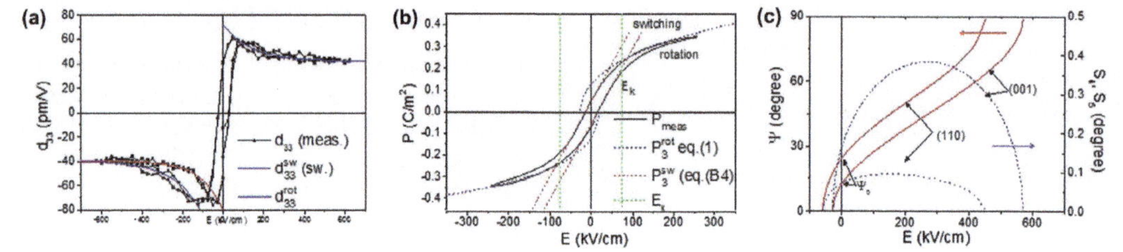

Figure 7. (a) $d_{33} - E$ of (110)-oriented PMN-PT film fitted with the rotation model. (b) Polarization hysteresis loop, fitted with polarization rotation at high fields and switching model at low fields. (c) Polarization angle and shear angle versus applied field of the (001) and (110) films. The shear of the (110) film is (somewhat arbitrarily) a factor of 4 smaller than that of the (001) film, reflecting a weaker coupling between shear strain and polarization.

structure at high fields) and for very large fields, when the polarization in all domains is parallel to the field the symmetry becomes tetragonal and the system is in a single domain state. In principle all the properties of the film can be derived by numerical minimization of the average free energy of the clamped ferroelectric film, $F = \tilde{G} - E_3P_3 = G_0 + \sum_i S_i\sigma_i - E_3P_3$ if the values of all stiffness and electrostrictive coefficients and elastic compliance were known (for example as done in [22]). $\tilde{G}$ is the zero field energy of the clamped film. G_0 is the standard Gibbs energy used in the thermodynamic theory of bulk ferroelectric crystals including elastic energy [23], S_i and σ_i are the components of the strain and stress vectors in Voigt notation, E_3 is the applied field in the out-of-plane direction and P_3 the polarization component in this direction. The model neglects the small energy contribution from the domain walls. The averaging is over the different domains. The unit cells in the four different domains have the same rhombohedral symmetry, therefore we assume that unit cells in each domain can be described with the same energy equation and thus that we may consider the film to be in a (quasi-) single domain state. We will use some results of the single domain model to obtain expressions for the field dependence of the film properties.

First we consider the (001)-oriented film, with in-plane 1- and 2-directions and the out-of-plane 3-direction. The direction indices also refer to the principle axes of the (001)-oriented pseudocube. In each of the four polarized domains the out-of-plane component of the polarization, P_3^{rot}, and the in-plane component, $P_{1,2}^{rot}$ can be written as:

$$P_3^{rot} = P_0 \sin\psi \; P_{12}^{rot} = \sqrt{P_1^2 + P_2^2} = P_0\cos\psi \quad (1)$$

where P_0 is the vector length and ψ the polarization angle with the film plane. For a clamped single domain film the strain in the out-of-plane direction of a (001)-oriented pseudo-cube is in general given by [24]:

$$S_3 = aS_m + b\left(P_1^2 + P_2^2\right) + cP_3^2 = aS_m + bP_0^2 + P_0^2(c-b)sin^2\psi \quad (2)$$

Here S_m is the misfit strain between film and substrate, $a = 2s_{13}/\left(s_{11} + s_{12}\right)$, $b = Q_{12} - (a/2)\left(Q_{11} + Q_{12}\right)$, and $c = Q_{11} - aQ_{12}$, with Q_{ij} and s_{ij} the electrostrictive and elastic compliances [25–27]. The right-hand side of Equation (2) is obtained after substitution of Equation (1). The values of all the parameters depend on the orientation of the unit cell. The shear of the clamped unit cell is [24]:

$$S_4 = Q_{44}P_2P_3 = \frac{Q_{44}P_0^2 \sin 2\psi}{2\sqrt{2}} = S_5 = Q_{44}P_1P_3 \quad (3)$$

The piezoelectric coefficient of the film d_{33} is obtained as:

$$d_{33} = \frac{\partial S_3}{\partial E_3} = P_0^2(c-b)\frac{\partial\psi}{\partial E_3}\sin 2\psi + 2P_0\frac{\partial P_0}{\partial E_3}\left[b + (c-b)sin^2\psi\right] = d_{33}^{rot} + d_{33}^{ext} \quad (4)$$

The first right hand term describes the effect of polarization rotation (index 'rot') on the piezoelectric coefficient and the second term that of polarization extension ('ext'). At very high voltages, when $\psi \approx 90°$, the unit cell is practically tetragonal ($S_4 = S_5 = 0$) with the polarization in the 3-direction, $\partial\psi/\partial E_3 \approx 0$. Thus Equation (2) reduces to $S_3 = aS_m + cP_3^2$ and Equation (4) to $d_{33} = d_{33}^{ext} = 2cP_0\frac{\partial P_0}{\partial E_3}$ and we can identify an effective electrostrictive constant for the clamped film at high fields, $Q_{33}^{eff} = c$. This corresponds to the usual formula $d_{33} = d_{33}^{ext} = d'_{33} - 2s_{13}d'_{31}/\left(s_{11} + s_{12}\right)$ for a clamped film, where we have defined $d'_{33} = 2Q_{11}\varepsilon_0\varepsilon_{33}P_0$ and $d'_{31} = 2Q_{12}\varepsilon_0\varepsilon_{33}P_0$. Although the latter expressions have the same structure as that of the longitudinal and transverse piezoelectric coefficients of a free crystal of PMN-PT, it is only a formal similarity, but the parameter value of ε_{33} that enters these relations is still determined by the clamping. With the literature values for the Q_{ij} and s_{ij} parameters of the pseudocubic phase we obtain c = 0.0137.

Measurements (Figure A1 in Appendix A) show that the high field values of ε_r (which should be equal to ε_{33} in the theoretical formulas) saturate at about 250 at 300 kV cm^{-1} (for larger fields the ε_r determined from $C - V$ measurements is strongly influenced by leakage currents).

With these values the (high field) piezoelectric constant is found to be $d_{33} = 42$ pm V^{-1}, very close to the measured value of $d_{33} \approx 39$ pm V^{-1} (Figure 5(a)). Thus the experimental high field value is well predicted by the model for a clamped, (001)-oriented, tetragonal film, suggesting that at the maximum field values used in the $d_{33} - E$ measurements d_{33} is determined by the polarization extension mechanism. At lower voltages the rotation dependent term in Equation (4) determines the piezo-electric coefficient, because $\partial P_0/\partial E_3$ is small,

$$d_{33}^{rot} = P_0^2(c-b)\frac{\partial\psi}{\partial E_3}sin2\psi \tag{5}$$

This is the central relation from which we establish a link between the experiments and the semi-empirical model described in the following. One expects a smooth transition from the rotation to the extension mechanism and thus a constant or smoothly changing $d_{33} - E$ loop in the voltage range where the rotation mechanism fades out and the extension mechanism comes up. If the high field value of d_{33} is determined by extension, as the close match between theoretical and experimental values suggests, then the transition voltage range is below the maximum field applied in the $d_{33} - E$ measurements (about 650 kV cm^{-1}). Experimentally d_{33} is approximately constant over a wide voltage range and equal to $d_{33} \approx 39$ pm V^{-1}. Assuming that at relatively low voltages the measured polarization change is dominated by polarization rotation, the product of the angle dependent terms in d_{33}^{rot} take the constant value k_1

$$\frac{\partial\psi}{\partial E_3}sin2\psi = k_1 = \frac{d_{33}}{P_0^2(c-b)} \tag{6}$$

Through the term $(c-b) = (1+a/2)(Q_{11}-Q_{12})$ the parameter k_1 (units cm kV^{-1}) is a sensitive function of the material parameters, especially of s_{13}. It appears that a change of for example s_{13} by as little as 15% can change $(c-b)$ by a factor of 3. Because of this sensitivity we will not use the calculated value, but extract the value k_1 from the polarization hysteresis measurements instead. Integration of Equation (6) gives an expression for the field dependence of the angle ψ of the polarization vector in the (001)-oriented unit cell:

$$-2(k_2 + k_1E_3) = cos2\,\psi \tag{7}$$

The (dimensionless) integration constant k_2 defines the angle $\psi_0 = 0.5acos(-2k_2)$ at zero field. If the (001) (respectively (110)) oriented unit cell is unstrained the integration constant can be obtained from the assumption that at zero applied field the polarization vector is oriented in the body diagonal direction, at angle $\psi_0^{001} = atan(1/\sqrt{2}) = 35.3°$ ($\psi_0^{110} = 54.7°$). From this follows $k_2^{001} = -0.167$ ($k_2^{110} = 0.167$) and a remanent polarization $P_r^{001} = P_0 sin\psi_0$=0.25C/m^2($P_r^{110}$ =0.35C/m^2), using $P_0 \approx 0.43$ C m^{-2} [27]. However if the lattice is strained the polarization vector is tilted towards the film plane for tensile stress, resulting in a lower zero field angle ψ_0 and consequently lower P_r.

In short, assuming that polarization rotation dominates the $d_{33} - E$ loop at lower voltages described by Equation (5), the measurements imply Equation (6). This results in the field dependent polarization angle defined by Equation (7). The coefficients k_1 and k_2 are obtained by fitting the polarization loop with Equations (1) and (7), from which a value for the material parameter $(c-b)_{001}$ is obtained. The parameter c follows from the high-field section of the $d_{33} - E$ measurement where polarization extension dominates the piezoelectric effect. Differences in the k coefficients for (001)- and (110)-oriented clamped films then reflect the effect of the orientation of the film on the properties. Finally, we note that contrary to the $P - E$ and $C - V$ measurements the $d_{33} - E$ measurements can be taken up to much higher field values, far into the regime where field dependent leakage currents start to dominate the electrical measurements. This is because the leakage currents do not influence the mechanical deformation of the film as long as the field can be applied. This allows the determination of the angle dependence of the polarization vector over a much larger field range than from electrical measurements.

Figure 5(b) shows the measured $P - E$ loop of Figure 3(a) of the (001)-oriented device, corrected for a small constant leakage resistance and field-shifted over the small self-bias field. This is the largest field range in which the leakage resistance is low and constant. (For larger applied fields the leakage strongly increases, distorting the loop significantly.) The loop is fitted in the high field regions with Equation (1), using the field dependence of the polarization angle given by Equation (7), with k_1^{001} and ψ_1^{001} as fitting parameters. In the interval E = [−100, 100] kV cm^{-1} the hysteresis loop deviates strongly from the polarization rotation model. This is because we expect that in this field range polarization switching (the out-of-plane polarization component switches direction) is the dominant mechanism for polarization change. The $P - E$ loop branches due to switching are indicated by the red dashed lines in Figure 5(b). (In Appendix B this is discussed further and the consequences for the device properties in this field range are described.) The fit of the polarization rotation model to the experimental loop is very sensitive to the values of these parameters, which can therefore be determined accurately (k_1^{001} within 0.05×10^{-8} and ψ_1^{001} within 2°). k_1 determines the slope and curvature of the loop, while ψ_0 sets the P-axis crossing. Figure 7(c) shows the polarization angle of the falling branch of the hysteresis loop as function of field for the fit parameter values $k_2^{001} = -0.45$ (ψ_0=12.4°) and k_1^{001} =1.67x10^{-8} (m V^{-1}). This corresponds to a value $(c-b)^{001}$=0.0126 (m^4/C^2), which can theoretically be obtained using the literature values for Q_i and s_{ij}, but with $s_{13} = 0.87s_{13}^{lit}$. The model suggests that the polarization vector tilt angles varies approximately linearly over a wide field range, [0, 500] kV cm^{-1}. The angle rapidly drops to zero close to a critical value $E_3 = E_{3\min} = -(0.5 + k_2^{001})/k_1^{001} = -27.4$ kV cm^{-1} where the polarization is lying in the film plane. For $E_3 > E_{3\max} = (0.5 - k_2^{001})/k_1^{001} = 570$ kV cm^{-1} the polarization is normal to the film plane. This corresponds to the high field range of the $d_{33} - E$ measurements, where ψ saturates at 90° and polarization extension determines the piezoelectric coefficient. Thus d_{33} is determined by polarization extension only above $E_{3\max}$. This corroborates the close match between the experimental high field value of d_{33} and the value calculated from the extension mechanism. Figure 7(c) also shows the calculated shear angle $S_4 = S_5$ of the rhombohedrally deformed pseudocube as function of the field, as calculated from Equation (3). At $E_3 = E_{3min}$ the shear angle is zero, indicating that the pseudocube has been tetragonally deformed. At zero field the shear angle is about 0.16° and increases with increasing field to reach a maximum of 0.38° at a field strength of about 275 kV cm^{-1}. For larger fields the rhombohedral deformation reduces again and the pseudocube becomes more tetragonal. In principle one might be able to determine the shear angles from XRD data. However, the XRD reflections are not very sharp, probably because of the overlapping of the reflections arising from the four-fold symmetry of the crystallographic domain structure.

We started with the conjecture that the angle dependence of the polarization can be derived from the constant d_{33} at higher field, using Equation (5). This equation can be rewritten as $d_{33} = 2Q_{33}^{\text{eff}}\varepsilon_{33}P_3$, with $Q_{33}^{\text{eff}} = (c - b)^{001}$. The constant measured value of d_{33} for the (001) film seems fortuitous, since, according to the polarization rotation model, it arises from the cancellation of the angle dependences in the product $\varepsilon_{33}(E)P_3(E)$ in Equation (5) for all field values, which is a consequence of the angle-field dependence in Equation (7). The fitting results in a zero field polarization angle of about $\psi_0^{001} = 12.4°$, much less than of the bulk body diagonal direction. Together with the small observed tensile strain in the film ($S_m = 0.0013$ calculated from the difference in thermal expansion coefficients from film and substrate) this implies that the polarization easy axis direction is very easily changed under the influence of strain, $\psi_0^{001}(S_m) \approx \psi_0^{001}(0) + (\partial\psi_0^{001}/\partial S_m)S_m$, with $\psi_0^{001}(0) =$ 35.3° and $\partial\psi_0^{001}/\partial S_m$ equal to −17.6° per 10^{-3} in plane strain.

In the case of the (110)-grown film a similar analysis as for the (001)-oriented film can be followed. In this case the film is not polarized in the [001] direction, but in the [110] direction. For the (001)- and (110)-oriented films the polarization vector rotates in opposite directions in the pseudocubic unit cell under the influence of the applied field (see figure 6). In the case of an unstrained (001)-oriented unit cell, the tilt angle with the film plane increases from $\psi_0^{001} = 35.3°$ rotating the polarization vector away from the in-plane [110] towards the out-of-plane [001] direction. For the tensile strained film discussed above the zero-field easy axis angle is found to be strongly reduced and equals 12.4°. In the unstrained (110)-oriented unit cell $\psi_0^{110} = 54.7°$ and the polarization rotates in the opposite direction as for the (001) film. One may therefore expect a different dependence of the tilt angle on applied field. Specifically the values of the coefficients *a*, *b* and *c* in Equation (2) are different for the (110)-oriented, clamped pseudo-cube, but the describing expressions for d_{33}, P_3 and ε_{33} have the same functional form when considering the orthorhombic representation. (The 45° tilted pseudocube can be described in terms of an orthorhombic unit cell ($[100]_O \times [010]_O \times [001]_O$ with approximate dimensions $\sqrt{2}a_{pc} \times a_{pc} \times \sqrt{2}a_{pc}$).) Defining P_3 in the [110] direction and P_1 in the [001] direction, then P_2 in the in-plane [1−10] direction is zero. It follows that $S_4 = 0$ (no shear in the (101) plane), but S_5 is a function of the polarization angle. Thus in the orthorhombic presentation the unit cell is slightly sheared in the short axis in-plane direction.

At high fields the measured value d_{33} becomes constant and approximately equal to that of the (001) film, implying that the products $P_0^2(c - b)\frac{\partial\psi}{\partial E_3}sin2\psi$ are equal for both orientations. The factors $(c - b)^{001}$ and $(c - b)^{110}$ are constants, thus the larger d_{33} value of the (110) film at lower field values, as compared to the high field values, indicates that the product $d_{33} \sim \left(\frac{\partial\psi}{\partial E_3}sin2\psi\right)^{110}$ is now field dependent at low fields. This field dependence may be approximated by the functional form $d_{33} \approx \left[P_0^2(c - b)_{110}\right]\left[k_1^{110} + k_3^{110}\exp(-k_4^{110}E)\right]$. From fitting to the $P - E$ and $d_{33} - E$ loops the k_i parameters are again determined and the field dependence of the polarization angle is obtained, as shown in Figure 7(c). We find $k_1^{110} = 1.84 \times 10^{-8}$, corresponding to $(c - b)^{110} =$ 0.0123. $k_2^{110} = -0.22$ m V^{-1} giving $\psi_0^{110} = 17.7°$. k_3^{110} and k_4^{110} follow from the fit to $d_{33} - E$). The different functional dependence of $\psi(E)$ does not cause noticeable qualitative differences in the calculated $P - E$ (and $C - V$) loops. Indeed the measured loops (see Figures 7(a) and (A1(b)) are equally well described by the modelled curves as in the (001) case. We note that the asymmetry of the $d_{33} - E$ curves of the (110) film (broad peak of the falling branch as compared to the smaller peak of the rising branch at negative field bias and secondly high peaks at negative bias as compared to positive bias) also indicates that $\psi(E)$ depends on the direction of the field change, especially for polarization down (negative bias) and thus on the applied field history. The large fields applied in the $d_{33} - E$ loop giving rise to the asymmetry, could not be used for the $P - E$ and $C - V$ measurements. Therefore we have used the $\psi(E)$ dependence derived from the positive field branch of d_{33}, for which the rising and falling curves are approximately equal, to calculate the curves in Figure 7(a), (c) and (d). Figure 7(b) shows again the good fit obtained with the rotation model. At high voltages the fitted rotation curves deviate somewhat from the measured loop. This is possibly due to the increased leakage current at these voltages in the (110) device.

The derived field dependence of the polarization angle (Figure 7(c)) in the positive field branch is steeper than for the (001) film, thus it appears that in the clamped film the polarization rotates somewhat easier in the (110) film towards the [110] direction than in the (001) film towards the opposite [001] direction under the influence of the applied field. The easier polarization rotation suggests also that there is less lattice distortion than in the (001) case. This implies a weaker coupling between the polarization components and the shear strain, thus a smaller value of Q_{44}^{110} (in the orthorhombic coordinate frame) than Q_{44}^{001}. We think that the difference in Q_{44} values between the (001) and (110) grown films, which couples the polarization rotation to the shear of the crystal, may be an important reason for the differences in properties, which are associated with the more or less easier rotation of the polarization vector under the influence of the applied field. In the (110) sample with less induced shear, the shear of domains with different in-plane polarization orientations counteracts the shearing of neighbouring domains less than in the (001) film. Thus with increasing field the unit cell can more easily adapt its shape to the rotated polarization angle or, vice versa, if the unit cell cannot shear easily the polarization cannot rotate that much.

We conclude that the simple rotation model quantitatively couples the measured field dependence of polarization and piezoelectric coefficient fairly well, assuming coherent rotation of the polarization as the dominant mechanism for polarization change in the voltage range in which the polarization loops are measured. This is plausible for a good epitaxial film with homogeneous properties and strain state. In Appendix A we discuss consequences of the rotation model for the film permittivity and find reasonable correspondence between the model and experiment. As mentioned in the introduction at applied fields below the value E_k, polarization switching dominates the polarization loop. In Appendix B a phenomenological model is developed to describe this behaviour and its consequences for the $C-V$ and $d_{33}-E$ loops.

5. Conclusions

In conclusion, relaxor PMN-PT thin films with (001) or (110) orientation were epitaxially grown on Si substrates with a SRO/CeO_2/YSZ buffer layer using PLD. The PMN-PT films are phase pure and relax to bulk lattice parameters over a thin layer. The (110)-oriented films show larger polarization and dielectric permittivity values. The effective longitudinal piezoelectric coefficients d_{33} of both types of film approach the same value for large bias fields, about 39 pm V^{-1}, significantly smaller than that of bulk PMN-PT. This is attributed to clamping by the substrate. The (001)-oriented film shows a relatively large self-bias voltage, which is ascribed to a strain gradient layer at the interface with the bottom electrode. On long term cycling the self-bias is slightly reduced, probably due to the introduction of charged defects in this strained layer. In the case of the (110)-oriented film the self-bias is small and has opposite sign and increases further on cycling.

The ferroelectric, dielectric and piezoelectric properties are described by a model, assuming that polarization rotation is the dominant mechanism for polarization change in the high electric field range. The rotation rate is larger for the (110)-oriented film than for the (001)-oriented film, which results in enhanced ferroelectric and dielectric properties, as is also observed experimentally. It is suggested that the difference in the value of the shear electrostriction parameter Q_{44} may play an important role in this. At low applied fields polarization change is dominated by polarization switching and domain wall motion. However, also at low fields the polarization rotation process appears to determine largely the dielectric properties.

The developed model makes it possible to disentangle different mechanisms for polarization change in clamped thin relaxor films (rotation, switching and extension), which dominate in different applied field regimes, and their effects on ferroelectric, dielectric and piezoelectric properties.

Acknowledgements

This work is supported by the Engineering Doctorate School of Roma TRE University and by NanoNextNL, a micro and nanotechnology consortium of the Government of the Netherlands and 130 partners.

Disclosure statement

No potential conflict of interest was reported by the authors.

References

[1] Do DH, Evans PG, Isaacs ED, et al. Structural visualization of polarization fatigue in epitaxial ferroelectric oxide devices. Nat. Mater. 2004;3:365.

[2] Nguyen MD, Vu HN, Blank DHA, et al. Epitaxial Pb(Zr,Ti)O[3] thin films for a MEMS application. Adv. Nat. Sci.: Nanosci. Nanotechnol. 2011;2:015005.

[3] Park SE, Shrout TR. Ultrahigh strain and piezoelectric behaviour in relaxor based ferroelectric single crystals. J. Appl. Phys. 1997;82:1804.

[4] Baek SH, Park J, Kim DM, et al. Giant piezoelectricity on Si for hyperactive MEMS. Science. 2011;334:958.

[5] Wang W, Zhu QX, Li XM, et al. Effects of ferroelectric/metal interface on the electric properties of PMN-PT thin films epitaxially grown on Si substrates. J. Mater. Sci. Mater Electron. 2013;24:3782.

[6] Tsang WS, Chan KY, Mak CL, et al. Spectroscopic ellipsometry study of epitaxially grown $Pb(Mg_{1/3}Nb_{2/3})O_3$-$PbTiO_3$ /MgO/TiN/Si heterostructures. Appl. Phys. Lett. 2003;83:1599.

[7] Dekkers M, Nguyen MD, Steenwelle R, et al. Ferroelectric properties of epitaxial Pb(Zr,Ti)O3 thin films on silicon by control of crystal orientation. Appl. Phys. Lett. 2009;95:012902.

[8] Jiang J, Hur SG, Yoon SG. Epitaxial PMN-PT Thin Films Grown on $LaNiO_3/CeO_2$/YSZ Buffered Si(001) Substrates by Pulsed Laser Deposition. J. Electrochem. Soc. 2011;158:G83.

[9] Chopra A, Pantel D, Kim Y, et al. Microstructure and ferroelectric properties of epitaxial cation ordered $PbSc_{0.5}Ta_{0.5}O_3$ thin films grown on electroded and buffered Si(100). J. Appl. Phys. 2013;114:084107.

[10] Krishna L, Sunder M, Moran PD. Growth of Epitaxial (110) $0.7Pb(Mg_{1/3}Nb_{2/3})O_3$-$0.3PbTiO_3$ Thin Films on r-Plane Sapphire Substrates by RF Magnetron Sputtering. J. Electro. Mater. 2010;39:132.

[11] Wang SJ, Ong CK, You LP, et al. Epitaxial growth of yttria-stabilized zirconia oxide thin film on natively oxidized silicon wafer without an amorphous layer. Semicond. Sci. Technol. 2000;15:836.

[12] Copetti CA, Soltner H, Schubert J, et al. High quality epitaxy of $YBa_2Cu_3O_{7-x}$ on silicon-on-sapphire with the multiple buffer layer YSZ/CeO_2. Appl. Phys. Lett. 1993;63:1429.

[13] Kondo M, Maruyama K, Kurihara K. Epitaxial Ferroelectric Thin Films on Silicon Substrates for Future Electronic Devices. Fujitsu Sci. Tech. J. 2002;38:46.

[14] Nguyen MD. Ferroelectric and piezoelectric properties of epitaxial PZT films and devices on silicon PhD Thesis. The Netherlands: University of Twente; 2010. ISBN 978-90-365-3047-7.

[15] Nguyen MD, Karakaya K, te Riele PM, et al. Piezoelectric materials for MEMS applications. Proceedings of the Third IEEE-NEMS; 2008 Jan 6–9; Sanya, China:315.

[16] Hou SY, Kwo J, Watts RK, et al. Structure and properties of epitaxial $Ba_{0.5}Sr_{0.5}TiO_3/SrRuO_3/ZrO_2$ heterostructure on Si grown by off-axis sputtering. Appl. Phys. Lett. 1995;67:1387.

[17] Boota M, Houwman EP, Dekkers M, et al. Epitaxial $Pb(Mg_{1/3}Nb_{2/3})O_3$-$PbTiO_3$ (67/33) thin films with large tunable self-bias field controlled by a $PbZr_{1-x}Ti_xO_3$ interfacial layer. Appl. Phys. Lett. 2014;104:182909.

[18] Zhang Z, Park JH, Trolier-McKinstry S. (001)-oriented $LaNiO_3$ bottom electrodes and (001)-textured ferroelectric thin films on $LaNiO_3$. MRS Proc. Ferroelectric Thin Films VIII. 2000;596:73.

[19] Substrate induced thermal strain causes tetragonal deformation of the pseudocube deposited at temperature T_{dep} = 600 °C. The out-of-plane lattice parameter of the clamped film at temperature T, $c^f(T)$, is related to the in-plane lattice parameter as af(T) as , assuming unit cell volume conservation upon deformation and fully relaxed film growth at deposition. αf and αs are the average thermal expansion coefficients of film and substrate respectively over the temperature range ΔT = Tdep - T. For PMN-PT (67/33) we deduced α^f=8.1 ppm/K from R.Wongmaneerung, R. Guo, A. Bhalla, R. Yimnirun and S. Ananta, Journal of Alloys and Compounds 461 (2008) 565. For the Si substrate αs = 2.8 ppm/K. With these numbers we find a room temperature c/a ratio of 0.994, hence small tensile strain Sm = 0.0020 in the film plane.

[20] Zhang S, Li F. High performance ferroelectric relaxor-$PbTiO_3$ single crystals: Status and perspective. J. Appl. Phys. 2012;111:031301.

[21] Fu H, Cohen RE. Polarization rotation mechanism for ultrahigh electromechanical response in single-crystal piezoelectrics. Nature. 2000;403:281.

[22] Khukhar VG, Pertsev NA, Kohlstedt H, et al. Polari zation states of polydomain epitaxial Pb(Zr1−*x*Ti*x*)O3 thin films and their dielectric properties. Phys. Rev. B. 2006;73:214103.

[23] Haun MJ, Furman E, Jang SJ, et al. Thermodynamic theory of the lead zirconate-titanate solid solution system, part I-V. Ferroelectrics. 1989;99:13, Secs. I–V

[24] Pertsev NE, Kukhar VG, Kohlstedt H, et al. Phase diagrams and physical properties of single-domain epitaxial Pb(Zr1-*x*Ti*x*)O3 thin films. Physical Review B. 2003;67:054107.

[25] The compliances of the cubic phase (under constant field in the [001] direction) with X,Y, and Z along [100], [010] and [001] respectively, are s_{11}= 69, s_{12}=−11.1, s_{13}=−55.7, s_{44}=14.5, (x10^{-12} m^2/N) for the multidomain single crystal poled in the [001] direction [26]. The electrostrictive constants are Q_{11}= 0.056, Q_{12}= −0.022, Q_{44}= 0.019 (m^4/C^2) [27].

[26] Zhang R, Jiang B, Cao W. Elastic, piezoelectric, and dielectric properties of multidomain $0.67Pb(Mg_{1/3}Nb_{2/3})O_3$– $0.33PbTiO_3$ single crystals. J. Appl. Phys. 2001;90:3471.

[27] Li F, Jin L, Wang D, et al. Electrostrictive effect in $Pb(Mg_{1/3}Nb_{2/3})O_{3-x}PbTiO_3$ crystals. Appl. Phys. Lett. 2013;102:152910.

[28] Neel L, Pauthenet R, Rimet G, et al. On the Laws of Magnetization of Ferromagnetic Single Crystals and Polycrystals. Application to Uniaxial Compounds. J. Appl. Phys. 1960;31:S27.

[29] Stephenson GB, Elder KR. Theory for equilibrium 180° stripe domains in $PbTiO_3$ films. J. Appl. Phys. 2006;100:051601.

[30] Pertsev NA, Arlt G. Forced translational vibrations of 90° domain walls and the dielectric dispersion in ferroelectric ceramics. J. Appl. Phys. 1993;74:4105.

Appendix A – dielectric constant due to polarization rotation

The relative dielectric constant in the rotation part of the hysteresis loop is obtained from Equation (1) as:

$$\varepsilon_{33}^{rot}(E) = \frac{1}{\varepsilon_0}\frac{\partial P_3}{\partial E_3} = \frac{P_0}{\varepsilon_0}cos\psi\frac{\partial \psi}{\partial E_3} = \frac{k_1 P_0}{\varepsilon_0}\frac{1}{2sin\psi} \quad \text{(A1)}$$

where we obtained $\partial\psi/\partial E_3 = k_1/sin2\psi$ from Equation (6). The zero field relative dielectric constant, assuming that polarization rotation determines the dielectric properties of the ferroelectric, results in a relation between the fitting parameters $\varepsilon_{33}^{rot}(0) = k_1P_0/2\varepsilon_0 sin\psi_0$. The high field value is $\varepsilon_{33}^{rot}(\infty) = k_1P_0/2\varepsilon_0$. In Figure A1 the measured and modelled curves for the relative dielectric constant as function of field are shown. At high fields Equation (A1) describes the experimental values ε_{33} quantitatively fairly well, without any

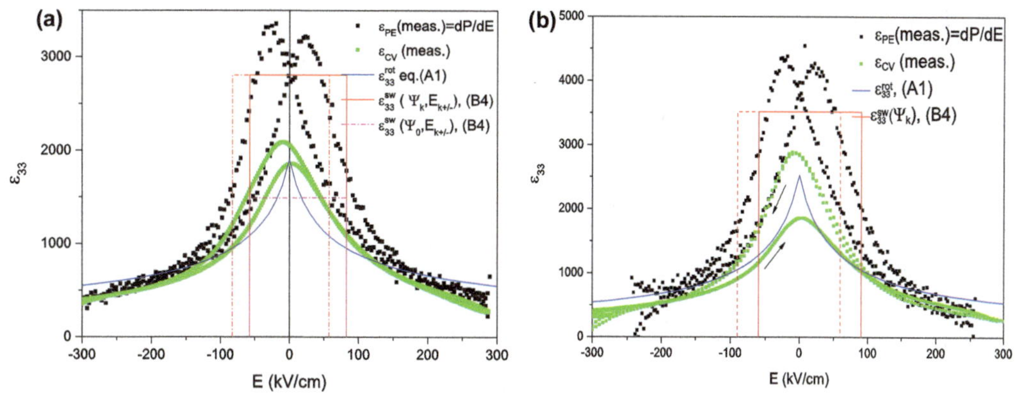

Figure A1. Measured relative dielectric constants versus field obtained from differentiation of the P – E loop ($\varepsilon_{PE}(meas)$) and from C – V measurements ($\varepsilon_{CV}(meas)$). The calculated curves follow from the model for the rotation (ε_{33}^{rot}) and switching (ε_{33}^{sw}) branches of the P – E loop. (a) (001)-oriented PMN-PT device. (b) (110)-oriented device.

additional fitting. At low fields there are larger deviations from the rotation model, due to the effects of polarization switching. This is discussed in Appendix B below.

Appendix B – polarization switching

At low fields the $P-E$ loops (Figure 3(a)) are slightly hysteretic. We expect that in this part of the loop both rotation and switching takes place so that the average polarization can be written as:

$$P_3^{rot+sw} = \phi p_u + (1-\phi)p_d = P_0(2\phi - 1)\sin\psi \quad \text{(B1)}$$

$$P_{12}^{rot+sw} = P_0(2\phi - 1)\cos\psi$$

Here ϕ is the volume fraction of material in which the polarization is in the (+3) direction with the value $p_u = P_0 \sin\psi$ and $p_d = -P_0 \sin\psi$ is the value of the polarization in the (-3)-oriented domains. For $\phi = 1$ (0) all domains have polarization up (down), and for $\phi = 1/2$ the polarizations in the up and downwards oriented domains cancel each other. The reason for the switching is not that the coercive field is compensated by the applied field, which it is evidently not the case at the high fields at which switching sets in, but that the total energy of the system with polarization up and down domains is lower than that of a (quasi-) monodomain state. A strong analogy with the polarization rotation and switching mechanisms is observed in uniaxial ferromagnetic systems with low anisotropy energy and negligible coercive field [28]. In the magnetic system the polydomain state arises from the reduction of the demagnetization energy term in the total energy. Stephenson and Elder [29] described the appearance of 180° stripe domains in compressively strained $PbTiO_3$ thin films, due to reduction of the depolarization energy. In analogy with the model of [28] the polarization switching is between polarization vectors with the same in-plane polarization components but opposite oriented out-of-plane components. It may not only be the reduction in polarization energy that initiates and drives the switching. The results of Kukhar et al. [22] on clamped ferroelectric polydomain PZT films indicate that the ferroelectric *c/a* oriented tetragonal domain formation arises due to a reduction of the overall elastic energy in the film. We assume that a similar mechanism is possible in the PMN-PT films, in which the rhombohedrally distorted monodomain splits up in domains with opposing distortion angles and thus out-of-plane polarization directions, to reduce the overall elastic energy. To simplify the analysis we assume that at low fields the switching process dominates and that the tilt angle of the polarization in the different domains is equal to a value ψ_k, with plus sign for upwards oriented polarization domains and minus sign for downwards oriented domains. ψ_k is approximately equal to the tilt angle at which polarization rotation changes into switching, occurring at a field E_k. Considering only switching Equation (B1) becomes:

$$P_3^{sw} = P_0(2\phi - 1)\sin\psi_k \; P_{12}^{sw} = P_0(2\phi - 1)\cos\psi_k \quad \text{(B2)}$$

The switching branches of the $P-E$ loops are approximately linear for low field strength, thus one can approximate $\phi = (E - E_{c\pm})/2E_k + 1/2$, for the rising (with E_{c+}) and falling (with $E_{c-} = -E_{c+}$)branches. (Here we removed any self-bias field by shifting the experimental loops along the field axis accordingly.) The rising 'switching' branch (index r) cuts the 'rotation branches' for (negative) ψ_{k-} at $E_{k-,r} = -E_k + E_{c+}$ and (positive) ψ_{k+} at $E_{k+,r} = E_k + E_{c+}$. The width of the switching branch is $\Delta E_k = E_{k+,r} - E_{k-,r} = 2E_k$. For the falling branch (index f) $E_{k+,f} = E_k + E_{c-}$ and $E_{k-,f} = -E_k + E_{c-}$. In Figures 5(b) and 6(b) the switching branches are indicated by the dashed red lines. We estimate $E_k \approx \left(|E_{k+,r}| + |E_{k+,f}|\right)/2$ to be 75 kV cm^{-1} for both the (001)- and (110)-oriented films. Since we have no model underlying the field dependence of the domain fraction yet, it is not clear what determines this value quantitatively, although in analogy with the model of [28] it is likely to be related to the strength of the polarization easy axis and depolarization energies. With that the polarization of the switching branches can be written as:

$$P_{3,r/f}^{sw} = \left(\frac{E - E_{c\pm}}{E_k}\right) P_0 \sin\psi_k \quad \text{(B3)}$$

The relative susceptibility due to switching follows as:

$$\varepsilon_{33,r/f}^{sw} = \frac{P_0 \sin\psi_k}{\varepsilon_0 E_k} \quad \text{(B4)}$$

and is, in this approximation of constant polarization angle ψ_k, constant in the field range $E_{k-,r} < E < E_{k+,r}$ for the rising branch and $E_{k-,f} < E < E_{k+,f}$ for the falling branch. The peaks of the rising and falling branches of $\varepsilon_{33,r/f}^{sw}$ are shifted with respect to each other over $E_{c+} - E_{c-} = 2E_{c+}$.

(Figure A1 shows the measured relative susceptibility ε_{CV} (corrected for the self-bias) from $C-V$ measurements of the (001)- and (110)-oriented devices, characterized by Figures 5 and 7, and the curves corresponding to the derivative $\varepsilon_{PE} = \partial P_3/\varepsilon_0 \partial E_3$ of the measured polarization loops (after self-bias and leakage correction). There is a large difference between ε_{CV} and ε_{PE} at low field bias, the latter being about 1.5× larger, whereas at high fields ε_{CV} and ε_{PE} are approximately equal. The susceptibility $\varepsilon_{33}(E)$ calculated from Equation (A1) with the same fitting parameter values as used for Figure 5(a) and (b), valid for the rotation part of the polarization branch, reproduces the measured high field susceptibility quite accurately, as well as the maximum at zero field. At low fields one would expect a contribution from the switching to the susceptibility and $\varepsilon_{33}(E)$ is estimated from Equation (B4). It reproduces the low-field values of the experimental $\varepsilon_{PE}(E)$ fairly well, as well as the peak separation in $\varepsilon_{PE}(E)$, but overestimates the measured maximum ε_{CV} significantly. On the other hand the peak width predicted by rotation polarization only is much smaller than the observed width, which is better reproduced by the switching model. Using ψ_0 instead of ψ_k in Equation (B4) gives a maximum value of the switching susceptibility that is more comparable to that of the measured ε_{CV}-loop. These are indications that polarization rotation still takes place in the switching part of the loop, causing the reduced peak height. However the very small separation (and unequal heights) of the two peaks in ε_{CV} cannot be explained by the simple switching model, given above. We think that the negligible peak separation in the ε_{CV} curves is related to the inertial and viscous forces acting on the moving domain walls (DWs). The $P - E$ loop and therefore ε_{PE} are measured with a field bias changing at a constant rate, $E(t) = at$. The DWs, present in the switching part of the $P - E$ loop, therefore move at a constant speed determined by the change in domain fraction and not subject to inertial forces, although they may lag behind the change in the field, due to the viscous forces. This explains why the ε_{PE} is estimated better by the switching model at low fields and by slow rotation of the polarization at high fields where there are no DWs anymore. The $C - V$ loop on the other hand is measured with a field bias of the form $E(t) = at + E_{ac} sin\omega t$. The polarization response on the oscillatory field is measured with a lock-in amplifier, supplying the in-phase signal, proportional to ε_{CV} and the out-of-phase signal from which the loss tangent $tan\delta_{CV}$ is obtained. This method therefore measures the oscillation frequency (and amplitude) dependent response of the domain-wall movement, $\partial\phi/\partial E = f(\omega, E_{ac}, \psi)$, as well as the polarization rotation within the domains. Considering that the $C - V$ curve is quantitatively fairly well described by Equation (A1), taking only polarization rotation into account, suggests that rotation is the dominant mechanism determining ε_{CV}. The broadening of the zero-field peak may then be ascribed to the additional contribution from domain wall motion to ε_{CV}. The high losses measured in the switching regime indeed indicate that the DW displacement is subject to a substantial loss mechanism and inertial reaction of the DW self-mass [30].

Considering only polarization rotation d_{33} is given by d_{33}^{rot}, producing a square $d_{33} - E$ loop: for the falling branch $d_{33,f}(E > -|E_{3\min}|) = P_0^2(c - b)k_1$ and for the rising branch $d_{33,r}(E < |E_{3\min}|) = -P_0^2(c - b)k_1$. These curves are shown in Figure 5(a) (blue lines) together with the experimental loop (shifted over the self-bias field). If only switching causes the change in the polarization in the 'switching' branch, then the unit cell dimensions do not change, hence $S_{3u} = S_{3d}$ due to the quadratic dependence of $S_{3u/d}$ on the polarization components. $S_{3u/d}$ is the strain in the upward/downward polarized domain. The average strain $S_3 = \phi S_{3u} + (1 - \phi)S_{3d} = aS_m + (bcos^2\psi_k + csin^2\psi_k)P_0^2$ is then constant and $\partial S_3/\partial E_3 = 0$. Hence the pure switching model predicts that the piezoelectric coefficient in a regime where only switching takes place is zero in the field ranges $E_{k-,r/f} < E < E_{k+,r/f}$ for the rising (falling) branch (red dashed lines in Figures 5(a) and 7(a)). If both switching and rotation are contributing then $\partial S_3/\partial E_3$ again becomes equal to Equation (5). The experimental $d_{33} - E$ data in the switching regime show a steep slope over the approximate field range in which switching takes place $\{-E_k \mp E_c^{\pm}, E_k \pm E_c^{\pm}\}$ Considering that the displacement is measured by a lock-in technique using a drive field bias of the form $E(t) = at + E_{ac} sin\omega t$, it is expected that, similarly as for the ε_{CV} measurements, there is a significant effect of the lossy and inertial DW motion in the switching range. The analysis indicates that field driven polarization rotation also takes place in the switching branches. Hence the simple model using either switching or rotation should incorporate the polarization rotation in the switching branch. Further the analysis indicates that the finite self-mass and viscous motion of the DWs influence the low-field susceptibility peaks and the piezoelectric constant in the switching regime.

9

Effects of fabrication methods on spin relaxation and crystallite quality in Tm-doped $Y_3Al_5O_{12}$ powders studied using spectral hole burning

Thomas Lutz[a], Lucile Veissier[a], Charles W. Thiel[b], Philip J. T. Woodburn[b], Rufus L. Cone[b], Paul E. Barclay[a] and Wolfgang Tittel[a]

[a]Department of Physics & Astronomy, Institute for Quantum Science and Technology, University of Calgary, Calgary Alberta T2N 1N4, Canada;
[b]Department of Physics, Montana State University, Bozeman, MT 59717, USA

ABSTRACT

High-quality rare-earth-ion (REI) doped materials are a prerequisite for many applications such as quantum memories, ultra-high-resolution optical spectrum analyzers and information processing. Compared to bulk materials, REI doped powders offer low-cost fabrication and a greater range of accessible material systems. Here we show that crystal properties, such as nuclear spin lifetime, are strongly affected by mechanical treatment, and that spectral hole burning can serve as a sensitive method to characterize the quality of REI doped powders. We focus on the specific case of thulium doped $Y_3Al_5O_{12}$ (Tm:YAG). Different methods for obtaining the powders are compared and the influence of annealing on the spectroscopic quality of powders is investigated on a few examples. We conclude that annealing can reverse some detrimental effects of powder fabrication and, in certain cases, the properties of the bulk material can be reached. Our results may be applicable to other impurities and other crystals, including color centers in nano-structured diamond.

KEYWORDS
Rare earth ion doped powders; powder processing; powder synthesis; annealing; spectral hole burning

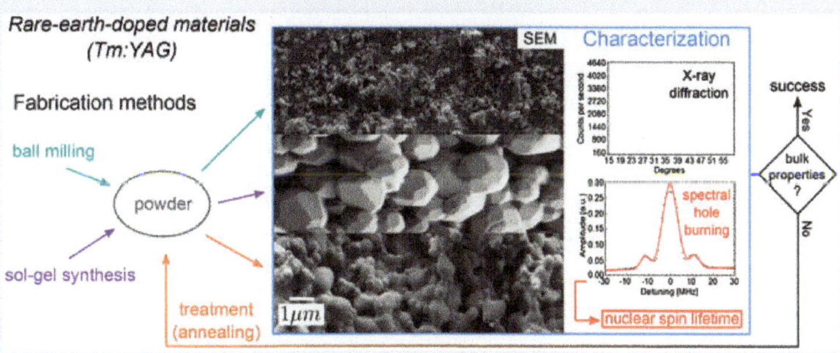

1. Introduction

Rare-earth-ion (REI) doped bulk crystals cooled to cryogenic temperatures are used for a multitude of applications. Examples are quantum memories [1], quantum information processing [2], and ultra-high-resolution optical spectrum analyzers [3]. In contrast to REI doped bulk materials, powders offer low cost and rapid prototyping. Furthermore, the understanding of powders constitutes a first step towards nanofabrication of devices from these materials [4–7]. However, despite much effort [8–11], producing monodisperse powders with properties comparable to those of bulk materials remains challenging.

Fabrication or manipulation of REI doped powders can induce stress, especially during grinding or milling, that creates strain in the crystal lattice [12,13]. In addition, impurities can contaminate the host matrix during synthesis. Strain and impurities often significantly impact the performance of the powders in both signal processing and more general luminescence applications. The goal of this work is to study REI doped powders at temperatures near 1.6 K and improve their properties to reach those of bulk materials. Towards that goal, we compared properties of powders that were synthesized chemically or milled down from larger crystals using either high-energy planetary ball mills or low-energy tumbling mills. We used scanning electron microscopy (SEM) to determine the shape and size of the particles, x-ray diffraction (XRD) to analyze the composition and phase of the crystalline structure, and we employed sensitive spectral hole burning (SHB) methods to probe variations in the optical decoherence dynamics, deduced from the spectral hole width, and the ^{169}Tm nuclear spin relaxation dynamics, deduced

CONTACT Thomas Lutz thomasl@ucalgary.ca

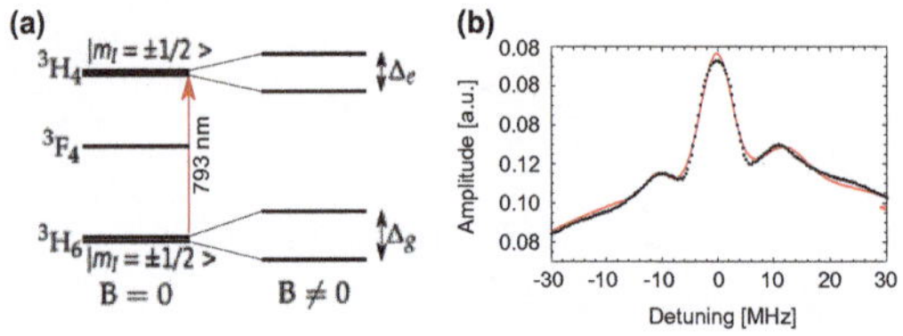

Figure 1. (a) Level structure of Tm:YAG without and with an applied magnetic field. (b) Hole burning spectrum of the 1% Tm:YAG bulk crystal together with a fit (red line).

from the spectral hole lifetime. We find that induced damage and strain in the crystal lattice, which does not affect XRD or SEM measurements, can produce large variations in the measured low-temperature dynamics of the powders that are observed using SHB techniques. Thus, SHB can serve as a quantitative characterization tool, complementing traditional techniques such as XRD, SEM or Raman scattering. Our results also demonstrate that mechanical and thermal treatment of REI doped crystals influences properties, such as the lifetime of nuclear spins, in a surprisingly strong way. Consequently, SHB is a well-suited technique to reveal the presence of residual damage in powders due to fabrication and to evaluate the effectiveness of methods used to reduce strain and improve material quality, such as thermal annealing.

2. Experiment

2.1. *Tm:YAG*

Our investigations employ crystalline $Y_3Al_5O_{12}$ doped with thulium impurities (Tm:YAG), whose relevant electronic level structure is shown in Figure 1. Under an external magnetic field, both the $^3H_6$ ground state and the $^3H_4$ excited state split into two non-degenerate states through the hyperfine interaction with the ^{169}Tm nuclear spin ($I = 1/2$), allowing persistent atomic population storage with lifetimes as long as hours in bulk crystals at liquid helium temperatures through optical pumping of the nuclear spin states.

The 20 GHz wide inhomogeneously broadened line of the $^3H_6 \leftrightarrow {}^3H_4$ transition in Tm:YAG is centered at 793.156 nm [14]. The samples were mounted in an Oxford Instruments Spectromag cryostat (Oxford Instruments: Abingdon, UK) and all powders were held in unsealed glass cuvettes. All samples were 0.5 mm thick, and, for the nominal Tm concentration of 1%, featured an optical depth of $\approx$1. This allowed for direct transmission detection of spectral holes while keeping optical scattering at an acceptable level. To prevent scattered light from reaching the detector, the cuvette was placed inside a copper box with two pinholes to allow light to enter and exit. For all measurements, the samples were cooled in helium vapor to 1.6 K and the applied magnetic field was set to 1 T.

2.2. *Spectral hole burning*

To assess the quality of different powders, we used persistent spectral holes. We extracted the lifetime of the spin states from the hole decay, and the magnitude of optical decoherence from the width of the hole [15]. As we describe later, we found that both quantities strongly depend on the crystal quality in terms of strain as well as impurities. In this approach, a spectral hole is burnt into the inhomogeneously broadened absorption line by using a laser to optically pump population between the nuclear spin states. After a certain waiting time, the absorption line is scanned using a tunable laser and the area of the spectral hole is measured. From the exponential decay of the hole area with waiting time, the spin lifetime is extracted. Note that the width of the spectral hole ideally corresponds to twice the homogeneous linewidth; however, the measured width is generally larger due to laser frequency jitter, spectral diffusion, and power broadening.

We used a Coherent 899-21 Ti:sapphire laser (Coherent Inc.: Santa Clara, California, USA) emitting at a wavelength of 793.38 nm and with a linewidth of less than 1 MHz. For the bulk single crystal measurement that served as a reference for the best material properties, the laser power was set to 5 μW. For the powder samples, more power was necessary to overcome loss due to scattering and achieve a sufficiently high signal-to-noise ratio. Otherwise, the experimental conditions were kept the same for all materials. The burn and read pulses were generated from the CW laser beam using two acousto-optic modulators (AOMs) in series. This arrangement gave an extinction ratio of >90 dB, ensuring that no unintentionally leaked light reached the sample. During the reading pulse, the laser sweep was implemented using a double-passed AOM scanned in frequency. A New Focus model 2051 photo-receiver (New Focus: Santa Clara, California, USA) was used to detect the transmitted light.

2.3. *Bulk single crystal reference*

First, as a reference against which we compare the properties of our Tm:YAG powders (the different methods used to create these powders are described below), we assessed the properties of a 1% doped Tm:YAG crystal from Scientific Materials Corporation (SMC) (Scientific Materials: Bozeman, Montana, United States) (growth number 3-8). The crystal is 0.5 cm thick and features an optical depth of about 0.5. Depending on the crystal orientation, we found spin lifetimes between 5 and 16 h [16]. Furthermore, as shown in Figure 1(b), the observed spectral hole width was limited to 6 MHz – much wider than the intrinsic kHz-wide homogeneous linewidth of Tm:YAG [17] under these conditions – primarily due to power broadening effects. The figure also shows clear side holes, split by 10 MHz T^{-1}

Table 1. Hole widths (Γ) (bold fonts indicate visible side holes), and lifetimes (T_a) of all measured materials at 1.6 K and $B = 1$ T. *ball milled.

Material	Size [μm]	Γ [MHz]	T_a [min]
SMC bulk		**6**	300–960
SMC thermal crushing	500	$\lesssim$ **10**	60
annealed	500	$\lesssim$ **10**	420
SMC low energy BM*	< 0.1	28.6	20
annealed	< 1	17.2	60
Crytur	30-50	$\lesssim$ 10	60
annealed	30-50	$\lesssim$ **10**	60
Crytur low energy BM	< 0.1	23	15
annealed	< 1	$\lesssim$ **10**	20
Crytur high energy BM	< 2	$\lesssim$ 10	10
Chemical synthesis	< 0.1	26.6	2

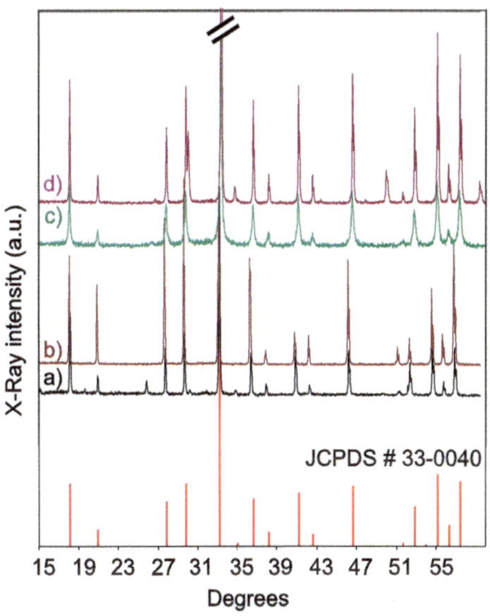

Figure 2. XRD spectra of selected powders together with the reference spectrum (JCPDS # 30-0040) of YAG. (a) Chemical synthesis; (b) Crytur non-annealed; (c) SMC low energy ball milled; (d) SMC low energy ball milled and annealed.

and caused by super-hyperfine coupling between the thulium ions and the nuclear spin of the ^{27}Al present in the host matrix [15,18]. Because powders are composed of randomly oriented crystallites, measurements of powders probe all possible orientations at once and we expect to observe lifetimes that span the range of those observed in the bulk material.

3. Results

In the following, we study the impact of fabrication and annealing methods on nuclear spin lifetimes T_a as well as on hole linewidths Γ. Due to the strong scattering caused by the powders, higher laser power was required for these measurements, leading to additional power broadening. As a consequence, hole linewidths were measurable only down to approximately 10 MHz. A summary of the properties of all investigated materials is shown in Table 1.

In addition, we performed SEM imaging and powder XRD analysis to further characterize each sample. Example powder XRD spectra are shown in Figure 2 for a selected subset of our samples. The spectra were obtained with a Scintag Inc X-1 Advanced Diffraction System (Acintag Inc.: Cupertino, California, USA). We find that all the analyzed samples formed the expected crystalline structure of YAG. Small deviations from the reference spectrum (JCPDS # 30-0040) can be observed for some samples. Differences in peak heights can be assigned mostly to sample preparation issues, while additional peaks correspond to the presence of residual amounts of impurity phases [19].

3.1. Crushing and ball milling

Our first approach to obtain small powders was "top-down" fabrication starting from a 1% doped Tm:YAG bulk single crystal originating from the same growth as the one studied above. To pre-crush the crystal, we heated the bulk crystal to ≈ 500°C and then immersed the hot crystal in water (21°C) to thermally shock the crystal, causing it to crack into millimeter and larger sized pieces. This procedure was repeated on all pieces with sizes greater than ≈ 5 mm until all were in the few millimeter or smaller size range. We then ground the small pieces in a mortar and pestle to produce a powder composed of crystallites with sizes of less than 0.5 mm. With spectral hole burning spectroscopy, we measured a hole width of $\lesssim$ 10 MHz, limited by power broadening. Side holes due to superhyperfine coupling with ^{27}Al nuclear spins were still resolvable, as in the bulk material. The spin lifetime was on the order of 1 h – significantly shorter than the values obtained in the bulk material at the same temperature and magnetic field. Since this powder originated from a bulk crystal whose properties were well characterized, and was composed of relatively large particles of ≈ 0.5 mm size that had been ground only for a short time, it is unlikely that any change in chemical composition or in-diffusion of impurities could have occurred (as might be a factor for much smaller particles when processed using high-energy methods). The only explanation for

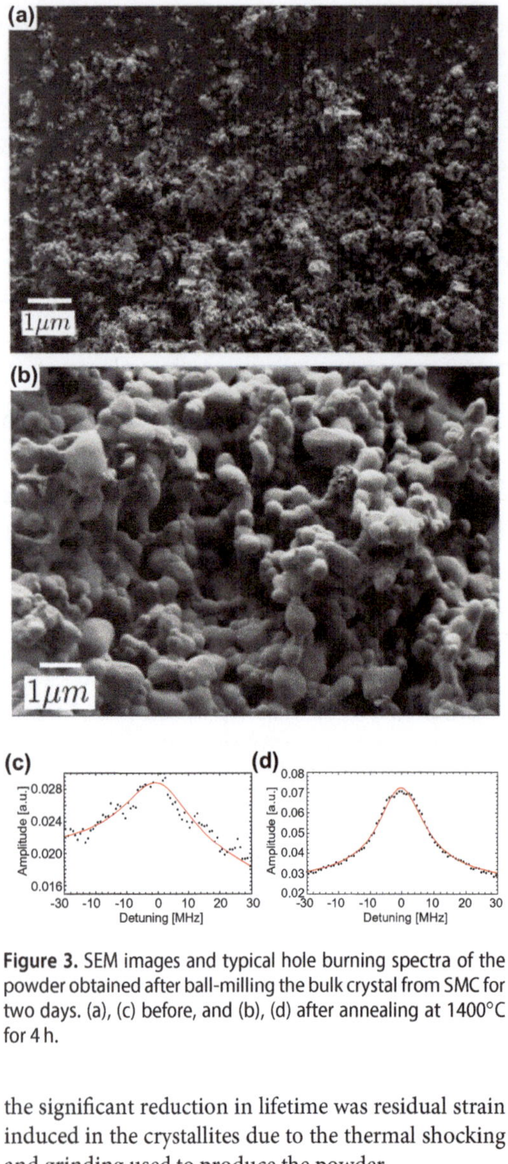

Figure 3. SEM images and typical hole burning spectra of the powder obtained after ball-milling the bulk crystal from SMC for two days. (a), (c) before, and (b), (d) after annealing at 1400°C for 4 h.

the significant reduction in lifetime was residual strain induced in the crystallites due to the thermal shocking and grinding used to produce the powder.

With the < 0.5 mm sized crushed powder as a starting material, we produced a smaller powder using a low-speed tumbling mill. The Tm:YAG powder was dispersed in ethanol and milled for 48 h using a mix of zirconia balls with sizes ranging from 1 mm to 1 cm. Figure 3(a) shows an SEM image of the obtained powder. The average size of the irregular particles was below 100 nm. The XRD spectrum, shown in Figure 2(c) displays the expected YAG structure with no observable peak broadening or amorphous background. A spectral hole burning trace, shown in Figure 3(c), revealed a hole width of 28.6 MHz. The hole was about five times larger than that observed in the bulk, which could not solely be attributed to additional power broadening and reveals a significant difference in the material not detectable from our XRD measurement. No sideholes were visible, which was expected since their splitting (10 MHz at $B = 1$ T) is less than the spectral hole width. Moreover, we found a further reduction of the spin lifetime to 20 min. Overall, as opposed to the XRD analysis, the SHB results suggest that a significant amount of damage (strain) was induced in the crystallites during the two days of milling.

The surprisingly strong effect of mechanical strain on coherence and spin relaxation dynamics may be understood within the framework of the two-level system (TLS) model first developed to describe dynamic disorder modes in amorphous materials [20,21]. These very low-frequency modes are enabled by the disordered structure and involve groups of atoms tunneling between local configurations with nearly equivalent energies. It has been proposed that low densities of TLS may also be enabled in crystalline materials by large inhomogeneous lattice strains [22], and optical decoherence of rare-earth ions due to interactions with TLS has been observed in both bulk crystals [23,24] and powders [25]. Furthermore, it is known that TLS can also be effective in causing rapid electron and nuclear spin relaxation [26–28]. Here, both the increase in spectral hole widths and the decrease in nuclear spin lifetimes that we observe in powders are likely to result from the creation of TLS from strain induced during the mechanical fabrication process. This interpretation is also consistent with past observations of increased electronic spin-lattice relaxation rates of Nd^{3+} doped into bulk YAG crystals with greater densities of structural defects in the lattice [29].

3.2. Effects of annealing

Next, we investigated to what degree annealing of the powders can repair the induced damage. After we confirmed that low annealing temperatures around 600°C were unable to improve material properties, and since higher temperatures generally lead to better properties [30], we annealed both the 0.5 mm and the 100 nm powders at our highest accessible temperature of 1400°C. The annealing was performed for 4 h in an oxygen atmosphere (using a tube furnace) to minimize the outdiffusion of oxygen from the YAG matrix. The 0.5 mm annealed crystals appeared unchanged under the SEM. The spin-state lifetime improved significantly to 7 h and became comparable to that extracted from measurements on the bulk. This suggests that annealing can mostly repair the damages induced by stress during the grinding via mortar and pestle.

The SEM image of the annealed 100 nm powder, shown in Figure 3(b), reveals that the size of the particles increased to around 1 μm. The particles were almost spherical, with close to uniform size distribution, and appeared to be agglomerated. Compared to the same powder before annealing (panel (a)) this is a major improvement. The XRD spectrum, shown in Figure 2(d), shows no significant change compared to

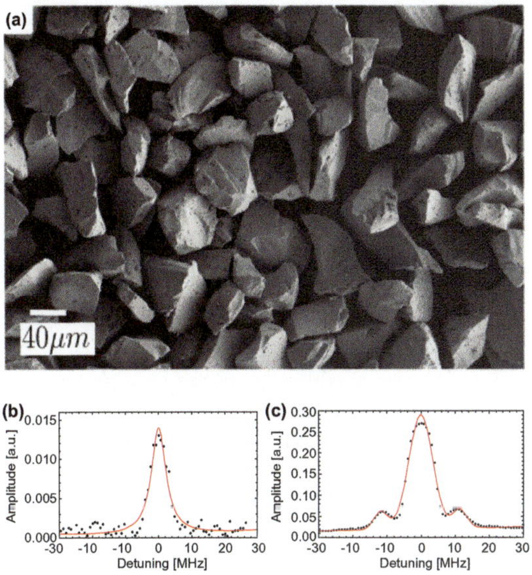

Figure 4. (a) SEM image of the 1% Tm:YAG powder provided by Crytur, and typical hole burning spectrum (b) before and (c) after annealing.

the non-annealed powder. The signal-to-noise ratio in this spectrum is better than the one in the spectrum for the non-annealed powder since more material was available for analysis. This indicates that, in contrary to the SHB measurement, the current XRD analysis is not sensitive enough to detect the changes caused by annealing for this powder. Indeed, the hole burning spectrum in Figure 3(d) shows a hole width of 17.2 MHz, which is 30% smaller than the value measured before annealing. Furthermore, the spin lifetime increased by a factor of three to 1 h. Hence, our method of annealing also improved the properties of the ball-milled powder, but it was not sufficient to completely re-establish the properties of the bulk. We anticipate that optimized annealing procedures, in particular involving temperatures closer to the melting point, will lead to further improvements.

3.3. Comparison of high- and low-energy ball-milling methods

The above results indicated that the grinding methods used here, namely mortar and pestle and ball-milling, caused damage to the host material. To better understand this surprisingly large effect, and to determine the degree to which damage can be repaired through annealing, we continued studying ball-milling methods starting with a powder from a commercial supplier (Crytur). The powder was created from outer fragments of a large crystal, grown from 99.999% pure starting materials using the Czochralski method. Those fragments were then pre-crushed using a mechanical jaw crusher and finally milled down to a size of 30–50 μm

Figure 5. SEM image of the 1% Tm:YAG powder after 4 h of high-energy planetary ball-milling.

in a tumbling mill. The resulting powder was cleaned using mineral acids.

First, we characterized the "as purchased" micro meter-sized Crytur powder using SEM, XRD, and SHB, as shown in Figures 4(a), (b), and 2(b). The SEM image reveals that the powder has sharp edges and few particles with dimensions below 30 μm. The XRD spectrum shows good YAG crystalline structure. From the hole burning trace we extracted a hole width of $\lesssim$ 10 MHz; however, the lifetimes were only on the order of one hour and comparable to the thermally crushed bulk crystal. This indicates that the crushing method employed by Crytur also induced a comparable amount or strain that is not detected in the XRD spectrum.

Since our annealing procedure was successful for the 0.5 mm-sized powder made from the SMC bulk crystal, we annealed the Crytur powder using the same method. The SEM image taken after annealing resembles that of the powder before annealing (see Figure 4(a)), and the hole width as well as the spin lifetime also remained comparable to those of the powder before annealing. Hence, in contrast to the powder created from the SMC crystal, besides an improvement in signal strength (see Figure 4(c)), our annealing method did not succeed in improving the properties of the powder. A possible explanation for this observation is the different way of creating the bulk crystal from which the powders were made. While both crystals were grown in a Czochralski process using starting materials of similar purity (99.999% in the case of Crytur, and 99.995% for SMC), the original bulk crystal from SMC has been cut from the center of a boule and annealed before shipping. In contrast, the starting material at Crytur was outer fragments of a large crystal that were not annealed before crushing. These differences in the procedures employed in the growth process and sample preparation may affect the density of intrinsic defects and strain in the initial bulk crystal, potentially explaining the differences between the samples that are observed here.

In addition to jaw crushing and low-energy ball milling, we also used a high-energy planetary ball mill to obtain small powders. This technique requires shorter milling times compared to the low energy approach, but the energy of the impacts during milling is higher. For a direct comparison between high-energy and low-energy ball milling, we used the same starting material, i.e. the non-annealed, 30–50 μm-size crystals from Crytur, and milled the powder for 4 h using 1 cm-diameter balls in a high-energy ball-mill. The SEM image in Figure 5 showed that the particles were not agglomerated and on average about 2 μm large. This is the smallest average size we could achieve using this mill with balls of that size. Figure 5 also reveals the sharp, irregular surfaces of the particles, suggesting significant strain, as well as a broad size distribution. For this powder, we measured a spectral hole width $\lesssim$ 10 MHz (no side holes are visible) and a lifetime of around 10 min – considerably shorter than any other milled material. We conclude that despite the reduced milling time, the crystallites contained an even greater amount of strain and defects due to the high-energy impacts during milling.

Finally, we employed a low-energy tumbling mill for 48 h to reduce the size of the original Crytur powder. The SEM image reveals agglomerated particles on the order of 100 nm. Furthermore, we found a spectral hole width of 23 MHz (no side holes are visible) and spin lifetimes of around 15 min. This new powder was very similar in all its properties to the one obtained from the SMC bulk crystal after the same ball-milling procedure (see Figure 3(a) and (c)).

The same annealing procedure as described before was then applied to the low-energy ball-milled powder. As observed in the case of the powders originating from the SMC bulk crystal, the particles grew from 100 nm to 1 μm, exhibiting clear garnet dodecahedral habit. Improvements in terms of spectral hole burning were more pronounced, with a measurement limited hole width of less than 10 MHz and visible side holes. However, the obtained lifetimes were still only around 20 min (as opposed to 1 h in the case of the SMC powder), which might be explained by a larger number of defects or impurities in the original bulk crystal from which the Crytur powder was obtained.

3.4. Direct chemical synthesis

Apart from milling larger crystals, small-sized powders can also be obtained using "bottom-up" chemical synthesis. In addition to not requiring initial growth of a high-quality crystal – a substantial advantage for fast prototyping – this method does not suffer from potential damage induced during grinding. We therefore examined the performance of a 1% Tm doped YAG powder that was prepared using a co-precipitation method [31–33]. This 1% Tm:YAG powder was synthesized from starting materials with at least 99.9% purity. An aqueous nitrate solution was prepared by dissolving

Figure 6. SEM image of the synthesized 1% Tm:YAG powder.

$Tm(NO_3)_3 \cdot 5H_2O$, $Y(NO_3)_3 \cdot 6H_2O$, and $Al(NO_3)_3 \cdot 9H_2O$ with a molar ratio of 0.03:2.97:5 in HPLC-grade water. The precipitant solution was prepared with a 1 M concentration of aqueous ammonium bicarbonate and ≈1 mM concentrations of sodium dodecyl sulfate and polyethylene glycol 400 added as anionic and nonionic surfactants, respectively. The solutions were mixed by stirring for 1 h and then the nitrate solution was added drop-wise to an excess of the precipitant solution while stirring at room temperature to form the precipitate. The resulting suspension was aged for several hours while stirring to allow the solutions to fully react, and the resulting gelatinous precipitate was centrifugally separated from the solution. The precipitate was then washed by vortex mixing and sonication using deionized water, ethanol, and finally n-butanol, and it was then allowed to dry overnight in a vacuum desiccator. A YAG precursor powder was obtained by subsequently drying the precipitate at 300°C for 4 h and then grinding in a mortar to obtain a fine powder. The precursor was then mixed with 5% Li_2CO_3 by weight as a flux to promote growth of large crystallites, placed in a covered alumina crucible, and then crystallized in a tube furnace at a temperature of 1300°C for 8 h in air. Analysis of the resulting powder using XRD, shown in Figure 2(a), revealed complete formation of single-phase crystalline YAG. The quality of the XRD spectrum, in terms of signal-to-noise ratio and peak widths, is comparable to the spectrum of the purchased Crytur powder (see Figure 2(b)). An SEM image of the synthesized 1% Tm:YAG powder is shown in Figure 6, revealing that it consisted of uniform, non-agglomerated crystallites with diameters of 500 nm to 2 μm that exhibited the characteristic dodecahedral crystal habit of the garnets.

Although the XRD and SEM results for the synthesized powders appeared promising, spectral hole burning measurements revealed the poor quality of the material. We observed a very broad hole width of 26.6 MHz with no visible side holes. Furthermore, the measured nuclear spin lifetime was limited to values of at most 2 min. Hence, the properties of the synthesized powder were significantly worse than the properties of all of the powders obtained using the top-down ball-milling approach. We expect that the poor performance of the synthesized crystallites may result from residual chemical impurities that can cause lattice strain from incorporation of the defects as well as the presence of paramagnetic impurities such as iron that can couple to the Tm^{3+} ions and induce nuclear spin relaxation and optical decoherence. Consequently, further studies are required on powders synthesized from higher-purity starting materials to investigate whether the properties show improvement and how they compare to powders of similar purity obtained from top-down fabrication.

4. Discussion of different characterization methods

The three methods used in this work, SEM, XRD and SHB, complement each other as they provide information about different aspects of the nanocrystal properties. We found that SHB was the most sensitive method to detect effects of crushing as well as annealing on the powder quality in our measurements.

SEM images constitute the most convenient method to assess the size of the particles, and also observe their shape, which can be an indication of the crystalline quality, provided one knows the crystal's habit. For instance, the powder obtained by chemical synthesis exhibits the characteristic dodecahedral crystal habit of YAG (see Figure 6). However, the SEM results do not allow one to identify the chemical composition, and thus cannot detect impurities present in the particles.

XRD analysis is the most reliable method to assess the composition and the phase of the nanocrystals. If the concentration of impurities is high enough, they can also be detected by this method. Potential amorphous character of the particles should also be visible as a broad background on the XRD spectrum. However, we were not able to observe the amorphous components or crystal strain that was clearly observed in our SHB measurements. In principle, by analyzing the peak widths in the spectrum, XRD can also give access to the particle size, but this requires a resolution beyond that of our instrument.

The SHB measurement gives access to the population lifetime of the nuclear spin states of Tm^{3+}:YAG, as well as to the homogeneous linewidth of the optical transition if the hole width is not broadened by laser power, which limited our resolution to 10 MHz. Both quantities are very sensitive to the crystal quality in terms of both strain and impurities. This sensitivity originates from the coupling between the ion's nuclear spin and the environment. In some cases, the SHB measurements reveal differences in the materials that are not observed in the SEM or XRD results. For example, the chemically synthesized nanocrystals exhibited very short spin lifetimes despite promising SEM and XRD results. The SHB measurement gives a quantitative, and sensitive method to determine the overall quality of the nanocrystals, even though it does not directly determine the nature of the limiting factors.

5. Conclusions and outlook

In summary, we studied properties of 1% Tm:YAG powders produced by different methods via SEM imaging, XRD analysis, and spectral hole burning. We found that SHB is a very sensitive and well-suited method to assess the quality of REI doped crystals in a quantitative way.

We found that any grinding or milling technique degraded the performance of the REI doped material in terms of homogeneous linewidth and especially spin-state lifetimes, most likely due to induced stress and deformation to the host matrix. Annealing is crucial and allowed us to recover the bulk spin-state lifetimes in the case of a soft-grinding method. In the case of more severe damage caused by ball-milling, bulk properties were partly recovered in powders originating from high-quality, annealed bulk crystals, whereas it remained poor in the case of powders obtained from non-annealed crystal fragments. This shows the importance of starting with high-quality crystals and the need for annealing. Finally, we found that the performance of the synthesized powder studied here was likely limited by the quality of the starting materials.

We anticipate our findings to be generalizable to other impurities and other hosts, in particular to color centers in nano-structured diamond, and to provide valuable insight towards nano-structuring REI doped crystals. Furthermore, SHB is a way to reveal variations in terms of quality and uniformity of powders that could have consequences for traditional applications where low levels of lattice defects can affect performance, such as luminescence efficiency and thermal quenching of phosphors.

Acknowledgements

W. T. is a Senior Fellow of the Canadian Institute for Advanced Research (CIFAR). This work is based upon research conducted in part at the Montana State University Imaging and Chemical Analysis Laboratory.

Disclosure statement

No potential conflict of interest was reported by the authors.

Funding

The authors acknowledge support from Alberta Innovates Technology Futures (ATIF), the National Engineering and Research Council of Canada (NSERC), and the National Science Foundation of the USA (NSF) [award nos. PHY-1212462, PHY-1415628, CHE-1416454].

References

[1] de Riedmatten H, Afzelius M. Quantum light storage in solid state atomic ensembles. ArXiv. 2015;1502.00307.

[2] Saglamyurek E, Sinclair N, Slater JA, Heshami K, Oblak D, Tittel W. An integrated processor for photonic quantum states using a broadband light-matter interface. New J. Phys. 2014;16:065019.

[3] Ménager L, Lorgeré I, Gouët J-LL, et al. Demonstration of a radio-frequency spectrum analyzer based on spectral hole burning. Opt. Lett.. 2001;26:1245.

[4] McAuslan DL, Longdell JJ, Sellars MJ. Strong-coupling cavity QED using rare-earth-metal-ion dopants in monolithic resonators: What you can do with a weak oscillator. Phys. Rev. A. 2009;80:062307.

[5] O'Brien C, Lauk N, Blum S, Morigi G, Fleischhauer M. Interfacing Superconducting Qubits and Telecom Photons via a Rare-Earth-Doped Crystal. Phys. Rev. Lett. 2014;113:063603.

[6] Walther A, Rippe L, Yan Y, Karlsson J, Serrano D, Nilsson AN, Bengtsson S, Kröll S. High fidelity readout scheme for rare-earth solid state quantum computing. ArXiv. 2015;1503:08447.

[7] Lutz T, Veissier L, Thiel CW, Cone RL, Barclay PE, Tittel W. Modification of phonon processes in nano-structured rare-earth-ion-doped crystals. ArXiv. 2015;1504:02471.

[8] Hong K, Meltzer R, Bihari B, et al. Spectral hole burning in crystalline Eu2O3 and Y2O3 : Eu3+ nanoparticles. Proceedings of the eleventh international conference on dynamical processes in excited states of solids. J. Lumin. 1998;7677:234.

[9] Malyukin YV, Masalov AA, Zhmurin PN. Single-ion fluorescence spectroscopy of a Y2SiO5:Pr3+ nanocluster. Phys. Lett. A. 2003;316:147.

[10] Macfarlane RM, Dejneka MJ. Spectral hole burning in thulium-doped glass ceramics. Opt. Lett. 2001;26:429.

[11] Perrot A, Goldner P, Giaume D, Lovrić M, Andriamiadamanana C, Gonçalves RR, Ferrier A. Narrow Optical Homogeneous Linewidths in Rare Earth Doped Nanocrystals. Phys. Rev. Lett. 2013;111:203601.

[12] Heitjans P, Masoud M, Feldhoff A, Wilkening M. NMR and impedance studies of nanocrystalline and amorphous ion conductors: lithium niobate as a model system. Faraday Discuss. 2007;134:67.

[13] Scholz G, Stösser R, Klein J, Silly G, Buzar JY, Laligant Y, Ziemer B. Local structural orders in nanostructured Al 2 O 3 prepared by high-energy ball milling. J. Phys. Condensed Matter. 2002;14:2101.

[14] Sun Y, Wang GM, Cone RL, Equall RW, Leask MJM. Phys. Rev. B. 2000;62:15443.

[15] Macfarlane RM, Shelby RM. Coherent Transient and Holeburning Spectroscopy of Rare Earth Ions in Solids. In: Kaplyanskii AA, RMMacfarlane editors. Spectroscopy of Solids Containing Rare Earth Ions. Elsevier; 1987. p. 51–184.

[16] Veissier L, Thiel CW, Lutz T, Cone RL, Barclay PE, Tittel W. In preparation.

[17] Liu G, Jacquier B. Spectroscopic Properties of Rare Earths in Optical Materials. Berlin, Heidelberg: Springer; 2005.

[18] Bonarota M, Ruggiero J, Gouët JLL, Chanelière T. Efficiency optimization for atomic frequency comb storage. Phys. Rev. A. 2010;81:033803.

[19] Kareiva A. Aqueous Sol-Gel Synthesis Methods for the Preparation of Garnet Crystal Structure Compounds. Mater. Sci. 2011;17:1392–1320.

[20] Anderson PW, Halperin BI, Varma CM. Anomalous low-temperature thermal properties of glasses and spin glasses. Philos. Mag. 1972;25:1.

[21] Phillips W. Tunneling states in amorphous solids. J. Low Temp. Phys. 1972;7:351.

[22] Watson SK. Tunneling States in Crystals with Large Random Strains. Phys. Rev. Lett. 1995;75:1965.

[23] Flinn GP, Jang KW, Ganem J, Jones ML, Meltzer RS, Macfarlane RM. Sample-dependent optical dephasing in bulk crystalline samples of Y2O3:Eu3+. Phys. Rev. B. 1994;49:5821.

[24] Macfarlane RM, Sun Y, Cone RL, Thiel CW, Equall RW. Optical dephasing by disorder modes in yttrium orthosilicate (Y2SiO5) doped with Eu3+. Proceedings of the 8th International Meeting on Hole Burning, Single Molecule, and Related Spectroscopies: Science and Applications. J. Lumin. 2004;107:310.

[25] Meltzer R, Jang K, Hong K, Sun Y, Feofilov S. Optical dephasing of rare earth ions in mixed crystalline and size-restricted systems. J. Alloys Compounds. 1997;250:279.

[26] Szeftel J, Alloul H. Nuclear spin lattice relaxation in the amorphous state: Towards an understanding. J. Non-Crystalline Solids. 1978;29:253.

[27] Askew TR, Stapleton HJ, Brower KL. Anomalous electron-spin relaxation in amorphous silicon. Phys. Rev. B. 1986;33:4455.

[28] Balzer-Jöllenbeck G, Kanert O, Steinert J, Jain H. Low-temperature properties of nuclear spin relaxation in inorganic glasses. Solid State Comm. 1988;65:303.

[29] Aminov L, Kurkin I, Lukoyanov D. Some peculiarities of spin-lattice relaxation of impurity rare-earth ions in crystals caused by the structure defects. Appl. Mag. Reson. 1998;14:447.

[30] Georgescu S, Constantinescu S, Chinie A, et al. 2007. Optical and Mossbauer studies on YAG: Eu nanocrystals synthesized by a sol-gel method, ROMOPTO 2006: Eighth Conference on Optics (International Society for Optics and Photonics); p. 67850A–67850A.

[31] Li X, Liu H, Wang J, Cui H, Zhang X, Han F. Preparation of YAG:Nd nano-sized powder by co-precipitation method. Mater. Sci. Eng. A. 2004;379:347.

[32] Pradhan A, Zhang K, Loutts G. Synthesis of neodymium-doped yttrium aluminum garnet (YAG) nanocrystalline powders leading to transparent ceramics. Mater. Res. Bull. 2004;39:1291.

[33] Xu G, Zhang X, He W, Liu H, Li H, Boughton RI. Preparation of highly dispersed YAG nano-sized powder by co-precipitation method. Mater. Lett. 2006;60:962.

10

A simple synthesis method for nanostructured Co-WC/carbon composites with enhanced oxygen reduction reaction activity

Jun Kang[a], Hye-min Kim[b], Nagahiro Saito[c] and Myeong-Hoon Lee[a]

[a]Division of Marine Engineering, Korea Maritime and Ocean University, Busan 606-791, Republic of Korea
[b]Graduate School of Engineering, Nagoya University, Nagoya 464-8603, Japan
[c]Department of Material Science and Engineering, Nagoya University, Furo-cho, Chikusa-ku, Nagoya 464 8603, Japan

ABSTRACT
Co nanoparticles (Co NPs) and nanoscale tungsten carbide (WC) are successfully synthesized simultaneously with mesoporous structured carbon black (C) using an innovative simple method, which is known as solution plasma processing (SPP), and NPs are also loaded onto carbon black at the same time by SPP. The introduction of Co NPs led to not only superior oxygen reduction reaction (ORR) activity in terms of onset potential and peak potential, but also to a more efficient electron transfer process compared to that of pure WC. Co-WC/C also showed durability for long-term operation better than that of commercial Pt/C. These results clearly demonstrate that the presence of Co NPs significantly enhanced the ORR and charge transfer number of neighboring WC NPs in ORR activities. In addition, it was proved that SPP is a simple method (from synthesis of NPs and carbon black to loading on carbon black) for the large-scale synthesis of NP-carbon composite. Therefore, SPP holds great potential as a candidate for next-generation synthetic methods for the production of NP-carbon composites.

KEYWORDS
Solution plasma processing; nanoparticle-carbon composite; oxygen reduction reaction; non-precious metal electrocatalysts

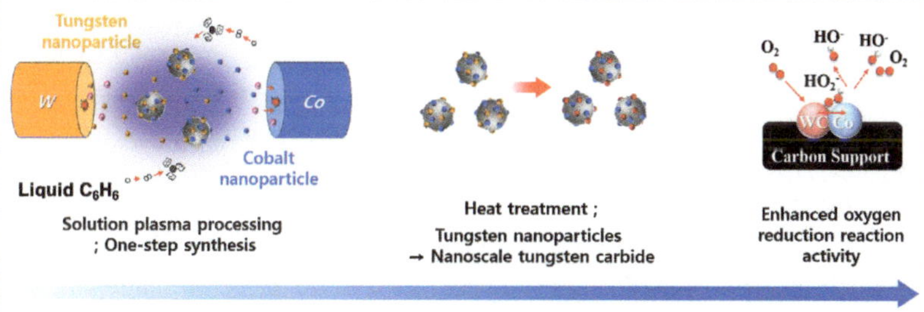

1. Introduction

The oxygen reduction reaction (ORR) is an important electrochemical reaction in many fields, including energy conversion (fuel cells, metal–air batteries), corrosion, biosensing and the production of hydrogen peroxide [1–4]. In particular, for fuel cell systems, the cathodic oxygen reduction is a major factor limiting performance because it is always a slow reaction in both acidic and basic solutions [5]. Currently the best and most frequently used catalyst for ORR is platinum because of its excellent ORR activity (i.e. low overpotential and large current density) and selectivity toward a direct four-electron pathway in both acidic and alkaline solutions. Aside from its activity, however, the Pt catalyst is disadvantageous in terms of its prohibitively high cost and limited supply, and its susceptibility to contamination by carbon monoxide (CO) [6]. As a result, the discovery of less expensive and more CO-tolerant alternatives to Pt electrocatalysts would greatly facilitate the commercialization of fuel cell systems. One of the most attractive ideas is the development of efficient non-precious metal electrocatalysts. Tungsten carbide (WC) is the most attractive material due to its high electrical conductivity, CO resistance, and corrosion resistance; it also has a platinum-like behavior for the chemisorption of hydrogen and oxygen [7–12]. However, the intrinsic catalysis of WC itself is very much lower than that of platinum, and thus its large over-potential for ORR

CONTACT Myeong-Hoon Lee leemh@kmou.ac.kr

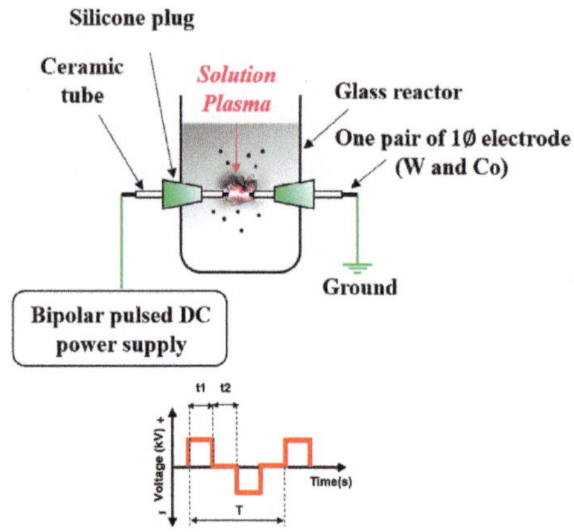

Figure 1. Schematic of the solution plasma process (SPP).

must be reduced in order to obtain high energy efficiency. In addition, the methods that can be used to obtain nanoscale WC are complicated and generally require multiple step processes, high temperature, long processing time, and a chemical agent. Therefore, it is necessary to find a simple method to synthesize WC with good dispersion.

To the best of our knowledge, there have been few attempts to enhance the catalytic activity of WC based materials for ORR with non-noble metals. Herein, accordingly, for improved catalytic potential and durability of a cathode, a combination of nanoscale WC with cobalt (Co) nanoparticles is considered as a potential alternative electrode material for fuel cell application. Recently Co based catalysts have been studied as non-noble metals to enhance ORR activity of carbonaceous materials, for economic reasons, and showed significantly improved electro-catalytic activity of ORR [13–15].

In addition, in this work, a simple and effective method to synthesize highly dispersed nanoscale Co-WC on carbon materials (Co-WC/C) in the form of nanocomposites using a one-step process (from synthesis of nanoparticles and carbon black to loading on carbon) by solution plasma processing (SPP) was reported. SPP has been applied for synthesis of carbon black directly from hydrocarbon solution and metal nanoparticles by electrode sputtering process of the working electrode within short processing time at atmospheric pressure. [16–20]. Moreover, synthesized metal nanoparticles can be uniformly dispersed on carbon black because the synthesis and loading processes can occur at the same time.

In this study the physical and chemical characteristics of the obtained Co-WC/C are investigated. Then, the enhancement of catalytic activity due to the presence of Co within the WC/carbon matrix is discussed in detail. The durability of Co-WC/C is compared to that of commercial Pt/C.

2. Experimental procedures

2.1. Synthesis of Co-WC/C

The experimental setup of SPP used for the production of Co-WC/C is schematically shown in Figure 1. Metal wires (W, Co) with diameter of 1.0 mm (W, Co, 99.95% Nilaco, Tokyo, Japan) and benzene (C_6H_6, 99.5% Kanto Chemical, Tokyo, Japan) were applied, respectively, as the precursors of Co, WC and carbon nanoparticles. To concentrate energy, each electrode was insulated by ceramic tubes; protruding length was 1.5 mm from the ceramic tube tips and tubes were inserted into a silicone plug. One pairs of metal electrodes (W, Co) was placed in a glass vessel (100 ml beaker, with a diameter of 5 cm and height of 7 cm); the distance between the tips of the electrodes was set at 0.5 mm. The electrodes were discharged in benzene using a bipolar–DC pulse power supply (Kurita, Kurita, Japan); the voltage, pulse frequency, and pulse width were controlled at 1.6 kV, 25 kHz, and 0.5 μs, respectively. Heat treatment was required to increase the conductivity of the carbon matrix and to facilitate the conversion from metastable carbide (cubic WC_{1-X}) to carbon rich carbide (hexagonal WC); this process was performed in a tube furnace with a 30 min dwell time, under a flowing Ar atmosphere, with a heat treatment temperature (HTT) of 1100°C (heating rate: 25°C min^{-1} and cooling rate: 7°C min^{-1}).

2.2. Characterization

Transmission electron microscopy (TEM, JEOL, JEM2500SE at 200 kV, JEOL Ltd., Tokyo, Japan) observations were conducted to study the microstructure, shape and size of the synthesized Co-WC/C nanocomposite. X-ray diffraction patterns (XRD, Rigaku Smart Lab Rigaku Corporation, Tokyo, Japan) were obtained using Cu Kα (λ = 1.54 nm) as a target over a scan range of 15–90° with 0.02° step size and 2° min^{-1} scan speed. The specific surface area, total pore volume, and mean pore size of all of the samples were calculated by N_2 adsorption–desorption isotherms using the Brunauer–Emmett–Teller (BET) method (Belsorp-mini II, Belsorp, USA). All samples were degassed at 200°C for 2 h prior to measurement of the BET.

2.3. Electrode preparation and electrochemical characterization

In order to evaluate the electrochemical properties of Co-WC/C, three electrode cells with a potentiostat (Hokuto Denko, Inc., HZ5000, Tokyo, Japan) were applied. The three-electrode cell consisted of a glassy carbon (GC) disk (3 mm in diameter) working electrode, Ag/AgCl (saturated KCl) reference electrode, and a platinum wire counter electrode. The samples were prepared

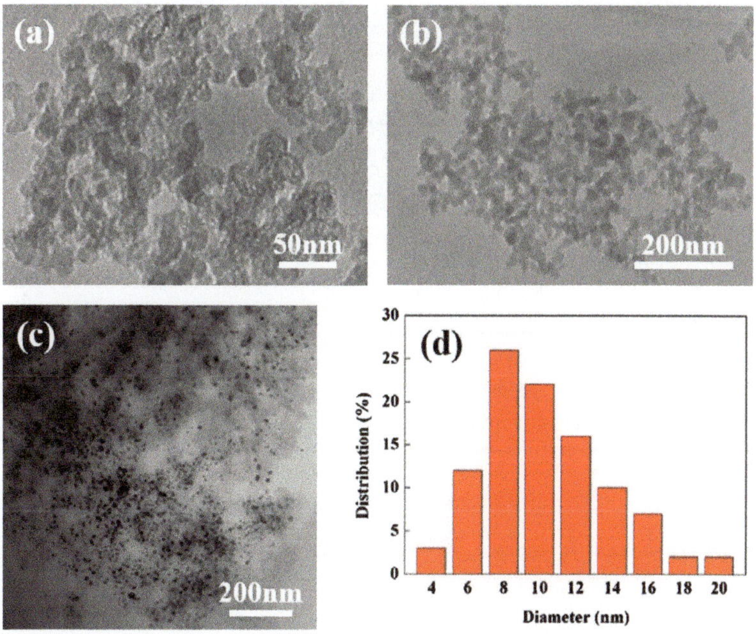

Figure 2. (a) High magnification TEM image of Co-WC/C; (b) low magnification TEM image Co-WC/C; (c) BF-STEM image of Co-WC/C; (d) size distribution of annealed Co-WC/C.

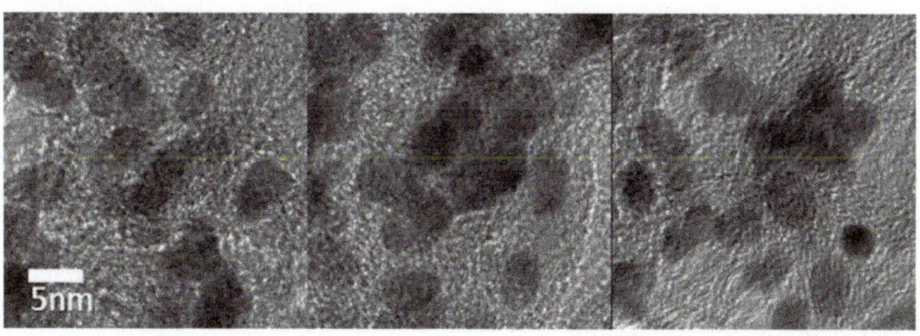

Figure 3. HR–TEM images of Co-WC/C.

by ultrasonicating a mixture of 5 mg of sample powder, 0.5 ml of ethanol, and 50 μl of Nafion solution (Sigma Aldrich, Germany 5 wt% Nafion) until a homogeneous suspension was formed. A droplet (10 μl) of the obtained suspension was placed on the glassy carbon and dried at room temperature prior to the measurement.

The cyclic voltammetry (CV) method was conducted in 0.1 M KOH electrolyte and O_2 gas was purged in the solution at a 50 mV s^{-1} scan rate from 0.3 V to −1 V (V versus Ag/AgCl). Moreover, we investigated the electron transfer kinetics during the ORR, and the linear sweep voltammetry (LSV) method was performed in 0.1 M KOH electrolyte and O_2 gas was purged in the solution with a 10 mV s^{-1} scan rate from 0.2 V to −1 V (V versus Ag/AgCl) at 900 to 3600 rpm rotation speed using a rotating disk electrode (RDE). A commercial 20 wt% Pt/C catalyst was selected as a performance benchmark. The electrochemical durability was measured in chronoamperometric responses in 0.1 M KOH with saturated oxygen concentration for 20,000 s at −0.4 V.

3. Results and discussion

Approximately 1000 mg of Co-WC/C was generated from one pair of electrodes in 100 ml of benzene after 30 min of SPP. The microstructure of the Co-WC/C was observed using TEM, as shown in Figure 2(a) and (b). These images show that the carbons were spherical-like in shape, with a uniform size ranging from 20 to 30 nm; the carbons were found to exist in a chain-like morphology. The principle that explains how the carbons can agglomerate was discussed in previous studies [16–18]. The morphology of the Co-WC was determined using bright field scanning TEM (BF-STEM) images (Figure 2(c)) and high resolution TEM (HR-TEM) images (Figure 3). These images clearly demonstrate that the nearly spherical nanoparticles (NPs) exist stably without any aggregation. The images also clearly reveal that the NPs were remarkably uniform, and were well dispersed over the entire surface of the carbons. The images show that the NPs had a narrow

Figure 4. STEM–BF and STEM–EDX mapping images of Co-WC/C.

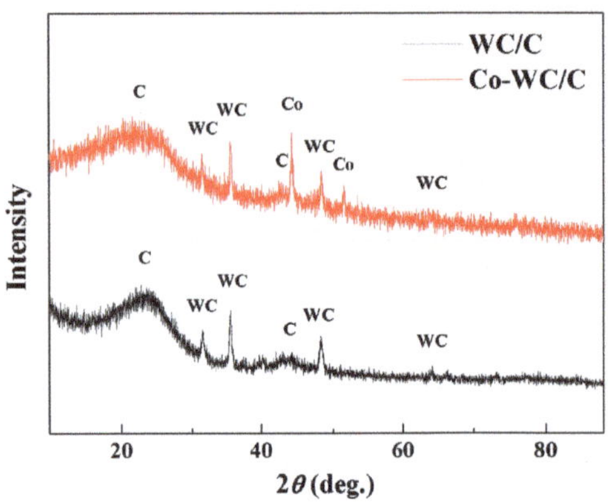

Figure 5. X-ray diffraction patterns for NP/C.

size distribution, with an average size of 10 nm (Figure 2(d)). The STEM energy-dispersive X-ray spectroscopy (STEM–EDX) mapping images show that Co and WC NPs are uniformly distributed on carbon surface and both NPs adjoin each other (Figure 4).

Inductively coupled plasma-atomic emission spectrometry (ICPAES, Perkin Elmer, USA, Optima 3300DV) was applied for a qualitative and quantitative chemical analysis of the composites. The elemental composition of Co and W in the sample was calculated as 57.97 and 43.03%, respectively and the weight percentage of entire metal NPs in composite was 14.01%.

Figure 5 shows the XRD patterns of the nanocomposites synthesized from different metal electrodes. As can be seen in the spectra, five peaks were observed from all of samples, corresponding to the 111, 200, 220, 311, and 222 reflections of the planes of WC (JCPDS #25-1047); other small peaks were also detected that indicated the formation of Co NPs (JCPDS #15-0806) from the electrodes. Based on the peak areas from the XRD pattern, the average particle size was calculated using Scherrer's formula $d = 0.9\lambda/\beta\cos\theta$, where 0.9 is the shape factor generally taken for a cubic system, λ is the X-ray source wavelength, which is typically 0.154 nm, β is the full width at half maximum intensity in radians, and θ is the Bragg angle [21]. As can be seen in Table 1, the nanoparticles were smaller than 10 nm; these results agreed with the TEM observations.

Table 1. Crystallite size of NPs (calculated from the XRD peak width).

2θ	Crystallite size (nm)	Average crystallite size (nm)
44.2	9.3	8.2
51.7	7.1	
76.1	8.1	

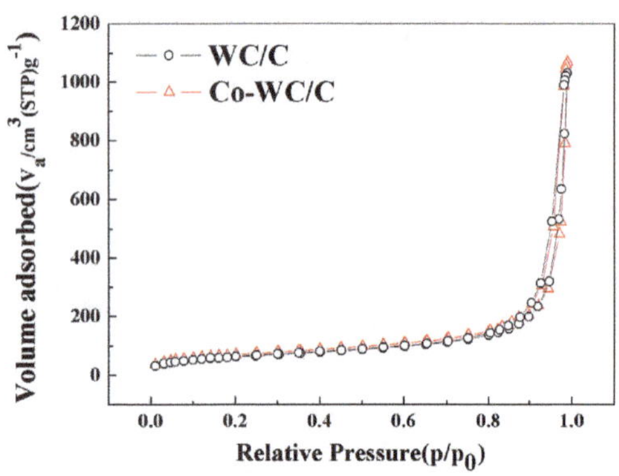

Figure 6. Nitrogen adsorption–desorption isotherms for Co-WC/C.

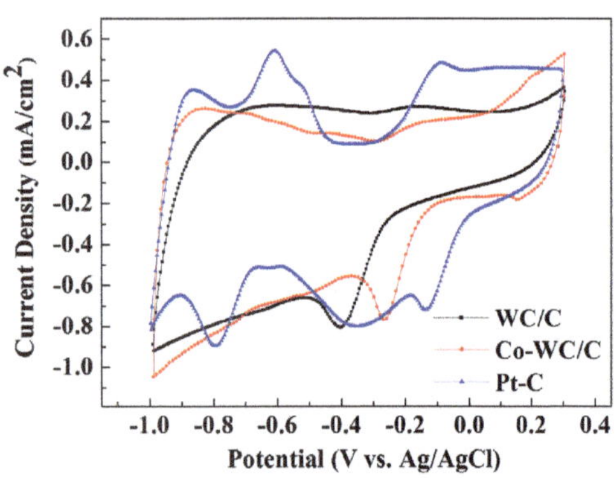

Figure 7. Cyclic voltammetry plots for WC/C, Co-WC/C and Pt/C.

To check the presence of NPs in solution (not onto carbon nanoparticles), discharged solution was filtered by syringe filter unit for gas chromatography and mixed with other carbon materials (Kejen black EC600Jd) which do not contain any metal particles. It was then

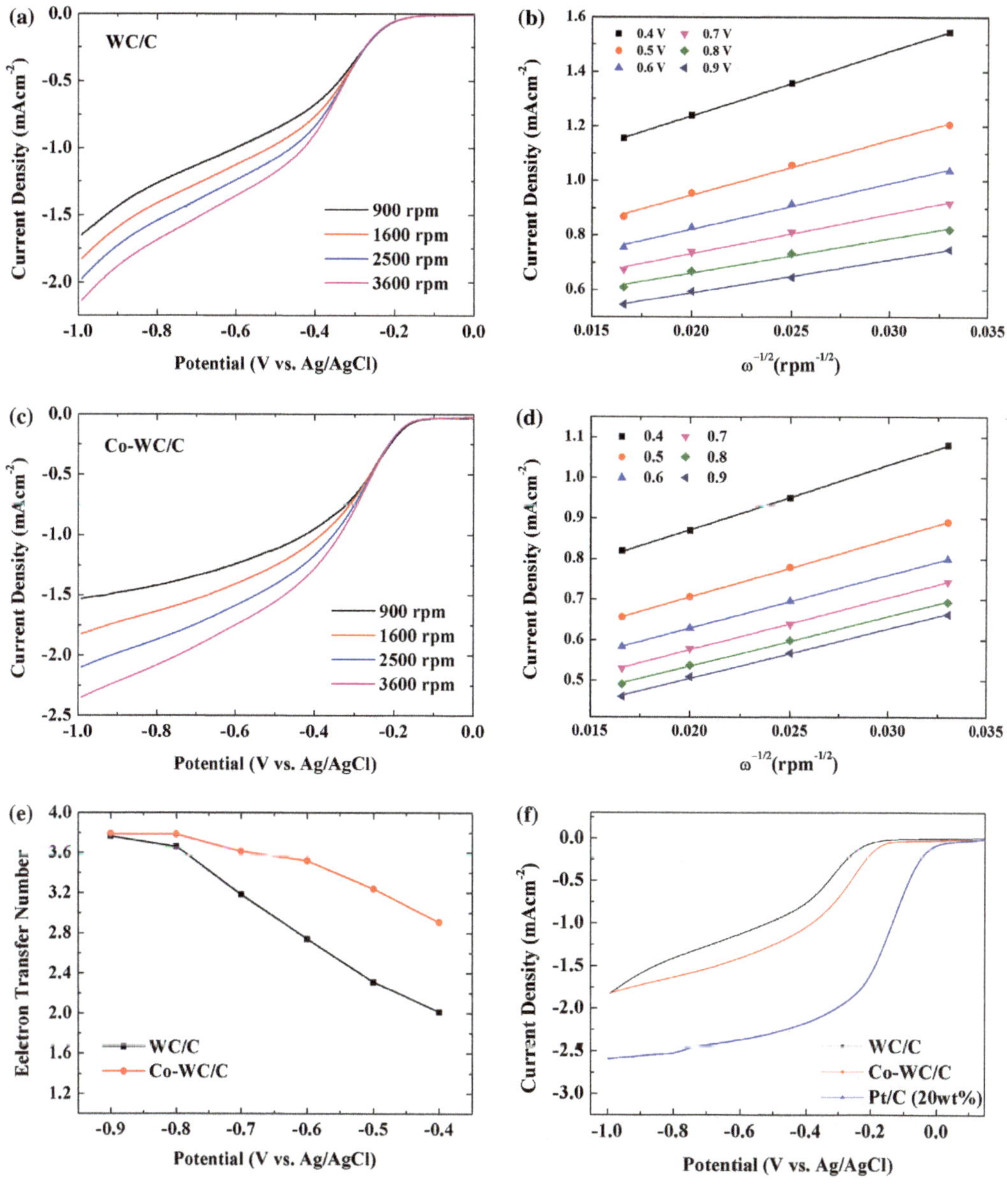

Figure 8. (a, c) Linear sweep voltammetry measurements; (b, d) Koutecky–Levich plot; (e) transferred electron numbers for WC/C and Co-WC/C derived from RDE measurement; (f) linear sweep voltammetry at a rotation rate of 1600 rpm and a scan rate of 10 mV s^{-1}.

dried and measured by XRD. Figure S1 shows X-ray diffraction patterns of filtered discharged solution, revealing the lack of NP peaks. It is supposed that all sputtered NPs from electrode were loaded on CNBs during discharging.

The pore structure properties of Co-WC/C, including the BET specific surface area, pore volume, and mean pore diameter were obtained by N_2 adsorption–desorption (Figure 6). The isotherms exhibited type IV and H1 type hysteresis loop characteristics according to the IUPAC classification. [22,23] The total surface area, total pore volume, and mean pore size were 367 m^2 g^{-1}, 1.56 cm^3 g^{-1}, and 8.56 nm, respectively. From these results, it is assumed that the synthesized carbon black presented a highly mesoporous structure; this structure promotes the diffusion of chemical species into the inner region of the carbon, leading to the active use of the catalyst on its surface.

The electrochemical performance of Co-WC/C was evaluated and discussed. Figure 7 shows the cyclic voltammetry (CV) data for the synthesized Co-WC/C, WC/C, and commercial Pt-supported (Pt/C) material under oxygen saturated conditions with 0.1 M KOH electrolyte.

The oxygen reduction potential of Co-WC/C was found to be in the range of −0.09 to −0.35 V. A detailed comparison first focused on WC/C and Co-WC/C. For WC/C, the maximum ORR peak and the on-set potential corresponding to ORR were observed at −0.40 V and −0.23 V, respectively. In the case of Co-WC, the maximum ORR peak and on-set potential were positively shifted to −0.26 V and −0.09 V, respectively. The results of cyclic voltammetry measurements showed that the cobalt nanoparticle promoted WC/C electrocatalyst was very active for ORR, with an onset potential of −0.09 V versus Ag/AgCl, which is over 140 mV better than the value for pure WC/C. However, in comparison with that of the Pt/C catalyst, the ORR activity of Co-WC/C was still incompatible with that of Pt.

To gain further insight into the ORR electrochemical process of the Co-WC/C sample, LSV measurements were carried out using a rotating disk electrode (RDE) in O_2 saturated 0.1 M KOH aqueous solution at a scan rate of 10 mV s^{-1}. This was performed at different rotation speeds ranging from 900 to 3600 rpm. Figure 8(a) and (c) display the RDE LSV for WC/C and Co-WC/C. The current density of Co-WC/C was significantly enhanced with increased rotation rate (from 900 to 3600 rpm); this enhancement can be ascribed to the increased diffusion rate at high speeds. The kinetic and diffusion mechanisms of the reaction can be explained by the Koutecky–Levich equations using the obtained parameters [24]:

$$1/j = 1/jL + 1/jk$$

$$1/j = 1/B\omega^{1/2} + 1/jk$$

$$B = 0.62nFCo * Do2/3v^{-1/6}$$

where j is the measured current density, j_k and j_L are the kinetic and diffusion-limiting current densities, B is the Levich current (A), ω is the rotation rate of the disk in radians ($\omega = 2\pi N$, where N is the linear rotation speed), F is the Faraday constant (F = 96 485 C), n is the number of electrons transferred per oxygen molecule, Co is the saturated concentration of O_2 in the electrolyte (1.2×10^{-6} mol cm^{-3}), Do is the diffusion coefficient of O_2 in the solution (1.9×10^{-5} cm^2 s^{-1}), and v is the kinematic viscosity of the electrolyte (1.1×10 cm^{-2} s^{-1}). The corresponding Koutecky–Levich plots (j^{-1} versus $\omega^{-1/2}$) at various electrode potentials (Figure 8(b) and (d)) were obtained from LSV. From the slope of the plot, the number of electrons (n) transferred per oxygen molecule of WC/C is calculated to be 2.3 at 0.5 V, suggesting that the ORR process proceeds through a dominant two-electron pathway. The Co-WC/C exhibits the highest n value of 3.23 at 0.5 V. This result indicates that the ORR process on Co-WC/C proceeds preferentially via parallel two-electron and four-electron pathways. The above results reveal that the incorporation of Co nanoparticles into the WC composite improves both the ORR activity and electron transfer in an alkaline medium.

It is predicted that Co nanoparticles neighboring WC/C increase the kinetic property by catalyzing the 2 × 2 serial oxygen reduction reactions in alkaline media

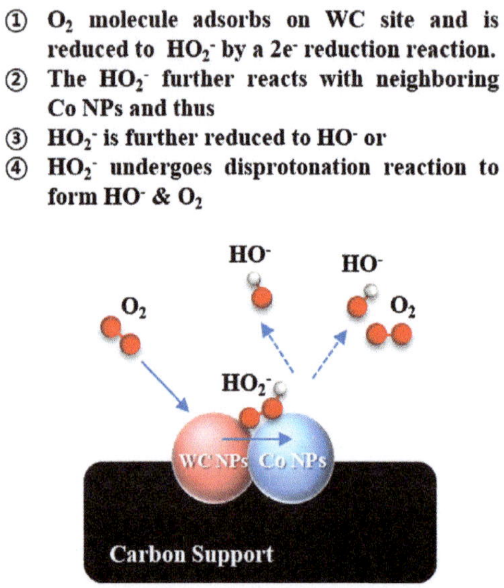

Figure 9. Proposed dual site ORR mechanism on Co-WC/C.

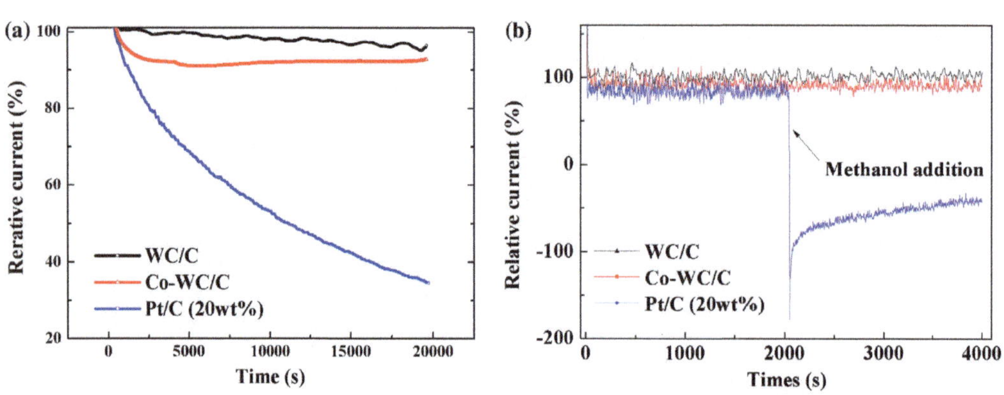

Figure 10. Chronoamperometric responses of Co-WC/C compared with that of Pt/C catalyst in O_2-saturated 0.1 M KOH at −0.4 V (a) without and (b) with methanol.

(Figure 9). Co addition in WC/C has beneficial effects on ORR activity by formation of active sites such as (1) WC NPs sites, and (2) Co particles, according to dual site mechanism in alkaline media [25].

The comparison of LSV curves at 1600 rpm for WC/C, Co-WC/C, and 20% Pt/C is shown in Figure 8(f). In comparison with WC/C, the half wave potential positively shifted from 0.43 V to 0.32 V in the presence Co NPs, which agreed with the examination from CV. A positive shift in the onset potential points out that the ORR of the Co-WC/C is more favorable than that of WC/C. However, it is still inferior to the state-of-art commercial 20% Pt/C in terms of both the onset potential and current density.

Another important property of catalysts for current fuel cell applications is their durability in long-term operation. Figure 10(a) shows the stability of WC/C, Co-WC/C, and commercial Pt/C catalysts, which were tested at –0.4 V (mass transport region) for 20,000 s in O_2 saturated 0.1 M KOH solution at a rotation speed of 1600 rpm. Evidently, WC/C and Co-WC/C showed a high durability for long-term operation after continuous reaction for 20,000 s, with relative current remaining at over 90% of the initial current. On the other hand, an obvious decrease in current was observed for the commercial Pt/C catalyst. To further assess the tolerance to the methanol crossover effect, 3 M methanol (CH_3OH, purity 99.8%, Kanto Chemical) was introduced at 2000 s into an O_2-saturated 0.1 M KOH solution during the chronoamperometric measurement at –0.4 V (Figure 10(b)). After the introduction of 3 M methanol, commercial Pt/C showed an obvious current drop corresponding to the methanol oxidation reaction, whereas that of WC/C and Co-WC/C remains unchanged and stable without degradation. From these results, we can conclude that Co-WC/C nanocomposites presented a higher durability than that of commercial Pt/C catalysts in an alkaline medium.

4. Conclusions

Co-WC/C nanocomposites were first synthesized by SPP as non-precious metal electrocatalysts for ORR. Electrochemical measurements in an alkaline medium revealed that the Co NPs affected the ORR activity. The introduction of Co NPs led to not only superior ORR activity in terms of onset potential and peak potential, but also to a more efficient electron transfer process compared to that of pure WC/C. Compared to commercial 20% Pt/C, Co-WC/C also exhibits long-term durability for ORR. Therefore, we believe that the introduction of Co NPs is an effective method for enhancing the ORR activity of pure WC; also, Co-WC/C could be a promising alternative ORR catalyst for application to alkaline fuel cell devices. In addition, SPP shows a potential that can be extended to the synthesis of other non-noble metal nanoparticles simply by changing the electrodes. We believe that further development using SPP in this field will facilitate and accelerate the application of non-noble metal ORR electrocatalysts for advanced energy conversion devices.

Disclosure statement

No potential conflict of interest was reported by the authors.

Notes on contributors

Jun Kang is an assistant professor and received his doctor of Engineering. The author's research interests include Nano Materials, Carbon Materials, and Plasma in Liquid. His recent publications include Synthesis of structure-controlled carbon nano spheres by solution plasma process. Carbon. 2013;60:292–298; Hierarchical meso–macro structure porous carbon black as electrode materials in Li–air battery. J Power Sources. 2014;261:156–161; A simple synthesis method for nano-metal catalyst supported on mesoporous carbon: the solution plasma process. Nanoscale. 2013;5:6874– 6882; In situ solution plasma synthesis of mesoporous nanocarbon-supported bimetallic nanoparticles RSC Advances 2015 ; 5(37): 29131-29134.

Hye-Min Kim is a doctoral student. The author's research interests include Carbon Material and Metal Oxide The author's recent publication includes Synthesis of colloidal MnO 2 with a sheet-like structure by one-pot plasma discharge in permanganate aqueous solution. RSC Advances. 2016; 6: 2826-2834.

Nagahiro Saito is a full professor and received his doctor of Engineering. The author's research interests include Plasma in Liquid and Nano materals. His recent publications are Synthesis process of gold nanoparticles in solution plasma. Thin Solid Films. 2009; 518 (3), 912-917; Size-controlled gold nanoparticles synthesized in solution plasma. The Journal of Physical Chemistry C. 2011; 115 (50), 24569-24576.

Myeong-Hoon Leeis a full professor and received his doctor of Engineering. The author's research interests are Electrochemistry, Corrosion control and Surface coating. His recent publications are Influence of annealing temperatures on corrosion resistance of magnesium thin film-coated electro-galvanized steel Modern Physics Letters B. 2015 ; 29,1540015-1~1540015-5; Polarization characteristics of four types of coating films by thermal spray in seawater solution. Modern Physics Letters B. 2015; 29, 1540019-1~1540019-6.

References

[1] Gamburzev S, Atanasov P, Wilkins E. Performance of glucose biosensor based on oxygen electrode in physiological fluids and at body temperature. Sensor Actuat B. 1996; 30: 179–183.

[2] Goux A, Pauporté T, Lincot D. Oxygen reduction reaction on electrodeposited zinc oxide electrodes in KCl solution at 70°C Electrochim. Acta. 2006;51: 3168–3172.
[3] Steele BCH, Heinzel A. Materials for fuel-cell technologies. Nature. 2001;414:345–352.
[4] Weidenkaff A, Ebbinghaus SG, Lippert T. $Ln_{1-x}A_xCoO_3$ (Ln = Er, La; A = Ca, Sr)/carbon nanotube composite materials applied for rechargeable Zn/Air batteries. Chem Mater. 2002;14:1797–1805.
[5] Basu S. 2007. Recent trends in fuel cell science and technology : New Delhi : Springer and Anamaya; p. 260.
[6] Stonehart, P. in Drake, J. 1994. Electrochemistry and Clean Energy : Cambridge : The Royal Society of Chemistry.
[7] Levy RB, Boudart M. Platinum-like behavior of tungsten carbide in surface catalysis. Science. 1973;181:547–549.
[8] Nikolov I, Nikolova V, Vitanov T. The effect of method of preparation on the catalytic activity of tungsten carbide. J Power Sources. 1979;4:65–75.
[9] Venkataraman R, Kunz HR, Fenton JM. Development of new CO tolerant ternary anode catalysts for proton exchange membrane fuel cells. J Electrochem Soc. 2003;150:A278–A284.
[10] Chen Z, Higgins D, Yu A, et al. A review on non-precious metal electrocatalysts for PEM fuel cells. Energ Environ Sci. 2011;4: 3167–3192.
[11] Oyama ST. Preparation and catalytic properties of transition metal carbides and nitrides. Catal Today. 1992;15:179–200.
[12] Oyama ST, Schlatter JC, Metcalfe JE, et al. Preparation and characterization of early transition-metal carbides and nitrides. Ind Eng Chem Res. 1988;27:1639–1648.
[13] Li M, Bo X, Zhang Y, et al. Cobalt and nitrogen co-embedded onion-like mesoporous carbon vesicles as efficient catalysts for oxygen reduction reaction, J Mater Chem A. 2014;2:11672–11682.
[14] Su Y, Zhu Y., Jiang H, et al. Cobalt nanoparticles embedded in N-doped carbon as anefficient bifunctional electrocatalyst for oxygen reduction and evolution reactions. Nanoscale. 2014;6:15080–15089.
[15] Wang S, Cui Z, Cao M. A template-free method for preparation of cobalt nanoparticles embedded in N-doped carbon nanofibers with a hierarchical pore structure for oxygen reduction. Chem-Eur J. 2015;21:2165–2172.
[16] Kang J, Li OL, Saito N. Synthesis of structure-controlled carbon nano spheres by solution plasma process. Carbon. 2013;60:292–298.
[17] Kang J, Li OL, Saito N. Hierarchical meso-macro structure porous carbon black as electrode materials in Li-air battery. J Power Sources. 2014;261:156–161.
[18] Kang J, Li OL, Saito N. A simple synthesis method for nano-metal catalyst supported on mesoporous carbon: the solution plasma process. Nanoscale. 2013;5:6874–6882.
[19] Kim DW, Li OL, Saito N. The role of the central Fe atom in the N 4-macrocyclic structure for the enhancement of oxygen reduction reaction in a heteroatom nitrogen-carbon nanosphere. Phys Chem Chem Phys. 2014;16:14905–14911.
[20] Kim DW, Li OL, Saito N. Enhancement of ORR catalytic activity by multiple heteroatom-doped carbon materials. Phys Chem Chem Phys. 2015;17;407–413.
[21] Azaroff LV. Elements of X-Ray crystallography. NewYork, USA: McGraw-Hill; 1968; p. 549.
[22] Rouquerol J, Avnir D, Fairbridge CW, et al. Physical and biophysical chemistry division commission on colloid and surface chemistry including catalysis. Pure Appl Chem. 1994;66:1739–1758.
[23] Kao HM, Shen TY, Wu JD, et al. Control of ordered structure and morphology of cubic mesoporous silica SBA-1 via direct synthesis of thiol-unctionalization, Microporous Mesoporous Mater. 2008;110:461–471.
[24] Liu R, Wu D, Feng X, et al. Nitrogen-doped ordered mesoporous graphitic arrays with high electrocatalytic activity for oxygen reduction, Angew. Chem Int Ed. 2010;49:2565–2569.
[25] Olson TS, Pylypenko S, Atanassov P. Anion-exchange membrane fuel cells: dual-site mechanism of oxygen reduction reaction in alkaline media on cobalt-polypyrrole electrocatalysts. J Phys Chem C. 2014;114:5049–5059.

11

Facile preparation of acid-resistant magnetite particles for removal of Sb(III) from strong acidic solution

Dong Wang[a,b], Kaiwen Guan[a,b], Zhiping Bai[a] and Fuqiang Liu[b]

[a]State Key Laboratory of Coordination Chemistry, School of Chemistry and Chemical Engineering, Nanjing University, Nanjing 210023, P.R. China
[b]State Key Laboratory of Pollution Control and Resource Reuse, School of the Environment, Nanjing University, Nanjing 210046, P.R. China

ABSTRACT
A new facile coating strategy based on the hydrophobicity of methyl groups was developed to prevent nano-sized magnetite particles from strong acid corrosion. In this method, three steps of hydrolysis led to three layers of protection shell coating Fe_3O_4 nanoparticles. Filled with hydrophobic methyl groups, the middle layer mainly prevented the magnetic core from strong acid corrosion. These magnetite particles managed to resist 1 M HCl solution and 2.5 M H_2SO_4 solution. The acid resistant ability was higher than those reported previously. After further modification with amino-methylene-phosphonic groups, these magnetite particles successfully adsorbed Sb(III) in strong acid solution. This new strategy can also be applied to protect other materials from strong acid corrosion.

KEYWORDS
Magnetite nanoparticles; acid-resistant; sol-gel; adsorption

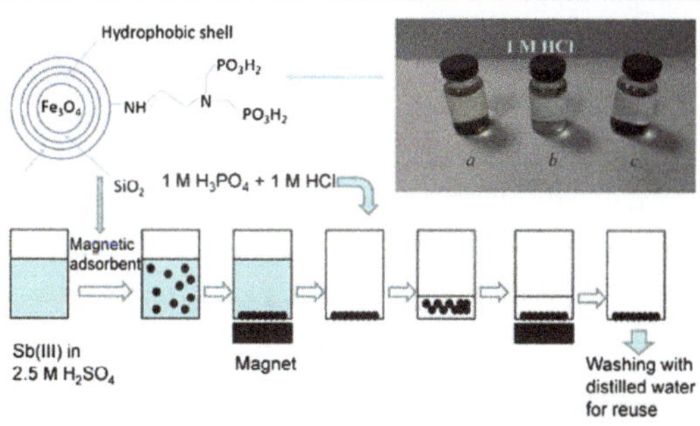

1. Introduction

In the past few years, magnetic solid phase extraction (MSPE) with magnetic particles as adsorbents has been applicable to metal removal and trace-metal analysis [1–5]. Compared with conventional solid phase extraction, sample pretreatment is substantially simplified by employing magnetic adsorbents because there is no need for packing a column with sorbents in batch operation. Instead, phase separation can be quickly and easily accomplished by an external magnetic field. Among the magnetic particles used in MSPE, micro- and nano-sized magnetite (Fe_3O_4) particles, which can be dissolved in strong acid solutions (pH < 1), are most popular [6].

Magnetite particles have been most commonly coated by silica gel, considering mechanical and chemical stabilities as well as thermal resistance under various conditions [7]. In addition, this method enjoys easy surface modification and control of interparticle interactions [8,9]. However, silica gel-coated Fe_3O_4 particles still cannot be applied in strong acid conditions because Fe_3O_4 cores are vulnerable to corrosion. Li et al. [10] and Liu et al. [11] prevented Fe_3O_4 cores from acid corrosion by employing dense liquid coating which required the reaction temperature to be maintained at 90°C. It is not convenient for coating Fe_3O_4 particles though, as nano-sized Fe_3O_4 particles are prone to conversion into Fe_2O_3 at high temperature when exposed to the atmosphere [12–14]. Besides, although these silica-coated materials can stand 0.1 M HCl, strong acid conditions are still troublesome. Therefore, the challenge remains to develop a suitable coating strategy that protects magnetite particles from strong acid corrosion by silica gel.

CONTACT Zhiping Fuqiang Bai Liu ✉ baizp@nju.edu.cn; jogia@163.com

In recent years, silica-coated Fe_3O_4 particles have been modified with various functional groups to remove different metal ions from wastes and water, such as Pd(II) [15], Cr(III) [16–18], Pb(II) [19,20], Ag(I) [21] and Cd(II) [22]. Since the existing coating technologies using silica gel fail to resist acids, corresponding magnetite particles have not been successfully used to adsorb dissociated metal ions under strong acid conditions.

On the other hand, some kinds of metal ion need to be removed from strong acid solutions or waste liquids simultaneously in industrial practice. For instance, removal of antimony plays an essential role in treating electrolytes of copper electro-refining [23,24]. The electrolyte is a solution of copper sulfate and sulfuric acid whose nominal composition is 40–45 g l^{-1} copper and 180–200 g l^{-1} sulfuric acid. In the past, numerous materials have been successfully developed to separate antimony species from water, such as hydroxyapatite [25], graphene [26], diatomite [27], goethite [28], bentonite [29], Fe-Mn binary oxide [30], raw and modified multi-walled carbon nanotubes [31] and hematite-coated magnetic nanoparticles [32]. However, these materials may not be suitable for removing Sb in copper refining due to the strong acid conditions of electrolyte and the demand for recycling. Meanwhile, silica gel-coated magnetic nanoparticles have not been applied in strong acid conditions either. In the industry, some ion-exchange resins possessing an amino-phosphonic acid functional group, including Eporus MX-2 (Miyoshi Oil and Fat Company, Tokyo, Japan), Unitika UR-3300 (Unitika Ltd, Osaka, Japan) and Duolite C-467 (Rohm and Haas Company, Philadelphia, USA), have been used to remove Sb and then recycled by being eluted with 6 M HCl [24]. Therefore, developing a new strategy that coats magnetite (Fe_3O_4) particles by silica gel is bound to expand the application of MSPE by preventing strong acid corrosion.

In this study, a new facile coating method based on the traditional sol-gel process was developed and operated at only 30°C, depending significantly on the hydrophobicity of methyl groups. After coating, this material managed to stand the corrosion by 1 M HCl solution for 24 h. To prove the practicability of the acid-resistant magnetite nanoparticles, amino-methylene-phosphonic groups were fixed onto their surfaces to adsorb Sb(III) in acid condition. These groups have been applied to remove Sb(III) from copper electro-refining electrolytes [33]. As we expected, this novel magnetic adsorbent successfully removed Sb(III) from 2.5 M H_2SO_4 solution, leaving no ferric ions in the supernatant and keeping the cores stable.

2. Experimental section

2.1. Materials

$FeSO_4 \cdot 7H_2O$, $Fe_2(SO_4)_3$, NaOH, HCl, H_2SO_4, formaldehyde and tetraethyl orthosilicate (TEOS) were all guaranteed reagents from Nanjing Chemical Reagent Co., Ltd. (Nanjing, China) Dimethyldiethoxysilane (DMDES), (3-aminopropyl China)triethoxysilane (APTES), 2-choroethylamine hydrochloride and Sb_2O_3 were purchased from Aladdin Industrial Corporation (Ontario, Canada) Sb_2O_3 was dissolved in concentrated sulfuric acid to prepare Sb(III) solution. All the chemicals were used directly without any further purification. Nano-sized ultrafine magnetic Fe_3O_4 powders were prepared by the coprecipitation method [14]. Referring to a previous work [34], N-(2-chloroethyl)iminobis(methylene)diphosphonic acid was synthesized from 2-choroethylamine hydrochloride with a yield of 87%.

2.2. Preparation of ultrafine magnetic Fe_3O_4 powders

$FeSO_4 \cdot 7H_2O$ (5.6 g) and $Fe_2(SO_4)_3$ (8 g) were dissolved into 100 ml of distilled water and stirred until the mixture became an orange colloid. Then 15 g NaOH was added into 200 ml of distilled water with strong stirring. When the NaOH solution was cooled down to room temperature, the orange colloid was dropwise added over one hour with continuous stirring. After the reaction, the black mixture was separated by being centrifuged at 5000 rpm. Then, the black precipitate was washed three times with distilled water (100 ml). Finally, the ultrafine Fe_3O_4 powders were dispersed in 150 ml of pure ethanol for later use.

2.3. Coating Fe_3O_4 powders by sol-gel process

In a typical experiment, a mixture of Fe_3O_4 powders dispersed in 150 ml of ethanol was sonicated for 30 min first. Then 25 ml of distilled water and 1 ml of aqueous ammonia were added into the mixture successively with strong stirring to adjust pH. Meanwhile, before the reagents were added, the temperature of the water bath was adjusted to 30°C and maintained until the experiment ended. Fifteen minutes later, 10 ml of TEOS was injected into the mixture, followed by reaction for one hour. Before further treatment, a small amount of particles were separated from the reaction system as sample *a*. After 1 ml of aqueous ammonia was injected again, 10 ml of TEOS and 6 ml of DMDES were added into the flask, and the reaction was maintained for 4 h. Then, by employing an external magnetic field, the black powders were easily separated from the mixture, and washed three times with ethanol (50 ml). Next, these cleaned magnetic powders were dispersed into 150 ml of ethanol and 25 ml of pure water again. After 1 ml of aqueous ammonia and 6 ml of DMDES were injected in sequence, the reaction was maintained under strong stirring for 24 h at 30°C. Finally, the magnetic Fe_3O_4 particles were successfully coated with a layer of hydrophobic silica. Magnetic field allowed easy phase separation and washing of these particles. After this coating process, the samples with a structure of Fe_3O_4@SiO_2@SiO-Me were marked as sample *b*.

2.4. Modification of amino group

Previous magnetic Fe_3O_4@SiO_2@SiO-Me particles were dispersed into 150 ml of ethanol and 25 ml of pure water again, into which was added 1 ml of aqueous ammonia under strong stirring at 30°C. Then 10 ml of TEOS was added into the flask. Twenty minutes later, 4 ml of APTES was poured into the flask and the reaction was maintained for 4 h. Last, the products marked as sample *c* with a structure of Fe_3O_4@SiO_2@SiO-Me@SiO-NH_2 were separated by external magnetic field and washed by ethanol (50 ml) twice and distilled water (100 ml) three times in order.

2.5. Modification of amino-methylene-phosphonic group

N-(2-chloroethyl)iminobis(methylene)diphosphonic acid (1 g) was dissolved in 50 ml of distilled water at below 10°C. Then, the acid solution was neutralized by adding 1 M NaOH solution dropwise until pH = 7. After this solution was poured into a flask, 5 g magnetic Fe_3O_4@SiO_2@SiO-Me@SiO-NH_2 was dispersed into the solution and heated to 70°C with stirring. One hour later, the flask was evacuated, with the temperature maintained at 70°C. After vacuum drying, the products were washed by 20 ml of 1 M HCl twice and distilled water (50 ml) three times in sequence. Finally, the products were dried in a vacuum drying oven at 50°C.

2.6. Characterization of coated magnetite particles

Morphology and size of coated magnetite particles were determined by transmission electron microscopy (TEM, JEM-200CX, JEOL Ltd, Tokyo, Japan). Zeta potentials of the prepared sorption material were measured by Zetasizer Nano-Z (Malvern Instruments, Worcestershire, UK). Element compositions of the shell were investigated by elemental analyzer (Heraeus CHN-O-Rapid Heraeus Group, Hanau, Germany). The X-ray diffraction (XRD) patterns were collected on a D8 ADVANCE diffractometer with Cu Kα radiation (Bruker Corporation, Billerica, USA). An ASAP2020 physisorption analyzer (Micromeritics, Norcross, USA) was employed to measure the specific surface areas.

2.7. Leaching test

To examine the coverage and property of the coating, 0.05 g Fe_3O_4@SiO_2@SiO-Me@SiO-NH_2 particles were dispersed into 25 ml of 1 M HCl acid solution. Twenty-four hours later, the suspension was separated by external magnetic field. The supernatant was collected and analyzed for ion concentration by atomic absorption spectroscopy (Hitachi 180-80, Hitachi Ltd, Tokyo, Japan).

2.8. Batch adsorption experiments

In batch adsorption experiments, 0.05 g Fe_3O_4@SiO_2@SiO-Me@SiO-NH_2 particles modified with amino-methylene-phosphonic groups were dispersed into 25 ml of Sb(III) solution with 2.5 M H_2SO_4 in a 100 ml conical flask under ultrasonic wave for 2 min. The flasks were shaken at 100 rpm and 25°C in a constant-temperature shaking table for 6 h to reach equilibrium. Adsorption isotherm at 25°C was obtained by varying the initial Sb(III) concentration from 0.01 to 0.5 g l^{-1}. After reaching the adsorption equilibrium, the mixture was separated by external magnetic field to collect supernatant. The Sb(III) concentration of supernatant was analyzed by atomic absorption spectroscopy.

2.9. Selective adsorption experiments

To investigate the adsorption performance in copper electro-refining electrolytes, a series of solutions containing 14.115 g l^{-1} Cu(II) and 2.5 M H_2SO_4 with different concentrations of Sb(III) were prepared. In a typical experiment, 0.05 g adsorbent was dispersed into 25 ml of simulated electrolytes in a conical flask. The flasks were shaken at 100 rpm and 25°C in a constant-temperature shaking table for 6 h to reach equilibrium. Then, after the suspension was separated by magnetic field, the supernatant was obtained to analyze the contents of Cu(II) and Sb(III) respectively.

3. Result and discussion

3.1. Characterization

To investigate the shell structure of Fe_3O_4@SiO_2@SiO-Me@SiO-NH_2, these particles were mixed with 6 M HCl at room temperature to remove the magnetic core, leaving white shell powders. The shells without cores were also determined by TEM and compared with normal samples. As shown in Figure 1(a), the magnetic cores with irregular morphologies are sized 10–20 nm. The morphologies of the shells without magnetic cores are exhibited in Figure 1(b). Obviously, the shell, with the core retained, was about 10 nm thick, which may partly be responsible for protecting the magnetic core from strong acid corrosion. In order to clarify the increase in particle size and to evidence the shell formation in different coating steps, samples *a* and *b* were also detected by TEM, as shown in Figure 1(c) and (d). The shell thickness of sample *b* (nearly 5 nm) resembled that of sample *a*, suggesting that coating of methyl groups did not significantly raise the particle size. SiO_2 shell coating and amino group modification markedly increased the

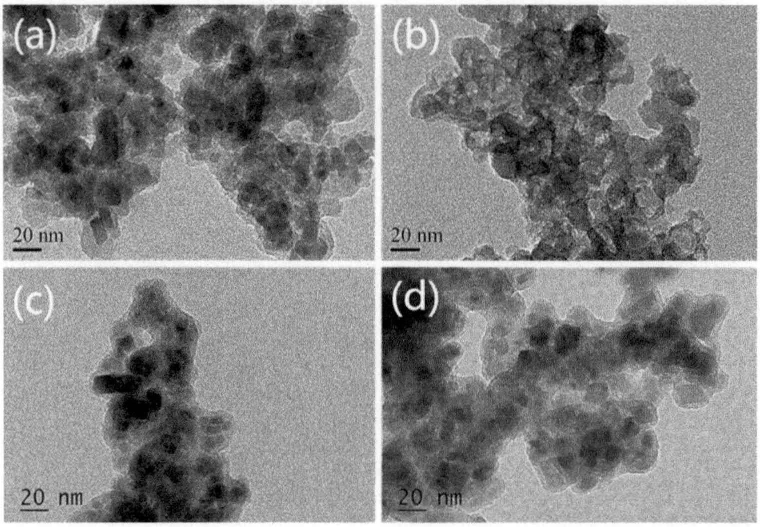

Figure 1. TEM images of (a) acid-resistant magnetic Fe_3O_4 nanoparticles; (b) shell of the sorbent; (c) sample *a*; (d) sample *b*.

Table 1. Elemental analysis (in wt%) of samples ***a-e***.

Sample	C	H	N
a	0.31	0.5	0.08
b	1.42	0.8	0.07
c	6.89	2.2	2.13
d	1.51	0.9	0.08
e	1.89	1.2	0.09

shell thickness and the particle size (Figure 1(a), (c) and (d)). Although the hydrophobic shell played a main role in resisting acids, it could hardly be distinguished by TEM. Therefore, whether the particles had been coated with the hydrophobic methyl groups remained uncertain, for which elemental analysis was performed. To investigate the structure of the coated magnetic Fe_3O_4 particles, samples *a–c* were collected successively from the reaction system after SiO_2 shell coating, hydrophobic shell coating and amino group modification. The elemental analysis results of samples *a–c* are listed in Table 1.

In sample *a*, the total content of elements C and H that may originate from residual ethanol and water was below 1%. Significant increase in the mass percent of element C in sample *b* demonstrated that methyl groups had been successfully modified on the surface of magnetic particles after coating of hydrophobic shell by using DMDES. In other words, the particles can be coated with the hydrophobic shell containing methyl groups by our method. But the increase range was not enough to reflect the increase in shell thickness. The C and N contents of sample *c* both obviously exceeded those of samples *a* and *b*. This fact implies the shell was rapidly thickened during modification of amino group, agreeing with the TEM result. According to the above analysis, the acid-resistant magnetic Fe_3O_4 nanoparticles were structurally identified (Figure 2).

To evaluate the effects of experimental operation on hydrophobic shell coating with methyl groups, other two comparison experiments were designed. In one of them, during the coating of hydrophobic methyl groups, 10 ml of TEOS and 6 ml of DMDES were added into the reaction system, 30 min after which another 6 ml of DMDES was added, and the reaction was then maintained for 4 h. The samples coated by this method were marked as sample *d*. In the other experiment, the magnetite particles were coated through hydrolysis of 10 ml of TEOS and 6 ml of DMDES in the same way twice during the coating of hydrophobic protection shell, giving sample *e*. Like samples *a–c*, samples *d* and *e* were also detected by elemental analysis.

During the coating of methyl groups, DMDES needs to be hydrolyzed with TEOS at the same time, because DMDES alone can hardly fix methyl groups on the surface of magnetite particles. On the other hand, the elemental analysis results of samples *b* and *d* showed that another 6 ml of DMDES did not markedly increase the carbon content. In other words, methyl groups were barely fixed on the particle surface when only DMDES hydrolyzed. Meanwhile, the elemental analysis results of samples *e* and *b* indicated that hydrolysis of DMDES with TEOS twice brought more hydrophobic methyl groups than those in sample *b*. However, the second hydrolysis did not increase the carbon content as the first one did.

As shown in Figure 1(a), all the coated magnetite particles seem to aggregate. In order to detect the dispersity of the particles, the specific surface area was measured before and after coating by BET method. In theory, the specific surface area of pure ultrafine Fe_3O_4 powders without coating with a diameter of 10 nm reaches 116 $m^2 g^{-1}$, and the experimental value was

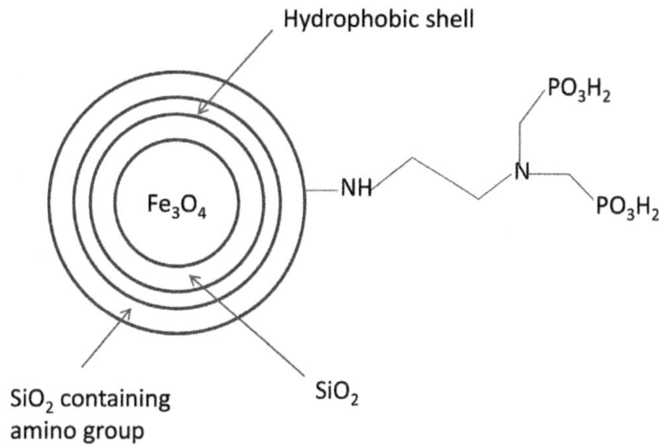

Figure 2. Schematic diagram of the structure of acid-resistant magnetic Fe_3O_4 nanoparticles.

107 $m^2 g^{-1}$, suggesting that the magnetite powders used in our study had a high dispersity. Meanwhile, the magnetite particles after coating had a specific surface area of 60 $m^2 g^{-1}$. Given that the surface area was inversely proportional to the particle diameter, the dispersity was still acceptable.

These acid-resistant magnetite particles were herein applied in a strong acid condition with 2.5 M H_2SO_4, so the zeta-potential of amino-methylene-phosphonic group-modified particles was measured in the same solution. As suggested by the actual value (+52.7 mV), these particles were quite stable in this condition.

3.2. Leaching test

After sample *c* was immersed in 1 M HCl solution for 24 h, atomic absorption spectroscopic results showed that the supernatant contained no ferric ions, verifying that these magnetite nanoparticles coated with hydrophobic methyl groups managed to resist 1 M HCl solution, which was notably superior to the outcomes of previous studies. Hence, our coating method may render magnetic nano- or micro-sized metal oxide particles applicable to strong acid conditions.

Samples *a*, *b* and *c* (0.01 g each) were respectively dispersed into 7 ml of 1 M HCl solution, 24 h after which color changes of the supernatants were observed (Figure 3). The supernatant of sample *a* turned yellow, indicating that ferric ions were leached into the solution. In other words, coating only one layer of SiO_2 failed to prevent the cores from corrosion by strong acid. Nevertheless, the innermost layer of SiO_2 was essential and indispensable for further coating. Furthermore, sample *b* kept floating above the acid solution, so the outside was full of hydrophobic methyl groups that hindered the particles from being dispersed into the acid solution. Since the supernatant of sample *c* was still colorless after 24 h, it had high acid resistance. In addition, the outmost layer, which was full of amino groups, benefited from further modification and generated a hydrophilic outside surface as well.

As mentioned above, copper electro-refining electrolyte usually contains 180-200 g l^{-1} sulfuric acid. Hence, we also used 2.5 M H_2SO_4 solution to test our acid-resistant magnetite particles. As expected, these magnetite particles stood the erosion of 2.5 M H_2SO_4 solution, because there were no ferric ions in the supernatant. For the sake of recycling, the coated magnetic Fe_3O_4 particles may have to resist the corrosion by stronger acid, such as 3 M HCl solution. The concentration of iron ions in the supernatant reached 1.7 mg l^{-1} one hour after these coated magnetite particles were immersed in 3 M HCl. Two hours later, such concentration rapidly increased to 15.6 mg l^{-1}, meaning that 3 M HCl penetrated the shell to erode the magnetic core. Three hours later, the color of the supernatant, as observed by naked eyes, turned green. Although these coated magnetite particles could not resist 3 M HCl solution, the leaching concentration of iron ions at one hour was acceptable, i.e. the leached ferric ions were negligible during short-term contact.

To prove the stability of coated magnetic particles in strong acid conditions, a powder X-ray diffraction study was performed to investigate the changes of the adsorbent before and after contacting with acid solution. As shown in Figure 4, the XRD pattern of acid-resistant magnetic Fe_3O_4 nanoparticles shows characteristic (220), (311), (400), (422), (511) and (440) peaks. The XRD pattern is in good agreement with the data for cubic Fe_3O_4 as reported in the JCPDS card (No. 88-315, $\alpha = 8.375$). Thus, this coating strategy did not undermine the property of the magnetic cores. On the other hand, the coated magnetite particles after contacting with 1 M HCl for 24 h have similar peaks to those before contact (Figure 4(a) and (b)), indicating that the magnetic Fe_3O_4 cores were well retained after contact, and that the hydrophobic shells successfully protected the cores from strong acid corrosion. In addition, the coated magnetite particles after contacting with 3 M for 24 h were also

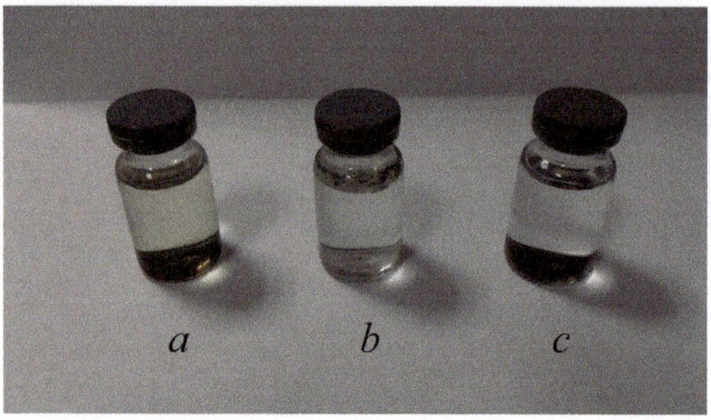

Figure 3. Samples *a*, *b* and *c* after being added in 1 M HCl solution for 24 h.

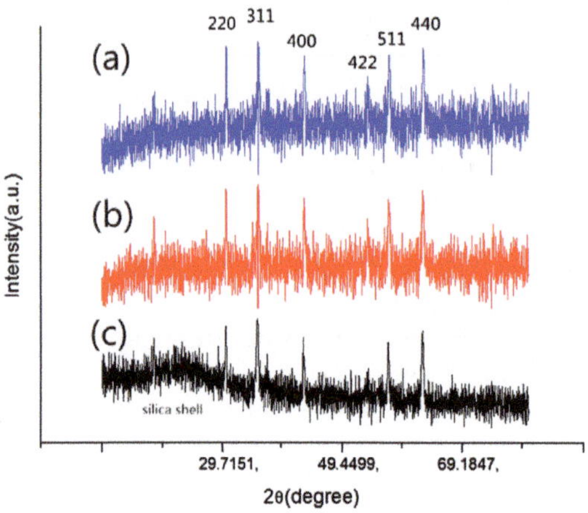

Figure 4. XRD patterns of (a) acid-resistant magnetic Fe_3O_4 nanoparticles; (b) acid-resistant magnetic Fe_3O_4 nanoparticles contacting with 1 M HCl for 24 h; (c) acid-resistant magnetic Fe_3O_4 nanoparticles contacting with 3 M HCl for 24 h.

detected by XRD (Figure 4(c)). Obviously, the pattern of the sample contacting with 3 M HCl is different from others. In other words, contacting with 3 M HCl solution for 24 h dissolved part of the magnetite cores. In fact, the black coated magnetite powders turned gray after such contact.

3.3. Effect of contact time

Fine powders are characterized by large specific surface areas and the resultant high adsorption speeds. In a previous work, however, the adsorbent with the same chelating group took 2 h to completely adsorb free metal ions in the acid solution. Thus, the effect of contact time was evaluated to clarify the adsorption equilibrium time of our magnetic Fe_3O_4@SiO_2@SiO-Me@SiO-NH_2 particles modified with amino-methylene-phosphonic groups. In a typical experiment, 0.15 g adsorbent was dispersed into 75 ml of Sb(III) solution with 2.5 M H_2SO_4 in a 300 ml conical flask. The initial concentration of Sb(III) was 133 mg l^{-1}. The conical flask was placed in a constant-temperature shaking table with a speed of 100 rpm at 25°C. At different time intervals, 1 ml of supernatant was collected to analyze the residual Sb(III) concentration (Figure 5).

In the beginning, the concentration of Sb(III) dropped quickly, implying the adsorption was fast initially. The adsorption of Sb(III) from 2.5 M sulfuric acid solution spent four hours to reach equilibrium. Given that the adsorption took 2 h of contact to remove most of antimony (III), a specific surface area of 60 $m^2\ g^{-1}$ did accelerate the adsorption of Sb(III) by using the nano-sized magnetite particles modified with amino-methylene-phosphonic groups.

3.4. Adsorption isotherm

Sb(III) adsorption isotherm of our acid-resistant magnetic nanoparticles in 2.5 M H_2SO_4 solution was obtained at 25°C by varying the initial concentration of Sb(III) from 0.01 g l^{-1} to 0.5 g l^{-1} (Figure 6). The adsorption isotherm was fitted with the Langmuir model and the Freundlich model as Equations (1) and (2), respectively.

$$q_e = \frac{q_m K_L C_e}{1 + K_L C_e} \tag{1}$$

$$q_e = K_F C_e^{1/n} \tag{2}$$

where C_e is the equilibrium concentration of Sb(III) in mg l^{-1}, and q_e is the amount of Sb(III) adsorbed at equilibrium in mg l^{-1}. In Equation (1), q_m (mg g^{-1}) indicates the adsorption capacity, whereas K_L (l mg^{-1}) is the Langmuir affinity constant related to the energy of adsorption. In Equation (2), K_F is defined as the Freundlich adsorption affinity, while n is an index related to the heterogeneity of the adsorbent surface.

The Langmuir isotherm [35] assumes that sorption occurs on a structurally homogeneous adsorbent and

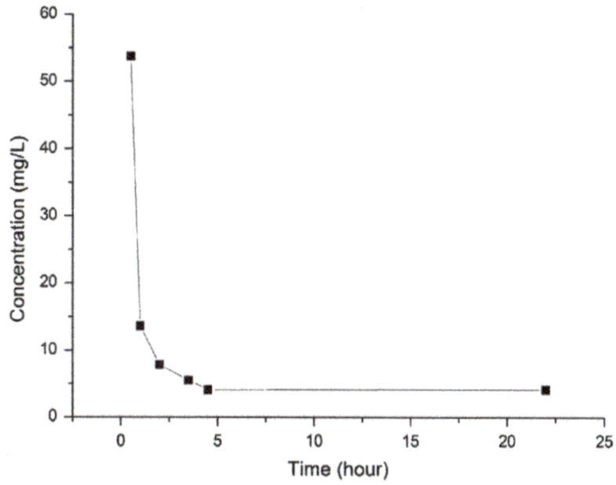

Figure 5. Removal of Sb(III) from 2.5 M H_2SO_4 solution. Initial concentration of Sb(III): 133 mg l^{-1}; adsorbent dosage: 2 g l^{-1}.

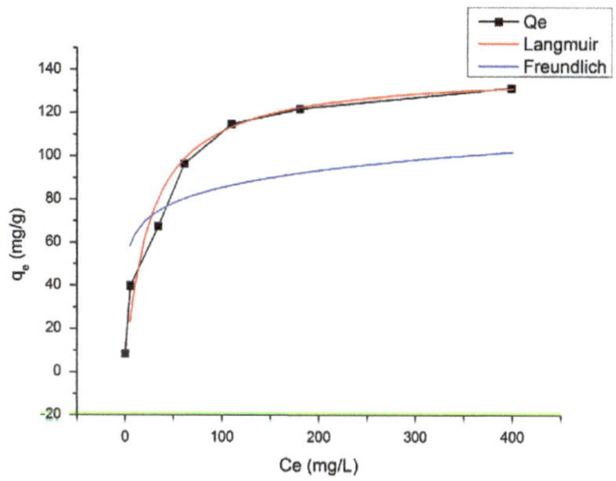

Figure 6. Sb(III) adsorption isotherms of acid-resistant magnetic Fe_3O_4 nanoparticles at 25°C in 2.5 M H_2SO_4 solution. Adsorbent dosage: 2 g l^{-1}; contact time: 6 h.

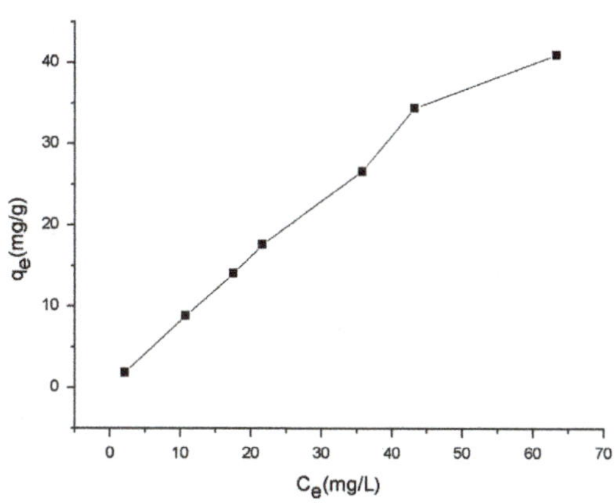

Figure 7. Selective removal of Sb(III) from cupric sulfate ([Cu(II)]: 14.115 g l^{-1}) solution with 2.5 M H_2SO_4 at 25°C. Adsorbent dosage: 2 g l^{-1}; contact time: 6 h.

all the adsorption sites are energetically identical. It is characterized by a decreasing slope as the concentration increases, since vacant adsorption sites decrease as the adsorbent is gradually covered. The Freundlich isotherm [36] is derived by assuming a heterogeneous surface with a non-uniform distribution of heat of adsorption over the surface and reversible adsorption. This model allows coexistence of several kinds of sorption sites on the solid and represents properly the sorption data at low and intermediate concentrations on heterogeneous surfaces. As shown in Figure 6, the Langmuir model can fit the experiment results more precisely than the Freundlich model does. Therefore, the adsorption process followed the Langmuir isotherm.

Table 2. Sb(III) concentrations after adsorption, Sb(III) concentrations in eluate and Fe concentrations in eluate for three recycles.

Round	Sb(III) concentration after adsorption(mg l^{-1})	Sb(III) concentration in eluate (mg l^{-1})	Fe concentration in eluate
1	55.4	146	nil
2	67.3	121	nil
3	75.6	107	nil

3.5. *Selective adsorption of Sb(III) from strongly acid $CuSO_4$ solution*

The adsorptive capacity of Sb(III) with different initial concentrations in a $CuSO_4$ solution ([Cu(II)]: 14.115 g l^{-1}) with 2.5 M H_2SO_4 was further studied. Referring to the batch adsorption experiment, 0.05 g sorbents were dispersed into 25 ml of this solution. Two hours later, the concentrations of Sb (III) and Cu^{2+} in the supernatant were both detected. The concentrations of Sb (III) with different initial concentrations were all reduced after adsorption, while those of Cu^{2+} remained unchanged. As shown in Figure 7, the acid-resistant magnetic sorbent modified with amino-methylene-phosphonic chelating group can selectively remove Sb(III) from strongly acid $CuSO_4$ solution. Hence, our sorbents are practically eligible for removal of Sb(III) from strongly acid copper electro-refining electrolytes.

3.6. *Recovery of the adsorbent*

Magnetic Fe_3O_4 nanoparticles have been utilized to remove metal ions from a water system, but they have seldom been reused because desorption of metal ions needs washing by using strong acids. For cost concerns, the acid-resistant magnetic Fe_3O_4 sorbent coated with hydrophobic shell was herein recycled and reused three times.

In a recycle, 0.1 g sorbents were dispersed into 25 ml of cupric sulfate ([Cu(II)]: 14.876 g l^{-1}) solution with 175.84 ppm Sb(III) and 2.5 M H_2SO_4, and the mixture was shaken at 100 rpm and 25°C. Two hours later, the sorbents were separated from the mixture by external magnetic field, and the supernatant was collected to measure the concentration of Sb(III). Then the sorbents were washed by 10 ml of distilled water three times. Next, these sorbents were dispersed into 10 ml of 1 M HCl and 1 M H_3PO_4 eluent, and shaken for 30 min. After the eluate was collected, the elution was repeated using 10 ml of the same eluent, and then the eluate was collected again. These two eluates were merged together to detect the concentrations of Fe and Sb (III) by atomic absorption spectrometry. Subsequently, the used sorbents were washed with 10 ml of distilled water twice before reusing. The recycling was repeated for another two rounds, and all the results are listed in Table 2, demonstrating that the acid-resistant magnetic Fe_3O_4 adsorbents may still function after being reused at least three times. Moreover, the adsorbed antimony could be eluted by using 1 M HCl and 1 M H_3PO_4, without iron ions leached. In other words, the acid-resistant magnetic Fe_3O_4 adsorbents prepared by our method are economical and potentially applicable to antimony (III) removal under strong acid conditions.

4. Conclusions

In this study, a new facile strategy has been successfully developed to protect nano-sized magnetite particles from corrosion by strong acid. This strategy was based on the hydrophobic property of methyl groups originating from the hydrolysis of DMDES. Through three different sol-gel processes at 30°C, the nano-sized Fe_3O_4 particles were coated by three corresponding shells with a structure of Fe_3O_4@SiO_2@SiO-Me@SiO-NH_2. The middle shell was full of methyl groups that resisted acid corrosion, and abundant amino groups, which were convenient for further modification, existed in the outmost shell. To adsorb Sb(III) in the acid solution, amino-methylene-phosphonic groups were modified on the adsorbent surfaces as chelating groups. As expected, the acid-resistant magnetic Fe_3O_4 nanoparticles selectively removed Sb(III) from cupric sulfate solution with 2.5 M sulfuric acid, and also could be reused at least three times. These findings prove that the coated magnetite particles modified with amino-methylene-phosphonic group can be applied in the purification of copper refining electrolytes by the MSPE process. The adsorption of Sb(III) from 2.5 M sulfuric acid solution needed four hours to reach equilibrium. Our coating strategy, with the chelating groups changed, may also be suitable for the adsorption of other metal ions in acid condition. Furthermore, it is feasible to protect other types of materials from strong acid corrosion with this method.

Disclosure statement

No potential conflict of interest was reported by the authors.

References

[1] Giakisikli G, Anthemidis AN. Magnetic materials as sorbents for metal/metalloid preconcentration and/or separation. A review. Anal. Chim. Acta. 2013;789:1.

[2] Robinson PJ, Dunnil P, Lilly MD. The properties of magnetic supports in relation to immobilized enzyme reactors. Biotechnol Bioeng. 1973;15:603.

[3] Wikstrom P, Flygare S, Grondalen A, Larsson P –O. Magnetic aqueous two-phase separation: A new technique to increase rate of phase-separation, using dextran-ferrofluid or larger iron oxide particles. Anal. Biochem. 1987;167:331.

[4] Towler PH, Smith JD, Dixon DR. Magnetic recovery of radium, lead and polonium from seawater samples after preconcentration on a magnetic adsorbent of manganese dioxide coated magnetite. Anal Chim Acta. 1996;328:53.

[5] Safaríkõvá M, Safarík I. Magnetic solid-phase extraction. J. Magn. Magn. Mater. 1999;194:108.

[6] Aguilar-Arteaga K, Rodriguez JA, Barrado E.Magnetic solids in analytical chemistry: A review. Anal. Chim. Acta. 2010;674:157.

[7] Poole CF. New trends in solid-phase extraction Trends Anal. Chem. 2003;22:362.

[8] Chen LG, Wang T, Tong J. Application of derivatized magnetic materials to the separation and the preconcentration of pollutants in water samples. Trends Anal. Chem. 2011;30:1095.

[9] Camel V. Solid phase extraction of trace elements. Spectrochim. Acta B. 2003;58:1177.

[10] Li SF, Liu M, Sun L. Preparation of acid-resisting ultramarine blue by novel two-stepsilica coating process. Ind. Eng. Chem. Res. 2011;50:7326.

[11] Liu QX, Xu ZH, Finch JA, Egerton R. A novel two-step silica-coating process for engineering magnetic nanocomposites. Chem. Mater. 1998;10:3936.

[12] Wu W, He QG, Jiang CZ. Magnetic iron oxide nanoparticles: synthesis and surface functionalization strategies. Nanoscale Res. Lett. 2008;3:397.

[13] Chen LY, Xu ZX, Dai H, Zhang ST. Facile synthesis and magnetic properties of monodisperse Fe3O4/silica nanocomposite microsphere with embedded structure via a direct solution-based route. J. Alloy. Compd. 2010;497:221.

[14] Kang YS, Risbud S, Rabolt JF, Stroeve P. Synthesis and characterization of nanometer-size Fe3O4 and γ-Fe2O3 particles. Chem. Mater. 1996;8:2209.

[15] Bagheri A, Taghizadeh M, Behbahani M, Asgharinezhad AA, Salarian M, Dehghani A, Ebrahimzadeh H, Amini MM. Synthesis and characterization of magnetic metal-organic framework (MOF) as a novel sorbent, and its optimization by experimental design methodology for determination of palladium in environmental samples. Talanta. 2012;99:132.

[16] Mashhadizadeh MH, Amoli-Diva M. Atomic absorption spectrometric determination of Al3+ and Cr3+ after preconcentration and separation on 3-mercaptopropionic acid modifiedsilica coated-Fe3O4 nanoparticles. J. Anal. At. Spectrom. 2013;28:251.

[17] Cheng GH, He M, Peng H, Hu B. Dithizone modified magnetic nanoparticles for fast and selective solid phase extraction of trace elements in environmental and biological samples prior to their determination by ICP-OES. Talanta. 2012;88:507.

[18] Suleiman JS, Hu B, Peng HY, Huang CZ. Separation/preconcentration of trace amounts of Cr, Cu and Pb in environmental samples by magnetic solid-phase extraction with Bismuthiol-II-immobilized magnetic nanoparticles and their determination by ICP-OES. Talanta. 2009;77:1579.
[19] Jiang HM, Yan ZP, Zhao Y, Hu X, Lian HZ. Zincon-immobilized silica-coated magnetic Fe3O4 nanoparticles for solid-phase extraction and determination of trace lead in natural and drinking waters by graphite furnace atomic absorption spectrometry. Talanta. 2012;94:251.
[20] Mandil A, Idrissi L, Amine A. Stripping voltammetric determination of mercury(II) and lead(II) using screen-printed electrodes modified with gold films, and metal ion preconcentration with thiol-modified magnetic particles. Microchim. Acta. 2010;170:299.
[21] Karimi MA, Hatefi-Mehrjardi A, Mohammadi SZ, Mohadesi A, Mazloum-Ardakani M, Kabir AA, Kazemipour M, Afsahi N. Solid phase extraction of trace amounts of silver (I) using dithizone-immobilized alumina-coated magnetite nanoparticles prior to determination by flame atomic absorption spectrometry. Int. J. Environ. Anal. Chem. 2012;92:1325.
[22] Wang Y, Tian T, Wang L, Hu XY. Solid-phase preconcentration of cadmium(II) using amino-functionalized magnetic-core silica-shell nanoparticles, and its determination by hydride generation atomic fluorescence spectrometry. Microchim. Acta. 2013;180:235.
[23] Hoffmann JE. The purification of copper refinery electrolyte. J. Miner. Metal. Mater. Soc. 2004;56:30.
[24] Wang SJ. Impurity control and removal in copper tankhouse operations. *JOM-US*. 2004;56:34.
[25] Leyva AG, Marrero J, Smichowski P, Cicerone D. Sorption of antimony onto hydroxyapatite. Environ. Sci. Technol. 2001;35:3669.
[26] Leng YQ, Guo WL, Su SN, Yi CL, Xing LT. Removal of animony (III) from aqueous solution by graphene as an adsorbent. Chem. Eng. J. 2012;406:211–212.
[27] Sari A, Çitak D, Tuzen M. Equilibrium, thermodynamic and kinetic studies on adsorption of Sb (III) from aqueous solution using low-cost natural diatomite. Chem. Eng. J. 2010;162:521.
[28] Xi JH, He MC. Removal of Sb (III) and Sb (V) from aqueous media by goethite. Water Qual. Res. J. Can. 2013;48:223.
[29] Xi JH, He MC, Lin CY. Adsorption of antimony (III) and antimony (V) on bentonite: kinetics, thermodynamics and anion competition. Microchem. J. 2011;97:85.
[30] Xu W, Wang HJ, Liu RP, Zhao X, Qu JH. The mechanism of antimony (III) removal and its reactions on the surface of Fe-Mn binary oxide. J. Colloid Interface Sci. 2011;363:320.
[31] Salam MA, Mohamed RM. Removal of antimony (III) by multi-walled carbon nanotubes from model solution and environmental samples. Chem. Eng. Res. Des. 2013;91:1352.
[32] Shan C, Ma ZY, Tong MP. Efficient removal of trace antimony (III) through adsorption by hematite modified magnetic nanoparticles. J. Harzard. Mater. 2014;268:229.
[33] Navarro P, Simpson J, Alguacil FJ. Removal of antimony (III) from copper in sulphuric acid solutions by solvent extraction with LIX 1104SM. Hydrometallurgy. 1999;53:121.
[34] Tessler MM. Novel starch ether derivatives, a method for the preparation thereof and their use in paper. US patent 4243479(A). 1981.
[35] Langmuir I. The constitution and fundamental properties of solids and liquids. part i. solids J. Am. Chem. Soc. 1916;38:2221.
[36] Freundlich HMF. Uber die adsorption in losungen. Z. Phys. Chem. 1906;57:385.

12

The chemical, mechanical, and physical properties of 3D printed materials composed of TiO_2-ABS nanocomposites

Matthew R. Skorski[a], Jake M. Esenther[a], Zeeshan Ahmed[b], Abigail E. Miller[a,c] and Matthew R. Hartings[a]

[a]Department of Chemistry, American University, 4400 Massachusetts Ave., NW, Washington, DC, 20016, USA
[b]Thermodynamic Metrology Group, Sensor Science Division, Physical Measurement Laboratory, National Institute of Standards and Technology, Gaithersburg, MD 20899, USA
[c]Food and Drug Administration, Washington, DC, USA

ABSTRACT
To expand the chemical capabilities of 3D printed structures generated from commercial thermoplastic printers, we have produced and printed polymer filaments that contain inorganic nanoparticles. TiO_2 was dispersed into acrylonitrile butadiene styrene (ABS) and extruded into filaments with 1.75 mm diameters. We produced filaments with TiO_2 compositions of 1, 5, and 10% (kg/kg) and printed structures using a commercial 3D printer. Our experiments suggest that ABS undergoes minor degradation in the presence of TiO_2 during the different processing steps. The measured mechanical properties (strain and Young's modulus) for all of the composites are similar to those of structures printed from the pure polymer. TiO_2 incorporation at 1% negatively affects the stress at breaking point and the flexural stress. Structures produced from the 5 and 10% nanocomposites display a higher breaking point stress than those printed from the pure polymer. TiO_2 within the printed matrix was able to quench the intrinsic fluorescence of the polymer. TiO_2 was also able to photocatalyze the degradation of a rhodamine 6G in solution. These experiments display chemical reactivity in nanocomposites that are printed using commercial 3D printers, and we expect that our methodology will help to inform others who seek to incorporate catalytic nanoparticles in 3D printed structures.

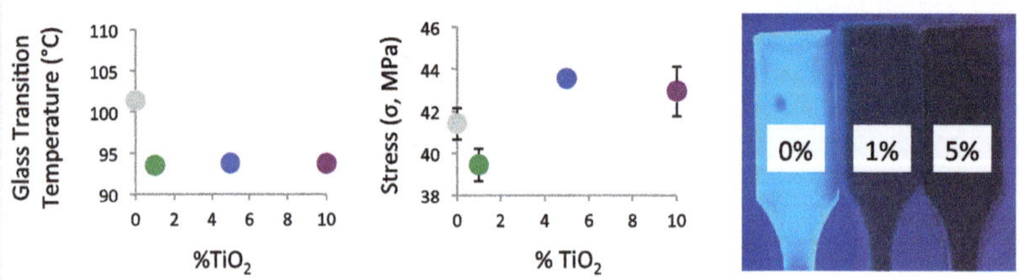

KEYWORDS
3D printing; nanocomposite; TiO_2 nanoparticle; ABS; photocatalysis

CLASSIFICATION
20 Organic and soft materials (colloids, liquid crystals, gel, polymers)

1. Introduction

Additive manufacturing, or three-dimensional printing, encompasses a number of techniques in which a material is deposited in a layer-by-layer fashion to produce a larger 3D structure.[1, 2] A diversity of materials (sugars,[3] thermoplastics, photopolymers,[4] glass,[5] metals,[6] metal oxides,[7] and ceramics,[8] to name a few) are used to produce 3D printed figures. Of these, thermoplastics are the most common for home enthusiasts and experimental laboratories due to the large number of commercially available printers that employ these materials, which results from the safety, ease of use, and cost of the equipment. Thermoplastics are extruded at temperatures above their glass transition temperature (T_g), at which they display increased flexibility and flow properties.

Commercial thermoplastic 3D printers have found their way into several research laboratories. Despite their popularity, their use has been mostly confined to producing tools that assist in performing research rather than producing structures that are subjects of research. This is especially true for the chemical sciences. Taken from a chemist's point of view, a printed item is primarily an inert object.

That is not to say that chemists and (bio)materials scientists have not found innovative ways to use 3D printing to advance their fields. Cronin and co-workers

CONTACT Matthew R. Hartings hartings@american.edu

have developed 3D printed reactionware using acetoxy silicone polymers that are normally used as household sealants.[9–11] Their reactionware has been used to facilitate syntheses, purifications, crystallizations, and analyses. Another subject where 3D printing has great potential is in the area of tissue engineering. Several groups of scientists have printed hydrogels that have been seeded with mammalian cells.[12–14] The hydrogels are fixed into place with UV light, and the cells are allowed to culture within the printed scaffold. In each of these examples, a polymer with optimal flow properties at room temperature is extruded and induced to set through some external stimulus (light, oxygen, etc.).

While 3D printing holds great promise for advancing chemistry and materials science, this promise has not been realized for the ubiquitous thermoplastic printers. We contend that one way to imbue active chemistry into 3D printed thermoplastic objects is to incorporate inorganic nanoparticles into the polymer filaments. Because organic dye molecules are infused into a polymeric matrix to create different colored filaments, inorganic nanoparticles should be able to be integrated in much the same way. The functionalities associated with different types of nanoparticles (catalysis,[15] luminescence,[15] thermal [16] and electric [17] conductivity, and gas storage,[18] to name a few) should retain their properties when incorporated into a thermoplastic. We have incorporated TiO_2 nanoparticles into acrylonitrile butadiene styrene (ABS) filaments. The TiO_2-ABS filaments were printed in multiple shapes and tested for their mechanical, physical, and chemical properties. TiO_2-polymer nanocomposites have been studied in detail.[19–24] Nanocomposites are materials in which nanoscale particles have been incorporated into a polymer matrix. In many nanocomposites, advantageous properties of the nanoparticles and the polymer are maintained in the bulk material. In some nanocomposites, enhancement of functional properties is observed. TiO_2 has been shown to increase the mechanical strength of the nanocomposite over the pure polymer.[20, 22, 24] TiO_2 can increase the dielectric properties of a polymer for use in insulating materials.[21] TiO_2 nanocomposites can also photocatalytically degrade environmental organic pollutants.[24] 3D printed TiO_2-ABS nanocomposites should follow similar trends, and 3D printing of this material could become a viable way to fabricate custom, on-demand parts that can meet any number of these technological needs. In the experiments presented here, we aim to determine if printed TiO_2-ABS composites maintain the advantageous properties of the nanocomposites as described above.

TiO_2 nanoparticles make an ideal test case for an inorganic nanoparticle filler within thermoplastic printing filaments. TiO_2 is a known photocatalyst, capable of generating free radical species in both aqueous and organic solvents and in the presence of oxygen upon irradiation.[25] The band gap of TiO_2 corresponds to UV wavelengths. Upon absorbance of light, electrons in the conduction band can form superoxide radicals ($O^{2-}\bullet$) from adsorbed oxygen, and holes in the valence band can form hydroxyl radicals ($\bullet OH$) from water. Because of these properties, TiO_2 has potential applications in the photocatalytic removal of pollution from air, water, and agricultural sources.[26, 27] It also has potential uses in solar energy [28] and solar hydrogen production.[29] The use of TiO_2 nanoparticles in many consumer products, such as paints, means that this material is abundant and can be acquired at low cost. All of these factors - low cost, ease of purchase, and readily testable chemical properties - make TiO_2 an ideal test case for producing functional nanocomposite ABS filaments.

We present here detailed observations of several chemical, physical, and mechanical properties of 3D printed TiO_2-ABS nanocomposites. We describe the preparation and evaluation of a range of TiO_2-ABS compositions (0, 1, 5, and 10% kg/kg). There is one previous study of the mechanical properties of printed 5% TiO_2-ABS structures.[22] Our measurements generally agree with the previous observations and expand upon the analysis of how TiO_2 affects the properties of printed ABS. As there are few studies of 3D printed TiO_2-polymer nanocomposites, we have attempted to understand how the different processing steps affect the chemical and physical properties of the polymer. We determined the effect of TiO_2 on the thermal transitions and the molecular weight of the processed (solvent cast, extruded, and printed) ABS. We measured the ability of the TiO_2 to quench the luminescence that arises from the styrene functional groups within ABS. Finally, we analyzed the photocatalytic degradation of rhodamine 6G by 3D printed structures of TiO_2-ABS nanocomposites. We expect that these experiments will inform future studies in which nanoparticles are incorporated into thermoplastic 3D printing filaments for developing chemically active 3D printed materials.

2. Materials and methods

2.1. Materials

Titanium(IV) oxide (nanopowder, 21 nm particle size by transmission electron microscopy, 80–90% anatase polyform with small percentage of rutile polyform) and rhodamine 6G were purchased from Sigma Aldrich (St Louis, Missouri, USA).[30] Commercial ABS filament was purchased from Octave Systems (Santa Clara, California, USA) (natural color, 1.75 mm width). High Performance Liquid Chromatography (HPLC)-grade water, dimethylformamide (DMF), and acetone were purchased from BDH Chemicals (Radnor, Pennsylvania, USA). ABS pellets (resin: GPA 100; color #: NC010; color: natural; lot #: C14–0702 K) was acquired from LTL Color Compounders, Inc. LiBr was purchased from Acros Chemicals (New Jersey, USA).

2.2. Production and printing of TiO_2-ABS filaments

ABS and TiO_2 mixtures (0, 1, 5, and 10% TiO_2 by dry weight) were suspended and dispersed in acetone (in a sealed container) over several hours at 35°C in a sonicator (VWR Symphony ultrasonic cleaner, VWR, Inc., Radnor, Pennsylvania, USA). The highest percentage of TiO_2 used in this study was 10% (kg/kg) because we found that nanocomposites produced at higher percentages were incompatible with our printing process. A typical batch included 20 total grams of ABS and TiO_2 in 200 ml of acetone. As an example, a 1% sample was made by suspending 0.2 g of TiO_2 and 19.8 g of ABS in 200 ml of acetone, held in an Erlenmeyer flask, and agitated in a heated sonicating bath. The dispersed samples were solvent cast in Teflon coated, aluminum frying pans (used for their large surface area). The pans were placed on hot plates set to 80 °C, and the solvent was evaporated. Solvent casting by heating samples in large surface area pans is necessary to reduce the amount of TiO_2 clumping observed in slower evaporation processes.

The films were cut into small squares (roughly 1 cm by 1 cm) and extruded with a DSM Xplore Micro 15 cc Twin Screw Compounder (DSM Xplore, Geleen, the Netherlands). The filaments produced using this compounder were too wide to be compatible with the 3D printer. These filaments were cut into small pieces using a wire cutter and extruded into a 1.75 mm-wide filament using a Filabot Wee Extruder from Filabot (Barre, Vermont, USA). These filaments were used to print various structures by a Flashforge Creator 3D Printer from Flashforge (Rowland Heights, California, USA) with dual extruders.

Two different shapes (dogbones and cylinders) were printed using the factory settings for ABS filaments. The dogbone shape and dimensions were printed to match the specification described by the American Society for Testing and Materials (ASTM) standard D638.[31] A diagram of these dimensions is shown in the Supporting Information. The cylinders were 2.5 cm in diameter, designed to fit in the solid sample holder used in fluorescence measurements. Both shapes were printed at a 10% fill volume and without a raft. The absence of a raft led to a smoother surface, which facilitated different spectroscopic measurements.

2.3. Presence of TiO_2 in ABS samples

Powder X-ray diffraction (XRD) measurements were used to confirm the presence of TiO_2 within the polymer nanocomposite. XRD measurements were recorded using a Rigaku Miniflex II equipped with an NaI scintillation counter detector, a 450 W Cu K α ($\lambda = 0.1540562$ nm) X-ray source, and a diffracted beam monochromator. Each sample was mounted on an aluminum holder.

2.4. Effect of TiO_2 and processing on ABS structure

To determine if processing (solvent casting, extruding, and printing) ABS in the presence of TiO_2 affects the polymer structure, we measured monomer content with Fourier transform infrared spectroscopy (FTIR), ABS thermal transitions with differential scanning calorimetry (DSC), and polymer size with gel permeation chromatography (GPC). FTIR spectra were recorded using a Bruker Alpha spectrometer (Billerica, Massachusetts, USA) with an attenuated total reflectance solid sample holder. DSC was performed using a TA DSC Q2000 from TA Instruments (New Castle, Delaware, USA). Roughly 3 mg of nanocomposite sample was placed in an aluminum Tzero pan (TA Instruments) and sealed with a hermetic lid (TA Instruments). Samples were exposed to the following heating cycle. The temperature was increased from 20 to 170 °C at a rate of 10 °C min^{-1}, held at 170 °C for 2 min, decreased from 170 to 20 °C at 10 °C min^{-1} and held at 20 °C for 2 min. These cycles were repeated for a total of three full heating and cooling cycles.

Polymer sizes were measured using a Styragel HR 4 THF GPC column (7.8 mm × 300 mm, Milford, Massachusetts, USA) attached to a Waters 1525 HPLC system (Standards SM-105 from Shodex, Tokyo, Japan). Polymer molecular weights were calibrated with polystyrene standards (Shodex Standards SM-105). Samples were prepared for GPC by dissolving the polymer in 0.05 M LiBr in DMF and filtering through a 0.45 μm-pore syringe filter.

2.5. Fluorescence data

All fluorescence measurements were recorded using a Perkin Elmer LS-55 luminescence spectrometer from Perkin Elmer (Watham, Massachusetts, USA). The ability of TiO_2 to quench the natural fluorescence of the styrene within the printed polymer was tested. Cylinders (2.5 cm diameter, 50 cm height) were printed with no raft. The side of the cylinder that was in contact with the print bed during fabrication was the smoother of the two sides. These procedures do not result in a cylinder that is flat enough for fluorescence measurements. A single drop of acetone was placed on the top of the cylinder and the window from the sample holder was set on top of the drop for 30 min before fluorescence measurements to flatten the cylinder. Performing this step increased the reproducibility of the data. Multiple cylinders were printed for each per cent TiO_2 in ABS, and each cylinder was measured six times. The cylinder was rotated within the sample holder in between each measurement. Emission spectra were collected with an excitation wavelength of 280 nm with a slit width of 10 nm and the emission slit width was set to 4.5 nm. Emission was scanned from 300 nm to 550 nm. The scan speed was 200 nm min^{-1}. The emission spectra were corrected such that the emission goes to zero at 550 nm.

The total integrated emission intensity was calculated for each measurement.

To test the photocatalytic properties of TiO_2 in the printed nanocomposites, we placed one half of a printed dogbone in a scintillation vial and immersed it in 10 ml of a 1 μM aqueous rhodamine 6G solution. The fluorescence of the solution was recorded before and after the sample was exposed to direct sunlight for 4 h. To measure fluorescence, the solution was placed in a quartz cuvette. Emission spectra were collected with a 480 nm excitation with a 10 nm slit width and an emission slit width of 10 nm. The emission was scanned from 500 nm to 650 nm. The total integrated intensity was calculated for each spectrum to determine if the TiO_2 in the printed materials could photocatalytically degrade the rhodamine 6G. As photogenerated radicals might also affect the ABS polymer, we recorded the FTIR spectra of printed dogbones before and after 4 h of exposure to direct sunlight.

3. Results and discussion

While developing our system we performed a number of preliminary experiments and observations. First we had to choose the polymer for making nanocomposites. The two polymers that are typically used in consumer 3D printers are ABS and polylactic acid (PLA). We chose to work with ABS as opposed to PLA for this work because we observed that PLA is more prone to decomposition during extrusion (experimental observation). One observation that we made while printing the TiO_2-ABS composites is that the standard printing speed (30 mm s^{-1}) led to poorly printed structures and often led to clogged print heads. We found piano wire to be the best tool for cleaning these clogs. Reducing the printing speed to 10 mm s^{-1} resulted in consistent, quality structures. We also would like to note the importance of allowing the print heads and printing bed to remain at operating temperature for at least a half an hour before starting a print job. This increased the temperature inside of the printing chamber and reduced the chances that the structure would curl up from the print bed during the print job.

3.1. TiO_2 in ABS

Figure 1 shows the XRD patterns of TiO_2, 0% TiO_2 in ABS (1%), 1% TiO_2 in ABS (1%), 5% TiO_2 in ABS (5%), and 10% TiO_2 in ABS (10%). The 1, 5, and 10% samples show the presence of TiO_2. The intense peak (at $2\theta = 25.4°$) from the anatase polymorph of TiO_2 is present in all TiO_2-ABS samples along with the signal (at $22\theta = 27.5°$) from the rutile polymorph of TiO_2. The signal of all TiO_2-associated peaks increases (with respect to the broad ABS signal at $2\theta \approx 20°$) with increasing percentage of TiO_2 within the ABS composite. Full spectra (from $2\theta = 3°$ to $2\theta = 90°$ are shown in the Supporting Information).

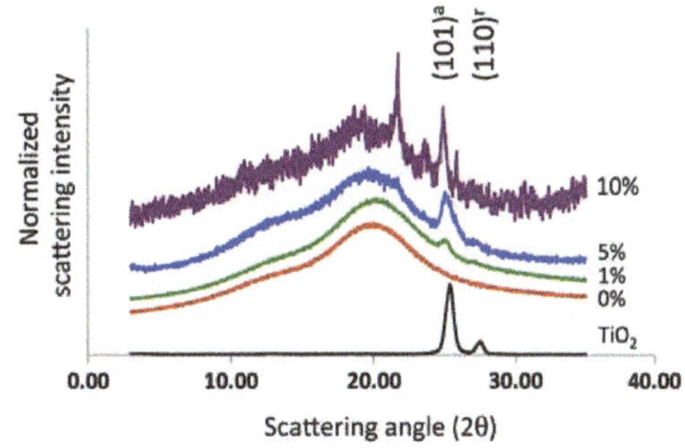

Figure 1. X-ray diffraction spectra of TiO_2 powder (black, no vertical offset), 0% TiO_2 in ABS (red, smallest vertical offset), 1% TiO_2 in ABS (green, small vertical offset), 5% TiO_2 in ABS (blue, medium vertical offset), and 10% TiO_2 in ABS (purple, large vertical offset). The scattering signals at $2\theta = 25.4°$ and 27.5° correspond to scattering from the anatase and rutile polymorphs, respectively (International Centre for Diffraction Data reference file for anatase TiO_2: 21–1272 and rutile TiO_2: 21–1276). The $2\theta = 25.4°$ scattering arises from the (101) plane of the anatase polymorph and corresponds to a d-spacing of 0.3504 nm. The $2\theta = 27.5°$ scattering arises from the (110) plane of the rutile polymorph and corresponds to a d-spacing of 0.3241 nm.

3.2. Effect of TiO_2 and processing on ABS structure

ABS is formed when styrene and acrylonitrile are polymerized in the presence of pre-formed polybutadiene creating a network of polybutadiene and poly(styrene-co-acrylonitrile) polymers. The solvent casting, extrusion, and printing steps place a significant amount of strain on the ABS. The presence of TiO_2 during these processes can exacerbate any damage to the polymer. Because TiO_2 can photocatalytically produce free radicals,[25] ABS can be degraded and altered throughout the progression from solvent casting to printed structure. To understand how each preparation step affected the polymeric structure, we monitored the FTIR spectra, glass transition temperature, and polymer molecular weight at different points in the processing for all of the TiO_2-ABS nanocomposites used.

FTIR analysis assesses how the individual components (acrylonitrile, butadiene, and styrene) are affected during the processing steps. Figure 2 displays the FTIR spectra for 3D printed samples (dogbones and cylinders). Multiple spectra were recorded for each TiO_2 percentage. The averaged spectra are shown in the top panel while the bottom panels show the average transmittance with standard deviations for the wavelengths associated with different components of the ABS-TiO_2 nanocomposite.

The data corresponding to TiO_2 absorption (absorbance feature at 465 cm^{-1}) show increasing amounts of TiO_2 with increasing TiO_2 percentage within the nanocomposite. These data are in agreement with the data in Figure 1. The spectra in Figure 2 are taken for printed samples, therefore the data show that TiO_2 remains

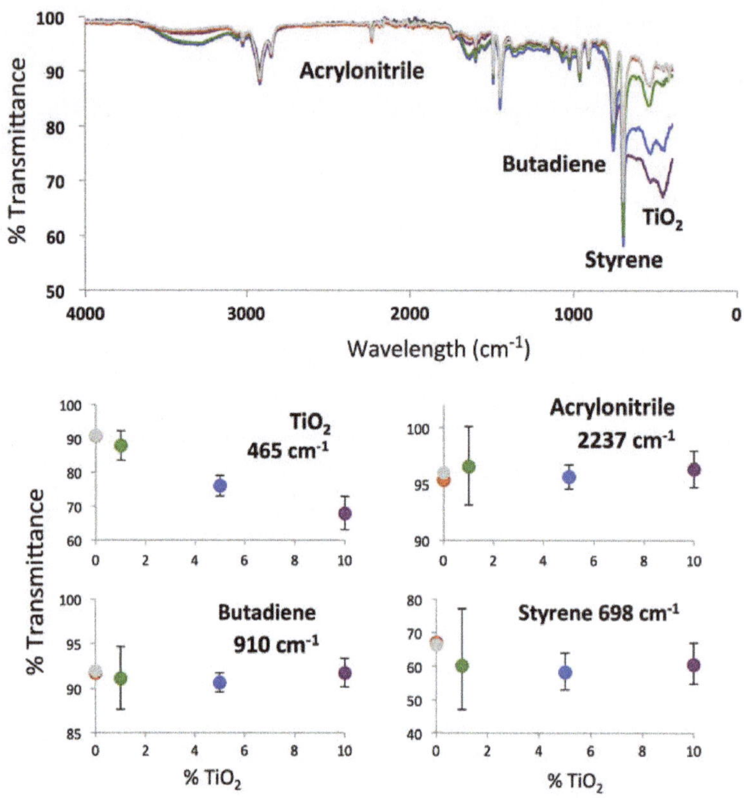

Figure 2. FTIR spectra for printed TiO_2-ABS composites. The top panel includes the full spectrum for each composite, averaged over 10 printed samples. The bottom panel displays the % transmittance at wavelengths associated with the different components for the TiO_2-ABS composites. The error bars correspond to the standard deviation from the averaged measurement over all spectra acquired. In these spectra, different colors (for the lines in the upper panel and circles in the bottom panels) correspond to samples printed from different composites. Gray: commercial ABS filament. Red: extruded 0% TiO_2-ABS composites. Green: extruded 1% TiO_2-ABS composites. Blue: extruded 5% TiO_2-ABS composites. Purple: extruded 10% TiO_2-ABS composites.

within the ABS matrix during the extrusion and printing processes.

The data for acrylonitrile (λ_{abs} = 2237 cm^{-1}), butadiene (λ_{abs} = 910 cm^{-1}), and styrene (λ_{abs} = 698 cm^{-1}) paint a different picture. None of these absorption features differ significantly from the samples printed with commercial ABS filament and lab-extruded (0% TiO_2) ABS filament. There are no trends observed in these samples with increasing TiO_2 concentration. This result is to be expected, as the sample matrices are still primarily ABS.

The standard deviations for the acrylonitrile, butadiene, and styrene features associated with the 1% TiO_2 samples are markedly higher than those of the other samples. The deviation in the 1% samples (±1.7, 9.2, and 16.9% for acrylonitrile, butadiene and styrene, respectively) is clearly larger than the deviation observed for the 5% and 10% TiO_2 samples. We argue that, as the amount of TiO_2 is increased in the nanocomposite, the amount of nanoparticle clumping increases. As the clumping increases, the direct net interactions between ABS and TiO_2 decreases. That is, because some TiO_2 nanoparticles are sequestered within aggregates, these nanoparticles do not contact the polymer matrix. If TiO_2 is capable of facilitating polymer degradation, samples with less clumping will increase degradation. We contend that TiO_2 dispersity in the 1% sample leads to increased polymer–nanoparticle interactions. The increased number of nanoparticle-polymer interactions in the 1% sample lead to changes in absorption features and, potentially, increased polymer degradation during processing. Increased deviations observed for the acrylonitrile, butadiene, and styrene FTIR peaks are due to increased polymer–nanoparticle interactions. This description (increased clumping leading to fewer interactions between the polymer and the nanoparticle) is consistent with other experimental observations, which will be discussed below.

ABS is naturally amorphous and displays no melting or crystallization features. ABS has a glass transition temperature (T_g) at roughly 105°C. The T_g values of the different nanocomposites are shown in Figure 3. The T_g of an unprocessed ABS pellet (unaltered from the supplier) and of the lab-extruded 0% TiO_2 ABS filament are both near the published value of 105°C. The

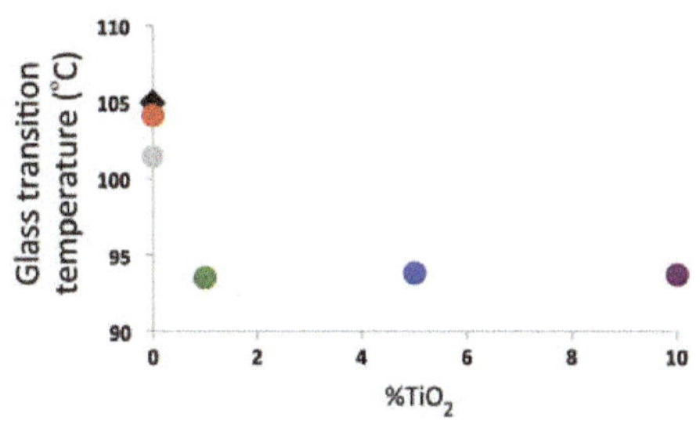

Figure 3. Glass transition temperatures of different samples from printed TiO_2-ABS nanocomposite filaments. Unprocessed ABS pellet (black diamond). Commercial ABS filament (gray circle). Lab extruded ABS filament (red circle). 1% TiO_2-ABS filament (green circle). 5% TiO_2-ABS filament (blue circle). 10% TiO_2-ABS filament (purple circle).

commercial ABS filament has a lower T_g (101°C). The T_g values for the 1, 5, and 10% TiO_2-ABS nanocomposites are all significantly lower than the pure ABS samples. The reduction in T_g upon TiO_2 addition is consistent with other experiments.[32]

The predominant cause for decreased T_g in the ABS that contains TiO_2 is the nanoparticle–polymer interaction. It has been observed that TiO_2 reduces the glass transition temperatures in polymeric systems where there is poor interfacial adhesion, leading to an increase in free volume and polymer chain mobility near the particle.[32–34] The T_g data are consistent with the FTIR spectra. Increasing the dispersion of the nanoparticles within the polymer matrix will increase the total surface area of TiO_2 available to ABS. If the 5% and 10% samples contained dispersed nanoparticles, the T_g values would decrease for these nanocomposites consistent with the amount of TiO_2 added. We contend that the T_g for the 5% and 10% samples do not decrease in comparison to the 1% sample due to nanoparticle aggregation.

The molecular weight of the ABS at different stages of processing was analyzed using GPC. Polystyrene standards were used to correlate retention time to molecular weight. The data for these standards were fit using a quadratic equation (shown in the Supplemental Information) to determine the molecular weight of the processed polymer. Table 1 shows the peak molecular weight (M_n) and polydispersity index (PDI) data for the TiO_2-ABS nanocomposites after: solvent casting, extruding through the compounder, extruding into a 1.75 mm filament, and 3D printing.

The retention time was determined by selecting the peak maximum in the recorded GPC trace. Molecular weight data were determined with the calibration curve (described above) using the Breeze software that controls the Waters HPLC system. Several samples for each processing step were acquired. The PDI, which is calculated by dividing the molecular weight average (M_w) by the number average molecular weight (M_n), is indicative of the molecular weight distribution of the polymers in solution. M_n decreases from the value observed for the ABS pellet for each ABS-TiO_2 sample except for the solvent cast 10% TiO_2 sample. We argue that the mass difference between the polymer from the ABS pellet and the solvent cast 10% TiO2 sample are statistically insignificant; a variation of 10 kDa is within the expected error of the measurement for the column used here. The change in molecular weight is largest for the 1% sample. This observation is consistent with the result showing that greater dispersion of TiO2 in the 1% sample leads to more ABS decomposition (Figure 2). From the data in this table, it would appear that most of the polymer degradation occurs during the solvent casting process. In this step, ABS is sonicated with TiO_2 at 37°C in acetone, poured into a Teflon coated pan, and heated at 80°C to evaporate the solvent. The effects of the other processing steps on polymer size seem to be less than the effect of the casting step. To support these data, we have also measured the FTIR spectra of the nanocomposites after each processing step. These spectra are presented in the Supporting Information.

3.3. *Mechanical properties of the printed TiO_2-ABS nanocomposites*

TiO_2 has previously been shown to alter the mechanical properties in printed ABS.[22] This type of observation is consistent with other experiments on TiO_2-polymer nanocomposite materials.[20, 23] As opposed to other nanocomposite structures, 3D printed structures have natural striations resulting from the printing resolution. It is possible that the printing process affects the homogeneity of the nanocomposite. We measured the mechanical properties of the printed materials to test the effect of these qualities and to directly study the effect of TiO_2 on ABS. ASTM D638 standard dogbones were printed and analyzed using a Series 5 Universal Testing Machine from Mark-10 (Copiague, New York, USA) for stress-strain and three-point bending measurements.

Figure 4 shows the stress-strain curves measured for 5% TiO_2-ABS dogbones, and Figure 5 shows the flexural testing results of these same materials. Several mechanical properties calculated from these experiments are shown

Table 1. Molecular weight and polydispersity index (PDI) determination for ABS after each processing step.

ABS pellet	M_n in kDa (PDI) = 210 (1.9)			
	Solvent cast	Extrusion 1	Extrusion 2	Printed
1% TiO_2	170 (1.9)	170 (1.8)	180 (2.0)	180 (1.9)
5% TiO_2	180 (1.8)	190 (1.7)	190 (1.8)	190 (1.7)
10% TiO_2	220 (1.7)	200 (1.6)	180 (1.8)	190 (1.9)

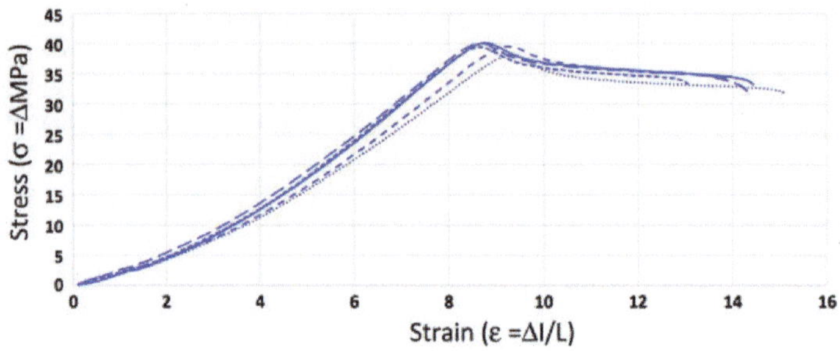

Figure 4. The stress-strain curves for the 5% TiO_2 ABS nanocomposites. The stress-strain curves for the other composites are shown in Figures S7–S9. The analyzed data for all composites are show in Table 2.

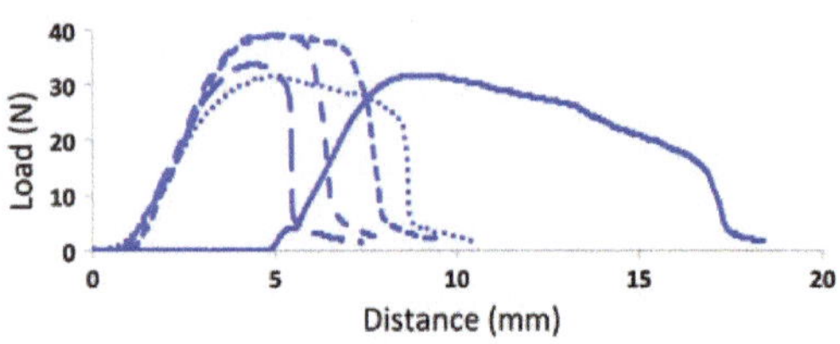

Figure 5. The three-point flexural measurements for the 5% TiO_2-ABS composite dogbones. The flexural measurements for the other composites are shown in Figures S10–S11. The analyzed data for all TiO_2-ABS composites are shown in Table 2.

in Table 2. Among these properties are the ultimate tensile strength (UTS), the stress measurement at UTS, Young's modulus, and the flexural strength. For each of these properties, the average measurements for each nanocomposite percentage are shown along with their deviations.

The UTS and flexural strength are both measurements of the extreme forces that a material can handle before it fails. The nanocomposites show similar trends for these values with respect to the amount of TiO_2 added. The 1% TiO_2 nanocomposites show a decrease in both UTS and flexural strength as compared to the 0% sample. The 5 and 10% TiO_2 nanocomposites show an increase in UTS and flexural strength in comparison to the 0 and 1% samples. Young's modulus values and the stress at material failure do not show any significant variations as a function of TiO_2 percentages.

We have argued that nanoparticle clumping within the polymer matrix is the cause for variations in glass transition properties as a function of TiO_2 percentages. We contend that clumping also plays a role in affecting the mechanical properties. The 1% TiO_2 samples show better dispersion than the 5% and 10% samples, and the 1% samples show lower mechanical strength in both tensile and flexural measurements, therefore the interaction between the polymer and the nanoparticles (21 nm) must have an adverse effect on the strength of the material matrix. As the nanoparticles clump together, there are several possible mechanisms that might lead to an increase in material strength. The first of these possibilities is that the effective size of the TiO_2 plays a role in determining tensile and flexural strength. As the effective particle gets larger (and the surface area to volume decreases), the total number of particle–polymer interactions decrease. If the interaction itself has an adverse effect on the strength of the matrix, decreasing the number of interactions will increase the strength. This possibility is consistent with the increased polymer flexibility near the nanoparticles that leads to a decrease in T_g.[33] Another possibility is that individual polymer chains can intercalate between nanoparticles within a single clump. The intercalation might make the polymer matrix stronger. We expect that the strength (justified by either of the two descriptions) will eventually stop increasing as a function of particle clumping, which has also been shown to induce fracture points in nanocomposites with high loading percentages.[20] We should note that in the previous study on 3D printed TiO_2-ABS structures, the researchers only prepared a 5% TiO_2 nanocomposite.[22] They observe an increase in tensile strength over the 0% nanocomposite. This observation is consistent with our experiments.

3.4. Photocatalytic degradation of rhodamine

TiO_2 and TiO2 nanocomposites have been shown to generate reactive oxygen-based radicals when exposed to sunlight.[25] These radicals can take part in reactions with organic compounds in close proximity to

Table 2. Mechanical properties and measurement uncertainties for average stress, average strain, Young's modulus, and flexural strength of each nanocomposite.

	Mechanical testing			
	Average stress (MPa)	Average strain (%)	Young's modulus (MPa)	Flexural strength (MPa)
0% TiO_2	41.4 ± 0.7	9.0 ± 0.2	6.1 ± 0.2	70 ± 2
1% TiO_2	39.4 ± 0.8	8.9 ± 0.3	6.1 ± 0.5	62 ± 5
5% TiO_2	43.6 ± 0.2	9.7 ± 0.8	6.2 ± 0.7	72 ± 8
10% TiO_2	43 ± 1	9.1 ± 0.9	6.5 ± 0.4	72 ± 2

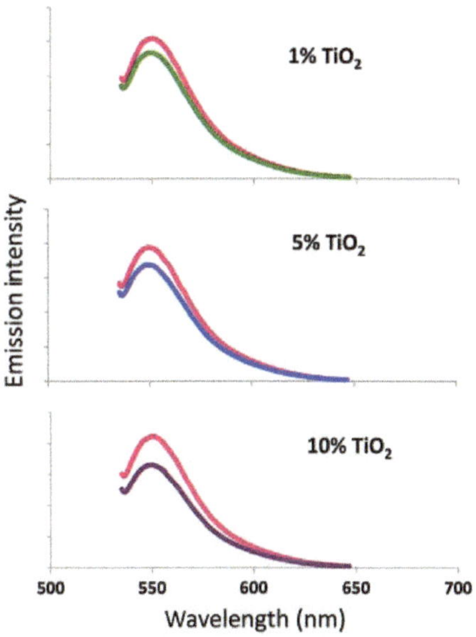

Figure 6. Fluorescence emission spectra of a rhodamine 6G solution before and after exposure to 4 h of sunlight in the presence of different samples of printed TiO_2-ABS nanocomposites. The pink curve in each plot corresponds to the rhodamine emission at the start of the experiment. The green curve (1% TiO_2-ABS), blue curve (5% TiO_2-ABS), and purple curve (10% TiO_2-ABS) correspond to the spectra recorded after 4 h in direct sunlight.

the nanoparticles. It has been proposed that TiO_2-containing materials can assist with the degradation of environmental pollutants in both wastewater streams and in the atmosphere. One of the most common tests of the photocatalytic properties of nanocomposites includes immersing the nanocomposite in a solution of fluorescent dye, exposing the mixture to light, and measuring the fluorescence of the solution as a function of time.[24]

Figure 6 shows the fluorescence spectra of rhodamine 6G solutions before and after exposure to sunlight in the presence of printed TiO_2-ABS samples. These spectra show a marked decrease in fluorescence emission over time; and the magnitude of this change increases as a function of TiO_2 concentration. The integrated intensity for each sample decreases as in the following manner. The 1% sample shows a decrease of integrated fluorescence intensity of 10% over time. The 5% sample shows a decrease of integrated fluorescence intensity of 13% over time. And the 10% sample shows a decrease of integrated fluorescence intensity of 22% over time.

It should be noted that these experiments used 3D printed dogbones. This geometry is hardly ideal for facilitating TiO_2 photocatalysis, which should include a high surface area to volume structure to increase the interaction of nanoparticles with the dye in solution. In fact we would only expect to see photogenerated radicals from the surface of the printed material. As the matrix that surrounds the TiO_2 can also be degraded by the photogenerated radicals, we measured the FTIR spectra before and after a 6 h solar exposure. Over this time period, we observe some minimal degradation of the ABS polymer (data shown in the Supporting Information). Because of the promising results described above, we have begun to experiment with different printed geometries to determine an optimal printed shape for applications that involve photocatalytic removal of environmental pollutants.

4. Conclusions

We have developed a process for generating TiO_2-ABS nanocomposite filaments for application in 3D printing. We have reported on the physical, mechanical, and chemical properties of these materials after they have been printed. The analysis of rhodamine 6G degradation characterizes the chemical and catalytic activity of a material that was produced using a commercial thermoplastic 3D printer. We expect to expand upon these studies by incorporating other, chemically active nanoparticles within ABS and bring new chemistry to 3D printed materials.

Acknowledgements

The authors would like to thank Douglas M. Fox of American University for helpful conversations and assistance in using twin screw compounder and Arec Jamgochian's assistance in the operation of the UTM at the National Institute of Standards and Technology (NIST).

Disclosure statement

No potential conflict of interest was reported by the authors.

Notes on contributors

At the time of publishing, Skorski and Esenther are both undergraduates at American University. Ahmed is on staff at the National Institutes of Standards and Technology whose primary research focus is on materials for photonics sensing. Miller has recently moved from a faculty position at American University to the United State's Food and Drug Administration. Hartings is a faculty member at American University who studies a broad range of polymer-transition metal nanoparticle nanocomposites, which includes areas that vary from biomineralization to synthesizing the types of industrial polymer nanoparticle systems described in this manuscript.

References

[1] Huang SH, Liu P, Mokasdar A, Hou L. Additive manufacturing and its societal impact: a literature review. Int J Adv Manuf Tech. 2013;67:1191–203.

[2] Ivanova O, Williams C, Campbell T. Additive manufacturing (AM) and nanotechnology: promises and challenges. Rapid Prototyping J. 2013;19:353–64.

[3] Miller JS, Stevens KR, Yang MT, et al. Rapid casting of patterned vascular networks for perfusable engineered three-dimensional tissues. Nat Mater. 2012;11:768–74.
[4] Tumbleston JR, Shirvanyants D, Ermoshkin N, et al. Continuous liquid interface production of 3D objects. Science. 2015;347:1349–52.
[5] Klein J, Stern M, Franchin G, et al. Additive Manufacturing of Optically Transparent Glass. *3D Printing and Additive Manufacturing*. 2015.
[6] Lewis GK, Schlienger E. Practical considerations and capabilities for laser assisted direct metal deposition. Mater Des. 2000;21:417–23.
[7] Yadoji P, Peelamedu R, Agrawal D, Roy R. Microwave sintering of Ni-Zn ferrites: comparison with conventional sintering. Mat Sci Eng B-Solid State Mater Adv Tech. 2003;98:269–78.
[8] Travitzky N, Bonet A, Dermeik B, et al. Additive manufacturing of ceramic-based materials. Adv Eng Mater. 2014;16:729–54.
[9] Kitson PJ, Rosnes MH, Sans V, et al. Configurable 3D-Printed millifluidic and microfluidic 'lab on a chip' reactionware devices. Lab on a Chip. 2012;12:3267–71.
[10] Kitson PJ, Symes MD, Dragone V, et al. Combining 3D printing and liquid handling to produce user-friendly reactionware for chemical synthesis and purification. Chem Sci. 2013;4:3099–103.
[11] Symes MD, Kitson PJ, Yan J, et al. Integrated 3D-printed reactionware for chemical synthesis and analysis. Nat Chem. 2012;4:349.
[12] Mannoor MS, Jiang Z, James T, et al. 3D Printed Bionic Ears. Nano Lett. 2013;13:2634–9.
[13] Duan B, Hockaday LA, Kang KH, et al. 3D Bioprinting of heterogeneous aortic valve conduits with alginate/gelatin hydrogels. J Biomed Mat Res Part A. 2013;101:1255–64.
[14] Hockaday LA, Kang KH, Colangelo NW, et al. Rapid 3D printing of anatomically accurate and mechanically heterogeneous aortic valve hydrogel scaffolds. Biofabrication 4. 2012.
[15] Narayanan R, El-Sayed MA. Catalysis with transition metal nanoparticles in colloidal solution: Nanoparticle shape dependence and stability. J Phys Chem B. 2005;109:12663–76.
[16] Link S, El-Sayed MA. Shape and size dependence of radiative, non-radiative and photothermal properties of gold nanocrystals. Int Rev Phys Chem. 2000;19:409.
[17] Cho MS, Park SY, Hwang JY, et al. Synthesis and electrical properties of polymer composites with polyaniline nanoparticles. Mater Sci Eng C-Biomim Supramol Syst. 2004;24:15–8.
[18] Li JR, Kuppler RJ, Zhou HC. Selective gas adsorption and separation in metal-organic frameworks. Chem Soc Rev. 2009;38:1477–504.
[19] Santos RM, Botelho GL, Machado AV. Development of acrylonitrile-butadiene-styrene composites with enhanced UV stability. J Mater Sci. 2014;49:5108.
[20] Evora VMF, Shukla A. Fabrication, characterization, and dynamic behavior of polyester/TiO_2 nanocomposites. Mater Sci Eng A-Struct Mater Prop Microstruct Process. 2003;361:358–66.
[21] Nussbaumer RJ, Caseri WR, Smith P, et al. Polymer-TiO_2 nanocomposites: A route towards visually transparent broadband UV filters and high refractive index materials. Macromol Mater Eng. 2003;288:44–9.
[22] Torrado Perez A, Roberson D, Wicker R. Fracture surface analysis of 3D-printed tensile specimens of novel ABS-based materials. J Fail Anal Prev. 2014;14,343–53.
[23] Luo YB, Li WD, Wang XL, et al. Preparation and properties of nanocomposites based on poly(lactic acid) and functionalized TiO_2. Acta Mater. 2009;57:3182–91.
[24] Kim J, Van der Bruggen B. The use of nanoparticles in polymeric and ceramic membrane structures: Review of manufacturing procedures and performance improvement for water treatment. Environ Pollut. 2010;158:2335–49.
[25] Nakata K, Fujishima A. TiO_2 photocatalysis: Design and applications. J Photochem Photobiol C-Photochem Rev. 2012;13:169–8.
[26] Ohko Y, Ando I, Niwa C, et al. Degradation of bisphenol A in water by TiO_2 photocatalyst. Environ Sci Tech. 2001;35:2365–8.
[27] Peral J, Domenech X, Ollis DF. Heterogeneous photocatalysis for purification, decontamination and deodorization of air. J Chem Tech Biotechnol. 1997;70:117–40.
[28] Oregan B, Gratzel M. A low-cost, high-efficiency solar-cell based on dye-sensitized colloidal TiO_2 films. Nature. 1991;353:737–40.
[29] Khan SUM, Al-Shahry M, Ingler WB. Efficient photochemical water splitting by a chemically modified n-TiO_2 2. Science. 2002;297:2243–5.
[30] Certain equipment or materials are identified in this paper in order to specify the experimental procedure adequately. Such identification is not intended to imply endorsement by the National Institute of Standards and Technology, nor is it intended to imply that the materials or equipment identified are necessarily the best available.
[31] Giordano RA, Wu BM, Borland SW, et al. Mechanical properties of dense polylactic acid structures fabricated by three dimensional printing. J Biomater Sci Polym Ed. 1996;8:63–75.
[32] Santos RM, Botelho GL, Machado AV. Development of acrylonitrile–butadiene–styrene composites with enhanced UV stability. J Mater Sci. 2014;49:510–8.
[33] Hamming LM, Qiao R, Messersmith PB, et al. Effects of dispersion and interfacial modification on the macroscale properties of TiO(2) polymer-matrix nanocomposites. Compos Sci Tech. 2009;69:1880–6.
[34] Hanemann T, Szabo DV. Polymer-Nanoparticle Composites: From Synthesis to Modern Applications. Materials. 2010;3:3468–517.

13

Silicon etch with chromium ions generated by a filtered or non-filtered cathodic arc discharge

Daniele Scopece[a], Max Döbeli[b], Daniele Passerone[a], Xavier Maeder[d], Antonia Neels[e], Beno Widrig[c], Alex Dommann[a], Ulrich Müller[d] and Jürgen Ramm[c]

[a]nanotech@surfaces laboratory, Empa, Swiss Federal Laboratories for Materials Science and Technology, Überlandstrasse 129, CH-8600, Dübendorf, Switzerland
[b]Ion Beam Physics, ETH Zürich, Otto-Stern-Weg 5, CH-8093, Zürich, Switzerland
[c]Oerlikon Surface Solutions AG, Oerlikon Balzers, Iramali 18, 9469 Balzers, Liechtenstein
[d]Laboratory for Nanoscale Materials Science, Empa, Swiss Federal Laboratories for Materials Science and Technology, Überlandstrasse 129, CH-8600 Dübendorf, Switzerland
[e]Center for X-ray Analytics, Empa, Swiss Federal Laboratories for Materials Science and Technology, Überlandstrasse 129, CH-8600 Dübendorf, Switzerland

ABSTRACT
The pre-treatment of substrate surfaces prior to deposition is important for the adhesion of physical vapour deposition coatings. This work investigates Si surfaces after the bombardment by energetic Cr ions which are created in cathodic arc discharges. The effect of the pre-treatment is analysed by X-ray diffraction, Rutherford backscattering spectroscopy, scanning electron microscopy and in-depth X-ray photoemission spectroscopy and compared for Cr vapour produced from a filtered and non-filtered cathodic arc discharge. Cr coverage as a function of ion energy was also predicted by TRIDYN Monte Carlo calculations. Discrepancies between measured and simulated values in the transition regime between layer growth and surface removal can be explained by the chemical reactions between Cr ions and the Si substrate or between the substrate surface and the residual gases. Simulations help to find optimum and more stable parameters for specific film and substrate combinations faster than trial-and-error procedure.

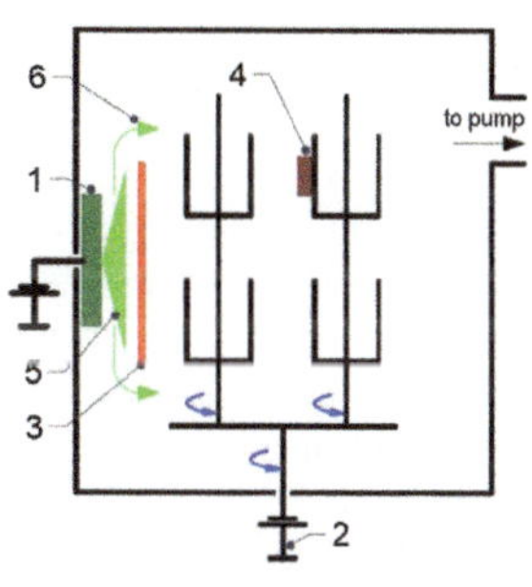

1. cathode of Cr
2. DC bias supply connected to the two-fold rotating sample holder
3. shutter to filter the macroparticles
4. samples of Si subject to a two-fold rotation
5. metal vapor produced from the cathode
6. filtered metal vapor

KEYWORDS
Adhesion; metal ion etch; filtered arc; nucleation; surface binding energy; TRIDYN

CLASSIFICATION
40 Optical; magnetic and electronic device materials; 402 Multi-scale simulation; 212 Surface and interfaces; 305 Plasma/Laser processing; 503 TEM, STEM, SEM; 504 X-ray/Neutron diffraction and scattering

1. Introduction

In physical vapour deposition (PVD) the bombardment of substrate surfaces with ions before deposition is an established process to improve the coating adhesion. For conductive substrates, the bombarding energy of the positively charged ions can be controlled by a negative bias applied to the substrates. A surface treatment with inert gas ions such as argon supports the desorption of gaseous species and removes surface contaminants by sputtering. The situation is different if, instead of inert gases, ionized metal vapour is utilized. Depending on the magnitude of the applied bias, the treatment can result in layer growth or surface removal by sputtering. The removal of surface atoms by metal ions is known as metal ion etch (MIE) [1,ch.8.5,2,3,4]. This denomination does not usually distinguish if only metal ions were utilized or if argon was added to increase the efficiency of sputtering. Generally, MIE is used to remove surface atoms.

CONTACT Daniele Scopece daniele.scopece@empa.ch; daniele.scopece.science@gmail.com

In industrial deposition technology, the ionized metal vapour for the substrate pre-treatment is frequently generated by cathodic arc sources. The arc discharge produces from the cathode material (target) highly ionized metal vapour with different charge states and neutrals [1]. In addition, the arc discharge generates macroparticles, so called droplets. These droplets are undesirable for some applications, because they may reduce the mechanical and thermal stability of the coatings. The droplets also influence the result of the MIE because they carry no charge, are different in size and contribute to layer growth, i.e. reduce the efficiency of sputtering. For the pre-treatment of the substrates with ions, the substrates can either be exposed directly to an arc source or, if droplets have to be avoided, the vapour from the arc source can be filtered. The first case will be named here *non-filtered MIE* (NF-MIE) and the second *filtered MIE* (F-MIE).

If substrate surfaces are bombarded with pure ionized metal vapour (ions, neutral vapour, macroparticles), there is a range of low bias voltages for which layer growth can be observed, while at higher (more negative) bias the substrate surface is removed. The transition bias range between layer growth and surface removal depends on the arc discharge (filtered, non-filtered), the composition of the ionized vapour, the residual gas components and pressure, and the substrate material. Better understanding of the basic physical and chemical processes at the substrate surface governing material deposition and removal will support a faster process development in industrial PVD coating applications.

The present work investigates this transition range in a PVD production system for Si substrates bombarded with Cr ions. The results of the pre-treatment are investigated for NF-MIE and F-MIE configurations as a function of substrate bias. The experimental data are compared with TRIDYN simulations [5–7]. These simulations are not able to describe the full complexity of the experiments but are extremely useful for a discussion of trends and a more detailed understanding of the transition region.

2. Experimental details

2.1. Sample preparation

The samples are prepared in an INNOVA batch-type production system of Oerlikon Surface Solutions AG (Figure 1). The deposition chamber is evacuated to a pressure of 2×10^{-4} mbar before the cathodic arc source, which produces the ionized metallic vapour for the substrate pre-treatment, is ignited. These vacuum conditions are typical for the deposition of temperature-sensitive components requiring temperatures below 300°C. This is also the reason why it is resigned to heat the vacuum chamber to accelerate desorption of gases and water from the inner chamber walls. Elemental Cr targets (cathodes) are utilized to produce the ionized metallic vapour with a DC discharge current of 140 A, resulting in a Cr evaporation rate of 12.2 g h^{-1} in both NF-MIE and F-MIE. The cathodes are installed at the inner chamber wall as schematically shown in Figure 1.

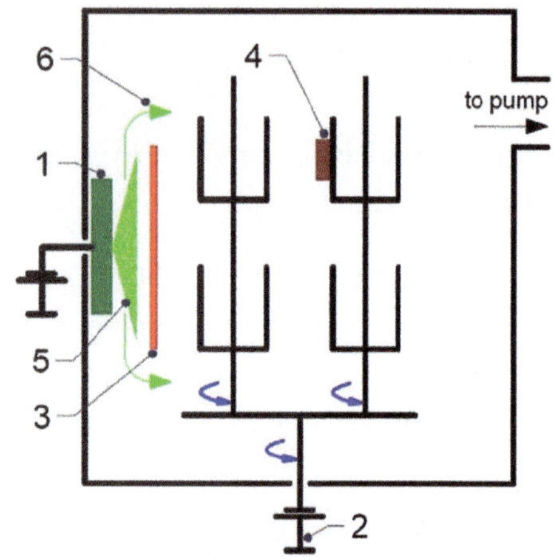

Figure 1. Schematics of the experimental set-up for the MIE in the deposition system. The picture shows: the cathode of Cr (1), the DC bias supply connected to the twofold rotating sample holder (2), the shutter to filter the macroparticles (3), the samples of Si subject to a twofold rotation (4), the metal vapour produced from the cathode (5) and the filtered metal vapour (6). The shutter (3) is fixed during F-MIE and removed for NF-MIE.

In the case of NF-MIE the substrates are directly exposed to the arc source. For the filtered case (F-MIE), there are different implementation approaches described in the literature [1,ch.7.4]. In our experiments, a fixed water-cooled shutter is installed in front of the cathode, to protect the substrate surfaces from the deposition with metallic droplets and from the condensation of neutral metallic vapour. This is a simple design which can be realized in every deposition system equipped with arc sources. The fixed shutter does not, however, completely block the ion current. Bias voltages between −50 V and −1200 V are applied to the substrates in order to modify the bombardment energy of the Cr ions. No additional gases (such as Ar) are added to the chamber, i.e. the arc discharge is self-sustaining and produces Cr vapour only. Although there are no additional gases fed to the vacuum chamber, the plasma created by the arc discharge current initiates gas desorption, especially of water vapour, from the inner chamber walls and fixtures. These gases are dissociated and activated in the arc discharge. Simultaneously, the metallic vapour "pumps" these gases efficiently due to gettering which is manifested in the drop of total chamber pressure right after arc ignition. The substrates are mounted on standard substrate holders and experience a twofold rotation during ion bombardment (Figure 1). The total substrate ion current (I_{Bias}, Table 1) measured for the substrate including substrate holder in F-MIE is between 0.7 and 0.8 A,

Table 1. Process parameters of the NF-MIE and F-MIE experiments.

NF-MIE	U_{Bias} [-V]	I_{Arc} [A]	I_{Bias} [A]	Cr [10^{16} at/cm^2]
	50	140	5	300
	200	140	5.3	135
	400	140	5.5	35
	600	140	6	5.7
	800	140	6.5	21
F-MIE				
	50	140	0.7	20.0
	200	140	0.7	13.6
	400	140	0.7	3.3
	600	140	0.8	11.6
	800	140	0.8	9.1
	1000	140	0.8	4.1
	1200	140	0.8	3.3

while it amounts to about 6 A for NF-MIE. The substrates are immersed in the plasma of the arc discharge. They charge to about −5 to −10 V against ground if they are left at floating potential, indicating that a plasma sheath is formed. The Si substrates are wet-chemically cleaned before loading to the deposition system. For each bias value, the Si substrates are bombarded for 30 min with the ionized metal vapour. At the end of this treatment, their temperature does not exceed 100°C for all bias voltages in F-MIE, but can exceed 500°C for direct exposure for NF-MIE.

A summary of the experimental conditions is given in Table 1.

2.2. Coating characterization

For a qualitative comparison of NF-MIE with the F-MIE treatments, the surface and fracture cross-sections of the coatings are investigated in a LEO 1530 scanning electron microscope (SEM). In this work we use a cross-section scanning electron microscopy and will define it as X-SEM.

In addition, the samples are characterized by Rutherford backscattering spectrometry (RBS) [8] at the ETH Laboratory for Ion Beam Physics. Measurements are performed using a 2 MeV ^{4}He beam under normal incidence to the sample surface with a beam spot size of 1 mm. Backscattered particles are detected at an angle of 168° towards the incident beam direction by a silicon PIN diode detector with an energy resolution of 12 keV (FWHM). The collected RBS data are fitted by simulations using the RUMP software [9]. For the thinnest deposits the resolution of the RBS measurement is not sufficient to resolve the exact depth profile of the material composition and the degree of implantation or sub-plantation of the layer. Therefore only the net areal density in number of metal atoms per Å^2 is determined and this is the quantity compared with our simulations with TRIDYN. In the case of NF-MIE some layers are thick enough for RBS to reveal resolved compositional depth profiles, which will be discussed in more detail below.

In order to check possible crystal phases created during deposition, the samples are analysed with X-ray diffractometry (XRD) on a PANalytical X'Pert PRO MRD instrument located at Empa Thun (CH) manufacturer: PANalytical using Cu-Kα radiation in a 2θ mode with 1° grazing incidence angle. This small angle is used to get maximum information of the surface of the coating and to avoid the signal from the substrate. The crystallography open database (COD) is used to identify the crystallographic phases [10,11].

The F-MIE samples are characterized with X-ray photoemission spectroscopy (XPS) in depth analysis. Depth profiles are acquired on a Physical Electronics Inc Chanhassen Instrument located in Empa Dubendorf (CH) [12] using monochromatized Al Kα radiation (hν = 1486.7 eV) and a hemispherical capacitor electron-energy analyser equipped with a channel plate and a position-sensitive detector. The electron take-off angle is 45° and the beam diameter used is around 150 μm. The analyser is operated in the constant pass energy mode at 58.7 eV and for the spectra a step size of 0.25 eV is used for all measurements. The binding energy is calibrated using Cu $2p_{3/2}$, Ag $3d_{5/2}$ and Au $4f_{7/2}$ at 932.62 eV, 368.21 eV and 83.96 eV, respectively to within ± 0.1 eV [13]. To compensate for possible surface charging, built-in electron and argon ion neutralizers are used. The base pressure of the system was below 5×10^{-7} Pa. Sputtering was performed with Ar^+-ions with 1 kV acceleration voltage and scanned over an area of 2 × 2 mm which results in a typical sputter rate on Ta_2O_5 of 2 nm min^{-1}. The real sputter rate on chromium, chromium oxide or silicon may differ from the actual one by up to a factor of 2; however, the direct comparability is certainly better than 10%. The Cr 2p, Si 2p, O 1s and C 1s peaks were measured. Depth profiles are analysed using the software MultiPak 8.2B [12] and peak areas are determined after a Shirley background subtraction. The atomic concentrations are calculated using the corrected relative sensitivity factors as given by the manufacturer and normalized to 100 at.%.

Table 2. Parameter utilized for TRIDYN simulations. The beam compositions for NF-MIE are the ones actually used with the neutrals included.

Parameter	Non-filtered MIE	Filtered MIE
Beam composition	Cr^{4+} (0.6%), Cr^{3+} (12.60%), Cr^{2+} (40.8%), Cr^{1+} (6.0%), Cr^{0} (40.0%), from [1]	Cr^{1+} (100%)
Final fluence [atoms/Å^2]	300	20
Number of histories	200,000	20,000
SBV (Cr,Si) [eV]	6.69 ($CrSi_2$)	6.69 ($CrSi_2$)
SBV (Cr,Si) [eV] test to interpolate the outliers	X	6.69 ($CrSi_2$)
SBV (Cr,O) [eV] Test to interpolate the outliers	X	10.20 (Cr_2O_3)
SBV (Cr,Cr) [eV]	4.12	
SBV (Si,Si) [eV]	4.70	

2.3. *TRIDYN simulations*

In this work we adopt the Monte Carlo code TRIDYN (version HZDR) [5–7], used successfully in previous works by other authors [14,15]. In a previous investigation [16], our attempts to simulate the results of F-MIE were not satisfactory and this is the reason to investigate this topic in greater detail in this work. This code uses a binary collision approximation to model the multiple and cascade collisions between incoming ions and the atoms of the substrate. We intentionally neglect non-kinetic effects such as diffusion or chemical reactions [17,18] and focus on the redistribution of energy and atoms via collisions. The parameters used in the simulations in both cases are reported in Table 2, justified in section 3 and briefly discussed here.

The total final fluence is extracted from the RBS measurements for the bias of −50 V for which sputtering has little influence and sticking probability is assumed to be 1. The kinetic energy of the ions is determined by the bias voltage applied in experiments and by the charge of the ions imposed in the simulations with no further broadening considered. The surface binding energy (SBE) of each atomic species is calculated from the local material composition via the binding energy matrix SBV describing mutual binding energies between all types of atoms present. The diagonal SBV components (Cr–Cr and Si–Si) are adopted from the TRIDYN code [10]. The off-diagonal SBV component is computed from the formation enthalpy of different compounds (Table 2) observed in our XRD results discussed below (section 4) and other values for comparison in selected cases. It has been checked that the number of simulated particle histories in the Monte Carlo calculation is large enough to exclude any significant statistical variation.

The TRIDYN simulations executed for NF-MIE (Figure 6(a)) are based on the ion charge state distribution given in Table 2, based on the closest conditions reported in the literature [1,p.508]. In addition to that, it is assumed that about 40 at.% of the Cr projectiles are neutrals, without a specific classification between droplets and atoms. This assumption is justified by the comparison of the evaporation rate of Cr (12 g h^{-1}) with the total substrate current at the substrate holder (about 6 A) and assuming that some of the neutral Cr will not be deposited at the substrate surface but at the chamber walls. The simulations of F-MIE are performed by utilizing a pure Cr^{1+} beam. However, slight modifications of this assumption do not alter the trend considerably.

Based on these parameter sets, the areal density of Cr as a function of substrate bias, the surface recession (i.e. etching or deposition depth) as a function of Cr fluence for different substrate biases, and the fraction of Cr as a function of depth for different substrate biases are calculated and utilized for the comparison with experimental results.

4. Results and discussion

Figure 2(a) and (b) shows the cross section of substrates treated in NF-MIE for substrate bias of −400 V and −800 V. At −400 V (Figure 2(a)), a coating can be detected. However, the determination of the thickness is difficult because the layer is not uniform in thickness. Values between 30 nm and 70 nm can be estimated. No coating is visible at the substrate which was biased with −800 V (Figure 2(b)). Instead, the substrate surface is damaged to a depth of 500 nm and more. The top view of the sample surface obtained at −400 V (Figure 2(c)) shows a fissured coating including circular areas with droplet residues. The appearance of these areas suggests that the droplets were removed during the NF-MIE process. The structure of the surface for −800 V (Figure 2(d)) shows a similar porosity, however, with more pronounced etching effect for the droplet regions. These regions do not contain Cr residues anymore. The findings indicate the difficulties to determine the end of layer growth in the NF-MIE process and the detrimental influence of droplets not only for the coating uniformity but also for a defined transition to etching. Given the occurrence of the damaged substrate surface at −800 V, no investigations of higher bias voltages were performed.

The substrate ion current is one order of magnitude lower for F-MIE, resulting in a coating thickness which cannot be determined by X-SEM investigations. The top view of the substrate surface shows a non-continuous layer with weak indications of island growth for −400 V bias (Figure 3(a)). For −800 V (Figure 3(b)) the surface is featureless, with the exception of few defects ascribed to particles. The transition between layer growth and etching is difficult to determine in the case of NF-MIE because of droplet induced etching effects. In F-MIE, it

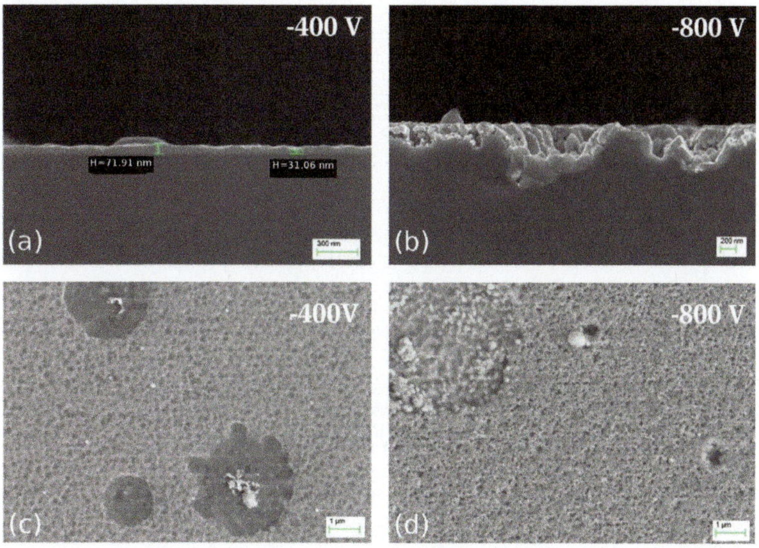

Figure 2. SEM micrographs of the Si samples obtained in NF-MIE. (a) and (b) show the cross section of the samples for –400 V bias voltage and –800 V bias, respectively. For both bias voltages, the surface appearance is shown in (c) and (d), respectively. See discussion in section 4.

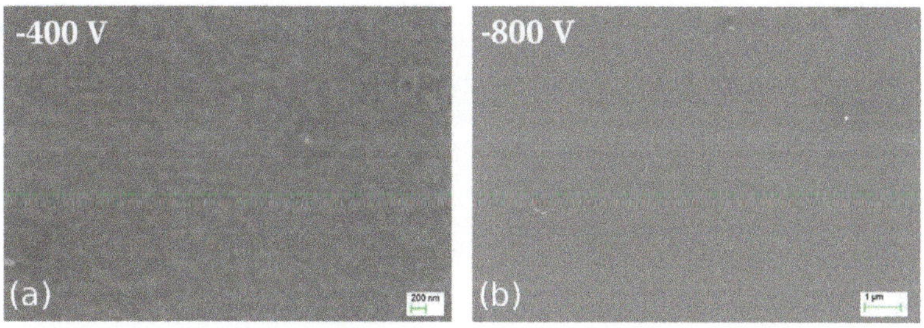

Figure 3. SEM micrographs of the Si surfaces obtained after F-MIE for (a) –400 V and (b) –800 V. The surface does not show crater formation like it does for samples of NF-MIE.

is difficult as well because a thickness measurement is not possible by standard methods like X-SEM. For thin film development, however, it would be of interest to determine the parameters for which the substrate material first appears at the surface and may influence the layer nucleation and the chemistry of the film-substrate interface.

Therefore, it was investigated if RBS could provide a better quantification of the transition region between growth and etching. In Figure 4, a typical RBS spectrum is given. The spectrum was measured for the sample which was exposed to F-MIE at –800 V bias. All spectra of the samples prepared in F-MIE show a Cr signal which is well separated from the Si substrate. This allows an accurate estimation of the total Cr areal density. The depth resolution of the technique, however, does not permit a distinction if the Cr layer is above the Si surface (layer growth) or intermixed with the substrate, which is also the case for the oxygen which may be present in the spectra.

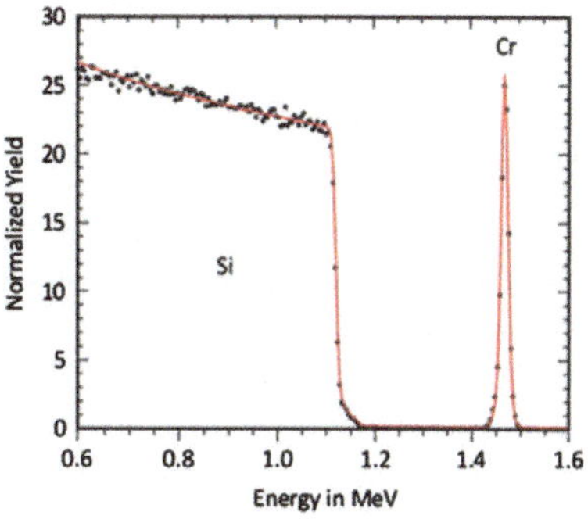

Figure 4. Typical RBS spectrum for F-MIE sample (–800 V bias).

All RBS spectra obtained from the samples of NF-MIE between –50 V and –600 V bias have also well separated Cr signals. An example is shown in Figure 5(a) for the experiment with –400 V bias. This sample was tilted during analysis by 57° to increase depth resolution.

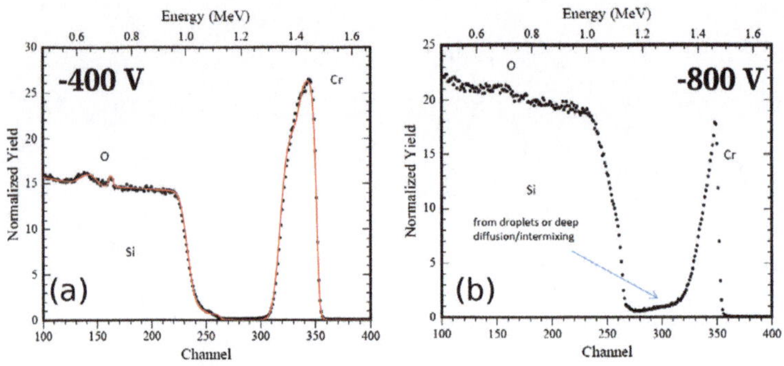

Figure 5. RBS spectra obtained from NF-MIE. For the bias of –400 V (a), the spectrum indicates oxygen accumulation at the interface and at the surface and additional Si intermixing in the interface. For the bias of –800 V (b), enhanced intermixing of Si and Cr is observed. The Cr low energy tail is either caused by droplets on the surface or Cr diffusion into the substrate.

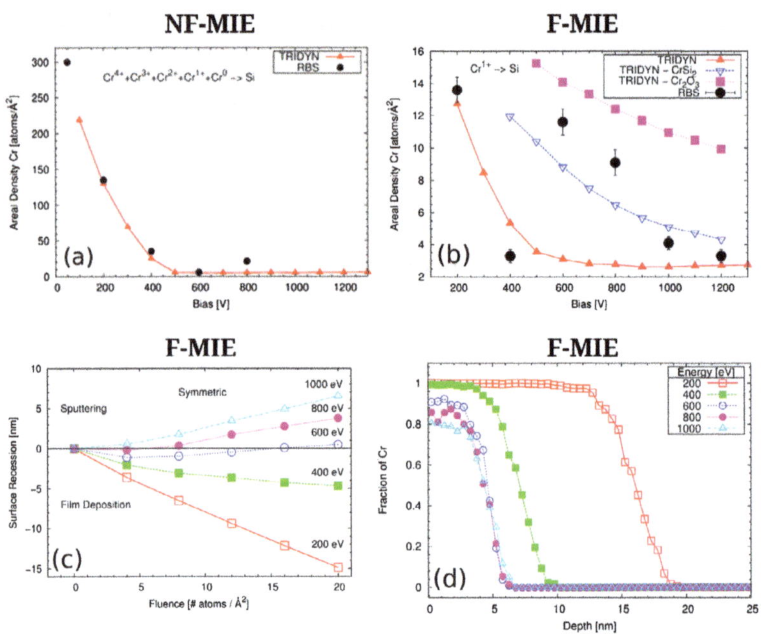

Figure 6. TRIDYN simulations (red curves) with the parameters specified in Table 2 superimposed on the RBS experimental data (black points) for NF-MIE (a) and F-MIE (b), respectively. In (b) the blue and purple curves are obtained with modified values of the SBE for the Cr silicide and the Cr oxide reported in the key, representing the limit of the possible mixed phases created (see text for discussion). (c) Surface recession in F-MIE for the different substrate bias (see section 4 for discussion). (d) Cr concentration profile versus depth for different substrate bias for F-MIE at the final fluence of 20 atoms/Å^2.

The spectrum indicates a thin oxide at the surface of the sample and an oxide in the interface. In addition, the comparison with RUMP simulations suggests a Si profile in the coating with a Si/Cr ratio of about 0.15 at the interface and zero at the surface, indicating that a Si–Cr intermixed region is formed before the Cr layer growth. For the bias of –800 V (Figure 5(b), sample not tilted), the Cr peak has a distinctive tail towards lower energies. This indicates Cr diffusion into the Si substrates or can result from increased surface roughness e.g. by droplets (see Figure 2(b)). The spectrum also shows in the Si signal at the surface a lower Si value than at greater depth, which might be explained by silicide or oxide formation. For all NF-MIE samples, oxide contaminations are detected in the interface. This indicates that at least in the beginning of the MIE the residual gas, mainly water vapour, influences the result of the MIE.

The measured Cr areal density (black dots) as a function of the substrate bias is given in Figure 6 for NF-MIE (a) and F-MIE (b) experiments. If, for sake of simplicity, we consider ionized vapour and kinetic processes only, some general trends should be reflected in these curves [19]. For energies up to a few hundred eV, sputtering is negligible and the bombarding atoms can penetrate only a few nm into the material. Disregarding chemical effects and volatility of certain elements and compounds, a film grows with virtually the same elemental composition as the impinging beam with an intermixed interface layer between substrate and film. Depending on the flux, the time span between intermixed interface formation and complete layer formation is different, i.e. the exposure time of the sample surface to vacuum ambient is different. For higher energies, layer growth turns into layer removal and the erosion of surface atoms starts due to sputtering. Simultaneously, the ion implantation depth increases with bombarding energy and most of the substrate atoms are removed at the beginning of the bombardment. After a layer with a thickness corresponding to the implantation range has been removed, more and more previously implanted ions are also sputtered and finally an equilibrium state is established.

To confirm these general trends for our experiments, TRIDYN simulations are performed with the parameters discussed in section 2.3 and reported in Table 2. The parameters of our simulations were chosen to obtain a reasonable agreement for the regime of layer growth at the lowest bias and are not based on measurements, as was discussed above. It was not the intention to completely describe NF-MIE and F-MIE under our conditions with TRIDYN, but to become able to better separate effects purely induced by energetic atomic collisions from those surface modifications that cannot be described by the binary collision approximation (on which TRIDYN is based) alone, such as diffusion, chemical processes and phase formation. This set of simulations is thought to extend and complement the one reported in [16] to take into account the two possible explanations highlighted above as suggested from experiments: oxides or silicides. In [16] we investigated the presence of oxygen in the beam due to the water vapour in the chamber. In this work we consider the option of no oxygen with formation of silicides or oxides in the compounds (but not in the beam explicitly), as discussed in detail below.

The results from TRIDYN simulations are given in Figure 6(a) and (b) (red line) for NF-MIE and F-MIE, respectively. With the chosen parameters, the simulations are brought in fair agreement with the experiments for layer growth.

Figure 6(c) reports the trend of the surface recession in function of substrate bias and for different fluences. Negative values of this quantity correspond to deposition (growth of coating), and positive values to sputtering (recession of the coating). For small fluences, it indicates dynamical equilibrium conditions at –800 V, shifting to –600 V for higher fluences. In Figure 6(d), the distribution of the Cr versus depth at the final fluence of 20 atoms/Å^2 for different substrate bias is shown. According to these simulations, Si will appear at the substrate surface at –600 V bias and may contribute to reactions. This is also the range for which the RBS data disagree with TRIDYN simulations (Figure 6(b)).

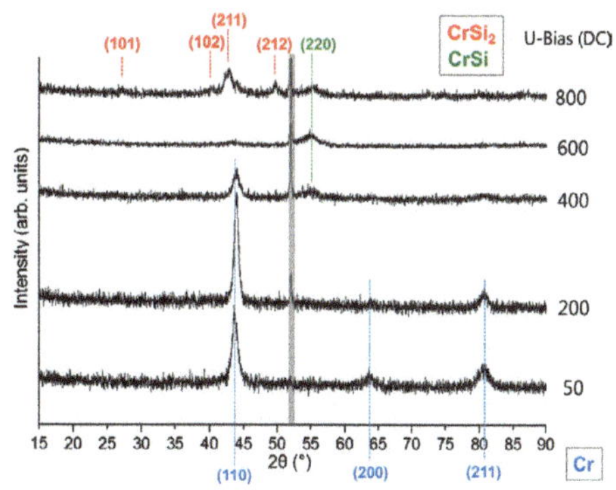

Figure 7. Grazing incidence XRD spectra of the substrate surfaces treated with NF-MIE for different substrate bias. This is a 2θ measurement with a grazing angle of 1°.

Indeed, based on our discussion (and from TRIDYN), we would expect a monotonic decrease of the atomic density with increased bias. On the contrary, an increase in areal densities is measured for NF-MIE at –800 V and for bias voltages of –600 V and higher for F-MIE.

This discontinuity in the curves can be understood as a jump in surface binding energy (by using the "TRIDYN" jargon) indicating chemical reactions at the surface, as outlined below for the case of F-MIE. To understand the reason behind the unexpected increase of the Cr areal densities, XRD grazing incidence measurements are performed to explore the phase formation close to the surface of the substrates obtained in NF-MIE (Figure 7). The analysis shows only the Cr Bragg peaks in the bias range between –50 V and –400 V which is in accordance with the net layer growth also observed in SEM analysis. At –400 V an additional peak appears which can be attributed to CrSi (COD 1010034) [20] and at –800 V $CrSi_2$ (COD 1010033) [20] is formed at the substrate surface. The XRD analysis supports the idea that for small Si concentrations at the substrate surface CrSi is formed and for higher Si concentrations (–800 V) enough Si is present at the surface to form $CrSi_2$. However, it is likely that in NF-MIE also thermal effects play a role. This means that chemical reactions occur during the bombardment of the Si surface with Cr vapour produced by the non-filtered arc source. In the spectrum, no oxide phases can be detected, despite the fact that oxygen is found in the RBS spectra. This indicates that possible formed oxides are amorphous.

The layers formed during F-MIE are too thin for XRD analysis. The performed XRD grazing incidence analysis (not shown here) could not identify a silicide. However, assuming a detection limit of about 3.5 nm, a silicide formation of a few monolayers cannot be excluded.

To find the reason for the increase of the Cr areal density measured by RBS and the depth profile of oxygen, XPS was performed for these samples. The results are shown in Figure 8 for bias voltages between –400 V and –1000 V. In all XPS spectra, an oxygen surface peak

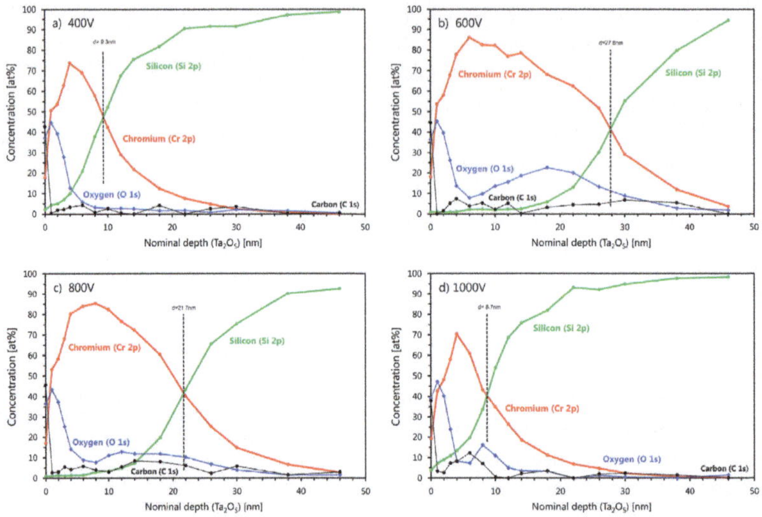

Figure 8. XPS measurements on the F-MIE samples. From panel (a) to (d): 400, 600, 800, and 1000 V bias. The depth is not calibrated. The amount of oxygen in the outliers 600 and 800 V is considerably higher than the other two: this supports the idea of formation of oxides against the one on silicides. See text for discussion.

is visible. The peaks have similar shapes and thickness. The shoulders in all Cr signals are correlated with the oxygen, indicating a surface oxidation of Cr. This surface oxide is most probably due to the exposure of the sample to ambient after the deposition process and therefore not related to F-MIE. At –400 V (Figure 8(a)) and –1000 V (Figure 8(d)) substrate bias, a small amount of Si can also be detected at the substrate surface, indicating that not only chromium oxide is formed. The thickness for the Cr containing regions in the two substrate surfaces is about the same. However, only at –1000 V do we see a clear indication of oxygen in the interface region to the silicon substrate. For –600 V (Figure 8(b)) and –800 V (Figure 8(c)), the Cr depth profile shows a larger thickness with a Cr concentration of about 70 at.%. In addition, a higher oxygen concentration between 10 and 20 at.% was measured, indicating that an oxide was formed. These results indicate partial oxidation of the sample surface during F-MIE, especially in this bias range. A qualitative explanation for this is not straightforward. However, the comparison with TRIDYN simulations can be of help in this case as well.

We can conclude that the higher areal densities obtained in the transition between layer growth and surface removal can be simulated with higher surface binding energies. If we assume silicide or oxide formation and therefore (partially) replace the surface binding energy (SBV) of the Si–Si, Cr–Cr or Si–Cr by the SBV values of silicides or oxides we can easily reduce the sputter yield and reproduce the higher areal densities. This is performed in Figure 6(b) for the –600 V and –800 V values for the $CrSi_2$ and Cr_2O_3 compounds. In this case the higher SBV and SBE values generate a higher amount of Cr in the sample and a better agreement with the points at 600 and 800 V bias. The two curves in Figure 6(b) (blue and purple) represent the extreme cases of a complete coverage of the surface by pure silicide or pure oxide (TRIDYN allows only the simulation of complete coverage and not mixed phases). The idea of the simultaneous formation of both Cr silicides and Cr oxides (both likely under-stoichiometric), as suggested by our XPS data, implies that the corresponding amount of Cr in this mixed phase lies between the two curves depicted in Figure 6(b), thus in fair agreement with the experimental RBS data.

TRIDYN is not only able to provide a good estimation for the substrate bias at which zero growth equilibrium has to be expected (i.e. the moment when chemical reactions are more likely to take place). It can also reveal details in the temporal evolution of the surface composition which may well be a possible cause for the suspected enhanced oxide formation for F-MIE. If a value for the particle flux is given, the curves of surface position versus fluence in Figure 6(c) can be transformed into surface position versus absolute time. The curves for 600 and 800 V bias are of special interest here. They demonstrate that even for a bias with net surface removal there is always an initial deposition of Cr due to implantation. The total time span of this transition phase shown in the figure corresponds to 30 min. Thus, surface atoms have a considerable amount of time to interact with the environment and create compounds that are more resistant to sputtering.

4. Conclusions

The transition between layer growth and surface removal for the treatment of Si surfaces with pure metallic Cr vapour from a cathodic arc source is difficult to control in production systems. The direct exposure of the substrates to the ionized vapour and high flux of ions may result in chemical reactions between the ionized metal and the atoms in the substrate surface which can turn in pronounced surface roughening. Silicides are formed already at a substrate bias (-400 V) for which net layer growth takes place. Increased reactivity is observed for the regions where droplets occur, indicating that excess Cr accelerates surface removal. This fact and a rather abrupt transition between layer growth and surface damage restrict the application of this approach and make the stable tuning of NF-MIE for tiny layer thickness impossible.

A better control can be obtained if the ions are produced by a filtered arc source. Such a source does not only generate droplet-free ionized vapour, but also reduces the ion flux and total fluence which in turn allows better control of the transition between growth and surface removal. It also reduces the thermal budget in the MIE process. This was shown for F-MIE with typical ion currents about an order of magnitude lower than corresponding currents in NF-MIE. Under these conditions, the filtered Cr vapour does not damage the Si surface. However, oxide formation is observed in the transition region probably due to longer residence times of atoms at the surface and indicates that the control of the residual gases is essential for a reliable F-MIE process operating at low fluences. An F-MIE can be critical if the ion flux is comparable with the impingement rate of the residual gas. A comparison with TRIDYN helps predict the bias region between layer growth and etching in which the substrate surface is changing slowly and becomes more susceptible to the influence of residual gases, diffusion and chemical reactions with substrate atoms. It is the region for which the atoms of the substrate appear at the surface. At this bias, the surface is not yet covered by growth nor experiences high etch rates and reactions forming compounds with larger SBV are possible. This region is very important to engineer interfaces. The fit of the experimental data in the growth regime with TRIDYN and the agreement or disagreement for the remaining curve indicate whether chemical reactions must be considered. Once F-MIE is described in this way, TRIDYN simulations can be extended to other substrate materials to allow the prediction of zero growth conditions.

Acknowledgements

This work is a part of the projects: "DD-Coat" (CTI-No. 13080.1 PFIW-IW) and the Marie-Curie Cofund Project @ Empa "Multi-Scale Computational Approach to Intermetallic Oxide Formation: Towards a Knowledge-Based Coating Design (MultiMOx)" (PSP 5211.00501.101.09). All authors acknowledge W. Moeller for an initial fruitful discussion to understand the code TRIDYN and Jörg Patscheider for a critical reading of the manuscript.

Disclosure statement

No potential conflict of interest was reported by the authors.

References

[1] Anders A. Cathodic Arcs. New York: Springer; 2008. chapter 2.
[2] Munz WO, Schulze D, Hauzer FJM. "A New Method for Hard Coating - Arc Bond Sputtering" Surf Coat Technol. 1992;50:169–178.
[3] Schonjahn, C, Ehiasarian AP, Lewis DB, et al. J Vac Sci Technol. 2001;19:1415–1420.
[4] Ehiasarian AP, Anders A, Petrov L. "Optimization of in situ subsate surface treatment in a cathodic arc plasma: A study by TEM and plasma diagnostics" J Vac Sci Technol. 2007;25:543–550.
[5] Möller, W., Eckstein W. "Tridyn - A Trim Simulation Code Including Dynamic Composition Changes." Nuclear Inst Meth Physics Research B. 1984;2:814
[6] Möller W, Eckstein W. Biersack JP. "Tridyn - Binary Collision Simulation of Atomic Collisions And Dynamic Composition Changes in Solids" Comput Phys Commun. 1988;51:355–368.
[7] Möller, W, Posselt M. Manual TRIDYN HZDR. 2012;1.
[8] Chu WK., Mayer JW, Nicolet MA. Backscattering spectroscopy. Academic Press: New York 1978.
[9] Doolittle, LR. "A Semiautomatic Algorithm For Rutherford Backscattering Analysis" Nucl Instr Meth B. 1986;15:227–231.
[10] Grazulis, S, Daskevic A, Merkys A, et al. "Crystallography Open Database (COD): an open-access collection of crystal structures and platform for world-wide collaboration" Nucleic Acids Res. 2012;40:D420–D427.
[11] Grazulis S, Chateigner D, Downs RT, et al. "Crystallography Open Database - an open access collection of crystal structures" J Appl Crystallogr. 2009;42:726.
[12] Physical Electronics Inc. Chanhassen, MN 55317 (USA). http://www.phi.com/
[13] ISO 15472, 2010-05
[14] Kubart T, Nyberg T, Berg S. "Modelling of low energy ion sputtering from oxide surfaces" J Phys D Appl Phys. 2010;43: 205204.
[15] Kubart T, et al. "Modelling of sputtering yield amplification effect in reactive deposition of oxides" Surf Coat Techn. 2010;204: 3882–3886.
[16] Döbeli M, Dommann A, Maeder X, et al. "Surface layer evolution caused by the bombardment with ionized metal vapor" Nucl Instr Meth B. 2014;332:337–340.
[17] Schmid K, Roth J. J "Erosion of high-Z metals with typical impurity ions" Nucl Mater. 2003;313:302.
[18] Schmid K, Roth J. J "Concentration dependent diffusion of carbon in tungsten" Nucl Mater. 2002;302:96.
[19] Sigmund P. "Mechanisms and Theory of Physical Sputtering by Particle Impact" Nucl Instr Meth. 1987;27:1.
[20] Boren B. "Roentgenuntersuchung der Legierungen von Silicium mit Chrom, Mangan, Kobalt und Nickel" Arkiv foer Kemi, Mineralogi och Geologi, A. 1934;11:1–28.

14

Luminescence of delafossite-type $CuAlO_2$ fibers with Eu substitution for Al cations

Yin Liu[a], Yuxuan Gong[a], Nathan P. Mellott[a], Bu Wang[a], Haitao Ye[b] and Yiquan Wu[a]

[a]New York State College of Ceramics, Alfred University, Alfred, NY, USA
[b]School of Engineering and Applied Science, Aston University, Birmingham, UK

ABSTRACT

$CuAlO_2$ has been examined as a potential luminescent material by substituting Eu for Al cations in the delafossite structure. $CuAlO_2$:Eu^{3+} nanofibers have been prepared via electrospinning for the ease of mitigating synthesis requirements and for future optoelectronics and emerging applications. Single-phase $CuAlO_2$ fibers could be obtained at a temperature of 1100 °C in air. The Eu was successfully doped in the delafossite structure and two strong emission bands at ~405 and 610 nm were observed in the photoluminescence spectra. These bands are due to the intrinsic near-band-edge transition of $CuAlO_2$ and the f-f transition of the Eu^{3+} activator, respectively. Further electrical characterization indicated that these fibers exhibit semiconducting behavior and the introduction of Eu could act as band-edge modifiers, thus changing the thermal activation energies. In light of this study, $CuAlO_2$:Eu^{3+} fibers with both strong photoluminescence and p-type conductivity could be produced by tailoring the rare earth doping concentrations.

KEYWORDS
$CuAlO_2$; electrospinning; luminescent materials

CLASSIFICATION
40 Optical; magnetic and electronic device materials; 105 Low-Dimension (1D/2D) materials; 100 Materials; 204 Optics / Optical applications; 200 Applications

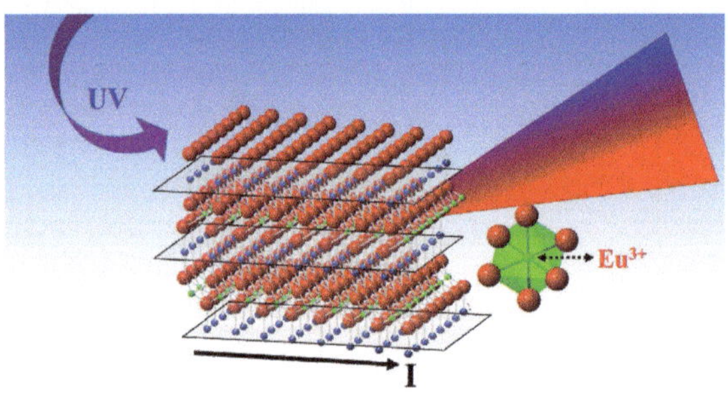

1. Introduction

Delafossite-type CMO_2 ternary oxides (C represents monovalent ion and M represents the trivalent ion) have been widely studied as p-type transparent conducting oxides and exhibit promise for various optoelectronics applications.[1–5] This class of materials has a rhombohedral space group of $R\bar{3}m$. The c-direction of the layered delafossite is occupied by a linearly coordinated O-Cu-O dumbbell. The edge-sharing MO_6 octahedra constitute the MO_2 layer along the *ab* plane, which is connected by the Cu^+ ions. $CuAlO_2$ is a typical *p*-type transparent oxide with a wide band-gap (>3 eV) and room temperature photoluminescence through its UV near-band-edge emission due to the recombination of free excitons.[1,6,7] The delafossite structure allows for chemical bond stretching from either the Cu-O bonds (*xy*-plane) or the Al-O bonds (*z*-direction) and the local lattice relaxations are crucial to electrical properties. By substituting the Al cation site with other trivalent ions, the conductivity could be enhanced due to the change of the electronic density via ligand field modification.[8,9]

While most researchers focus on the electrical properties of $CuAlO_2$ as a p-type transparent conducting oxide, there are few reports of $CuAlO_2$ as a potential phosphor material. In addition, with the development of field emission display and other flat display technologies,

CONTACT Yiquan Wu wuy@alfred.edu

new generation of phosphors with good luminescence, conductivity and stability are required. $CuAlO_2$ could be a promising host material in which the Al site could be substituted with various trivalent rare earth dopants, without changing the hole transport within the Cu^+ plane.[9] Since the main conduction path in delafossite crystals is close-packed Cu^+ layers,[10] the electrical properties could be sustained in addition to photoluminescence properties. Conventional solid-state synthesis of $CuAlO_2$ powders or thin films requires high temperature sintering and repeated thermal treatments. In this study, we prepared $CuAlO_2$ fibers via a cost-effective electrospinning method. The wire-like $CuAlO_2$ nanostructures possess higher surface area and sintering activity, which could lower the annealing temperature. Additionally, the one-dimensional material may also present extraordinary effectiveness in light emitting and transparent conducting.[11–13] The dopant cation, Eu^{3+}, was chosen as the emission activator and intense red emission from f-f transition of Eu^{3+} was identified. The Eu^{3+} activator center was successfully doped into the Al^{3+} site and this delafossite-type material could be used as potential host for luminescence application.

2. Experimental details

The 0.01 mol $CuAl_{1-x}Eu_xO_2$ (x = 0.001, 0.003, 0.01, 0.03, 0.05 and 0.1) was prepared through stoichiometric mixing of copper nitrate, aluminum nitrate and europium nitrate in deionized water. Poly(vinylpyrrolidone) (PVP, molar weight ~140,000 g mol^{-1}) was then added into the aqueous solution to obtain desirable viscosity for electrospinning. The viscous solution was then electrospun onto aluminum foil via a self-made electrospinning stage. The as-spun polymer fibers were collected and annealed in air at 1100 °C for 3 h.

The thermo-gravimetric (TG) and differential thermal analyses (DTA) were carried out on Q600 TG/DTA (TA Instruments, New Castle, DE) in flowing air. The powder X-ray diffraction measurement was performed on Bruker D2 Phaser (Bruker-AXS, Madison, WI) using Cu Kα radiation line. The Rietveld refinement was carried out by TOPAS software kits (Bruker-AXS, Madison, WI). The microstructure of the electrospun fibers was observed through an FEI Quanta 200F environmental scanning electron microscope (ESEM) (Fei, HIllsboro, OR) combined with energy dispersive spectroscopy (EDS). Steady state photoluminescence (PL) spectrum measurements were performed on a Spex FluoroLog Tau-3 (HORIBA Jobin Yvon, Edison, NJ) at 300 K. A Fourier-transform Raman spectrometer (Thermo Nicolet 6700, Thermo Nicolet Corp., Madison, WI) was used to record room-temperature Raman spectra of the delafossite samples. X-ray photoelectron spectroscopy (XPS) (PHI Quantera, ULVAC-PHI, Chanhassen, MN) was used to examine the local chemical environment in the doped samples (x = 0.001, 0.003, 0.01). The X-ray source was a monochromated Al Kα line at 1486.6 eV. The beam sweeps for each high resolution scan were adjusted (three sweeps: Al_{2p}/Cu_{2p}; one sweep: O_{1s} and C_{1s}) to yield a signal-to-noise ratio of >100:1 with exception of doping agents Eu_{4d} (three sweeps), which were adjusted to yield a signal-to-noise ratio of >50:1. The Hall measurement (MMR Technologies, San Jose, CA) was conducted at room temperature to determine the carrier type and density. The fibrous mat was cold pressed onto a square glass substrate and coated with silver electrodes on the four corners.

3. Results and discussion

Thermal treatment of electrospun fibers involves the burnout of organics and simultaneous crystallization. The thermo-gravimetric and differential thermal analyses are shown in Figure 1. The TG curve consists of three stages of weight loss. The first weight loss event occurred at ~100 °C, which could be attributed to the volatilization of adsorbed water and surface organics. The second event, at ~225 °C, is accompanied by a strong thermopositive peak in the DTA signal. This is due to the decomposition of PVP lateral chains and the nitrate salts. The third stage of weight loss begins at 400 °C, with a small thermal positive peak, which could correspond to the further burnout of PVP fundamental chains. [14,15] At above 800 °C, the TG curve exhibits a broad endothermic peak in the DTA curve. The TG plateau indicates that the thermal decomposition process was complete and the decomposed copper and aluminum nitrate undertook a combination reaction to finally form $CuAlO_2$. The overall weight loss is nearly 80%, given the large amount of PVP addition into the precursors required to yield ideal elasticity for electrospinning.

The X ray diffraction (XRD) patterns together with RWP (residual weight pattern) values from the refinements are shown in Figure 2 along with the standard $CuAlO_2$ ($R\bar{3}m$) phase for comparison. All samples show a single $CuAlO_2$ phase with no other crystal structure identified. This shows that the post annealing condition is sufficient to transform the polymeric as-spun fibers into a single, crystalline delafossite structure. In addition, samples (not shown in Figure 2) with a higher annealing temperature of 1200 °C show undesired $CuAl_2O_4$ phase due to the oxidation of cuprous ions. Annealing at lower temperatures than 1040 °C, both $CuAlO_2$ and CuO phases result, due to the insufficient reaction between CuO and $CuAl_2O_4$.[16,17] It is noteworthy that crystallinity differs among samples with different Eu doping. Samples with the 0.001 and 0.003 Eu doping show lower degree of crystallinity corresponding to broader diffraction peaks. However, when the doping concentration exceeds 0.01, well-defined $CuAlO_2$ delafossite peaks are exhibited. According to the bonding environment shown later, the substituted Eu and resultant Eu-O bond may alter the bonding environment, allowing for higher mass

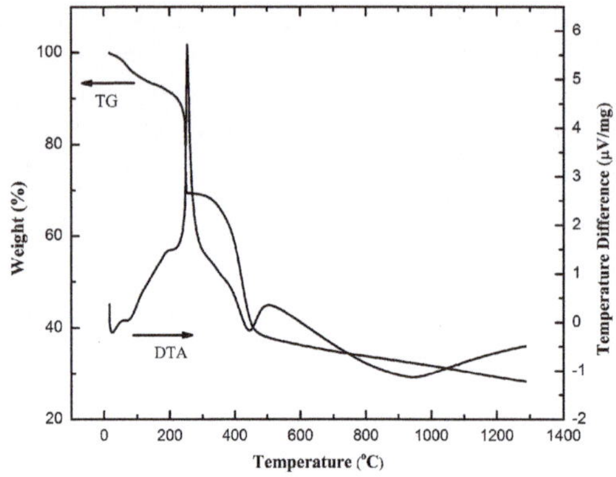

Figure 1. TG-DTA curve of as-spun polymeric fibers.

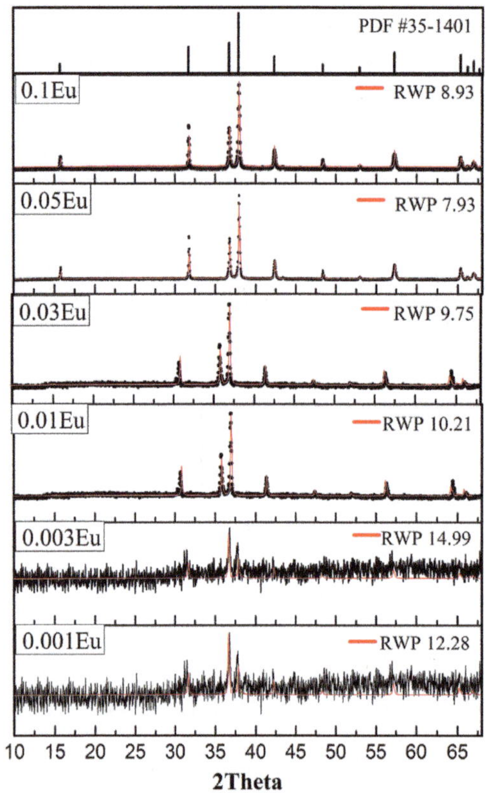

Figure 2. XRD patterns and profile fitting of $CuAl_{1-x}Eu_xO_2$ (x=0.001, 0.003, 0.01, 0.03, 0.05, 0.1) electrospun fibers annealed at 1100 °C for 3 h, with standard PDF#35-1041 corresponding to 3R-polytype delafossite $CuAlO_2$.

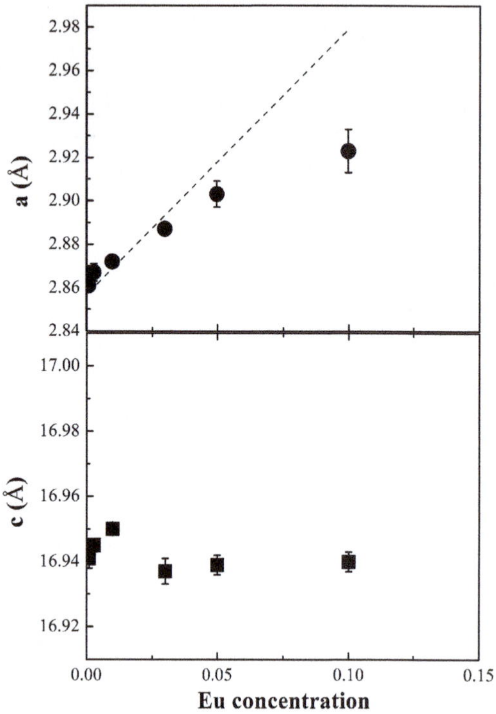

Figure 3. The lattice parameters *a* and *c* of the 3R-polytype structure delafossite $CuAl_{1-x}Eu_xO_2$ as a function of Eu concentration. The dashed line shows the calculated a values based on $a = 2.784 \times r_R + 1.437$, where $r_R = (1 - x)r_{Al} + xr_{Eu}$ (Vegard's law).

diffusivity and more sufficient crystallization, which might explain the promotional effect of Eu on $CuAlO_2$ crystallization. In summary, the nanofiber-derived single-phase $CuAlO_2$ via single step annealing and relatively short dwelling time implies that one-dimensional ceramic fibers have higher sintering activity than powders or thin films.[18]

The lattice parameters extracted from the Rietveld refinement of the delafossite phase $CuAl_{1-x}Eu_xO_2$ are shown in Figure 3. The lattice parameter a increases with increasing Eu concentration while c remains nearly constant. This behavior could be explained by a pseudo-linear relationship between lattice constant and trivalent ion radius r_R in a delafossite structure.[19] According to our host material, $CuAlO_2$, the lattice constant could be estimated by $a = 2.784 \times r_R + 1.437$, in which the constant 2.784 was calculated from the *a* value at zero doping level from the standard powder diffraction file (PDF#35-1401). The trivalent ion radius r_R follows a weighted sum of both Al^{3+} radius and Eu^{3+} radius ($r_R = (1 - x)r_{Al} + xr_{Eu}$).[20] By employing the Vegard's law, the Eu^{3+} ions are assumed to exclusively occupy the Al^{3+} sites, due to their equivalent charge and more difficulties in impurity doping of Cu-O dumbbells with strong covalency between Cu^+ d^{10} levels and O-2p orbitals. Based on the two equations above, the calculated value is shown as a dashed line in Figure 3. It can be seen that both experimental and calculated parameters follow a similar trend as a function of Eu concentration. The deviation could be attributed to the stoichiometric variation in starting precursors and the difference between nominal and doped Eu ion concentrations. The c value is independent of r_R and at various doping levels, the lattice parameter c remains almost constant. The trends in both *a* and *c* in terms of dopant concentration imply that Eu^{3+} substitutes Al^{3+} site in this delafossite structure.

The SEM micrographs are shown in Figure 4. After thermal annealing, the smooth polymeric as-spun fibers become coarse and rough, with nano-sized polygonal grains observed on the fibers. The microstructure evolution with the increasing doping concentration is also prominent, from well-sustained non-woven fibers to a disconnected porous structure. According to the significant radius difference between Eu^{3+} and Al^{3+} ions, the introduction of Eu^{3+} in the Al^{3+} site may cause lattice distortions and unrelaxed strains. Additionally, the radial crystallization on self-supported fibers imposes excess surface strain on the crystallites. As a result, excess enthalpy was produced for crystallization and a higher crystallinity is expected with a further increasing in Eu level, in a similar manner revealed from XRD patterns. When the Eu concentration increased with further increase in crystallinity to a degree that the bonding to crystallites cannot hold the accumulated stress in the cylindrical fibrous structure, the non-woven fibers break up and the crystallites start growing as isolated particles.

During the annealing process of the electrospun fibers, the organics derived from the PVP addition must be eliminated in order to achieve well-crystallized ceramic fibers suitable for luminescence application. EDS was employed to confirm the burn-out of the polymers shown in Figure 5. Before thermal annealing, the energy peaks associated with C and N were prominent because of the existence of PVP ($(C_6H_9NO)_n$). After 1100 °C annealing, there were no C or N peaks identified; only Cu, Al, O, and Eu (enlarged inset in Figure 5) peaks were observed. The polymer burn-out, thermal decomposition and crystallization proceeded simultaneously during the annealing, which shows that

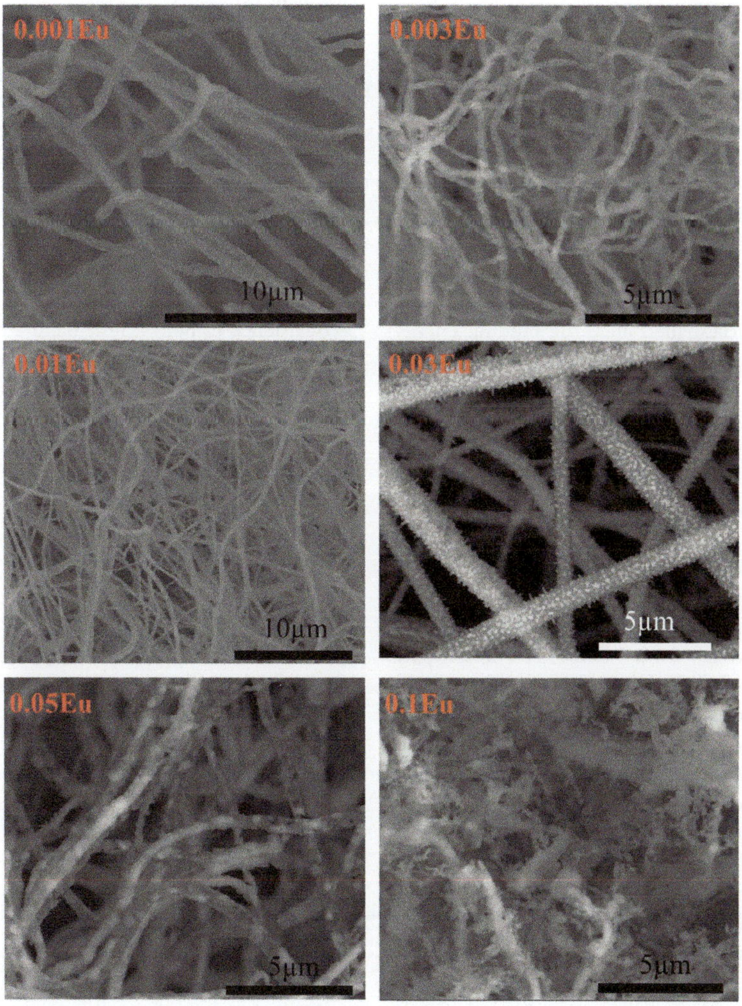

Figure 4. SEM microstructure of $CuAl_{1-x}Eu_xO_2$ (x=0.001, 0.003, 0.01, 0.03, 0.05 and 0.1) electrospun fibers annealed at 1100 °C for 3 h.

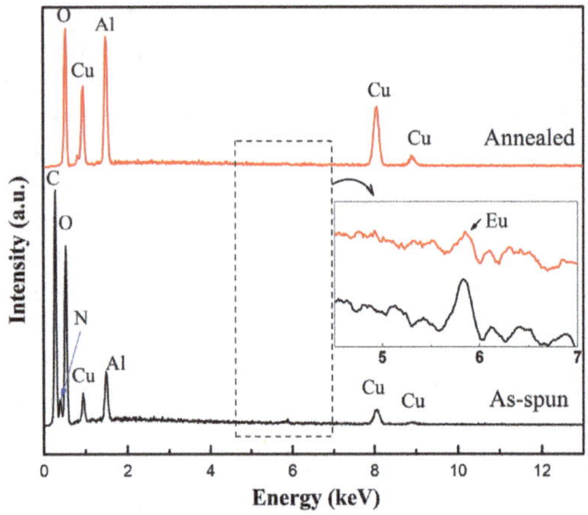

Figure 5. EDS spectra of as-spun polymeric fibers and annealed $CuAl_{0.99}Eu_{0.01}O_2$ ceramic fibers. The enlarged inset shows the presence of Eu.

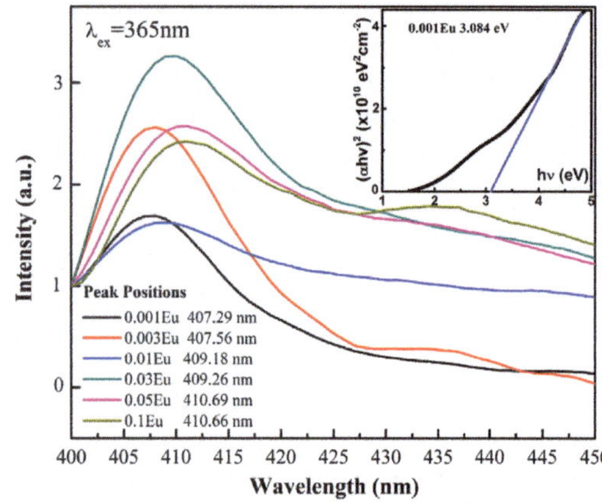

Figure 6. Near-band-edge photoluminescence of $CuAl_{1-x}Eu_xO_2$ excited at 365 nm. Inset shows the Tauc plot used for evaluating the optical band gap of $CuAl_{0.999}Eu_{0.001}O_2$.

this delafossite-type phosphor could be synthesized as one-dimensional fibers.

The photoluminescence of $CuAl_{1-x}Eu_xO_2$ (x=0.001, 0.003, 0.01, 0.03, 0.05 and 0.1) electrospun fibers has been studied at two excitation wavelengths (365 and 420 nm). Delafossite-type $CuMO_2$ has room temperature luminescence due to its UV near-band-edge emissions [21–24]. The Cu-O bonds in the O-Cu-O dumbbell layers determine the electronic structure near the band-gap and lead to strong localization of excitons in the *x–y* plane as well as larger binding energy. This causes hybridization between the $3d_z^2$ and 4s orbitals in the $CuMO_2$ delafossite. Since the binding energy exceeded the room temperature thermal energy (kT ≈ 0.025 eV), room temperature PL emissions were presented.[6,7] We herein reported the room temperature near-band-edge emissions from the Eu-doped $CuAlO_2$ fibers with slightly different peak energies compared to other studies.[7,25] The peak positions in terms of relative intensity shown in Figure 6 mainly centered at ~407 nm, indicating a 0.3 eV decrease in the direct band-gap compared to a normal 370 nm emission in thin films or bulks. Since the reported direct band-gap values of $CuAlO_2$ vary from 3.0 to 3.8 eV and highly depend on fabrication method, the near-band-edge emission is supposed to shift accordingly. The oxygen vacancies, surface defects and crystallinity associated with fabrication methods in low dimensional $CuAlO_2$ thin films or nanoparticles often yield dispersed band-gap values. In this particular study, the peak energy shift is also attributed to a fiber-derived band-gap energy change. A similar phenomenon has been observed in TiO_2 ceramic fibers,[26] i.e. that band-gap increases with the excess pressure on the fiber circular cross section. The inset in Figure 6 shows the measured optical band gap in our study, the absorption edge (3.084 eV corresponds to ~402 nm) of which is similar to the PL peak energy, which leads to the conjecture that the emission with the excitation wavelength of 365 nm results from the CuAlO2 near-band-edge transition. We also need to consider other factors regarding to the violet emission. The defect origin from V_{Cu} in the delafossite material is assumed to form an acceptor level above the zero vibrational level,[27,28] leading to a Stokes shift in the spectrum. Since our experiments were performed at room temperature and no other peaks associated with Eu^{3+} activators were observed from the emission spectra, the broad violet emission is unlikely from transition to defect levels. Byrne et al. [25] recently observed a coexistence of both a near-band-edge UV emission and a blue emission at ~430 nm at 14.5 K, with almost 100 nm Stokes shift. Revisiting the study by Jacob et al. [29] on $CuYO_2$ and $CuLaO_2$, the prolific hybridization from Cu 4p/3d orbitals and the resultant $3d_z^2+4s$ and $3d_z^2$ - 4s splitting only give rise to an asymmetric and overlapped peak feature rather than large Stokes shift. In addition, the effect of defect states on room temperature excitation still remains unclear, which definitely requires further investigation. The one-dimensional fibrous microstructure with major surface defects may also cause a band-gap narrowing in some oxide materials.[30] However, since there have been no reports on the photoluminescence of $CuAlO_2$ in the form of fibers to the best of our knowledge, at the current stage we may conjecture that the decrease in the band-gap associated with lower excitation energy (365 nm in our case) leads to the red shift in the PL emission spectra. At different dopant levels, the band-gap extracted from the PL peak energies in Figure 6 decreases with the increase of Eu level, indicating the further band modifications as a result of trivalent ion substitution and change of degree in Cu 4p/3d hybridization.

We also compared the peak energies in terms of doping levels. Regardless of the variation in relative PL emission intensities, the spectrum shifts to longer wavelength as the doping level goes up, roughly from 407.29 nm in $CuAl_{0.999}Eu_{0.001}O_2$ to 410.66 nm in $CuAl_{0.9}Eu_{0.1}O_2$. Based on previous study,[29,31] this monotonic shift to lower energies is due to the change of hybridization between Cu $4p_{x,y}$ to $3d_z^2 \pm 4s$, under the influence of trivalent partial substitution on Cu-O bond. As the Cu-O covalency increase from $CuAlO_2$ to $CuAl_{1-x}Eu_xO_2$, hybridization between the $3d_z^2$ and 4s orbitals became more prolific which would lift the non-bonding $3d_z^2$-4s orbital energy in order to compensate for the antibonding $3d_z^2+4s$ orbitals. Therefore the transition energy between $3d_z^2$-4s and $4p_{x,y}$ is lowered upon introducing Eu^{3+} ions. Our observation also confirmed the prediction that this Stokes shift would increase as the size of trivalent metal ion increases.[25] For all the doped samples at excitation wavelength of 365 nm, only the broad near-band-edge emission was identified and no peaks associated with Eu f-f transitions were observed above 500 nm. At higher excitation energies, the excitons near the band edge may undergo fast recombination across the band-gap and non-radiative relaxation to the ground state while the transitions to Eu activators become absent.

PL excitation and emission spectra from $CuAl_{0.99}Eu_{0.01}O_2$ are shown in Figure 7. The excitation peaks between 300 and 500 nm correspond to Eu^{3+} intra-4f transitions. The $^7F_0$-$^5D_2$ line at 465 nm is the strongest line in the excitation spectrum. The peak at 365 nm was identified as the intrinsic excitation peak from $CuAlO_2$, according to Ahmad's work on $CuAlO_2$ nanoparticles [32] and our previous work on nanofibers.[31] The emission spectrum with 465 nm excitation shows broad bands at 587, 610, 654 and 690 nm, which are due to the f-f transition of $^5D_0 \rightarrow {}^7F_J$ of Eu^{3+}, respectively. For the 0.01Eu sample, $^5D_1 \rightarrow {}^7F_1$ transition was also observed. The two-channel emission diagram is depicted in Figure 8(a). The strongest $^5D_0 \rightarrow {}^7F_2$ red emission is the result of hypersensitive transitions with $\Delta J = 2$ indicating the environment around the activator centers. The increasing $^5D_0 \rightarrow {}^7F_2$ emission intensity from samples with x = 0.001–0.01 indicates a higher symmetry local site. Upon the occupancy of Eu^{3+} ions in the octahedral center, the lattice was distorted and the local symmetry of Eu^{3+} continued to change with increasing Eu^{3+} doping level. As a consequence, the probability of each $^5D_0 \rightarrow {}^7F_J$ transition channels may change, which leads to the variation of intensity of the six $^5D_0 \rightarrow {}^7F_2$ emissions shown in Figure 8(b). The overall red emission intensity increases with the increase of doping concentration up to 0.01Eu. Above this doping level, the intensity significantly drops due to the concentration quench effect [33,34] shown in Figure 8(c). With the increase of Eu^{3+} concentration, the cross-relaxation becomes dominant due to the reduced distance between two Eu^{3+} activator ions. Therefore, the luminescence emission intensity almost vanishes at higher Eu^{3+} concentration. Since the activator is introduced solely on the Al^{3+} site, the critical transfer distance (R_c) could be estimated by the following equation:

$$R_c \approx 2\left(\frac{3V}{4\pi x_c N}\right)^{\frac{1}{3}} \quad (1)$$

where V is the unit cell volume, x_c is the critical concentration and N is the number of Al^{3+} ions per unit cell. For $CuAlO_2$ host ($N=3$), x_c is roughly equal to 0.01 at which concentration the maximum luminescent intensity was observed. The cell volume from XRD refinement is 120.75 Å^3 and the R_c was calculated to be approximately 20 Å. For a Eu^{3+}–Eu^{3+} distance larger than 5 Å, the multipolar interaction is dominant and the exchange interaction is negligible.[35] Therefore, the main mechanism for concentration quenching in the $CuAlO_2$:Eu^{3+} phosphor is the multipolar interaction.

Room temperature Raman spectra are shown in Figure 9. It has been widely acknowledged that in delafossite structure with $\Gamma = A_{1g}+E_g+3A_{2u}+3E_u$, only the first

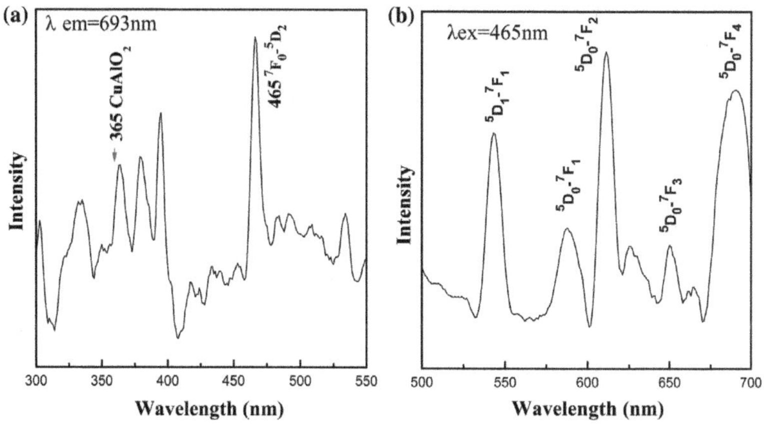

Figure 7. Photoluminescence excitation spectrum (a) showing the direct band-gap transition in $CuAlO_2$ host, and the emission spectrum (b) of $CuAl_{0.99}Eu_{0.01}O_2$.

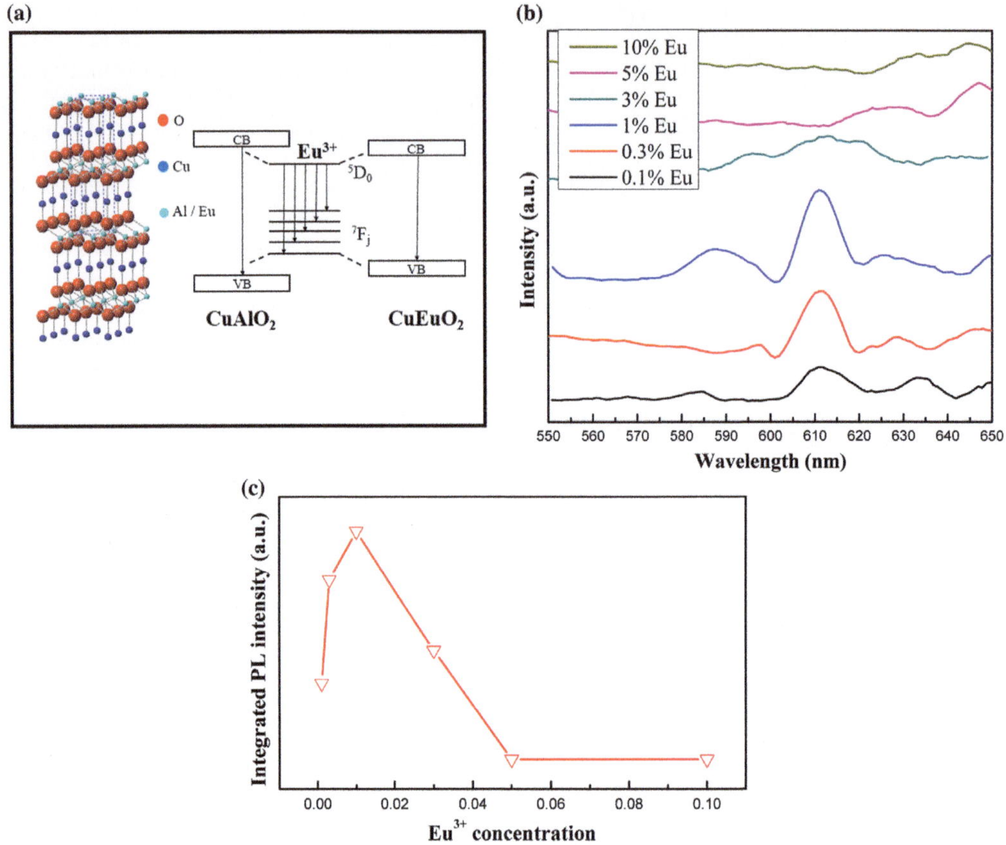

Figure 8. Schematic diagram of photoluminescence emission channels in the rare-earth doped delafossite $CuAlO_2$ (a), $^5D_0 \rightarrow {}^7F_1$ and $^7F_2$ transitions with respect to doping concentration (b) and integrated PL intensity of 610 nm emission as a function of doping concentration (c).

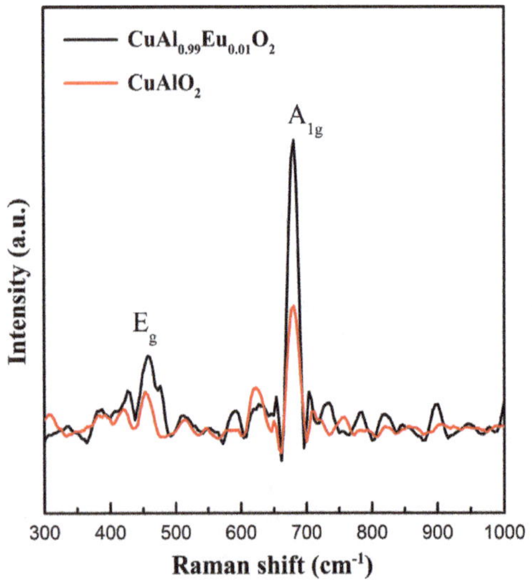

Figure 9. Room temperature Raman spectrum under 1064 nm excitation provided by a Nd:YVO_4 laser.

two modes are Raman active.[36,37] The A modes correspond to the vibrational movements in the direction of Cu-O bonds and the E modes represent the vibrations in the perpendicular direction. A previous report on particulate $CuAlO_2$ [25] shows two sharp peaks at 418 and 767 cm^{-1}, respectively. However, the Raman scattering peaks could shift to a large extent depending on the trivalent ion species. $CuLaO_2$ with the same crystal geometry has two peaks centered at 318 and 652 cm^{-1}. [38] The lowered frequencies could be due to a higher M atom mass and weaker M-O bond. The room temperature Raman spectrum that we obtained here shows a similar E_g position but a slight lower frequency of A_{1g} mode, which might provide supplementary information on trivalent substitution. Upon the introduction of larger Eu^{3+} ions, the MO_6 along the *ab* plane would go through a decrease in bonding energy, which is also confirmed by XPS in later discussion. Therefore the E_g characterized by the vibration along the *ab*-plane mode should shift to lower frequency. However from the experimental data, the E_g shows a typical $CuAlO_2$ peak position whereas the A_{1g} peak shifts to lower frequency for both undoped and doped samples, which might lead to the conjecture that the decrease of the A_{1g} frequency might due to the electrospinning synthesis during which the MO_6 octahedral

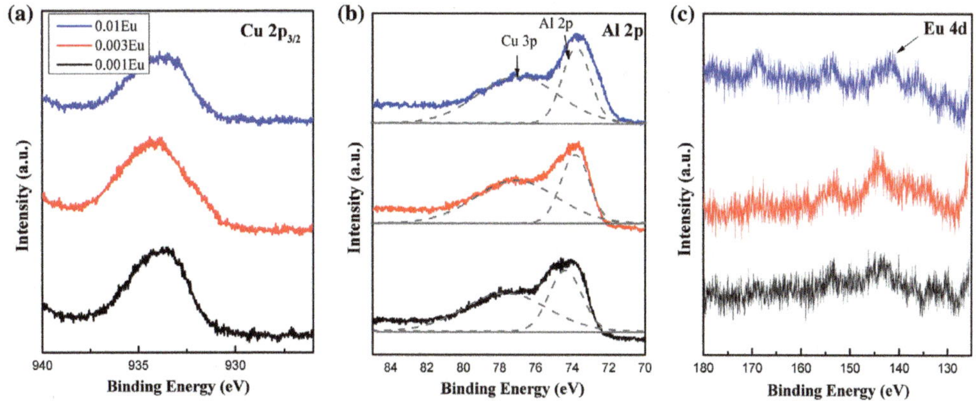

Figure 10. XPS spectra of (a) Cu $2p_{3/2}$, (b) Al 2p, and (c) Eu 4d levels.

bonding energy is lowered. Another possible explanation for the decrease of A_{1g} mode could be extrapolated from our near-band-edge emission observation. Since the Cu-O covalency is predicted to increase in the doped sample, the higher degree of hybridization might cause the A_{1g} vibration mode to move to lower frequency. In addition, the band width of E_g mode becomes broader after Eu doping, indicating that the E_g mode is more sensitive to the substitution of trivalent ions.

The binding energies (BE) of the atoms were examined by XPS and calibrated with respect to the C 1s BE (284.6 eV). The XPS high resolution Cu 2p, Al 2p and Eu 4d spectra are shown in Figure 10. The BE for Cu $2p_{3/2}$ is centered at ~933.4 eV for all three samples with no shift observed with Eu doping. It has been widely acknowledged that in the delafossite structure the strong two-dimensional confinement of Cu-O bonds along the *z*-axis restrains the electronic structure.[6] Therefore the substitution of Eu^{3+} ions along the *ab* plane has little effect on the charge density around Cu^{+} ions. For the spectra shown in Figure 10(b), the peaks can be deconvoluted into Al 2p and Cu 3p components. In contrast to Cu $2p_{3/2}$ peaks, the Al 2p peaks shift to lower binding energies at higher dopant concentrations (74.52 eV for x=0.001, 74.09 eV for x=0.003 and 73.54 eV for x=0.01) due to the environment change around the MO_6 octahedra along the *ab* plane. The different Cu $2p_{3/2}$ and Al 2p BE responses to trivalent ion substitution leads to the conjecture that the introduced rare earth within the MO_2 layer would insignificantly disturb the hole conduction path in the Cu^{+} layer. Therefore, the p-type conductivity might not be impaired. In Figure 10(c), the presence of Eu 4d spectra indicates the successful doping of europium ions into the delafossite lattice. The trivalent europium compounds with orbit-spin splitting Eu^{3+} $4d_{5/2}$ and Eu^{3+} $4d_{3/2}$ have been shown to have a binding energy of 135 and 142.5 eV,[39,40] which is similar to what we observed in our study.

The temperature dependence of electrical conductivity of the electrospun fibers is shown in Figure 11. Arrhenius

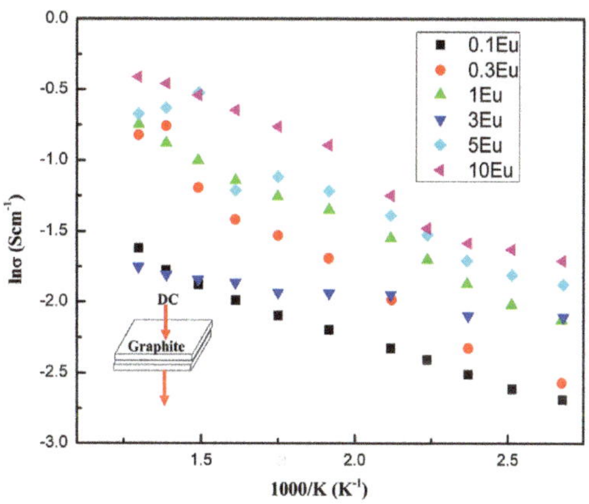

Figure 11. Arrhenius plots of DC electrical conductivity in $CuAlO_2$ fibers.

plots were used to identify the thermal-activation type of these wide band-gap materials. DC current was applied through a piece of fiber mat sandwiched by two rectangular graphite foils, which were used to minimize the contact resistance. The conductivity increases as the temperature increases, showing semiconduction. At ~300 K, the electrical conductivity for 0.1% Eu doped $CuAlO_2$ fibers is ~0.05 S cm^{-1} with a thermal activation energy of ~0.09 eV. The room temperature conductivity increased with elevated Eu doping levels, from 0.05 to 0.17 S cm^{-1}. This could be attributed to the substitution of Eu cations which induced the lattice distortion and increased the hole concentrations. On the other hand, the substitution of Eu on the Al site may introduce a smaller band-gap associated with $CuEuO_2$ and this modification at the near-band-edge may also contribute to the hole conductivity enhancement. Based on the plot shown in Figure 11, when the doping level is as small as 0.1%, logσ decreases linearly with the reciprocal of temperature. However as the doping concentration increased, a variable-range hopping mechanism occurred,[41] which could be seen from the segmented slopes at different temperature ranges. The slope at higher temperature range between 570 and 800 K is lower than that at a lower temperature range between 300 and 400 K. A revisit to PL energy peaks shown Figure 6 could lead to the conjecture that the change in thermal activation may result from the various band-gaps induced by Eu partial substitution, since the band-gaps decrease with the increase of Eu concentration. In addition to the DC temperature-dependent conductivity measurement, the Hall coefficient (R_H) for the 0.1%Eu doped sample determined by Hall measurement is + 5.48 cm^3/C, corresponding to a carrier density of 1.14×10^{18} cm^{-3}, while the heavily doped 10% sample has a RH = +6.94 cm^3/C and a carrier density of 8.99×10^{17} cm^{-3}, both showing a p-type semiconductivity.

4. Conclusions

$CuAl_{1-x}Eu_xO_2$ (x = 0.001, 0.003, 0.01, 0.03, 0.05 and 0.1) fibers have been fabricated through a combination of a chemical solution method and electrospinning technique. This method could effectively synthesize single-phase $CuAlO_2$ via one-step annealing in air. The lattice parameter *a* follows a pseudolinear relationship with Eu concentration while the lattice parameter *c* remains almost constant with Eu concentrations. This implied that the delafossite-type $CuAlO_2$ could be used as a potential host material for the rare-earth Eu^{3+} partial substitution on trivalent site. Further photoluminescence measurements indicated that upon two excitation wavelengths of 390 and 465 nm, the $CuAl_{1-x}Eu_xO_2$ exhibits two emission behaviors from both intrinsic near-band-edge violet emission and Eu^{3+} $^5D_0 \rightarrow {}^7F_J$ red emission, which shows that this material might be versatile in multi-channel light emitting. All fibers show semiconducting behavior as a function of temperature, and the room temperature conductivity increased monotonically with an increase of Eu concentration. The variation in thermal activation energy as a result of Eu substitution indicates different hopping mechanisms occur at inhomogeneous near-band-edges in the Eu doped delafossite structure.

Disclosure statement

No potential conflict of interest was reported by the authors.

Funding

Y. Liu and Y. Q. Wu gratefully acknowledge the US Air Force Office of Scientific Research (contract FA9550-14-1-0155) for funding and supporting the research. This work was partially supported by FP7 Marie Curie Action (project No: 295208) sponsored by European Commission to Dr. H. T. Ye.

References

[1] Yu R-S, Liang S-C, Lu C-J, et al. Characterization and optoelectronic properties of p-type N-doped $CuAlO_2$ films. Appl Phys Lett. 2007;90:191117–191117.

[2] Ong CH, Gong H. Effects of aluminum on the properties of p-type Cu–Al–O transparent oxide semiconductor prepared by reactive co-sputtering. Thin Solid Films. 2003;445:299–303.

[3] Fang M, He H, Lu B, et al. Optical properties of p-type $CuAlO_2$ thin film grown by rf magnetron sputtering. Appl Surf Sci. 2011;257:8330–8333.

[4] Chen HY, Tsai MW. Delafossite-$CuAlO_2$ thin films prepared by thermal annealing. J Nano Res. 2011;13:81–86.

[5] Mo R, Liu Y 2011 Synthesis and properties of delafossite $CuAlO_2$ nanowires J Sol-gel Sci Techn 57 16–19.

[6] Banerjee AN, Joo SW, Min B-K. Quantum size effect in the photoluminescence properties of p-type semiconducting transparent $CuAlO_2$ nanoparticles. J Appl Phys. 2012;112:114329.

[7] Banerjee A, Chattopadhyay K. Size-dependent optical properties of sputter-deposited nanocrystalline p-type transparent $CuAlO_2$ thin films. J Appl Phys. 2005;97:084308.

[8] Jiang HF, Zhu XB, Lei HC, et al. Effect of Cr doping on the optical–electrical property of $CuAlO_2$ thin films derived by chemical solution deposition. Thin Solid Films. 2011;519:2559–2563.

[9] Tsuboi N, Hoshino T, Ohara H, et al. Control of luminescence and conductivity of delafossite-type $CuYO_2$ by substitution of rare earth cation (Eu, Tb) and/or Ca cation for Y cation. J Phys Chem Solids. 2005;66:2134–2138.

[10] Lee M, Kim T, Kim D. Anisotropic electrical conductivity of delafossite-type $CuAlO_2$ laminar crystal. Appl Phys Lett. 2001;79:2028–2030.

[11] Bu W, Hua Z, Chen H, et al. Epitaxial synthesis of uniform cerium phosphate One-Dimensional nanocable heterostructures with improved luminescence. J Phys Chem B. 2005;109:14461–14464.

[12] Jia G, Zheng Y, Liu K, et al. Facile surfactant-and template-free synthesis and luminescent properties of one-dimensional Lu_2O_3: Eu^{3+} phosphors. J Phys Chem C. 2008;113:153–158.

[13] Yang J, Li C, Cheng Z, et al. Size-tailored synthesis and luminescent properties of one-dimensional Gd2O3:Eu3+ nanorods and microrods. J Phys Chem C. 2007;111:18148–18154.

[14] Zhang X, Xu S, Han G. Fabrication and photocatalytic activity of TiO_2 nanofiber membrane. Mater Lett. 2009;63:1761–1763.

[15] Zhang Z, Li X, Wang C, et al. ZnO hollow nanofibers: fabrication from facile single capillary electrospinning and applications in gas sensors. J Phys Chem C. 2009;113:19397–19403.

[16] Jacob K, Alcock C. Thermodynamics of CuAlO2 and CuAl2O4 and phase equilibria in the system Cu2O-CuO-Al2O3. J Am Ceram Soc. 1975;58:192–195.

[17] Xiong D, Zeng X, Zhang W, et al. Synthesis and characterization of $CuAlO_2$ and $AgAlO_2$ delafossite oxides through low-temperature hydrothermal methods. Inorg Chem. 2014;53:4106–4116.

[18] Liu Y, Olson TL, Wu Y. Luminescence and microstructure of Nd Doped $Y_2Si_2O_7$ electrospun fibers. J Am Ceram Soc. 2014;97:2390–2393.

[19] Isawa K, Yaegashi Y, Komatsu M, et al. Synthesis of delafossite-derived phases, $RCuO_{2+\delta}$ with R= Y, La, Pr, Nd, Sm, and Eu, and observation of spin-gap-like behavior. Phys Rev B. 1997;56:3457.

[20] Chong M, Pita K, Kam C. Photoluminescence of sol–gel-derived Y_2O_3: Eu^{3+} thin-film phosphors with Mg^{2+} and Al^{3+} co-doping. Applied Physics A. 2004;79:433–437.
[21] Shahriari DY, Barnabe A, Mason TO, et al. A high-yield hydrothermal preparation of $CuAlO_2$. Inorg Chem. 2001;40:5734–5735.
[22] Tsuboi N, Ohara H, Hoshino T, et al. Luminescence properties of delafossite-type $CuYO_2$ doped with calcium, oxygen or rare earth Tb. Jpn J Appl Phys. 2005;44:765.
[23] Kumekawa Y, Hirai M, Kobayashi Y, et al. Evaluation of thermodynamic and kinetic stability of $CuAlO_2$ and $CuGaO_2$. J Therm Anal Calorim. 2010;99:57–63.
[24] Liu Q-J, Zhang N-C, Sun Y-Y, et al. Structural, mechanical, electronic, optical properties and effective masses of $CuMO_2$ (M= Sc, Y, La) compounds: first-principles calculations. Solid State Sci. 2014;31:37–45.
[25] Byrne D, Cowley A, Bennett N, et al. The luminescent properties of $CuAlO_2$. Journal of Materials Chemistry C. 2014;2:7859–7868.
[26] Kumar A, Jose R, Fujihara K, et al. Structural and optical properties of electrospun TiO_2 nanofibers. Chem Mater. 2007;19:6536–6542.
[27] Scanlon DO, Walsh A, Morgan BJ, et al. Effect of Cr substitution on the electronic structure of $CuAl_{1-x}Cr_xO_2$. Phys Rev B. 2009;79:035101.
[28] Scanlon DO, Watson GW. Conductivity limits in $CuAlO_2$ from screened-hybrid density functional theory. J Phys Chem Lett. 2010;1:3195–3199.
[29] Jacob A, Parent C, Boutinaud P, et al. Luminescent properties of delafossite-type oxides $LaCuO_2$ and $YCuO_2$. Solid State Commun. 1997;103:529–532.
[30] Naldoni A, Allieta M, Santangelo S, et al. Effect of nature and location of defects on bandgap narrowing in black TiO_2 nanoparticles. J Am Chem Soc. 2012;134:7600–7603.
[31] Liu Y, Huang Y, Seo HJ, et al. Blueshift in near-band-edge emission in Y^{3+}-doped $CuAlO_2$ nanofibers. Opt Mater Express. 2014;4:2602–2607.
[32] Ahmad A, Jagadale T, Dhas V, et al. Fungus-based synthesis of chemically difficult-to-synthesize multi functional nanoparticles of CuAlO2. Adv Mater. 2007;19:3295–3299.
[33] Zhao J, Zhang W, Xie E, et al. Structure and photoluminescence of β-Ga_2O_3:Eu^{3+} nanofibers pre pared by electrospinning. Appl Surf Sci. 2011;257: 4968–4972.
[34] Chen J, Wang J, Zhang F, et al. Structure and photoluminescence property of Eu-doped SnO_2 nanocrystalline powders fabricated by sol–gel calcination process. J Phys D Appl Phys. 2008;41: 105306.
[35] Zhang J, Wang Y. Eu^{3+}-doped $Ba_3Bi(PO_4)_3$: A red phosphor for white light-emitting diodes. Electrochem Solid-State Lett. 2010;13:J35–J7.
[36] Yassin O, Alamri S, Joraid A. Effect of particle size and laser power on the Raman spectra of $CuAlO_2$ delafossite nanoparticles. J Phys D Appl Phys. 2013;46:235301.
[37] Pellicer-Porres J, Martinez-Garcia D, Segura A, et al. Pressure and temperature dependence of the lattice dynamics of $CuAlO_2$ investigated by Raman scattering experiments and *ab initio* calculations. Phys Rev B. 2006;74:184301.
[38] Salke NP, Rao R, Achary S, et al. Raman spectroscopic investigations on delafossite $CuLaO_2$ at high pressures. *J* Phys Conf Ser. 2012;377:012020.
[39] Lu D-Y, Sugano M, Sun X-Y, et al. X-ray photoelectron spectroscopy study on $Ba_{1-x}Eu_xTiO_3$. Appl Surf Sci. 2005;242:318–325.
[40] Schneider W-D, Laubschat C, Nowik I, et al. Shake-up excitations and core-hole screening in Eu systems. Phys Rev B. 1981;24:5422.
[41] Yanagi H, Inoue S-i, Ueda K, et al. Electronic structure and optoelectronic properties of transparent p-type conducting $CuAlO_2$. J Appl Phys. 2000; 88:4159–4163.

15

Microstructure study of a severely plastically deformed Mg-Zn-Y alloy by application of low angle annular dark field diffraction contrast imaging

Dudekula Althaf Basha[a], Julian M. Rosalie[c], Hidetoshi Somekawa[a], Takashi Miyawaki[a,b], Alok Singh[a], Koichi Tsuchiya[a,b]

[a]Structural Materials Unit, National Institute for Materials Science, Sengen 1-2-1,Tsukuba 305-0047, Japan;
[b]Graduate School of Pure and Applied Sciences,University of Tsukuba, Tsukuba, Ibaraki305-8577, Japan;
[c]Erich Schmid Institute of Materials Science,Austrian Academy of Sciences, Austria

ABSTRACT

Microstructural investigation of extremely strained samples, such as severely plastically deformed (SPD) materials, by using conventional transmission electron microscopy techniques is very challenging due to strong image contrast resulting from the high defect density. In this study, low angle annular dark field (LAADF) imaging mode of scanning transmission electron microscope (STEM) has been applied to study the microstructure of a Mg-3Zn-0.5Y (at%) alloy processed by high pressure torsion (HPT). LAADF imaging advantages for observation of twinning, grain fragmentation, nucleation of recrystallized grains and precipitation on second phase particles in the alloy processed by HPT are highlighted. By using STEM-LAADF imaging with a range of incident angles, various microstructural features have been imaged, such as nanoscale subgrain structure and recrystallization nucleation even from the thicker region of the highly strained matrix. It is shown that nucleation of recrystallized grains starts at a strain level of revolution $N = 1/4$ (earlier than detected by conventional bright field imaging). Occurrence of recrystallization of grains by nucleating heterogeneously on quasicrystalline particles is also confirmed. Minimizing all strain effects by LAADF imaging facilitated grain size measurement of 150 ± 25 nm in fully recrystallized HPT specimen after $N = 5$.

KEYWORDS
STEM; LAADF; diffraction contrast; severe plastic deformation (SPD); recrystallization; high pressure torsion (HPT)

CLASSIFICATION
10 Engineering and structural materials; 503 TEM, STEM, SEM; 106 metallic materials

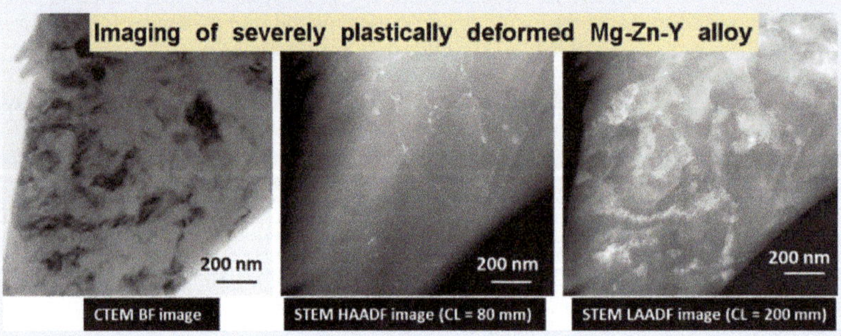

1. Introduction

In recent years severe plastic deformation (SPD) has been commonly employed to improve the mechanical properties, e.g. strength, ductility and toughness, of metallic materials. Mechanical properties improve because of grain refinement by dynamic recrystallization processes and strain accumulation. Submicron size grain can easily be achieved by SPD processes. SPD also modifies the crystallographic texture of the alloys, which can have direct impact on the mechanical (deformation) behavior. Some of the common SPD techniques include repeated rolling, such as equal channel angular extrusion (ECAE) [1], and caliber rolling [2] and high pressure torsion (HPT) [3].

In order to understand the deformation and recrystallization process during SPD, as well as the mechanical properties, it is important to study the microstructure such as grain structure and phase distribution, and deformation structures such as dislocations and twins. Because of the fine scale of the structures, as well as to obtain crystallographic information, it is necessary to use electron microscopy, especially transmission electron microscopy (TEM). However, the severely deformed microstructures are a challenge for conventional TEM (CTEM), which works well with

CONTACT Alok Singh alok.singh@nims.go.jp
Present address: Structural Materials Unit, National Institute for Materials Science, Sengen 1-2-1,Tsukuba 305-0047, Japan

defect-free annealed samples or with discrete lattice defects. Strain introduced by severe plastic deformation makes it difficult to look past the strong diffraction contrast.

STEM is a powerful imaging mode for detailed analysis of crystalline materials and for investigating electron beam-specimen interactions [4–8]. STEM has a considerable advantage over conventional TEM because various signals emitted from the same specimen can be collected simultaneously by utilizing different detectors having different shapes and sizes and thus effectively used for efficient imaging with high contrast. High angle annular dark field (HAADF) imaging is an important tool in STEM mode. HAADF images are formed by incoherent thermal diffuse scattering (TDS) electrons, utilizing a large inner angle detector [4,9]. Furthermore, in incoherent imaging, each atom independently contributes intensity to the resulting Z-contrast image. Howie suggested the use of an annular detector with the inner collection angle exceeding 40 mrads to ensure that all the scattering was thermally diffused incoherent scattering [9]. Apart from studies over several years using Z-contrast imaging techniques, diffraction-contrast STEM imaging methods have recently found much interest in the study of crystal defects, which is a major branch of CTEM [10–12]. If the inner detection angle of the annular dark field (ADF) detector is small, Bragg reflections from crystalline materials can significantly contribute to the ADF-STEM imaging. As the inner collection angle decreases, the scattering from crystals is no longer incoherent in nature, as other effects such as low-angle scattering from precipitates occur. Contrast reversal effects contribute to ADF imaging. At low scattering angles the diffraction contrast effectively becomes stronger than Z contrast and the imaging will become more similar to conventional dark-field TEM imaging. STEM imaging of crystalline defects can offer several practical benefits that are impossible to realize in CTEM imaging. It has been shown that STEM imaging allows for more uniform contrast of large sample areas with reduced thickness fringes, bend contours, and other auxiliary contrast effects and with greater signal-to-noise ratio and precisely resolved defects even in thick specimens which are difficult to resolve in traditional CTEM imaging [10]. Conventional techniques that are applicable in CTEM for the imaging of crystal defects, such as finding the Burgers vector of a dislocation or the displacement vector of a stacking fault, are also applicable in STEM [13].

Figure 1 shows a schematic diagram of STEM with BF and the ADF signal detection geometry. The bright field (BF) STEM image is formed by collecting the transmitted signal and it appears similar to the CTEM phase contrast image. The ADF detector is centered with respect to the optic axis of the microscope and

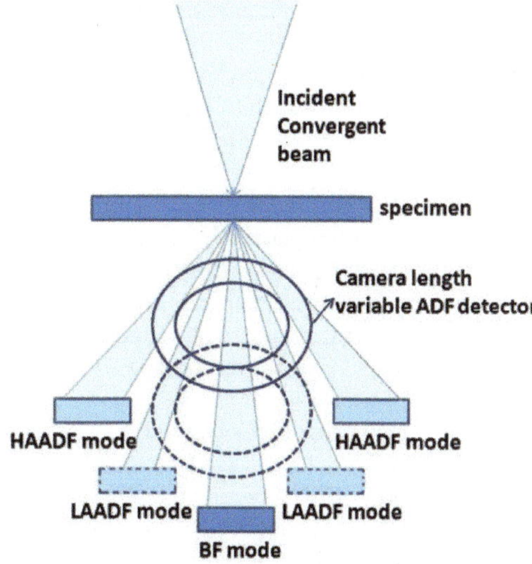

Figure 1. Schematic diagram of the virtual displacement of the ADF detector in STEM mode. The ADF detector is at the upper position in HAADF mode to collect high angle scattered electrons which contain Z-contrast information, and at the lower position in LAADF mode to collect low angle scattered electrons which contain both Z-contrast and diffraction contrast information. The unscattered rays can be collected by the BF detector.

the scattered electrons over an angular range can be collected by virtually (electronically) shifting the detector position along the optic axis with respect to sample, i.e. by changing the camera length [14]. The ADF detector is in the upper position in HAADF mode to collect high angle scattered electrons which contain Z-contrast information, and it is at the lower position in LAADF mode to collect low angle scattered electrons which contain both Z-contrast and diffraction contrast information. In general, the strength of the magnetic lenses is varied accordingly to project high- or low-angle scattered electrons onto the fixed annual detector. The displacement of the detector position in STEM mode corresponds to the beam tilt in CTEM. The image contrast is sensitive to the displacement of the ADF detector and thus certain features of the specimen can be enhanced by collecting a specific portion of the scattered signals, a technique which is extremely useful for imaging inhomogeneous and deformed samples. The degree of coherence decreases as the scattered angle of electrons increases for both the elastic and the inelastic imaging of thin objects.

In this work, for the first time, STEM-LAADF methodology has been applied to observe the overall deformation and recrystallization of grains by HPT. In the HPT process, a disk shaped sample is compressed between two dies under a high compressive stress of a few GPa, and then one of the dies is given a rotation to create torsion. The applied strain is proportional to the number of rotations, given by $\varepsilon = \frac{\pi N r}{\sqrt{3} t}$, where ε is the strain, N is the number of turns, r is the distance to the disk center and t is the sample thickness. In HPT a high strain can be applied even at low temperatures, such as room temperature, because the physical constraint on the sample by the dies does not let the sample break.

STEM imaging within a particular angular range or with a certain camera length to observe grain fragmentation and recrystallization initiation in severely plastically deformed microstructures is demonstrated with a number of illustrations. Although the principles of this kind of imaging are well known, they have not been covered by any specific theoretical paper in detail. It has also been shown that LAADF imaging would be a uniquely powerful tool to observe the sharp twin boundary or grain boundary contrast in deformed recrystallized regions of the specimen, and to estimate accurate recrystallized ultrafine grain size, especially in specimens containing secondary phase particles of similar size as that of the grains. The alloy studied here is a Mg-3Zn-0.5Y (at%) alloy, in which Mg-Zn precipitates create a distinct compositional contrast in the magnesium matrix. Yttrium-bearing ternary phases (mainly quasicrystalline icosahedral phase) display a higher level of compositional contrast. Thus the samples of this alloy provide diffraction as well as compositional contrast. In this alloy composition, when processed by extrusion, very high mechanical strength accompanied by ductility has been reported [15]. The strengthening and ductility is attributed to fine grain size obtained by dynamic recrystallization during extrusion and simultaneous modification of texture, aided by presence of the quasicrystalline phase. Thus the main interest of this study is observation of recrystallization by introduction of different amounts of strain by means of HPT. In an earlier work, we have also studied Mg-3.4at%Zn binary alloy processed by HPT [16,17]. In this study, early recrystallization of grains has been detected by imaging with LAADF. Segregation of zinc occurs on newly created boundaries. Grain boundaries of recrystallized grains have been distinguished by creating different contrasts. A systematic application of these imaging techniques to SPD alloys has been shown for the first time.

2. Experimental details

An alloy of nominal composition Mg-3Zn-0.5Y (at%) was prepared by melting pure metals under protective atmosphere in an electric furnace and cast in a steel mold. The cast material was extruded and then subjected to caliber rolling (eight passes) to form into a rod. The rod was annealed at 573 K for 24 h in order to anneal out the strain due to rolling, for grain growth and for dissolution of the Mg-Zn binary precipitates. After annealing, a grain size of about 10 μm was obtained. The annealed rods were machined into rods of 10 mm diameter, from which circular disks of 1 mm thickness were cut. These disks were ground down to 0.85 mm thickness and subjected to high pressure torsion (HPT). The HPT processing was performed at room temperature at a constant pressure of 5 GPa in which the lower anvil was rotated at a speed of 1 rpm. Disks were deformed through a total number of revolutions, N, of 0, 1/4, 1/2, 1 and 5 turns. Of these, results of only samples of $N = 0$, 1/4 and 5 are described, because of the significant changes in them. For microstructure characterization, samples were carefully extracted at a distance of 2–3 mm from the center of the disk.

TEM samples were prepared by ion milling (Gatan, model 691, Gatan Inc, Pleasanton CA, USA) after mechanical grinding and then examined using a FEI-Tecnai F30 (FEI Technologies Inc., Hillsboro, Oregon, USA) microscope operating at 300 kV. Microstructure observations were performed both in conventional CTEM and in STEM mode with a probe size of 2 nm. Energy dispersive X-ray spectroscopy (EDS) was performed in STEM mode. STEM-HAADF images were recorded by Fischione model 3000 HAADF detector (E.A. Fischione Instruments Inc., PA USA) using camera lengths ranging from 560 to 80 mm (occasionally 1200 mm), corresponding to the inner collection angle in the range of few mrad to 50 mrad. The inner and outer diameters of the ADF detector were 4 and 28 mm, respectively, and the beam convergence angle was 18.6 mrad. Table 1 shows a list of camera lengths for the Tecnai F30 microscope and corresponding acceptance angles. CTEM was also performed on a JEOL 2000FX microscope (JEOL Ltd., Tokyo, Japan).

For convenience, change in image contrast is represented with respect to change in camera length, instead of scattering angles, since each camera length of detector needs to be represented by two angles, the inner and outer acceptance angles. Based on experimental observations of diffraction effects, micrographs recorded with camera lengths of 80 mm are termed HAADF-STEM images, whereas those with camera lengths greater than 80 mm are termed LAADF images in this paper.

3. Results and discussion

Deformation by HPT was studied by observing samples with increasing strain from the compressive stress and $N = 0 - 5$ turns of torsion. The deformation started with twinning accompanied by dislocation slip. Subsequently, concentration of stress can occur at specific locations such as twin boundaries, leading to initiation of recrystallization. Nucleation of recrystallized grains also occurs on second phase, such as quasicrystal phase in the present alloy.

Table 1. ADF detector acceptance angles as a function of camera length.

	ADF detector acceptance angle (mrad)	
Camera length (mm)	Inner, θ_i	Outer, θ_o
80	50.0	336.6
100	40.0	273.0
120	33.3	233.3
150	26.6	184.4
200	20.0	139.0
250	16.0	111.5
300	13.3	93.0
380	10.5	73.5
560	7.1	50.0
1200	3.3	23.3

3.1. Twinning and dislocation slip in the $N = 0$ sample, and the specimen tilt dependent channeling contrast

As is well known for hexagonal metals, wrought processing introduces a strong texture, in which the grains tend to align with basal planes in the direction of mechanical working (such as rolling or extrusion). Hexagonal metals commonly tend to twin by $\{10\bar{1}2\}$ type of twinning (extension twin) when a tensile stress is applied along the hexagonal axis [18,19]. Such a situation occurs in the experimental setup of the present study. The disks subjected to HPT are sliced in cross-section from a rolled bar, in which grains are expected to be aligned preferentially with their basal planes along the compression axis. Therefore, massive twinning occurs on application of compressive pressure, observed in the $N = 0$ sample.

Figure 2a and c show BF-TEM images of the two types of deformation twins in $N = 0$ HPT (compressed) sample. The images are viewed along $[\bar{1}\bar{1}23]$ and $[\bar{1}2\bar{1}0]$ zone axis, respectively. The mirror images of diffraction spots with respect to $\{10\bar{1}1\}$ and $\{10\bar{1}2\}$ planes in selected area diffraction patterns (SADP) of Figure 2b and d clearly show the type of twins. Twinning under these conditions, characterized by $\{10\bar{1}1\}$ and $\{10\bar{1}2\}$ twin planes, cause a lattice reorientation of approximately 56° and 86°, respectively, around a $\langle 11\bar{2}0\rangle$ axis. In addition, strong strain contrasts are observed in these micrographs, which arise due to a high density of dislocations. These dislocations have evidently formed

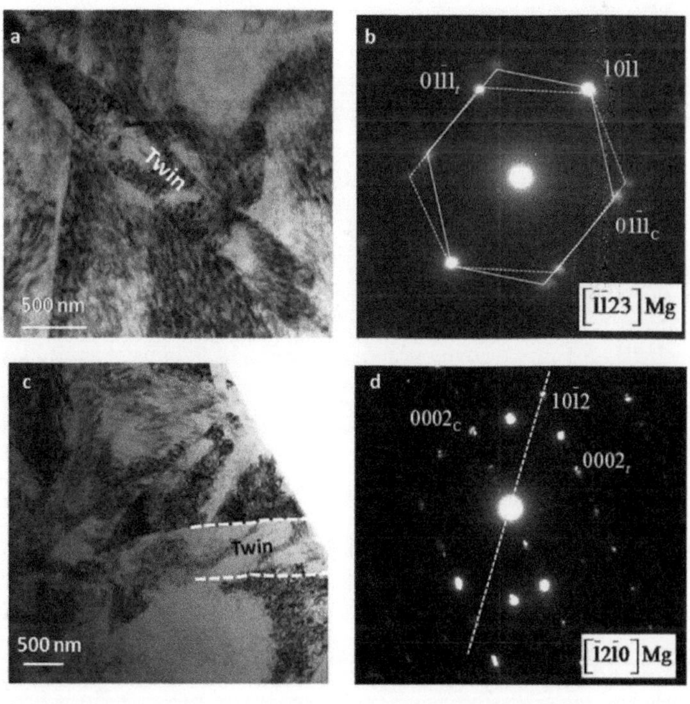

Figure 2. Conventional TEM bright field micrographs showing deformation twins in $N = 0$ HPT sample of Mg-3Zn-0.5 (at%) alloy. (a) A twin viewed along $[\bar{1}\bar{1}23]$ zone axis, and the corresponding composite diffraction pattern (b). The twin system is $\{10\bar{1}1\}\langle 10\bar{1}2\rangle$. (c) Another twin viewed along $[12\bar{1}0]$ zone axis and the corresponding composite diffraction pattern (d). The twin system is $\{10\bar{1}2\}\langle 10\bar{1}1\rangle$.

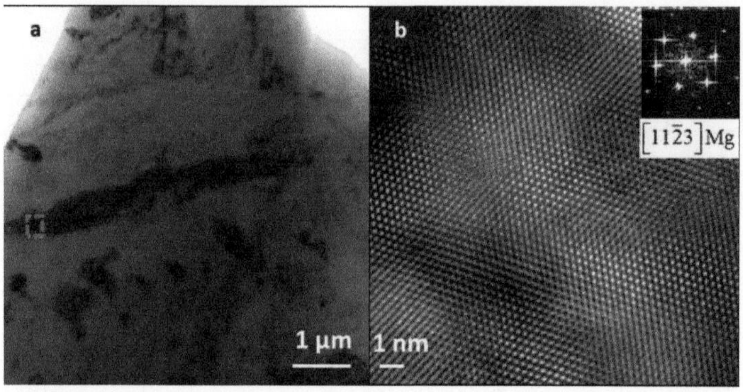

Figure 3. (a) Conventional TEM bright field image showing a twin oriented along [11$\bar{2}$3] zone axis. (b) HRTEM image with FFT (inset) of the same twin acquired from a small region shown in square box in (a).

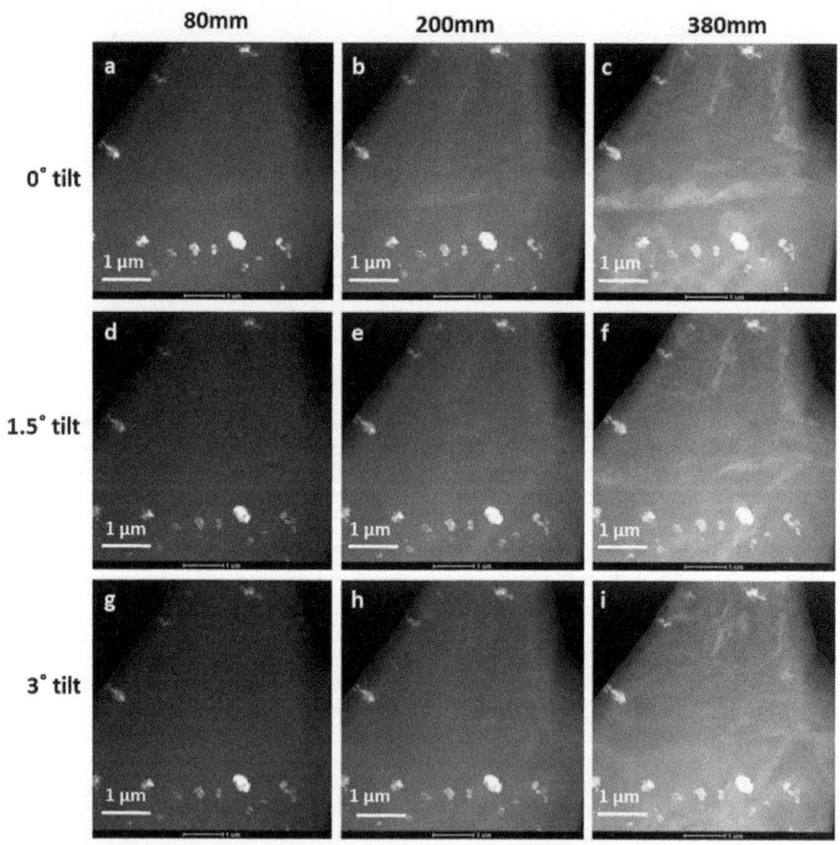

Figure 4. A series of STEM-ADF images showing variation of twin contrast from the same region as shown in Figure 3a, at different tilt positions and at different camera lengths. Rows from top to bottom: tilts of 0°, 1.5° and 3°, columns from left to right: camera lengths 80, 200 and 380 mm.

due to the high amount of stress of 5 GPa, as well as easy slip caused by reorientation of lattice due to twinning. For this reason of high strain, twin boundaries do not appear very well defined. $\{10\bar{1}1\}$ type twins are usually known to form at a high level of stress.

Next, the contrast of twins under HAADF-LAADF is examined. In the present investigations, channeling effects need to be avoided systematically, since they can prevent a direct interpretation of the HAADF and LAADF images in STEM mode in terms of chemical localization by Z-contrast and Bragg diffraction contrast respectively. Figure 3 shows BF-TEM and high resolution TEM (HRTEM) images of a typical twin found in the $N = 0$ HPT sample. The twin plate is oriented along the $[11\bar{2}3]$ zone axis as shown in the HRTEM image and its fast Fourier transform (FFT) inset. Local variations in lattice spot intensities indicates strain in the lattice

This sample region was then examined in ADF mode, as shown in Figure 4. The sample was tilted in STEM mode by about 0°, 1.5° and 3° away from the zone axis, and the effect of tilt on imaging was observed. At each tilting position, ADF images of the same twin were acquired with camera lengths of 80, 200 and 380 mm. It is assumed that upon tilting the sample a few degrees away from the zone axis, thickness remains unchanged, as ADF image intensity depends strongly on the mass thickness of the sample. When the sample is at 0° tilt, the HAADF image acquired with a camera length CL = 80 mm image (Figure 4a) shows bright contrast not only from the ternary phase particles due to Z-contrast, but also from twin plate due to channeling effects along the zone axis. In the same tilt conditions, LAADF images acquired with CL = 200 and 380 mm (Figure 4b and c) show much stronger bright contrast from the twin due to additional contributions from diffracted beams as well as channeling effects along the zone axis. These LAADF images still exhibit bright contrast from the particles as the Z-contrast contribution is still retained. When the sample is tilted 1.5° away from the zone axis, the twin intensity in HAADF image (Figure 4d) decreases significantly and does not exhibit any bright contrast, whereas the twin still exhibits a reasonable intensity in LAADF images (Figure 4e and f). At further tilt to 3°, the twin does not show any bright contrast in either HAADF (Figure 4g) or LAADF (Figures 4h and i) images (while other strain contrast are visible). In general, the HAADF image (Figure 4a) should not exhibit any bright contrast since at high scattering angles only Z-contrast will be expected. This demonstrates the contribution of channeling effects from well-aligned atomic columns in the zone axis oriented twin. These channeling effects will not be seen when the crystal is tilted even a few degrees away from the zone axis and hence the twin in the HAADF image (Figure 4d) loses its intensity by tilting just 1.5° away from zone axis. The higher brightness contrast from the twin in the LAADF image (Figure 4c) compared to the HAADF image (Figure 4a) indicates the strong dominance of those low order diffraction disks that fall on the ADF detector. At the same time, it is not possible to disregard the contributions from channeling effects at low scattering angles since twin intensity in the LAADF image (Figure 4e and f) slightly decreases when the twin is tilted 1.5° from the exact zone and it completely loses its intensity with an additional tilt of 3° from the zone axis (Figure 4h and i). The existence of weakly bright intensity of the twin in the LAADF image (Figure 4e and f) when it is in a 1.5° tilt condition from low index zone shows that the contribution of diffraction effects is not as sensitive as that of the channeling effects to the zone axis. From the above observations it can be seen that contributions of intensities for HAADF and LAADF imaging are not from the same source and that by tilting the sample and/or by changing the camera length, the source of contrast can be varied dramatically. These systematic tilting studies in HAADF and LAADF modes eliminate the possibility of presence of any recrystallized grains at the strain level in the $N = 0$ HPT specimen.

3.2. Subgrain structure formation in N=1/4 sample

Further slip of dislocations occurs when torsion is applied to the sample, in addition to the 5 GPa load. This leads to formation of subgrains, which occurs when dislocations align to form subgrain boundaries. Figure 5 shows the microstructure of one of the regions of the $N = 1/4$ sample. At this strain level, strain contrast at a fine scale is discerned in the CTEM image in Figure 5a. The accompanying diffraction pattern in Figure 5b shows spots belonging to the $[11\bar{2}0]$ zone axis, and a number of other spots indicative of grain fragmentation. However, it is difficult to observe grain fragmentation or formation of subgrain structures in the bright field image due to the high accumulated strain. The HPT deformation process with repeated revolutions leads to generation of excessive dislocations in grain interiors, which further leads to formation of dislocation cell structures and subgrains with low angle grain boundaries. Therefore the same region as shown in Figure 5 was imaged using STEM-ADF with several different camera lengths, shown in Figure 6. The HAADF image acquired with CL = 80 mm (Figure 6a) shows lines of Zn enrichment, which indicate formation of new boundaries, possibly low angle grain boundaries. Enrichment of zinc on boundaries by possible segregation of zinc is confirmed by EDS mapping shown in Figure 6e. In Figure 6a–d, a large region in the middle-left of the micrograph, marked 'A' in (a), does not show similar boundaries. In the LAADF images of Figure 6b acquired with higher camera lengths of 200 mm, regions of strain can be observed, which are seen to terminate at the boundaries. In the larger region A in the middle-left, there is a large ring of continuous strain contrast, indicated by an arrowhead, which suggests that this region is a single large grain containing lattice strain. On increasing the camera length to 300 mm (Figure 6c), the contrast becomes softer, and more subtle details become visible. More well-defined subgrains are observed, such as one marked by an arrow. The STEM-ADF image acquired with CL = 1200 mm (Figure 6d) appears similar to an annular bright field (ABF) image due to contributions from the peripheral transmitted beam at very high camera lengths [20]. By appearance

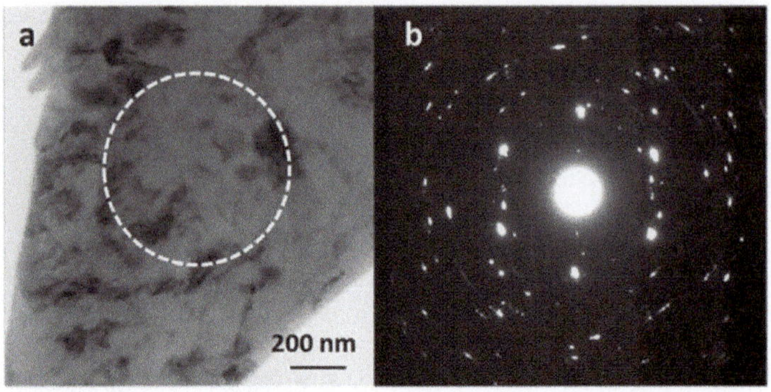

Figure 5. (a) Conventional TEM image recorded from a deformed region of $N = 1/4$ HPT processed sample. (b) SADP acquired from corresponding region marked in (a), depicting multiple diffraction spots and arcs, which indicate the formation of low angle boundaries.

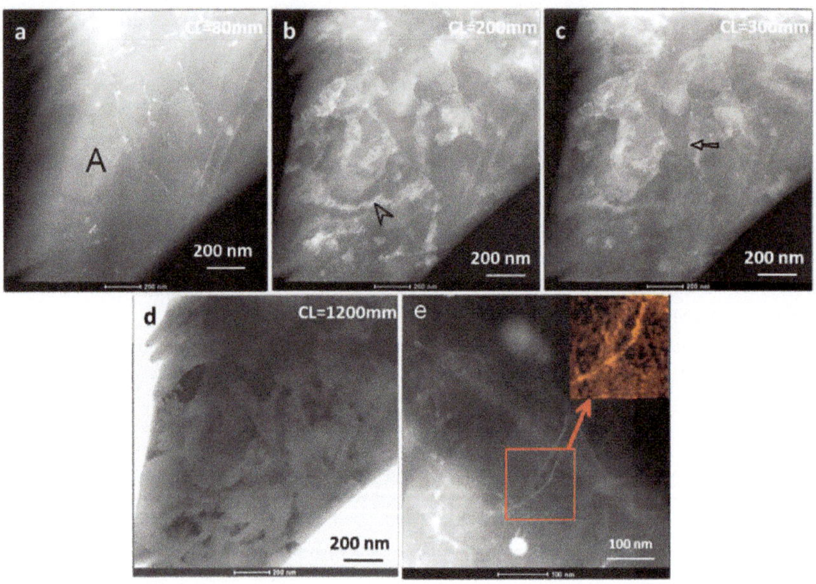

Figure 6. STEM-ADF images acquired from the same area as shown in Figure 5, in $N = 1/4$ HPT processed sample. (a) HAADF image (CL = 80 mm) showing only Zn segregation on low angle boundaries. (b–d) LAADF images (CL = 200, 300 and 1200 mm) exhibiting varying contrast from individual fragmented grains clearly, which is not observed either in CTEM or HAADF images. Furthermore, the STEM-ADF acquired with large camera length of 1200 mm (d) resembles an annular bright field (ABF) image. (e) STEM-ADF image from another region in the sample. The EDS map of zinc from an area marked by a square is shown in the inset.

it corresponds to the ADF image of Figure 6b. This example shows that STEM-ADF imaging becomes invaluable in observation of early deformation pathways before recrystallization occurs.

3.3. *Imaging of recrystallized grains nucleated in thick and highly strained region*

Figure 7a shows a CTEM dark field micrograph at another location in the $N = 1/4$ sample. Regions of strain concentration are imaged, which indicate local lattice distortion in Bragg condition. No grain boundaries are visible. Specimen thickness in this region was determined to be 226 nm by measuring fringe width in a disk of $\{10\bar{1}0\}$ convergent beam electron diffraction pattern in a two-beam condition. Two regions, 1 and 2, are marked by circles. The diffraction pattern acquired from region 1, shown in Figure 7b, exhibits diffraction spots from essentially the same zone axis, though the spots are in the form of arcs indicative of strain. This suggests that this region is still unrecrystallized. The diffraction pattern shown in Figure 7c, acquired from region 2, exhibits concentric rings, which suggest the presence of several grains in this small region, i.e. the formation of recrystallized grains (having large angle grain boundaries). Since the CTEM image does not show any recrystallized grains, further investigation was carried out by STEM-ADF imaging. Figure 8 shows a series of corresponding STEM-ADF micrographs acquired from the same region of the sample with different camera lengths. The HAADF image acquired with CL = 80 mm under the same sample tilt conditions as in Figure 7 is shown in Figure 8a. A distribution of sharp bright contrast appears, which can be attributed to boundaries decorated with segregation of zinc. The contrast of this region is changed quite dramatically in the LAADF image acquired with a camera length of 150 mm, shown in Figure 8b. Now strain contrast is the dominant feature, although it still shows the Z-contrast information. This contrast obscures any detailed information. With further increase in the camera length, Figure 8c, the strain contrast is softened and more details are visible in between. Figure 8d shows a magnified view of the square box region in Figure 8c, which corresponds to region 2 in Figure 7a. In this micrograph, grains with a size of about 100 nm with sharp grain boundaries are clearly observed. The distinction between regions 1 and 2 of Figure 7a, unrecrystallized versus recrystallized, is also confirmed from Figure 8a, in which region 1 does not show presence of any grain boundaries by zinc segregation contrast, while region 2 shows a high density of grain boundaries decorated with zinc contrast. However, grain boundaries can be observed such as in Figure 8d even in the absence of any Z-contrast. This example demonstrates that ADF imaging with conical illumination, using an appropriate camera length for the ADF detector to remove the strain effects, becomes extremely useful to retrieve a large amount of embedded structural information. Information from such buried structures is more difficult to access with traditional TEM imaging, leading to a quick and incorrect conclusion that recrystallization nucleation has not yet occurred. Furthermore, the visibility of recrystallized grains depends on beam convergence angle, camera length and specimen thickness. In this example the sample thickness of about 200 nm is comparable to the scale of the microstructure in which the grain size is in the range of 100–200 nm. This illustration shows the power of low-angular filtered ADF imaging as a unique tool for observing recrystallized grains and their formation mechanisms from very interior locations of the deformed sample at early periods of applied strain. With further increase in the camera length, Figure 8e, the composition contrast is at a minimum, and finer details of strain contrast are visible. At camera length of 1200 mm, Figure 8f, the contrast looks like a reversal of CL = 150 mm (Figure 8b). The contrast is now equivalent of ABF.

3.4. DRX nucleation on second phase particles

Secondary phase particles of hard phases are known to promote dynamic recrystallization (DRX) [21]. Early recrystallization occurs over secondary phase particles. The effect of fine particles on DRX and grain refinement mechanisms is also one of the major points in investigation on severely deformed magnesium alloys. The case of the secondary phase being the quasicrystal phase is of particular interest, because recrystallization on it can modify the grain texture [15,22]. The icosahedral quasicrystal phase has point group symmetry of $m\overline{35}$, with a large order of symmetry of 120. Assumption of only a single orientation relationship of $2f_{ico} \| (0001)_{hex}$ [23] can produce 15 variants of a hexagonal crystal. More orientation relationships are known to occur between magnesium and quasicrystal phase [24]. Figure 9a shows a DRX grain attached to a quasicrystal phase particle. This grain appears to be free of severe strain and exhibits a sharp grain boundary, and is therefore assumed to be recrystallized, by nucleating on the quasicrystal particle. Other than this grain (and the quasicrystal particle), only strain contrast is visible in this micrograph. This region was then imaged in STEM-ADF mode. The HAADF image acquired with CL = 80 mm displays only bright contrast from the quasicrystal particle. Figure 9b shows a micrograph recorded with camera lengths of 300 mm, which reveals bright contrast from the recrystallized grain attached to the quasicrystal particle. In this image, more small grains are seen attached to the quasicrystal phase particle, having recrystallized by nucleating over the particle. Thus it is established that quasicrystal particles effectively act as heterogeneous nucleation sites for dynamically recrystallized grains, even in a sample with relatively low amount of strain. Diffraction patterns to confirm crystallographic relationship were obtained (not shown here).

3.5. Observation of grain boundaries of ultrafine nanosize grains in recrystallized regions

It is very important to determine and to control grain size and secondary phase particles or precipitates size, as these have a significant effect on the mechanical properties of metals and alloys. The size of the grains can vary in a wide range, from very few tens of nanometers to up to a size of 1 μm in ultrafine grain alloys. From traditional CTEM imaging it can be challenging to distinguish between a matrix grain and a second phase particle because of the presence of both mass-thickness contrast and diffraction contrast. Misidentification of grain with particle can lead to incorrect estimation of grain size or particle size distribution, determination of which is very important for microstructural characterization and consequently the optimization of mechanical properties. To avoid such complications, LAADF/HAADF coupled STEM imaging can be effectively utilized to distinguish grains, at first from strain contrast, and then from second phase particles, and then to measure their size distribution.

Figure 10a shows a CTEM bright field image containing recrystallized grains and secondary phase particles from a recrystallized region in the $N = 5$ HPT sample. This image also shows that recrystallization has occurred over the entire region of the sample. However,

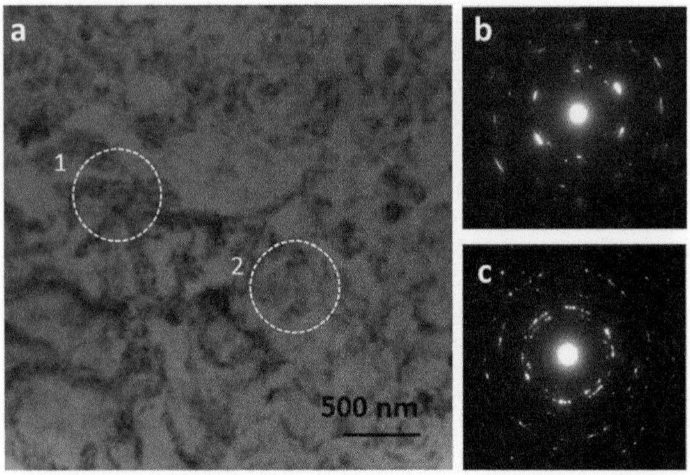

Figure 7. (a) Conventional TEM bright field image acquired from highly strained region of $N = 1/4$ HPT sample. (b) Spreading arc spots in SADP acquired from region 1 in (a), indicating an unrecrystallized region. (c) Concentric rings in SADP acquired from region 2 in (a), indicating recrystallization in that region even though recrystallized grains are not seen.

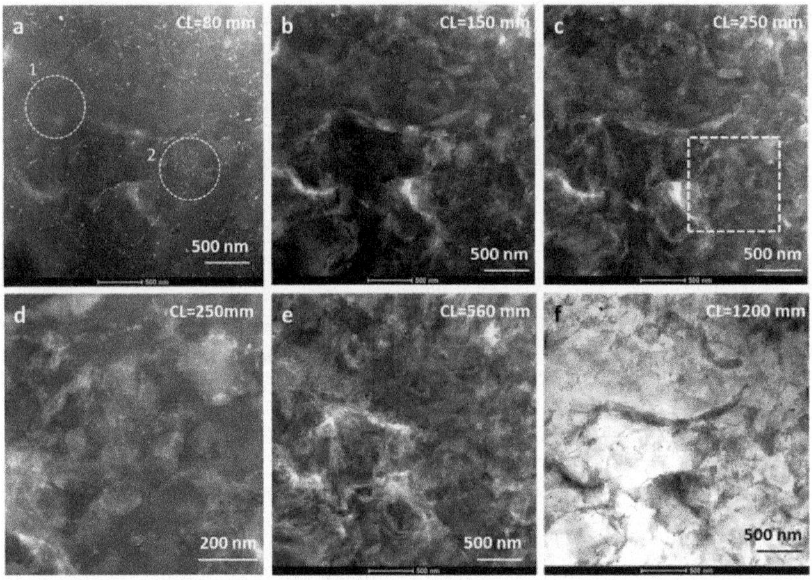

Figure 8. STEM-ADF images acquired from the same area as shown in Figure 7a in the $N = 1/4$ HPT sample. (a) HAADF image acquired with camera length 80 mm showing Z-contrast information. (b) LAADF image acquired with camera length of 150 mm showing not only Z-contrast information but also exhibiting varying contrast from lattice strain owing to diffraction contrast. (c) LAADF image acquired at camera length of 250 mm, additionally exhibiting a varying contrast from few recrystallized grains (shown in box) from high angle boundaries, clearly owing to a diffraction contrast not observed in either CTEM or HAADF images. The boxed region is shown magnified in (d). (e–f) STEM-ADF acquired with large camera lengths of 560 and 1200 mm, again exhibiting contrast only from lattice strain. This indicates that the visibility of grains depends on proper choice of camera length for a particular beam convergence angle and specimen thickness.

the grains cannot be distinguished clearly, in part due to uneven strain contrast within grains and partly because of uneven thickness of the region. The sample is thinner in a diagonal region from the bottom left to top right. Figure 10b–d shows three corresponding ADF images from the same location with three different camera lengths of 80, 380 and 560 mm. The advantage of LAADF/HAADF STEM coupled imaging for identifying grains and particles can be clearly observed. The HAADF image acquired at the shorter camera length of 80 mm shows bright contrast from only secondary phase particles. These particles are indistinguishable from the matrix grains in the bright field image of Figure 10a. In LAADF images acquired at higher camera lengths of 300 and 560 mm, the individual grains and grain boundaries are also clearly

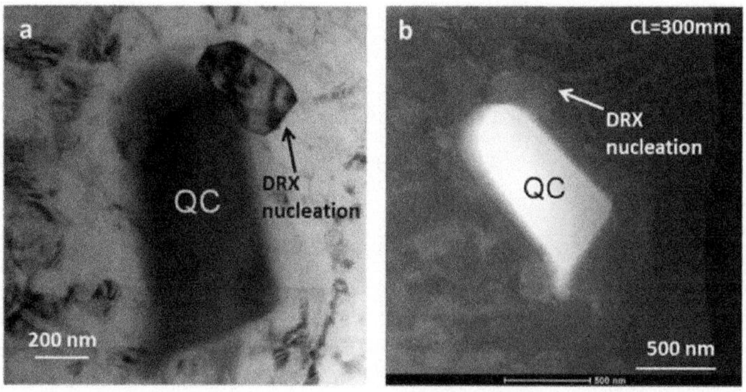

Figure 9. (a) Conventional TEM bright field image showing a DRX grain nucleated on Mg_3Zn_6Y quasicrystalline (QC) particle in $N = 1/2$ HPT processed sample. (b) STEM-LAADF images (CL = 300 mm) showing diffraction contrast from newly recrystallized grains nucleated on the quasicrystalline particle.

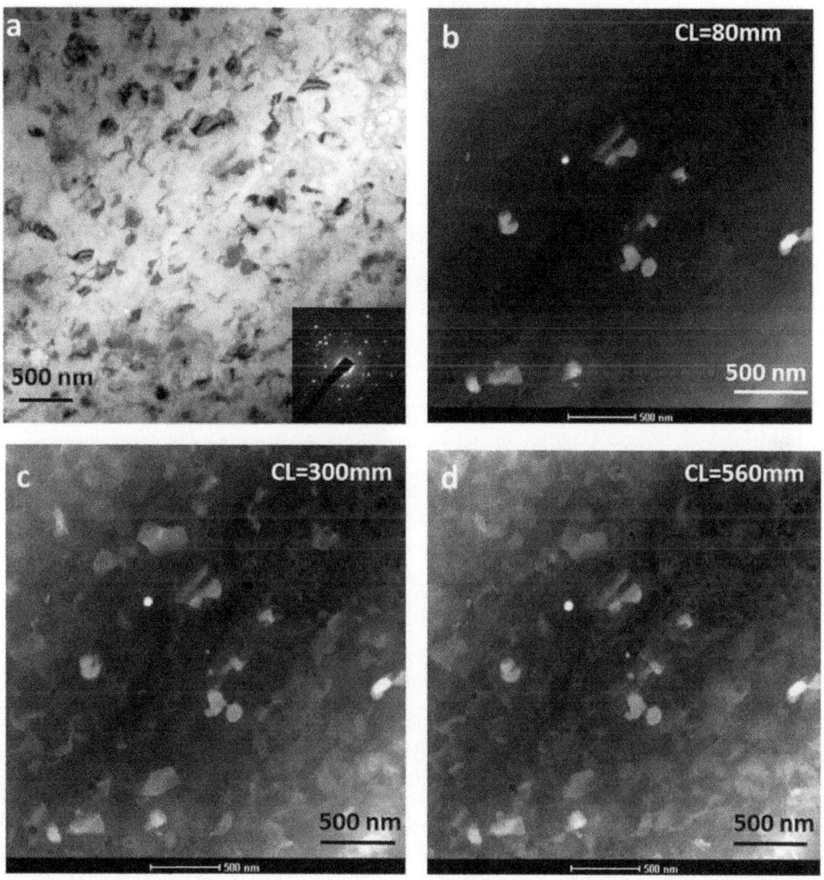

Figure 10. (a) Conventional TEM bright field image showing a recrystallized region in $N = 5$ HPT sample. The concentric rings in SADP as shown in the inset indicate recrystallization occurrence. (b) HAADF image acquired from same region with camera length 80 mm showing only Z-contrast. (c) LAADF image acquired from same region with camera length 300 mm showing grain refinement. (d) LAADF image acquired with camera length 560 mm. Both HAADF and LAADF imaging techniques together can be used to find grain size distribution and particle size distribution.

visible apart from the secondary phase particles. From a number of HAADF/LAADF images acquired from different locations of the sample, by measuring the farthest distance between recrystallized grain corners from the LAADF images of over 250 grains, the average grain size after five rotations of HPT is found to be 150 ± 25 nm. Thus, in this case both HAADF-STEM and LAADF-STEM investigations must be performed to gain information about grain size distribution and particle size distribution. Moreover, in the CTEM image (Figure

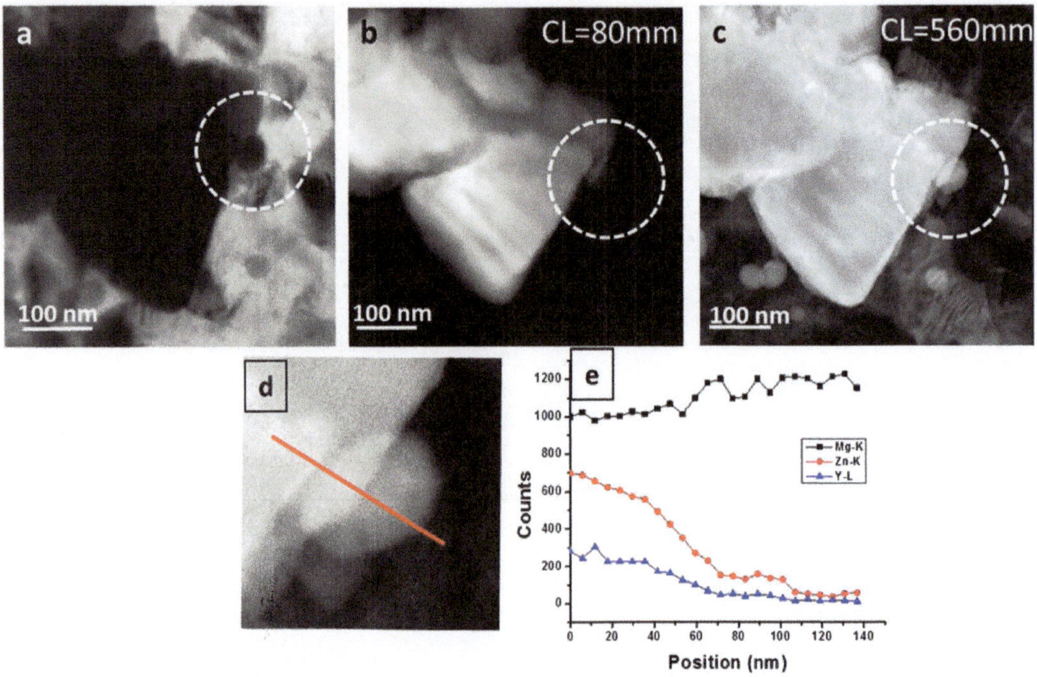

Figure 11. Observation of nucleation of a precipitate on a second phase particle. (a) Conventional TEM image showing a small precipitate seemingly attached to the second phase particle. (b) HAADF image (CL = 80 mm) acquired from the same location shows weak intensity from the precipitate. (c) LAADF image (CL = 380 mm) acquired from the same location shows bright intensity from the precipitate. (d) EDS line composition profile across the precipitate. Here both HAADF and LAADF imaging together confirm that this particle is a precipitate.

10a), only few grains show strong diffraction contrast, whereas in STEM-LAADF images (Figure 10c–d) acquired with CL = 300 and 560 mm more grains exhibit strong diffraction contrast. This is because the convergent beam used in STEM imaging facilitates the beam to incident on a sample over a range of angles, hence even the grains slightly away from Bragg condition with respect to the parallel beam also exhibit diffraction contrast. In the case of observation by CTEM, grain boundaries are not clearly visible, therefore dark field observation is employed to look at the grain size, such as in our previous report [16]. By this method, only a few grains are in contrast at any instance, and the grain boundaries are still not imaged due to strain contrast near the boundaries. Therefore the estimation of grain size is not accurate. In the micrographs of Figure 10, a qualitative difference in imaging between thin and thick regions can be observed. This is because, in general, in STEM-ADF (HAADF or LAADF) imaging, as the specimen thickness increases, the background intensity also increases due to the increase in elastic scattering. In LAADF imaging, the background intensity from thick regions of the specimen dominates. The diffracted beam intensities from thinner regions are weaker as compared to background intensity from thicker regions. It can be noticed in the micrographs that the contrast difference is reduced as the camera length increases.

Earlier study of Mg-Zn binary HPT alloy reported round particles of a Mg-Zn phase of about 10–20 nm in size at strain of $N = 20$ [16]. Similar particles were also observed in this study. Figure 11a shows a CTEM bright field image of specimen $N = 5$. Smaller dynamically nucleated precipitates are present in an Mg matrix. They are also observed on the ternary phase particle (yttrium rich) boundaries. From the CTEM image of the thick specimen, it is very difficult to identify whether a particular phase attached to it is a dynamically nucleated precipitate or broken ternary phase particle. The detection of small precipitated particles by HRTEM is also impeded if the particles are located in thicker regions of the specimen. Figure 11b and c show ADF images of the same location obtained in STEM mode with different camera lengths. The HAADF image acquired with CL = 80 mm (Figure 11b) shows strong brighter contrast from bigger quasicrystal particles and weak bright contrast from the smaller precipitates. At the same time these smaller precipitates appear brighter than the matrix in HAADF, which indicates that these precipitates are binary phase consisting of Mg and Zn atoms. In an LAADF image acquired at higher camera length of 560 mm (Figure 11c), the smaller precipitate marked with a circle is apparently directly attached to the bigger particle. A few neighboring precipitates also exhibit similarly strong bright contrast.

Thus HAADF/LAADF imaging together indicates that the particle could be newly heterogeneously precipitated particles from matrix formed at particle–matrix interfaces as well as inside the matrix during deformation. These new particles can further stabilize the grain refinement process because of creation of additional phase boundaries which can retard the grain growth process. In order to further confirm that this particle is precipitated, EDS line profile mapping in STEM mode was carried out. The composition profile shown in Figure 11d across the interface between ternary phase particle and the attached phase shows that there is sudden decrease in yttrium-L profile at the interface when it crosses the ternary phase particle, whereas Mg-K and Zn-K line profiles show almost no change in composition across the interface, which clearly indicates that the attached phase is a Mg-Zn binary phase.

This study establishes the grain size evolution by HPT processing. The grain size of the starting specimen (before deformation) was about 10 μm. The grain size after compression ($N = 0$) was also similar to that of the initial sample, with formation of twins. After $N = 1/4$ turns, the specimen shows two types of grains - one is subgrains with low angle boundaries, with size in the range of 300 nm to 2 μm. Another type of grains is recrystallized grains of size from 80 to 170 nm. The recrystallization is complete after $N = 5$ turns of HPT, with an average size of about 150 nm. As the strain increases, more subgrains are formed, but the recrystallized grain sizes remains nearly the same.

The microstructural evolution during HPT process as studied here shows similarities to the binary Mg-3.4at%Zn alloy studied by HPT, where the initial alloy was extruded (and consequently had a texture) and subsequently solution treated at 573 K, to obtain a grain size of about 28 μm [16,17]. In the binary Mg-Zn alloy, heavy twinning of matrix grains is observed initially, followed by subgrain formation. At $N = 3$, a bimodal microstructure is reported, of large grains with accumulated strain and fine recrystallized grains. Zinc segregation and consequent precipitation occurred at the grain boundaries, cited as the reason for very fine grain size by pinning of the boundaries, resulting in the final grain size of about 140 nm [16]. In the present study of the ternary Mg-Zn-Y alloy, a very small amount of zinc was available in the matrix, due to formation of the quasicrystal phase with a general composition Mg_3Zn_6Y. However, zinc segregation at the boundaries is still observed. The source of zinc is most likely the supersaturation of the matrix from the solution treatment, and a possible decomposition of the ternary phase under strain. Details of chemical compositional analysis will be published in a separate paper. A complete recrystallization is observed by $N = 5$, assisted by recrystallization on the quasicrystal phase particles.

4. Conclusions

Microstructural evolution, especially grain refinement by recrystallization during severe plastic deformation of a Mg-3Zn-0.5Y (at%) alloy processed by HPT at room temperature was studied by conventional TEM as well as STEM-low angle annular dark field (LAADF) imaging technique with an optimized annular detection angle. Suitability of LAADF for studying severe plastically deformed alloys is demonstrated, where grain boundaries can be imaged even in highly strained regions. The following are the main findings of this study.

(1) Extensive twinning was observed due to compression in the $N = 0$ sample. Specimen tilt-dependent channeling effect studies demonstrate that sources of contrast from the twin are not same in HAADF and LAADF imaging modes. Observation of twins is sensitive to sample tilt due to channeling effects, while strain contrast is not.
(2) No recrystallization nucleation is observed in twins.
(3) LAADF imaging alone permits a clear view of initial grain fragmentation and formation of sub grain structures in the $N = 1/4$ specimen.
(4) By LAADF imaging, nucleation of recrystallization was detected at an early deformation stage of $N = 1/4$. Severe strain contrast effects were subdued to observe the grain boundaries.
(5) Complete recrystallization was observed in the $N = 5$ sample. Clear observation of grain boundaries became possible in LAADF by eliminating strong strain contrast within grains and by reduction of the effect of uneven specimen thickness. Besides, by HAADF contrast, particles of second phase were clearly distinguished from the matrix grains. Average grain size in the $N = 5$ specimen was determined to be 150 ± 25 nm.
6. Nano-sized binary phase precipitates were analyzed by ADF imaging and composition profile, and distinguished from the fragments of the ternary phase.

Acknowledgements

The authors thank Dr. Akira Hasegawa and Ms. Noriko Isaka for fruitful discussions. This work was supported in part by a Grant-in-Aid for Scientific Research on Innovative Area 'Bulk Nanostructured Metals' through MEXT, Japan [Contract no. 22102004].

References

[1] Valiev RZ, Estrin Y, Horita Z, Langdon TG, Zehetbauer MJ, Zhu YT. Producing bulk ultrafine-grained materials by severe plastic deformation. J. Miner. Met. Mater. Soc. 2006;58:33–39.

[2] Aoki I, Yanagimoto S. Fundamental research on rolling of rod-shaped materials Part 1 (Strain distribution of oval-square-oval rolling method). J. Jpn. Soc. Technol. Plast. 1968;9:597.

[3] Zhilyaev AP, Langdon TG. Using high-pressure torsion for metal processing: Fundamentals and applications. Prog. Mater. Sci. 2008;53:893–979.

[4] Crewe AV, Eggenberger DN, Wall J, Welter LM. Electron gun using a field emission source. Rev. Sci. Inst. 1968;39:576-583.

[5] Crewe AV, Wall J, Welter LM. A High, Resolution scanning transmission electron microscope. J. Appl. Phys. 1968;39:5861-5868.

[6] Cowley J. Scanning transmission electron microscopy of thin specimens. Ultramicroscopy. 1977;2:3-16.

[7] Brown L. Scanning transmission electron microscopy: microanalysis for the microelectronic age. J. Phys. F: Metal Phys. 1981;11:1.

[8] Nellist PD, Pennycook SJ. Incoherent imaging using dynamically scattered coherent electrons. Ultramicroscopy. 1999;78:111-124.

[9] Howie A. Image contrast and localized signal selection techniques. J. Microsc. 1979;117:11-23.

[10] Phillips PJ, Mills MJ, De Graef M. Systematic row and zone axis STEM defect image simulations. Philos. Mag. 2011;91:2081-2101.

[11] Phillips PJ, Brandes MC, Mills MJ, De Graef M. Diffraction contrast STEM of dislocations: Imaging and simulations. Ultramicroscopy. 2011;111:1483-1487.

[12] Phillips PJ, De Graef M, Kovarik L, Agrawal A, Windl W, Mills MJ. Atomic-resolution defect contrast in low angle annular dark-field STEM. Ultramicroscopy. 2012;116:47-55.

[13] Humphreys CJ. Fundamental concepts of stem imaging. Ultramicroscopy. 1981;7:7-12.

[14] Glaisher RW, Spargo AEC, Smith DJ. A systematic analysis of HREM imaging of elemental semiconductors. Ultramicroscopy. 1989;27:35-51.

[15] Singh A, Osawa Y, Somekawa H, Mukai T. Effect of microstructure on strength and ductility of high strength quasicrystal phase dispersed Mg-Zn-Y alloys. Mater. Sci. Eng. A. 2014;611:242.

[16] Meng F, Rosalie JM, Singh A, Somekawa H, Tsuchiya K. Ultrafine grain formation in Mg-Zn alloy by in situ precipitation during high-pressure torsion. Scripta Mat. 2014;78-79:57-60.

[17] Meng F, Rosalie JM, Singh A, Tsuchiya K. Precipitation behaviour of an ultra-fine grained Mg-Zn alloy processed by high pressure torsion. Mater. Sci. Eng. A. 2015;644:386-391.

[18] Partridge PG. The crystallography and deformation modes of hexagonal close-packed metals. Metall. Rev. 1967;12:168.

[19] Reed-Hill RE, Robertson WD. The crystallographic characteristics of fracture in magnesium single crystals. Acta Metall. 1957;5:728.

[20] Okunishi E, Ishikawa I, Sawada H, Hosokawa F, Hori M, Kondo Y. Visualization of light elements at ultrahigh resolution by STEM annular bright field microscopy. Microsc. Microanal. 2009;15:164-165.

[21] Apps PJ, Bowen JR, Prangnell PB. The effect of coarse second-phase particles on the rate of grain refinement during severe plastic deformation. Acta Mater. 2003;51:2811-2822.

[22] Singh A, Somekawa H, Mukai T. Compressive strength and yield asymmetry in extruded Mg-Zn-Ho alloys containing quasicrystal phase. Scripta Mater. 2007;56:935.

[23] Singh A, Tsai AP, Nakamura M, Watanabe M, Kato A. Nano-precipitates of icosahedral phase in quasicrystal strengthened Mg-Zn-Y alloys. Philos. Mag. Lett. 2003;83:543.

[24] Singh A, Watanabe M, Kato A, Tsai AP. Twinning and orientation relationship of icosahedral phase with magnesium matrix. Acta Mater. 2005;53:4733.

16

Unconventional fractional quantum Hall effect in monolayer and bilayer graphene

Janusz Jacak and Lucjan Jacak

Institute of Physics, Wrocław University of Technology, Wyb. Wyspiańskiego 27, 50-370, Wrocław, Poland

ABSTRACT

The commensurability condition is applied to determine the hierarchy of fractional fillings of Landau levels in monolayer and in bilayer graphene. The filling rates for fractional quantum Hall effect (FQHE) in graphene are found in the first three Landau levels in one-to-one agreement with the experimental data. The presence of even denominator filling fractions in the hierarchy for FQHE in bilayer graphene is explained. Experimentally observed hierarchy of FQHE in the first and second Landau levels in monolayer graphene and in the zeroth Landau level in bilayer graphene is beyond the conventional composite fermion interpretation but fits to the presented nonlocal topology commensurability condition.

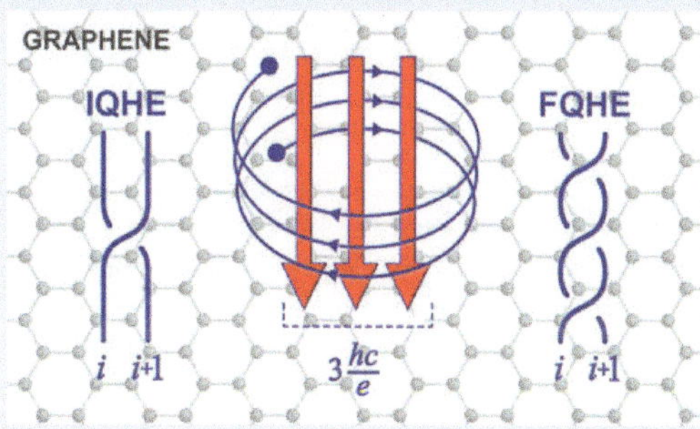

KEYWORDS
Monolayer graphene; bilayer graphene; FQHE; hierarchy of filling fractions

CLASSIFICATION
40 Optical, magnetic and electronic device materials; 104 Carbon and related materials; 105 Low-Dimension (1D/2D) materials

1. Introduction

Recent progress in Hall experiments with graphene has revealed many new features in longitudinal and transversal resistivity in Hall configurations characteristic for fractional quantum Hall effect (FQHE), both in suspended graphene scrapings [1–3], and in graphene samples on crystalline substrate of boron nitride [4,5]. The different structure of Landau levels (LLs) in graphene in comparison to the conventional semiconductor two-dimensional electron gas is the source of a distinct scheme for integer quantum Hall effect (IQHE) in graphene, referred to as the 'relativistic' version [6]. The Berry phase induced shift for chiral carriers in graphene together with the four-fold spin-valley degeneracy of LLs result in $\nu = 4\left(n + \frac{1}{2}\right)$ series for fillings at which IQHE plateaus occur in subsequent centers of LLs for the case of monolayer graphene. When $SU(4)$ spin-valley symmetry is broken by a larger magnetic field, the new features for IQHE appear, corresponding to removal of LL subbands degeneracy [7]. In the bilayer graphene an extra degeneracy of n = 0 and n = 1 LLs shifts the IQHE plateau positions to the ends of consecutive LLs, because of eight-fold degeneracy of the lowest LL (LLL) in this case. The higher LLs are still fourfold-degenerate in bilayer graphene. Simultaneously, more and more features at fractional fillings of LLs are observed related to FQHE, revealing also its specific character mostly connected with the different structure of subbands of LLs and particle-hole symmetry due to meeting of the conductivity and valence bands in Dirac points. The new filling fractions for FQHE are observed in the six first subbands of LLs with $n = 0$ and $n = 1$ in monolayer graphene [2–5]. Especially interesting is an observation of unusual even denominator fillings for FQHE in bilayer graphene, including the most pronounced feature at $\nu = -\frac{1}{2}$ [1].

In the present paper we analyze the hierarchy for fractional fillings linked to strongly correlated multiparticle states in graphene using the topological commensurability approach developed earlier for the ordinary 2DEG Hall systems [8–10]. In this way we explain the structure of fractional fillings of LL

CONTACT Janusz Jacak janusz.jacak@pwr.edu.pl

subbands and demonstrate its evolution with growing number of LL and its subband. This approach, described in [11,12], gives the hierarchy of FQHE in agreement with the available experimental data for monolayer and bilayer graphene. In particular the explanation of the even denominator filling ratios for bilayer graphene, with the FQHE state at $\nu = -\frac{1}{2}$, has been achieved by application of the commensurability topological method to the double layer Hall system.

2. Commensurability condition

The concept of the commensurability of cyclotron trajectories with interparticle spacing in a 2D charged system at strong magnetic field is born in relation to the braid group approach to multiparticle systems in the presence of a magnetic field, and employs the interaction of electrons, essential for formation of any correlated state, including states corresponding to FQHE. The collective behavior of a quantum multiparticle state must be assigned by the statistics phase shift acquired by the multiparticle wave function if one considers position exchanges of particle pairs [8,13]. The exchange of particles is understood as exchange of the position-arguments of the multiparticle wave function. This quantum feature is associated with the one dimensional unitary representation (1DUR) of the full braid group related to the system. The full braid group is the first homotopy group of the multiparticle configuration space, and topologically classifies all possible particle exchanges. This group is denoted as $\pi_1(\Phi) = \pi_1((M^N - \Delta)/S_N)$, where M^N is the N-fold normal product of the manifold $M = R^2$ for the plane, Δ is the diagonal point set in this product (when coordinates of two or more particles coincide, subtracted in order to assure conservation of the number of particles), S_N is the permutation group of N elements [14]. The quotient structure of the configuration space $\Phi = (M^N - \Delta)/S_N$ is related with the indistinguishability of quantum identical particles – the property necessary for the quantum reason, when renumeration of particles has no effect. The full braid group $\pi_1(\Phi)$ is a topological object collecting all classes of nonhomotopic trajectory loops in the configuration space Φ, where points which differ only by enumeration of particles are unified. Any details of the dynamics of an N particle system resulting in special shapes of trajectories are not important here, unless they cause the topological nonequivalence. Only topology of trajectories on M decides whether one trajectory loop can be continuously transformed into another one without a cut or not. If a smooth continuous transformation is impossible between two trajectories, they are topologically nonequivalent and fall into distinct classes of the full braid group. Thus, the full braid group does not reflect the dynamics details, but rather identifies the topology restrictions imposed on the multiparticle system, which are associated to some global features of the system including the manifold type and only those aspects of interaction that might cause topological nonequivalence of trajectories. The Coulomb repulsion of electrons on the plane is a central prerequisite for the braid group definition at magnetic field presence, because the scale of interparticle separation defined by the Coulomb repulsion must interfere with the planar cyclotron orbits, discriminating in this way possible correlation types. As the loops from $\pi_1(\Phi)$ describe exchanges of particles the full braid group contains information on quantum statistics of particles, although classical particles do not have any such statistics. It was proved [13] that the 1DURs of the full braid group serve as the determinant of the quantum statistics of particles. Therefore, if particle classical positions given by arguments of the multiparticle wave function $\Psi(x_1, ..., x_N)$ (x_i is the coordinate of i-th particle on the manifold M, i.e., the classical position of the i-th particle on M) are changing along a selected loop from the $\pi_1(\Phi)$, then this wave function acquires the phase shift $e^{i\alpha}$ defined by the 1DUR of this particular braid [15]. In this way the statistics of quantum particles can be identified. For the same classical particles the quantum different particles can be defined as assigned by distinct 1DURs of the related full braid group. For three dimensional space, $M = R^3$, the full braid group is simply the permutation group S_N, with only two different 1DURs [14]:

$$\sigma_i \rightarrow \begin{cases} e^{i0} = 1, \\ e^{i\pi} = -1, \end{cases} \qquad (1)$$

where σ_i is the generator of S_N, i.e. the elementary braid describing exchange of i-th and $i+1$-th particles, whereas other particles are left in their positions. These two 1DURs correspond to bosons and fermions, respectively. For two-dimensional manifolds M the braid groups differ considerably from the permutation group. For $M = R^2, \pi_1(\Phi)$ is an infinite group with the 1DURs [14]:

$$\sigma_i \rightarrow e^{i\alpha}, \ \alpha \in [0, 2\pi), \qquad (2)$$

corresponding to so-called anyons (including also bosons for $\alpha = 0$ and fermions for $\alpha = \pi$). In this way the unique topology of 2D space causes the substantial change in quantum statistics available for particles in comparison to the 3D space.

There are, however, also other important consequences of 2D topology, which were not accounted for by the anyon concept. The special feature of planar multiparticle systems manifests itself in the presence of the perpendicular magnetic field strong enough that the classical cyclotron orbit is shorter in comparison to the interparticle separation on the plane. Because the classical trajectories of charged particles are defined by cyclotron orbits at the presence of a magnetic field, thus exchanges of neighboring particles on the plane are possible only if the size of the cyclotron orbit fits to the separation between particles, as is illustrated in

Figure 1. Let us emphasize that the separation between particles on the plane is fixed by the Coulomb interaction between particles preventing one particle approaching another in the case of uniformly distributed N particles on the plane. This highlights the fundamental role of the electron interaction in formation of any correlated states, including those which can be characterized in topological commensurability terms. The commensurability between the cyclotron radius and the interparticle spacing is required for the definition of the generators σ_i of the full braid group, where σ_i describes exchange of i-th and $i+1$-th particles. If the cyclotron orbit is incommensurate with the separation between particles on the plane, then the classical trajectories describing σ_i elements of the full braid group are impossible and this braid group cannot be implemented. This means that the quantum statistics cannot be defined in this case and any correlated multiparticle quantum state cannot be organized. In this way the possibility for the definition of particle exchanges (i.e. the definition of the braid group and the quantum statistics) decides whether the quantum correlated collective state can be organized.

The situation when cyclotron orbits are shorter than interparticle separation happens at fractional filling of the LLL. For magnetic fields larger than this one which corresponds to the completely filled LLL, the classical cyclotron orbits are too short to match neighboring particles and the definition of the generators of the full braid group is precluded. Nevertheless, at some 'magic' fractional fillings of the LLL the correlated states are experimentally observed and referred to as FQHE [16–18]. This means that possibility of particle exchanges is recovered somehow. In the framework of the composite fermion (CF) approach [19] the enlargement of cyclotron orbit size is achieved by screening of the external magnetic field by auxiliary field flux quanta attached to particles. This artificial model allows for identification of the main line of fractional fillings for FQHE by mapping of the fractional state onto integer quantum Hall states in the resultant magnetic field reduced by the average field of fluxes pinned to CFs. The CF model does not explain neither the origin of auxiliary field fluxes nor the mechanism of creation of composite particles and the construction is rather formal.

The manifestation of FQHE can be explained, however, in the braid group terms [9]. Though the generators σ_i of the full braid group cannot be defined when the cyclotron orbits are shorter than interparticle separation, there exist other braids which in 2D fit to interparticle separations. This exceptional property of planar multiparticle systems possess multilooped cyclotron braids, i.e., σ_i^q, where q is an odd integer, and for e.g. $q=3$, σ_i^3 describes the braid for exchange of the i-th and $i+1$-th particles with one additional loop. Exclusively in 2D the multilooped cyclotron orbits

Figure 1. Schematic demonstration that commensurability (left) of cyclotron orbit with interparticle separation satisfies topology requirements for braid interchanges in uniformly and equidistantly (due to interaction) distributed 2D particles; for smaller cyclotron radii particles cannot be matched (center), for larger ones the interparticle distance cannot be conserved (right).

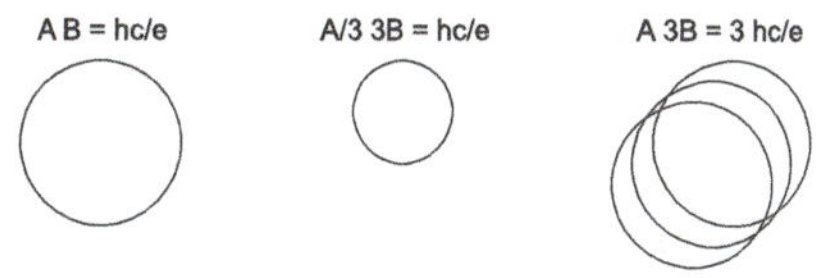

Figure 2. Schematic illustration of cyclotron orbit enhancement in 2D due to multi-loop trajectory structure (third dimension added for visual clarity).

have larger size which can fit to interparticle separation at the 'magic' fractional fillings of the LLL, the same ones at which FQHE is observed. The reason for the enhancement of the size of planar multilooped orbits is linked with constant surface spanned by 2D orbits despite its multilooped character (in opposition to the 3D case when each additional loop adds also a surface portion spanned by this loop). When the total external field flux is passing through the 2D multilooped orbit it must be shared among all loops. Thus the flux portion per each loop diminishes, resulting in their size growth. This is illustrated in Figure 2.

In Figure 2 (left) the scheme of the cyclotron orbit at magnetic field B is pictorially shown as accommodated to the quantum of the magnetic field flux, i.e. $BA=\frac{hc}{e}$. This serves as the definition of the cyclotron orbit size A because this surface A fits to the interparticle separation in the case of the completely filled LLL, $\frac{S}{N}$, S is the sample area, N is the number of particles. If only single-looped orbits are considered, then at, for instance, three times larger field, $3B$, the cyclotron orbit accommodated again to the flux quantum is too short in comparison to the interparticle separation $\frac{S}{N}$ (which fits to the B field orbits but not to $3B$ orbits). This is illustrated in the central panel of Figure 2. Nevertheless, if tree-loop orbits are considered, then in the flat geometry of 2D space, the external flux $3BA$ must be shared among three loops with the same surface A (i.e. BA for each loop). Thus, each loop accommodated to the flux quantum $\frac{hc}{e}$ has the orbit with the surface A and gives the contribution BA to the flux, resulting in the total flux $3BA$ per particle, as needed – this is schematically illustrated in Figure 2 (right). The size of A in the right panel is equal to A in the left panel, which means that the three-loop orbits fit to the interparticle separation.

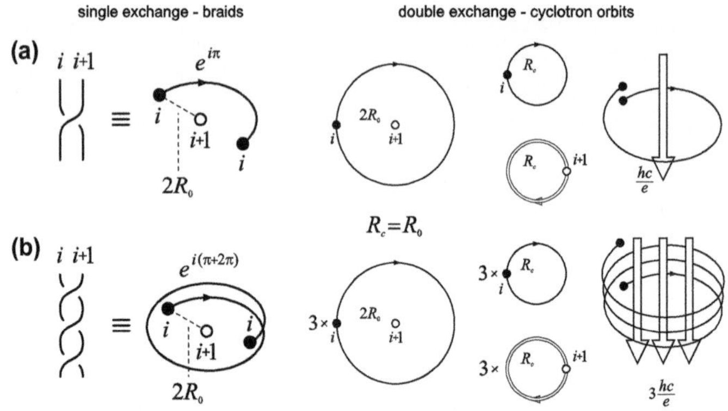

Figure 3. The braid generator σ_i corresponds to the single exchange of particles (left), the cyclotron orbit (relative) corresponds to the double exchange (right), for (a) $\nu = 1$ when single-looped cyclotron trajectory reaches neighboring particles, $R_c = R_o$; and (b) braid generator σ_i^3 for $\nu = \frac{1}{3}$ with additional loop needed for $R_c = R_o$ (R_c is cyclotron radius, $2R_o$ is particle separation, i.e. $\pi R_c^2 = \frac{hc}{eB}$, $\pi R_o^2 = \frac{S}{N}$).

As the braid group generator must be defined by the half of the cyclotron orbit (cf. Figure 3) thus the braid with one additional loop corresponds to the cyclotron orbits with three loops – such a generator has the form $b_i^{(3)} = (\sigma_i)^3$. The group generated by $b_i^{(3)}, i = 1, \ldots, N$ (new elementary braid exchanges) is the subgroup of the original full braid group. This subgroup is called the cyclotron braid subgroup and its 1DURs define statistics of 2D charged particles at strong magnetic field corresponding to fractional filling $\nu = \frac{1}{3}$ of the LLL, giving rise to the explanation of the Laughlin statistics for FQHE at this fractional filling (the exponent $p = 3$ in the related Jastrow polynomial in Laughlin function [17]). The generalization to more loops attached to the braid generator one by one, results in double increase of loop number in multilooped cyclotron orbits and thereby in fractions $\nu = \frac{1}{p}$, $p - odd\ integer$. This approach successfully reproduces the hierarchy of the experimentally observed filling fractions corresponding to FQHE in the LLL and in the higher LLs for 2DEG in conventional semiconductors [10].

If $\frac{S}{N} < A$, i.e. when cyclotron orbits are larger than interparticle separation (as in the right panel in Figure 1), some special commensurability opportunity important for braid group definition also occurs. For the filling fractions when $x\frac{S}{N} = A; x - integer$, the cyclotron orbits fit to every x-th particle separation ($x\frac{S}{N} = \frac{S}{N/x}$), which also allows for the definition of the generators σ_i in the form of ordinary single-looped braids, similarly as it happened for the completely filled higher LLs. The related statistics is the same as for IQHE though at some fractional fillings of higher LLs, which is demonstrated in [10] in good correspondence with the experimental observations available up to the third LL for 2DEG in conventional semiconductor Hall systems [20–22]. This situation of too large cyclotron orbits in comparison to particle separation may happen only in such LL subbands where the condition $x\frac{S}{N} = A\,(x - integer)$ can be fulfilled. It may happen only for $n \geq 1$ where n is the number enumerating LLs, because the cyclotron orbit size in n-th LL grows by factor $2n+1$, i.e., $A = \frac{(2n+1)hc}{eB}$ for particles with kinetic energy $(2n+1)\frac{eB}{2mc}$ in nth LL. Simultaneously, for $n \geq 1$ too short cyclotron orbits may be encountered only close to subbands edges, for sufficiently small planar density of particles in the subband, because in the higher LLs the cyclotron single-looped orbits are larger as accommodated to higher kinetic energy. This pushes FQHE(multiloop) features in higher LLs toward the edges of subbands in LLs with $n \geq 1$, whereas in the central regions of these subbands the new fractional features occur related with the single-looped cyclotron orbits and thus with IQHE-type of correlations but at fractional filling rates. These correlated states are referred to as FQHE(single-loop). The quantization of the transverse resistance R_{xy} related to these fractional filling rates ν of higher LLs is as for ordinary FQHE, $\frac{h}{e^2\nu}$, but the correlations of Laughlin type are with the exponent $p = 1$ in the Jastrow polynomial, displaying single-looped braid exchanges like in IQHE. The number of these new fractional filling features grows as $2n$ with the LL number.

3. Hierarchy for FQHE in graphene

In graphene one deals with the relativistic version of LLs [6,23]. This is due to the specific band structure in this material being a gapless semiconductor with points K and K' on the border of the hexagonal Brillouin zone where the valence and conduction bands meet [6]. Thus the low energy particle-hole excitations can be described by the effective Dirac Hamiltonian corresponding to cone-shape of both bands close to the meeting points. The quantization due to presence of the magnetic field has the form of degenerated LLs, though with spectrum not equidistant as for ordinary 2DEG but enumerated by $\sqrt{n}$, n is the number of the

LL [23]. This form of the self-energies results from the linear in momentum Hamiltonian close to K and K' points, whereas the degeneracy of each LL subband is the same as for conventional 2DEG and equals to $\frac{BS}{hc/e}$. The number of subbands per each LL (each n) is here 4. This corresponds to the ordinary Zeeman spin-splitting and to the so-called valley splitting (expressed often in terms of a pseudospin) due to doublet of inequivalent K points mixed with two sublattices for C atoms in crystal lattice of graphene [23]. Taking into account that the Zeeman splitting in graphene is small [6] and the valley splitting depending on structure imperfections and the external magnetic field is small as well, the fourfold approximate spin-valley additional degeneracy is assumed to determine the filling fractions for 'relativistic' IQHE in graphene, in the following form $\nu = 4(n+\frac{1}{2})$ in correspondence with the experimental observations [6]. The remarkable difference between this filling rate formula and that for the ordinary 2DEG (with factor 2 instead of 4, due to only approximate spin degeneracy) is the presence of the overall shift by the factor 2 ($\frac{1}{2}$ in the formula $\nu = 4(n+\frac{1}{2})$). This shift is due to the Berry phase manifestation in the LLL for graphene and resulting in the sharing of the LLL states between particles and holes from the conduction and valence band at the zero energy level. Due to this feature the bottom of the LLL is shifted by 2 (in terms of the filling factor) upward if one counts only negatively charged carriers and oppositely – downward for valence band positive holes. The corresponding filling rates for holes from the valence band are a negative mirror refection of those for electrons from the conduction band. Note that changing between particles and holes can be easily achieved in graphene by the shift of the Fermi level around the Dirac point by application of a lateral voltage.

The Berry phase contribution is linked to an additional π phase shift due to the chiral valley pseudospin when one adiabatically traverses with a selected particle a closed loop (e.g. along the semiclassical cyclotron loop) with momentum 2π convolution.

Similar analysis of the LLs can be done for the bilayer graphene [23,24]. Due to off-diagonal interlayer hopping the local Hamiltonian attains back the quadratic form with respect to the momentum. Thus the LL spectrum in bilayer graphene resembles that for the ordinary 2DEG with four subbands for each level except for the LLL which is eightfold degenerated. This extra degeneracy of the LLL arises from the action of the square of an annihilation operator on oscillator states with $n = 0$ and $n = 1$ [23,24]. Because the LLL subbands are distributed among particles and holes, the bottom for uniformly charged carriers is located in the center of the eightfold quasi-degenerated LLL. This property is associated again with the Berry phase shift for chiral particles, though in the case of bilayer graphene it gives additional 2π phase shift. Therefore, steps of the relativistic IQHE are located in bilayer graphene back at integer filling rates of the subsequent LLs whereas for monolayer graphene were located at half-fillings of the LLs [23,24].

3.1. FQHE in monolayer graphene

For the magnetic field strong enough that $\nu \in (0,1)$ and by the lateral voltage Fermi level shifted to the conduction band, one deals with fractionally filled first conduction subband $n = 0, 2 \uparrow$ (2 indicates the valley pseudospin down-orientation and $\uparrow$ indicates orientation of the ordinary spin). The degeneracy of each subband is $N_0 = \frac{BS}{hc/e}$ and for $N < N_0$ the filling rate, $\nu = N/N_0$, is fractional.

The cyclotron orbits in the LLL must be accommodated to the bare kinetic energy $T = \hbar\omega_c(n+\frac{1}{2})$ with $n = 0$, ($\omega_c = \frac{eB}{mc}$). The topology is not modified by the crystal field and to define the commensurability condition the braid cyclotron dimensions repeats those of the noninteracting gas. The cyclotron orbits have the same size for all particles due to the flat band quenching the kinetic energy competition. This results in the same averaged velocity and the same cyclotron orbit size for all particles in a perpendicular magnetic field (though in quantum mechanical treatment, velocity is not well defined since its coordinates do not commute). The cyclotron orbits restrict the braid topology of all trajectories uniformly in 2D, thus restrict the braid group structure despite particularities of the dynamics in the crystal field, which do not change trajectory topology. Therefore for graphene the cyclotron orbit structure is governed by ordinary (the same as for 2DEG) Landau levels restrictions despite the fact that a specific band structure with Dirac points, being the result of the crystal field in graphene, highly modifies LLs but not in terms of the bare kinetic energy. The specific for graphene quantum dynamics is included to the Feynman path integral, whereas the additional summation over topologically nonequivalent trajectory classes concerns the braid group structure the same as for 2DEG upon the magnetic field. The difference between the conventional 2DEG systems and graphene will be related in the regard to the statistics with distinct number of LL subbands in graphene compared to the conventional 2DEG and to the Barry phase shift of fillings in the LLL.

Thus, the cyclotron orbit size in the subband $n = 0, 2 \uparrow$ equals to $\frac{hc/e}{B} = \frac{S}{N_0}$ (S is the sample surface). As this orbit size is lower than the interparticle separation $\frac{S}{N}$ (as $N < N_0$), the multilooped braid structure is necessary. From the commensurability condition $q\frac{S}{N_0} = \frac{S}{N}$ one finds $\nu = \frac{N}{N_0} = \frac{1}{q}$, ($q$ – *odd integer* to protect the braid structure [9]). For holes in this subband one can expect the symmetric filling rates $\nu = 1 - \frac{1}{q}$.

Similarly as for the ordinary 2DEG one can generalize this simple series by assumption that the last loop of the multilooped cyclotron orbit can be commensurate with the interparticle separation as for some other filling ratio expressed by $l > 1$, whereas the former loops take away an integer number of flux quanta. For $l = 1, 2, 3, \ldots$ the last loop reaches every l-th particles. In this way one can obtain the hierarchy of fillings for FQHE in this subband of the LLL, $\nu = \frac{l}{l(q-1)\pm 1}$, $\nu = 1 - \frac{l}{l(q-1)\pm 1}$, where $l = 1, 2, \ldots$ and minus in the denominators indicates possibility of the eight-figure orientation of the last loop with respect to the antecedent one. The Hall metal states can be characterized by the limit $l \to \infty$ in the above formula (i.e., for zeroth flux taken away by the last loop, or equivalently, the last loops reaching infinitely distant particle, as for ordinary fermions without the magnetic field, which was the case in the Hall metal archetype for $\nu = \frac{1}{2}$ in the conventional 2DEG), which gives the hierarchy for the Hall metal states, $\nu = \frac{1}{q-1}$, $\nu = 1 - \frac{1}{q-1}$. Some other possibility may correspond with the case when in q-looped orbit $q - 1$ loops are accommodated to every x-th particle ($x = 1, 2, 3, \ldots$, whereas the last one fits to every l-th particle separation. Such an ordering is observed in ordinary 2DEG Hall systems in the LLL, for e.g. $\nu = \frac{4}{11}, \frac{5}{13}, \frac{3}{8}, \frac{3}{10}$ (out of the CF hierarchy), but is not observed in the LLL in monolayer graphene to date, though it is observed in the first LL in monolayer graphene (as will be analyzed below). To account for the Berry phase anomaly in graphene the overall shift of ν by -2 can be performed, but we use here the net filling fractions.

For the completely filled subband $n = 0, 2 \uparrow$, i.e., for $\nu = 1$, one arrives at IQHE. For a lower magnetic field (or a larger number of electrons), when the three first subbands (two of them belonging to valence band holes) of the LLL are filled and in the last subband of LLL $n = 0, 2 \downarrow$, the cyclotron orbit size $\frac{S}{N_0}$ is still lower than the interparticle separation $\frac{S}{N-N_0}$ (because $N - N_0 < N_0$). Thus multilooped structure is repeated from the previous subband. This results with the same FQHE hierarchy as for antecedent subband, only shifted ahead by 1. For the completely filled LLL (i.e., for all four of its subbands completely filled) one deals with IQHE according to its main-line $\nu = 4(n + \frac{1}{2})$.

Similarly one can consider fillings of the following LL with $n = 1$. This level also has four subbands, but in this level the bare kinetic energy is equal to $\frac{3\hbar\omega_c}{2}$ and the related cyclotron orbit size is $\frac{3hc}{eB} = \frac{3S}{N_0}$. For $N \in (2N_0, 3N_0]$ we deal with graduate filling of $n = 1, 1 \uparrow$ subband. Cyclotron orbits of size $\frac{3S}{N_0}$ must be compared here with interparticle separation scale $\frac{S}{N-2N_0}$. For a small number of electrons in this subband one deals with the multilooped structure (corresponding to the inequality $\frac{3S}{N_0} < \frac{S}{N-2N_0}$), when $q\frac{3S}{N_0} = \frac{S}{N-2N_0}$, q *– odd integer*, which gives the main series for FQHE(multiloop) in this subband, $\nu = 2 + \frac{1}{3q}$. Similarly as before the complete related hierarchy can be written as $\nu = 2 + \frac{l}{l3(q-1)\pm 1}$, $\nu = 3 - \frac{l}{l3(q-1)\pm 1}$ $l = i/3$, $i = 1, 2, \ldots$, with the Hall metal hierarchy in the limit $l \to \infty$. These series are located closer to the subband edges, whereas in the center of this subband the other commensurability conditions are possible. When $\frac{3S}{N_0} = \frac{xS}{N-2N_0}$ and $x = 1, 2, 3$ (fitting of the single-looped orbit with every x-th particle) we get $\nu = \frac{7}{3}, \frac{8}{3}, 3$, respectively, corresponding to single-looped cyclotron orbits similar as for IQHE. Thus, for $\nu = \frac{7}{3}, \frac{8}{3}$ one deals with FQHE(single-loop) – the new Hall feature manifesting itself only in higher LLs, where cyclotron orbits may be larger than the interparticle separation.

At the special case $\frac{3S}{N_0} = \frac{1.5S}{N-2N_0}$ one can arrive at $\nu = \frac{5}{2}$ with the paired particles (pairing does not change the cyclotron radius but twice diminishes the carrier number to $\frac{N-2N_0}{2}$, which gives the commensurability for pairs at $\nu = \frac{5}{2}$).

Moreover, for q-looped orbits, their size may be accommodated to every x-th particle in the subband, resulting in additional hierarchy in all subbands of the first LL, $\nu = 2(3, 4, 5) + \frac{xl}{l3(q-1)\pm 1}$, $\nu = 3(4, 5, 6) - \frac{xl}{l3(q-1)\pm 1}$ which for $q = 3$, $x = 2, 3$, $l = i/3$, $i = 1, 2, 3$ reproduces $\nu = \frac{7}{3}, \frac{8}{3}, \frac{12}{5}, \frac{13}{5}, \frac{17}{7}, \frac{18}{7}, \frac{22}{9}, \frac{23}{9}, \frac{10}{3}, \frac{11}{3}, \frac{17}{5}, \frac{18}{5}, \frac{24}{7}, \frac{25}{7}, \frac{13}{3}, \frac{14}{3}, \frac{22}{5}, \frac{23}{5}$ with FQHE features recently observed in the three first subbands of the $n = 1$ LL in monolayer graphene at ultra-low temperatures [5].

The following subbands are filled with electrons upon the similar scheme. For the subband $n = 1, 1 \downarrow$ the cyclotron orbit size is $\frac{3hc}{eB} = \frac{3S}{N_0}$ (as in all subbands with the same n), whereas the interparticle distances are measured with the plaque $\frac{S}{N-3N_0}$, where $N \in (3N_0, 4N_0)$. The commensurability condition $\frac{q3S}{N_0} = \frac{S}{N-3N_0}$ results in the main series for FQHE(multiloop), $\nu = 3 + \frac{1}{3q}$, which can be developed to the full hierarchy similarly as described above. The condition $\frac{3S}{N_0} = \frac{xS}{N-3N_0}$ with $x = 1, 2, 3$ results in fractions with single-looped correlations of FQHE(single-loop)-type for $\nu = \frac{10}{3}, \frac{11}{3}$ and IQHE for $\nu = 4$, correspondingly, whereas a paired state can be realized at $\nu = \frac{7}{2}$.

3.2. FQHE in bilayer graphene

The special topology of the bilayer graphene creates the opportunity to verify the braid group based concept of the commensurability of cyclotron orbit size with the interparticle spacing in the planar system of interacting particles. Let us emphasize once more that the interaction is essential for correlation manifestation – the interaction prevents particles from approaching one another closer than the particle separation resulting from the planar density. This separation is rigidly kept by the interaction allowing the commensurability notion to be useful. Exclusively when the cyclotron orbits fit accurately to the particle spacing the exchange of neighboring particles in the presence of the perpendicular magnetic field is possible along specific braids named as the cyclotron braids.

It must be emphasized that commensurability condition, as related to the braid group, concerns the classical archetype of the quantum system, thus is not addressed to real quantum particles which do not have any trajectories. This agrees with the spirit of the braid group approach in which classical trajectories in configuration space supply information on quantum statistics via 1DURs of the braid group. To organize any correlated collective state, the related statistics of particles must be determined first. In order to establish the statistics of quantum particles, the appropriate braid group must be defined whereas the statistics is governed by the selected 1DUR of this group [13].

The bilayer graphene is, however, not strictly two dimensional and for the bilayer graphene the topological situation changes considerably. Two sheets of the graphene plane lie close to each other and the hopping can change the electron positions among the planes. Thus, we deal here with double the number of electrons residing on a two-sheet structure simultaneously instead of the single sheet as was the case for the monolayer graphene.

The above requirements, to fulfill the commensurability condition in order to define the related braid group describing the correlated multiparticle state, also apply to the bilayer graphene, with a single distinction with respect to the monolayer case. The *doublelooped* cyclotron orbits may have in bilayer graphene the same size as the single-looped orbit in opposition to the monolayer case. This follows from the fact that the second loop can be located in the opposite sheet of graphene than the first one and the external field passing through such doublelooped orbit is twice larger than the flux passing through the single-looped orbit. Each loop has in this case its own separate surface in contrary to the multilooped cyclotron orbit located on the purely 2D plane. Taking into account that in the bilayer system loops of the multilooped orbit may be located partly in both 2D sheets, the contribution of the one loop must be avoided whereas the remaining loops must share the same flux as passing through a single-looped orbit, independently of how loops are distributed among two sheets. Thus, one can write out the commensurability condition in the bilayer graphene for the case of too short single-looped cyclotron orbits in the following form (for concreteness in the subband $n = 0, 2 \uparrow$ of the LLL – the first particle-type subband of the LLL):

$$\begin{aligned} &\frac{hc}{eB} = \frac{S}{N_0} < \frac{S}{N}, \\ &(p-1)\frac{hc}{eB} = (p-1)\frac{S}{N_0} = \frac{S}{N}, \\ &\nu = \frac{N}{N_0} = \frac{1}{p-1} = \frac{1}{2}, \frac{1}{4}, \frac{1}{6}, \ldots, \end{aligned} \tag{3}$$

where, N is the total number of particles in both graphene sheets, N_0 is the degeneracy counted for both sheets together, S is the surface of the sample (the surface of the single sheet), and p is an odd integer to assure that the half of the cyclotron orbit defines the braid.

The factor $p - 1$ in Equation (3) is caused by the fact that only orbits from the ideal 2D sheets of bilayer graphene contribute to the enlargement of the effective cyclotron orbits (no matter in which are located doubling loops) with exception of a single orbit which may be located in the opposite sheet to the first one. This sole loop contributes to the total flux with the additional flux quantum due to its own surface and this loop must be omitted. The next orbits must duplicate the former ones (in fact two) without rising to the surface and no matter in which sheet are they located, because in both they will duplicate loops already present there. Thus, in the enhancement of the effective p-looped cyclotron orbit take part only $p - 1$ loops.

Let us emphasize that for such multilooped orbits the total number of loops is still p – thus the generators of the corresponding cyclotron subgroup are of the form $b_i^{(p)} = \sigma_i^p$, resulting in the Laughlin correlations with the p exponent for the Jastrow polynomial. But due to the distinct commensurability of orbits with interparticle separation (Equation (3)) the related filling fractions are $\nu = \frac{1}{p-1}$ (p-odd) in the first particle-type subband of the LLL, i.e., in the subband $n = 0, 2 \uparrow$. This even denominator main series of the FQHE hierarchy for bilayer graphene pretty well coincides with the experimental observations [1].

For holes in this subband (let us emphasize that these holes are not holes from the valence band but correspond to unfilled states in the almost filled subband of particle type) one can write $\nu = 1 - \frac{1}{p-1}$, whereas the generalization to the full hierarchy of FQHE in this subband attains the form, $\nu = \frac{l}{l(p-2)\pm 1}$, $\nu = 1 - \frac{l}{l(p-2)\pm 1}$, where $l > 1$ corresponds to some filling factor for other correlated Hall state, including completely filled LLs with IQHE. In the next subbands of the LLL, $n = 0, 2 \downarrow$ (assuming that this subband succeeds the former one), the hierarchy is identical only shifted ahead by one, because commensurability condition has the same form for all subbands with the same n due to the same size of the cyclotron orbits.

Some novelty occurs, however, in the following two subbands of the LLL, $n = 1, 2 \uparrow$ and $n = 1, 2 \downarrow$. The FQHE main series in the first of these subbands of the LLL, $n = 1, 2 \uparrow$, has the form,

$$\begin{aligned} &\frac{3hc}{eB} = \frac{3S}{N_0} < \frac{S}{N-2N_0}, \\ &(p-1)3\frac{hc}{eB} = (p-1)\frac{3S}{N_0} = \frac{S}{N-2N_0}, \\ &\nu = \frac{N}{N_0} = 2 + \frac{1}{3(p-1)} = 2 + \frac{1}{6}, 2 + \frac{1}{12}, 2 + \frac{1}{18}, \ldots, \end{aligned} \tag{4}$$

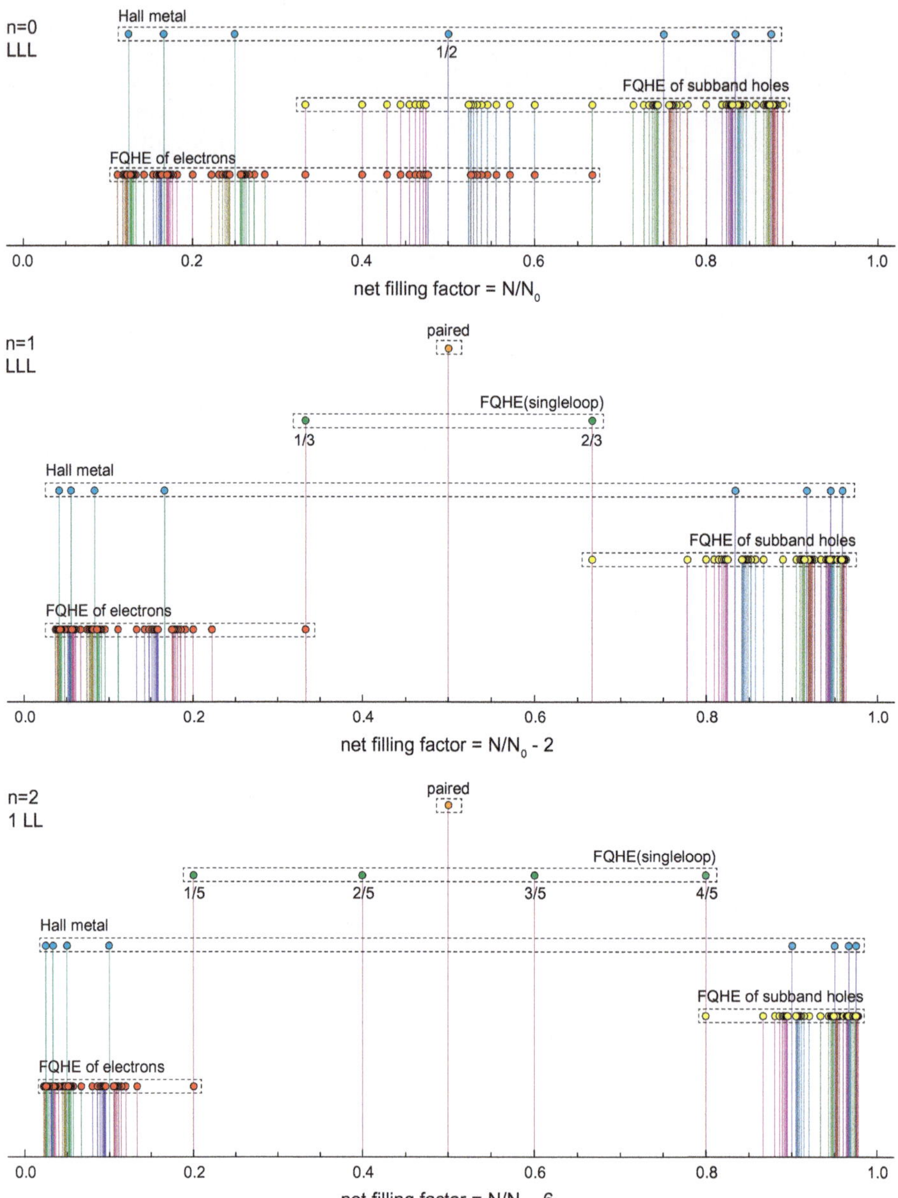

Figure 4. Evolution of fractional filling hierarchy in three first LLs of the monolayer graphene; for each LL the first particle subband is illustrated, the next subbands in each LL repeat the hierarchy from the first one. Different types of ordering are indicated with spikes of various height. Series for ordinary FQHE(multiloop), FQHE(single-loop), Hall metal and paired state are displayed according to the hierarchy described in Table 1 with $q = 3 - 9, 1 = 1 - 10$; only a few selected ratios from these series are explicitly written out.

The generalization of this main series for holes in the subband and to the full FQHE hierarchy in this subband is as follows: for subband holes, $\nu = 3 - \frac{1}{3(p-1)}$ and for the full FQHE hierarchy in this subband, $\nu = 2 + \frac{l}{l3(p-2)\pm 1}$, $\nu = 3 - \frac{l}{l3(p-2)\pm 1}$, $l = \frac{i}{3}$, $i = 1, 2, 3, \ldots$ (with Hall metal hierarchy in the limit $l \to \infty$).

In the subband $n = 1, 2 \uparrow$ of the LLL the new commensurability opportunity occurs (the one which appeared only in the first LL of the monolayer graphene): $\frac{3}{N_0} = \frac{x}{N-2N_0}$ for $x = 1, 2, 3$, which gives fillings ratios $\nu = \frac{7}{3}, \frac{8}{3}, 3$, correspondingly. All these rates are related with single-looped cyclotron trajectories, thus with single-loop correlations similar as for IQHE (though the first two for not integer filling rates). This new Hall feature, typical for LLs with $n \geq 1$, we called FQHE(single-loop). Moreover, for $x = 1.5$ one can consider twice diminishing of particle number $(N - 2N_0)/2$ due to the pairing, which gives perfect

Table 1. LL filling factors for FQHE determined by commensurability condition (*paired* indicates condensate of electron pairs), for the first particle subband in each of the three first LLs ($n = 0, 1, 2$) for the monolayer graphene.; subb. stands for subband.

LL subb.	FQHE(single-loop), paired, IQHE	FQHE(multiloop) ($q - odd, l = \frac{i}{2n+1}, i = 1,2,3,\ldots$)	Hall metal
$n = 0,2 \uparrow$	1	$\frac{1}{q}, 1 - \frac{1}{q}, \frac{l}{l(q-1)\pm 1}, 1 - \frac{l}{l(q-1)\pm 1}$	$\frac{1}{q-1}, 1 - \frac{1}{q-1}$
$n = 1,1 \uparrow$	$\frac{7}{3}, \frac{8}{3}$, ($\frac{5}{2}$ *paired*), 2, 3	$2 + \frac{1}{3q}, 3 - \frac{1}{3q}, 2 + \frac{l}{3l(q-1)\pm 1}, 3 - \frac{l}{3l(q-1)\pm 1}$	$2 + \frac{1}{3(q-1)}, 3 - \frac{1}{3(q-1)}$
$n = 2,1 \uparrow$	$\frac{31}{5}, \frac{32}{5}, \frac{33}{5}, \frac{34}{5}$, ($\frac{13}{2}$ *paired*), 6, 7	$6 + \frac{1}{5q}, 7 - \frac{1}{5q}, 6 + \frac{l}{5l(q-1)\pm 1}, 7 - \frac{l}{5l(q-1)\pm 1}$	$6 + \frac{1}{5(q-1)}, 7 - \frac{1}{5(q-1)}$

Table 2. LL filling factors for FQHE determined by commensurability condition (*paired* indicates condensate of electron pairs), for the first particle subband in each of the two first LLs ($n = 0, 1$ for the extra degenerated LLL and $n = 2$ for the first LL beyond the LLL) for the bilayer graphene.

LL subb.	FQHE(single-loop), paired, IQHE	FQHE(multiloop) ($q - odd, l = \frac{i}{2n+1}, i = 1,2,3,\ldots$)	Hall metal
$n = 0,2 \uparrow$ (LLL)	1	$\frac{1}{(q-1)}, 1 - \frac{1}{(q-1)}, \frac{l}{l(q-2)\pm 1}, 1 - \frac{l}{l(q-2)\pm 1}$	$\frac{1}{q-2}, 1 - \frac{1}{q-2}$
$n = 1,2 \uparrow$ (LLL)	$\frac{7}{3}, \frac{8}{3}$, ($\frac{5}{2}$ *paired*), 2, 3	$2 + \frac{1}{3(q-1)}, 3 - \frac{1}{3(q-1)}, 2 + \frac{l}{3l(q-2)\pm 1}, 3 - \frac{l}{3l(q-2)\pm 1}$	$2 + \frac{1}{3(q-2)}, 3 - \frac{1}{3(q-2)}$
$n = 2,1 \uparrow$ (first LL)	$\frac{21}{5}, \frac{22}{5}, \frac{23}{5}, \frac{24}{5}$, ($\frac{9}{2}$ *paired*), 4, 5	$4 + \frac{1}{5(q-1)}, 4 + \frac{l}{5l(q-2)\pm 1}, 5 - \frac{1}{5(q-1)}, 5 - \frac{l}{5l(q-2)\pm 1}$	$4 + \frac{1}{5(q-2)}, 5 - \frac{1}{5(q-2)}$

Table 3. Comparison of filling hierarchy in the LLL level in the bilayer graphene for two mutually inverted successions of two lowest subbands: $n = 0,2 \uparrow, n = 1,2,\uparrow$ (upper) and $n = 1,2 \uparrow, n = 0,2 \uparrow$ (lower).

LL subb.	FQHE(single-loop), paired, IQHE	FQHE(multiloop) ($q - odd, l = \frac{i}{2n+1}, i = 1,2,3,\ldots$)	Hall metal
$n = 0,2 \uparrow$	1	$\frac{1}{(q-1)}, 1 - \frac{1}{(q-1)}, \frac{l}{l(q-2)\pm 1}, 1 - \frac{l}{l(q-2)\pm 1}$	$\frac{1}{q-2}, 1 - \frac{1}{q-2}$
$n = 1,2 \uparrow$	$\frac{4}{3}, \frac{5}{3}$, ($\frac{3}{2}$ *paired*), 1, 2	$1 + \frac{1}{3(q-1)}, 2 - \frac{1}{3(q-1)}, 1 + \frac{l}{3l(q-2)\pm 1}, 2 - \frac{l}{3l(q-2)\pm 1}$	$1 + \frac{1}{3(q-2)}, 2 - \frac{1}{3(q-2)}$
$n = 1,2 \uparrow$	$\frac{1}{3}, \frac{2}{3}$, ($\frac{1}{2}$ *paired*), 1	$\frac{1}{3(q-1)}, 1 - \frac{1}{3(q-1)}, \frac{l}{3l(q-2)\pm 1}, 1 - \frac{l}{3l(q-2)\pm 1}$	$\frac{1}{3(q-2)}, 1 - \frac{1}{3(q-2)}$
$n = 0,2 \uparrow$	1,2	$1 + \frac{1}{q-1}, 1 + \frac{l}{l(q-1)\pm 1}, 2 - \frac{1}{q-1}, 2 - \frac{l}{l(q-2)\pm 1}$	$1 + \frac{1}{q-2}, 2 - \frac{1}{q-2}$

commensurability of cyclotron orbits of pairs with the separation of the particle pairs at $\nu = \frac{5}{2}$.

The last subband $n = 1, 2 \downarrow$ in the LLL in bilayer graphene is filled with electrons in the similar manner because for both subbands with $n = 1$ the cyclotron orbits have the same size. Thus the hierarchy of fractional filling for the last subband in the LLL is shifted by 1 from the antecedent subband without any modification. The situation changes, however, in the next LL (the first one beyond the LLL). In the first such LL (with $n = 2$) the cyclotron orbits suited to commensurability condition are determined by the bare kinetic energy for $n = 2$, and the corresponding cyclotron orbit size is equal to $\frac{5hc}{eB} = \frac{5S}{N_0}$. The similar analysis as in the previous LL gives here the main series and the full hierarchy for FQHE(multiloop) in the subband $n = 2, 1 \uparrow$, $\nu = 4 + \frac{1}{5(p-1)}$, $\nu = 4 + \frac{l}{l5(p-2)\pm 1}$, $l = \frac{i}{5}$, $i > 1$, respectively (inclusion of subband holes resolves itself to the substitution of 4+ by 5− in both above formulae). Similarly as previously, the limit $l \to \infty$ gives the Hall metal hierarchy. The difference in comparison to the previous LL consists here also in the presence of four (instead two) satellite FQHE(single-loop) states symmetrically located around the central paired state. In the subband $n = 2, 1, \uparrow$ the satellite states occur at $\nu = \frac{21}{5}, \frac{22}{5}, \frac{23}{5}, \frac{24}{5}$ and the central paired state at $\nu = \frac{9}{2}$. This hierarchy scheme is repeated in all four subbands of the first LL.

The evolution of the fractional filling hierarchy of subsequent LLs is illustrated in Figures 4 and 5, for the monolayer and bilayer graphene, correspondingly and is summarized in Tables 1 and 2.

For bilayer graphene the degeneration of $n = 0$ and $n = 1$ states results in eightfold degeneracy of the LLL, doubling fourfold spin-valley degeneracy. The degeneracy is not exact and with rising magnetic field amplitude both the Zeeman splitting and the valley splitting grows. Stress, deformation and structure imperfections also cause the increase of the valley splitting. Inclusion of the interaction plays a similar role. Coulomb interaction causes mixing of $n = 0, 1$ states lifting their degeneracy. Especially interesting is such a degeneracy lifting which admits inverted order of fillings of LLL subbands with distinct $n = 0, 1$. The inversion of orders $n = 0, 1$ to $n = 1, 0$ affects the filling rates hierarchy. Assuming that the LLL subband with $n = 1$ is filled earlier than the $n = 0$ subband, we get the following hierarchy for the first subband $n = 1, 2 \uparrow$: multilooped orbits for $\nu = \frac{l}{l3(p-2)\pm 1}$, $\nu = 1 - \frac{l}{l3(p-2)\pm 1}$, single-looped orbits for $\nu = \frac{1}{3}, \frac{2}{3}$ and a paired state for $\nu = \frac{1}{2}$. Assuming the next subband, $n = 0, 2 \uparrow$, we get the hierarchy of fillings for this subband in the form: multilooped orbits for $\nu = 1 + \frac{l}{l(p-2)\pm 1}$, $\nu = 2 - \frac{l}{l(p-2)\pm 1}$ and no single-looped orbits. The comparison of reverted orderings of two first LLL subbands is summarized in Table 3.

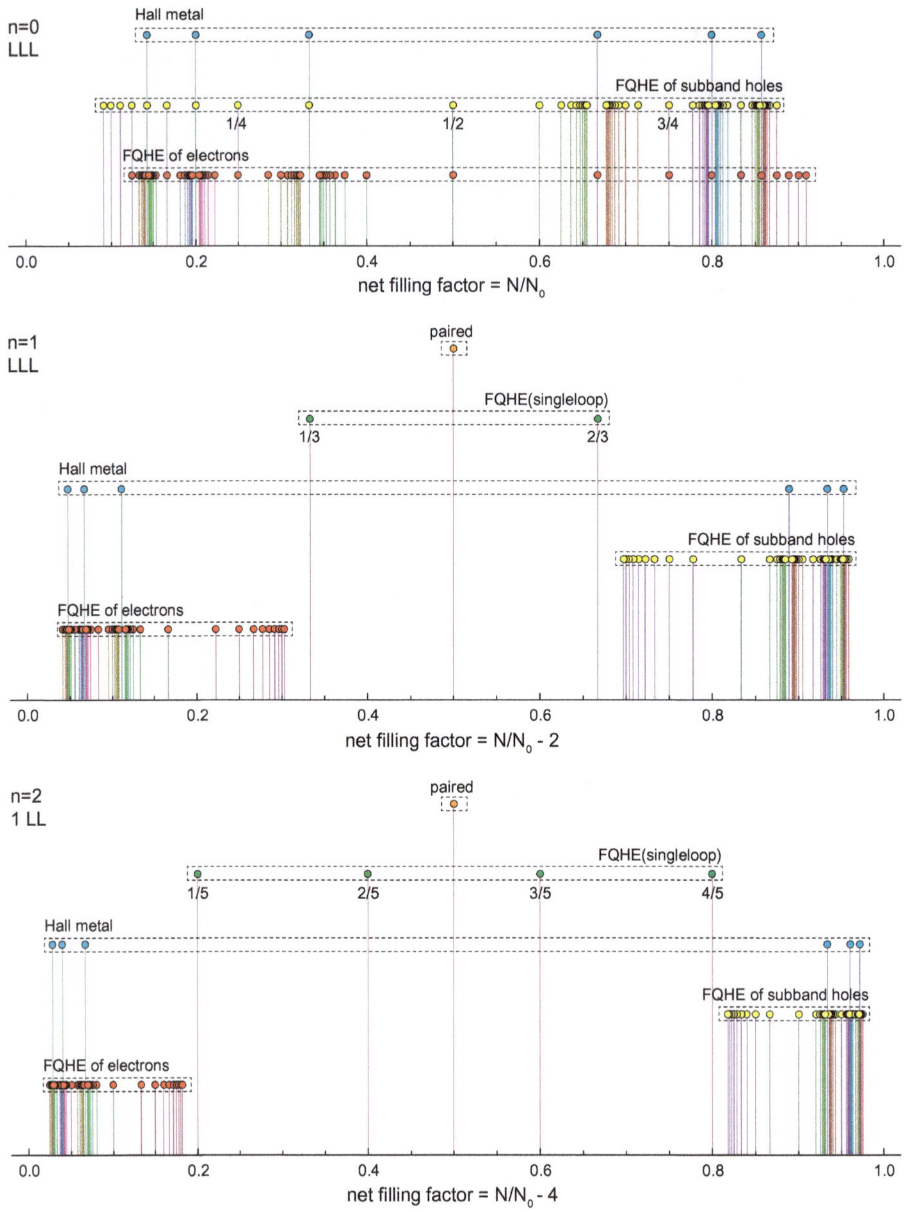

Figure 5. Evolution of fractional filling hierarchy in the two first LLs of the bilayer graphene; for the LLL two subbands with $n = 0$ and $n = 1$ are illustrated. Different types of ordering are indicated with spikes of various heights. Series for ordinary FQHE(multiloop), FQHE(single-loop), Hall metal and paired state are displayed according to the hierarchy described in Table 2 with $q = 3 - 9$, $i = 1 - 10$; only a few selected ratios from these series are explicitly written out.

One can consider also the situation in the LLL of bilayer graphene, when the degeneracy of $n = 0, 1$ states is lifted is such a way that both levels cross at certain filling factor $\nu^* < 1$ (cf. [25], where mixing between $n = 0, 1$ states is numerically analyzed for small models on torus or sphere). Let us assume for a model that first the $n = 1$ subband is filled ($n = 1, 2 \uparrow$) up to ν^*. At this filling the subband $n = 1, 2 \uparrow$ crosses with the subband $n = 0, 2 \uparrow$ and the latter is filled for $1 + \nu^* > \nu > \nu^*$. The related hierarchy of fractional fillings looks like an ordinary filling of the subband $n = 1, 2 \uparrow$, however, with an insertion of $n = 0, 2 \uparrow$ filling structure. Depending on the value of ν^* the various patterns are achievable by simple combination of hierarchy patterns listed in Table 3 (including also inverted ordering of $n = 1$ and $n = 0$ subbands).

4. Comparison with experiment

Searching for FQHE states in graphene is particularly challenging because of the different 'relativistic' structure of LLs, which is more complicated than the conventional semiconductor 2DEG. Moreover, the filling factor can be changed in graphene both by the external magnetic field and by the particle concentration via shifting of the Fermi level near the Dirac points by application of the lateral voltage. Due to spin-valley degeneracy and Berry phase contribution related to the chiral valley pseudospin, the IQHE is observed in graphene for fillings $\nu = 4(n + \frac{1}{2}) = 2, 6, 10, 14, \ldots$ for particles from the conduction band and for the mirror negative factors for holes from the valence band. Despite using very strong magnetic fields (up to 45 T), FQHE was, however, not detected in graphene samples deposited on a substrate of SiO_2 [7]. Instead, at these strong magnetic fields the emergence of additional plateaus of IQHE has been observed for the fillings $\nu = 0, \pm 1, \pm 4$, indicating spin-valley degeneracy lifting as a result of an increase of the mass of Dirac fermions [7]. Only after mastering the technology of the so-called suspended ultrasmall graphene scrapings with extreme purity and high mobility of carriers above 200,000 $cm^2V^{-1}s^{-1}$ (high mobility is necessary to observe FQHE also in the case of conventional semiconductor 2D heterostructures, which may be related to multi-looped quasi-classical cyclotron movement of wave packets in the case of multi-looped braids associated with FQHE [26]; note that in conventional semiconductor 2D heterostructures carrier mobility reaches even higher values of millions $cm^2V^{-1}s^{-1}$ [27]), it was possible to observe FQHE in graphene at net fillings $\nu = 1/3$ and $-1/3$ (the latter for holes, at the opposite polarization of the gate voltage, which determines position of the Fermi level, either in the conduction band, or in the valence band) [28,29]. Both these papers report the observation of FQHE in graphene for medium strong magnetic fields: in a field of 14 T, for electron concentration of 10^{11} cm^{-2} [28]; and in a field of 2 T, but for a concentration level smaller by one order of magnitude (10^{10} cm^{-2} and the mobility of 200,000 $cm^2V^{-1}s^{-1}$) [29], as shown in Figure 6.

FQHE in suspended graphene is observed at relatively high temperatures around 10 K [30], and even higher (up to 20 K) [31], which seems to be related with the stronger electric interaction in view of the lack, in the case of suspended samples, of a dielectric substrate (with the dielectric constant in case of SiO_2, ~ 3.9) and, on the other hand, with very high cyclotron energy in graphene (i.e., large energy gap between incompressible states).

The competition between the FQHE state with the insulator state near the Dirac point, corresponding to a rapidly decreasing concentration has also been demonstrated [32,33] (Figure 7).

From the perspective of cyclotron groups, experimental results on FQHE in graphene [28–31,34] seem to be compliant with the expectations of the braid description. In the case of graphene, the specific band structure with conical Dirac bands leads to simultaneous participation (in Dirac point) of both bands - of holes and of electrons, which combined with the massless character of Dirac fermions manifests itself through an anomalous 'relativistic' IQHE [7,32,35]. Controlling lateral gate voltage (within the range ca. 5 − 60 V [1,4,28]) allows regulation of the density of carriers at a constant magnetic field. One should therefore expect that at relatively small densities of carriers (electrons, or symmetrical holes at reverse voltage polarization), the cyclotron orbits will be too short to admit braid exchanges of particles at a sufficiently strong magnetic field – although weaker for smaller concentrations - and experimental observations exactly support this [28,29]. For low concentration, while closing on the Dirac point, one may expect that too strong fields would exceed the stability threshold of the FQHE state in competition with the Wigner crystal (assuming a similar character of this competition in the case of massless Dirac fermions in reference to conventional semiconductor 2D structures) and that it corresponds to the emergence of the insulating state near the Dirac point in a sufficiently large magnetic field as visible in experiment [36]. In the case of the hexagonal structure of graphene, electron (or hole) Wigner crystallization [37] may exhibit interference between the triangular crystal sublattices, and including of the resonance (hopping) between these two sublattices may cause blurring of the sharp transition to the insulator state, which seems compliant with observations (Figure 7).

The progress in the experiment allowed also for observation of FQHE in graphene on the crystal substrate of boron nitride (*BN*) in high magnetic fields of the order of 40 T (remarkably, FQHE features were noticed in this case up to $\nu = 4$) [4,5].

The mobility of carriers in graphene is lower than in traditional 2DEG, but taking into account that the carrier concentration in graphene can be lower in comparison to semiconductor heterostructure [27,38], the corresponding mean free path in both cases well exceeds the sample dimension (of μm order, as the mobility is proportional to the concentration and to the mean free path of carriers).

Energy gaps protecting incompressible FQHE states are larger in graphene than in traditional semiconductor materials, reaching a magnitude in order of 16 K (at $\nu = \frac{4}{3}$ and $B = 35$ T), which is referred to Dirac massless character of carriers. In conventional semiconductor heterostructure the corresponding gaps are much lower and the observed FQHE stability with temperature is much more fragile.

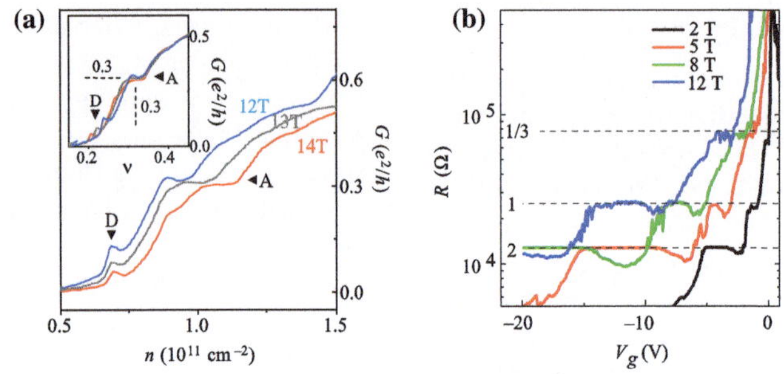

Figure 6. (a) FQHE observation in suspended graphene for the filling 0.3 (1/3) in a field of 12–14 T with the concentration of 10^{11} cm^{-2} and the mobility of 250,000 $cm^2V^{-1}s^{-1}$; (b) FQHE singularities in suspended graphene for the filling $\frac{1}{3}$ in a field of 2–12 T with the concentration of 10^{10} cm^{-2} and the mobility of 200,000 $cm^2V^{-1}s^{-1}$ (after [28,29]).

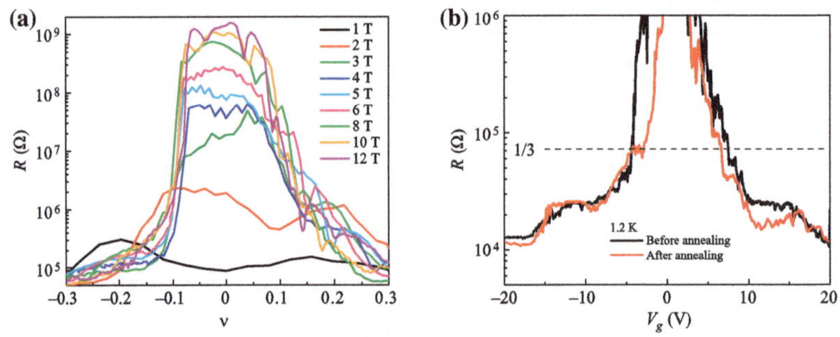

Figure 7. The emergence of an insulator state accompanying the increase in the strength of a magnetic field around the Dirac point; (b) competition between FQHE and the insulator state for the filling $-1/3$: annealing removes pollution – enhances mobility and provides conditions for the emergence of *plateau* for FQHE (after [29]).

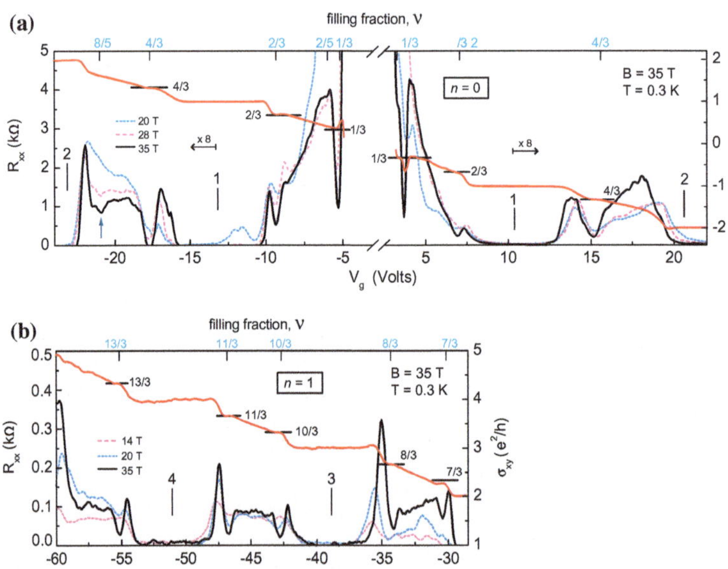

Figure 8. Fractional quantum Hall effect for graphene on BN. Magnetoresistance (left axis) and Hall conductivity (right axis) in the $n = 0$ and $n = 1$ Landau levels at $B = 35$ T and temperature ~ 0.3 K (after [4]). All filling ratios indicated in blue agree with the hierarchy given in Table 1.

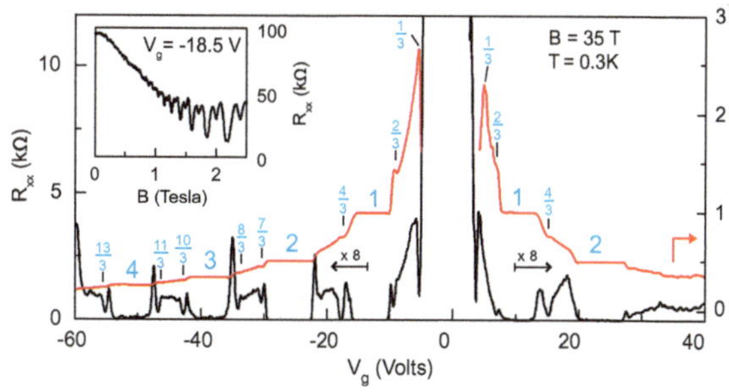

Figure 9. Magnetoresistance (left axis) and Hall resistance for graphene on BN (right axis) versus gate voltage acquired at B = 35 T. Inset shows Shubnikov-de Haas oscillations at Vg = −18.5 V (after [4]). All filling ratios indicated in the figure (in blue) agree with the hierarchy given in Table 1.

The recent development in experiments with monolayer graphene on BN substrate [4,5] and with suspended small sheets [2,3] allowed for observation of more and more Hall features at fractional fillings of subsequent subbands of two first LLs. While the sequence of fillings in the lowest subband of the LLL fits well to CF predictions (including CFs with two and four flux quanta attached), an explanation of the filling structure of next subbands strongly deviates from this simple picture. The pattern of filling rates repeated in the subbands of the first LL goes beyond the CF concept [2,3,5]. This phenomenon is referred in these papers to the various scenarios of breaking of the approximate SU(4) spin-valley symmetry in graphene. Because of the smallness of the Zeeman splitting E_Z in comparison to the Coulomb energy in graphene, $E_Z/E_C \sim 0.01\varepsilon$, similarly as of lattice scale in comparison to the magnetic length, $a/l_B \sim 0.06$ (where $l_B = \sqrt{\frac{\hbar}{eB}}$ and ε is the dielectric susceptibility ~ 3.2) [4], the subbands which differ with spin and valley-pseudospin orientation are closely located and can be regarded as approximately degenerated. This SU(4) spin-valley symmetry can be next broken by various factors and one can search arguments for unusual filling ratios hierarchy in related symmetry breaking and phase-like transitions. Despite there being many related ideas, no fully consistent picture has yet been attained in this way.

If one compares the experimentally observed fractions for characteristic FQHE features in longitudinal and Hall conductivities measured on a variety of samples of graphene with the pattern of fractional hierarchy for two lowest LLs, as illustrated in Figure 4, one notices the coincidence of this hierarchy with the measured data. All fractions found experimentally can be reproduced by this hierarchy (cf. Table 1). From this comparison it is visible why the CFs are efficient only in the LLL. This is linked with the fact that, exclusively in the LLL, cyclotron orbits are always shorter than the interparticle separation and additional loops are necessary. These loops can be modeled by fictitious field flux quanta attached to CFs. Though the analogy of CF flux quanta to additional loops is not exact, it allows one to get at least the similar main line of the filling hierarchy in the LLL as that given by the commensurability condition. The usefulness of the CF model is, however, limited in higher LLs because beginning from the first LL the multilooped commensurability is needed rather close to the subband edges, whereas the central regions of all subbands of the first LL are occupied by the following doublets of filling factors: $(\frac{7}{3}, \frac{8}{3})$, $(\frac{10}{3}, \frac{11}{3})$, $(\frac{13}{3}, \frac{14}{3})$, $(\frac{16}{3}, \frac{17}{3})$, corresponding to single-looped commensurability condition, not related to CF modeling, but visible in experiments as FQHE(single-loop) [2–5]. The number of centrally located filling rates for FQHE(single-loop) grows next with the LL number as $2n$. The repeating doublet of filling ratios for $n = 1$ is noticeable in Figure 9 and in more accurate measurements in suspended samples [2,3] besides of those on the BN substrate [4,5]. Worth noting is the observation [5] that stability of FQHE(single-loop) states is of similar order as of IQHE states and higher in comparison to FQHE(multiloop) state features as is visible in Figure 10. This might be associated with stronger correlations related to the single-looped braids as for IQHE states.

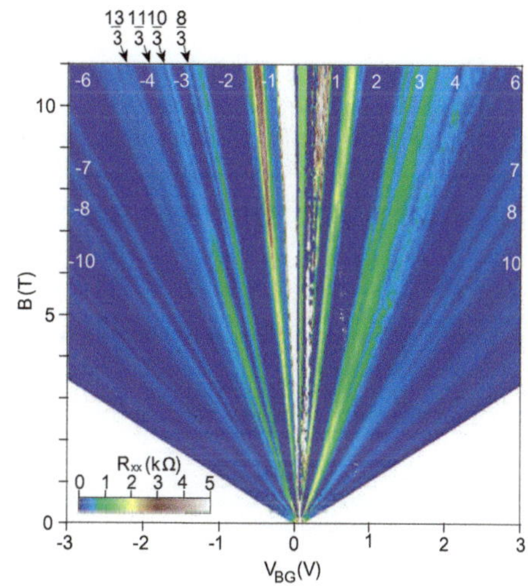

Figure 10. Fan diagram for $\rho_{xx}(\nu, B)$ for graphene up to 11 T (after [5]).

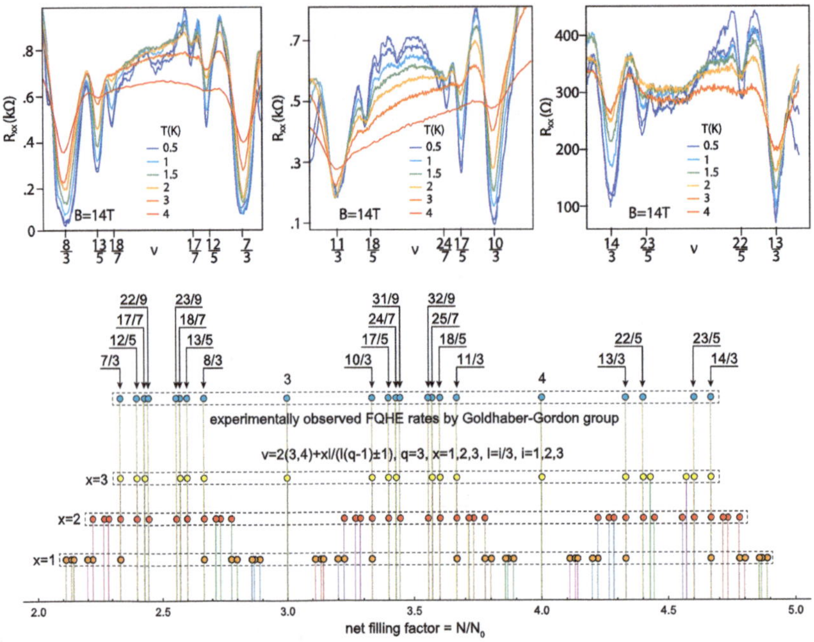

Figure 11. Not fully developed FQHE states with residual longitudinal resistance, corresponding to correlation of every second or every third particles at fractional rates reproduced by the commensurability series $\nu = 2(3,4) + \frac{xl}{l3(q-1)\pm1}$ with $q = 3$, $x = 2, 3$, $l = \frac{i}{3}$, $i = 1, 2, 3$ (upper panels visualize longitudinal resistivity measurements after [5]).

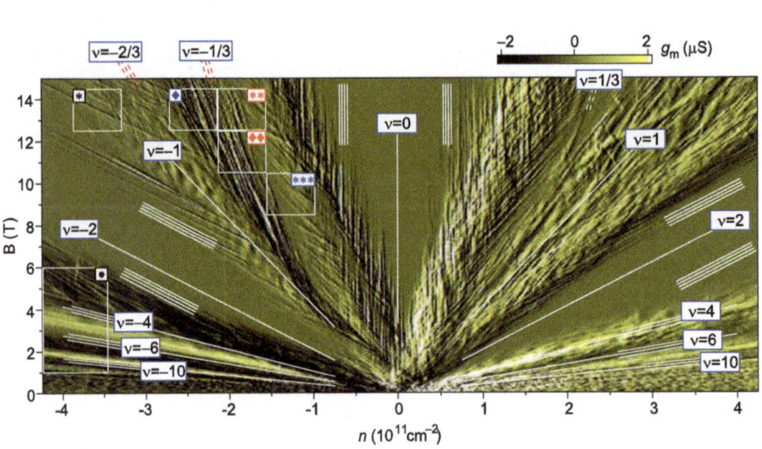

Figure 12. Color rendition of the transconductance in (N, B) plane – the tiny pattern agrees with the hierarchy for monolayer graphene given in Table 1 (after [39]).

The hierarchy induced by the commensurability condition reproduces the positions also of other observed features in two lowest LLs in the case of the monolayer graphene. The elongate plateaus at edges of subbands, with IQHE-rates in the centers, embrace minima related to closely located FQHE(multiloop)-rates indistinguishable at the observation resolution. In higher LL the new features, however, occur in between the above-mentioned doublets, but with nonzero longitudinal resistivity contrary to other FQHE states. This suggests that not all particles participate in these correlated states and they may correspond to the multilooped correlations of every second or every third particles according the appropriate cyclotron commensurability. And indeed these new features in the first LL of monolayer graphene, recently reported [5] at $\nu = \frac{7}{3}, \frac{8}{3}, \frac{12}{5}, \frac{13}{5}, \frac{17}{7}, \frac{18}{7}, \frac{22}{9}, \frac{23}{9}, \frac{10}{3}, \frac{11}{3}, \frac{17}{5}, \frac{18}{5}, \frac{24}{7}, \frac{25}{7}, \frac{13}{3}, \frac{14}{3}, \frac{22}{5}, \frac{23}{5}$, are reproduced one-to-one by the commensurability series $\nu = 2(3,4) + \frac{xl}{l3(q-1)\pm1}$ with $q = 3$, $x = 2, 3$, $l = \frac{i}{3}$, $j = 1, 2, 3$ as shown in Figure 11. One can notice that the rates $\frac{7}{3}, \frac{8}{3}, \frac{10}{3}, \frac{11}{3}, \frac{13}{3}, \frac{14}{3}$ are repeated in this hierarchy, though they correspond to single-looped

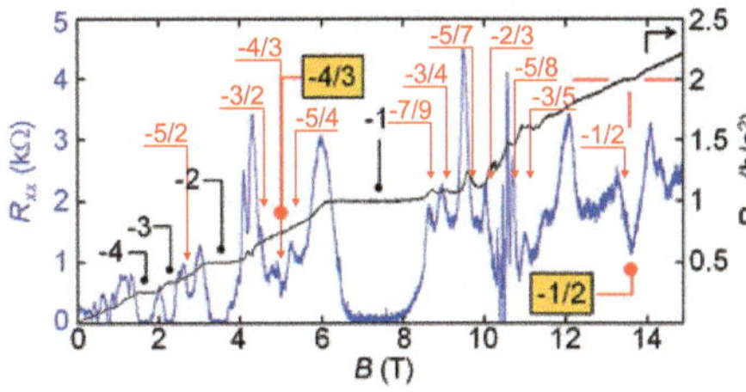

Figure 13. Observation of FQHE at T = 0.25 K in bilayer suspended graphene. Magneto-resistance Rxx (blue curve) and Rxy (black curve) at the lateral voltage −27 V (after [1]). Red symbols represent ratios from the hierarchy given in Table 2.

more stable correlation as is visible in Figure 10 and in the upper panel of Figure 11.

Another experimental evidence for the above hierarchy follows from the measurement [39] of the transconductance in the single electron field effect transistor configuration of small scrapings of graphene ($0.8 \times 3\ \mu$m) in a varying gate voltage $V_{bg} + \delta V_{bg}(t)$ ($f = 433$ Hz, $V_{bg} \sim 1$ V, $\delta V_{bg} \sim 10$ mV), which has revealed a fine structure of local correlated states. This structure is visible in typical transport experiments as noise-like oscillations but in the not-noisy regime. The detailed inspection of this fluctuations was done by visualization of the transconductance in the (N, B)-plane in Figure 12, which revealed a highly ordered pattern attributed to series of local correlated states closely accompanying fractions for the IQHE (and the FQHE as well) in monolayer graphene. The linear character of these new features was discovered, including several bunches collinear to directions of main IQHE/FQHE ratios in the (N, B)-plane. Straight lines, which lie in a (N, B)-plane and have a common point in a coordinate origin, are connected to a constant filling factor value; $B = \frac{hc}{ev}N$ (units are selected that $\frac{hc}{e} = 1$). An arbitrary line in a (N, B)-plane can be described as $B = \alpha N + \beta$. Its successive points are related to filling factors making up the hierarchy $\nu = \frac{N}{\alpha N + \beta}$. The latter matches quite well the FQHE hierarchy in consecutive subbands of LLs, according to the scheme arising from the commensurability conditions for monolayer graphene, $\nu = b \pm \frac{l}{l(2n+1)(p-1)\pm 1}$, with n enumerating LL, b depending on n and LL degeneracy, first $\pm$ arising from an electron-hole symmetry and second $\pm$ from the figure-of-eight shape of the trajectory. These fractions are gradually shifted toward subband edges with growing n and compressed to smaller periods for higher n, which also agrees with the details of the observation [39]. Similarly, vertical lines appearing in the picture ($\nu = \frac{const}{B}, N = const$) might be associated with a basic filling factor set $\nu = \frac{1}{p}$.

The most convincing evidence supporting the correctness of the commensurability condition is, however, the coincidence of the related predictions with the experimental observations in bilayer graphene. The pronounced feature of these observations is the occurrence of primary fractional features with even denominators also in the lowest subband of the LLL in bilayer graphene, oppositely to monolayer one [1]. The commensurability condition for bilayer graphene reproduces perfectly the observed experimentally hierarchy, including $\nu = -\frac{1}{2}$, cf. Figure 13; as is illustrated in Figure 5 and summarized in Table 2.

Note finally that the FQHE hierarchy in bilayer systems with the characteristic even denominators holds also for bilayer 2DEG conventional Hall setups and indeed the $\nu = \frac{1}{2}$ state has been discovered there [40,41], which also agrees with the commensurability predictions.

5. Conclusions

The condition for commensurability of cyclotron orbits building cyclotron braid subgroups with interparticle spacing in homogeneous 2D charged systems is formulated in order to verify possibility of arrangement of correlated multiparticle Hall states. Using this commensurability condition based on the braid group approach to statistics of interacting many particle systems, the hierarchy of fractional fillings for LLs in graphene has been determined. The FQHE evolution with growing numbers of LLs has been described. In higher LLs the new opportunities for commensurability occur leading to the different than the ordinary FQHE(multiloop) correlated states. The fractional fillings of LLs related with this new commensurability opportunity have been identified, starting from the first LL. They are referred to FQHE(single-loop) because the related correlations are described by single-looped braids. Both the monolayer and bilayer graphene have been considered and the essential difference of related hierarchy structures has been demonstrated and described. The even denominator main line of the fractional filling hierarchy in bilayer graphene is found in agreement with the experimental observations. The confirmation of the presented hierarchy for the monolayer and bilayer graphene can be found in experimental data for both, the graphene on BN substrate as well as the suspended samples including bilayer graphene available for fillings up to sixth spin-valley subband.

The success in description of hierarchy for FQHE using commensurability conditions is linked with the nonlocal and topological character of this approach. It is known that FQHE results from non-perturbative effects of the Coulomb interaction in 2D, which give rise to gapped many-body states at special filling factors. These states have long-range entanglement and cannot be explained in a simple single-particle picture. Similarly, the composite fermion theory, effective in the LLL of conventional 2DEG and in monolayer graphene, is in fact not strictly local, utilizing the auxiliary concept of flux tubes, or vortices, fixed to particles and allowing via this multiparticle effective trick to discriminate highly nonlocal effects. The commensurability approach presented in this paper bases on the topological braid group description of statistics in many body systems and falls to the same class of nonlocal multiparticle models. It elucidates artificial assumptions of composite fermion theory and allows for the generalization of the topological approach beyond the limit of the composite fermion model. In particular, the commensurability approach is effective also in higher LLs and allows the hierarchy of FQHE to be defined quite differently than that in the LLL - in agreement with experimental observations. Utilizing the energy minimization methods (exact diagonalization on finite models), the difference between FQHE states belonging to various Landau levels is referred to the different form of the effective Coulomb interaction (i.e. Coulomb interaction dressed by the form factor corresponding to a given Landau level). The energy minimization gives energy gaps for fractions related to special multiparticle correlations of FQHE. These fractions (and their hierarchy) can be, however, identified by a nonlocal effective-multiparticle topological method of commensurability braid approach, generalizing in this way the previous composite fermion approach from the LLL. The advantage of the commensurability condition is especially clearly visible in the case of bilayer graphene, where the states at fractions with even denominators cannot be explained by the simple composite fermion model, even in the LLL.

Acknowledgements

The support from the NCN Project UMO-2011/02/A/ST3/00116 is acknowledged.

References

[1] Ki DK, Falko VI, Abanin DA, Morpurgo A. Observation of even denominator fractional quantum Hall effect in suspended bilayer graphene. Nano Lett. 2014;14:2135.
[2] Feldman BE, Krauss B, Smet JH, Yacoby A. Unconventional sequence of fractional quantum Hall states in suspended graphene. Science. 2012;337:1196.
[3] Feldman BE, Levin AJ, Krauss B, Abanin DA, Halperin BI, Smet JH, Yacoby A. Fractional quantum Hall phase transitions and four-flux states in graphene. Phys. Rev. Lett. 2013;111:076802.
[4] Dean CR, Young AF, Cadden-Zimansky P, Wang L, Ren H, Watanabe K, Taniguchi T, Kim P, Hone J, Shepard KL. Multicomponent fractional quantum Hall effect in graphene. Nature Phys. 2011;7:693.
[5] F Amet, Bestwick AJ, Williams JR, et al. Composite fermions and broken symmetries in graphene. :arXiv:1410.3549v1
[6] Castro Neto AH. Guinea F, Peres NMR, Novoselov KS, Geim AK, The electronic properties of graphene. Rev. Mod. Phys. 2009;81:109.
[7] Zhang Y, Jiang Z, Small JP, Purewal MS, Tan Y-W, Fazlollahi M, Chudov JD, Jaszczak JA, Störmer HL, Kim P. Landau-level splitting in graphene in high magnetic fields. Phys. Rev. Lett. 2006;96:136806.
[8] Wu YS. General theory for quantum statistics in two dimensions. Phys. Rev. Lett. 1984;52:2103.
[9] Jacak J, Jacak L. Recovery of Laughlin correlations with cyclotron braids. Euorophys. Lett. 2010;92:60002.
[10] Jacak J, Jacak L. The commensurability condition and fractional quantum hall effect hierarchy in higher Landau levels. JETP Lett. 2015;102(1):19–25.
[11] Łydżba P, Jacak L, Jacak J. Hierarchy of fillings for the FQHE in monolayer graphene. Sci Rep (Nature). 2015;5:14287.
[12] Jacak J, Jacak L. Difference in hierarchy of FQHE between monolayer and bilayer graphene. Phys. Lett. A. 2015;379:2130.
[13] Laidlaw MG, DeWitt CM. Feynman functional integrals for systems of indistinguishable particles. Phys. Rev. D. 1971;3:1375.
[14] Birman JS. Braids, Links and Mapping Class Groups. Princeton UP: Princeton; 1974.
[15] Imbo TD, Imbo CS, Sudarshan CS. Identical particles, exotic statistics and braid groups. Phys. Lett. B. 1990;234:103.
[16] Tsui DC, Störmer HL, Gossard AC. Two-dimensional magnetotransport in the extreme quantum limit. Phys. Rev. Lett. 1982;48:1559.
[17] Laughlin RB. Anomalous quantum Hall effect: an incompressible quantum fluid with fractionally charged excitations. Phys. Rev. Lett. 1983;50:1395.
[18] Prange RE, Girvin SM. The Quantum Hall Effect. New York: Springer Verlag; 1990.
[19] Jain JK. Composite-fermion approach for the fractional quantum Hall effect. Phys. Rev. Lett. 1989;63:199.
[20] Eisenstein JP, Lilly MP, Cooper KB, Pfeiffer LN, West KW. New physics in high Landau levels. Physica E. 2000;6:29.
[21] Dolev M, Gross Y, Sabo R, Gurman I, Heiblum M, Umansky V, Mahalu D. Characterizing neutral modes of fractional states in the second Landau level. Phys. Rev. Lett. 2011;107:036805.
[22] Willett RL. The quantum Hall effect at 5/2 filling factor. Rep. Prog. Phys. 2013;76:076501.
[23] Goerbig MO. Electronic properties of graphene in a strong magnetic field. Rev. Mod. Phys. 2011;83:1193.
[24] McCann E, Falko VI. Landau-level degeneracy and quantum Hall effect in a graphite bilayer. Phys. Rev. Lett. 2006;96:086805.
[25] Papić Z, Abanin DA. Topological phases in the zeroth Landau level of bilayer graphene. Phys. Rev. Lett. 2014;112:046602.
[26] Jacak J, Jacak L. On triggering role of carrier mobility for Laughlin state organization. JETP Lett. 2014;98(11):684–688.
[27] Pfeiffer L, West KW. The role of MBE in recent quantum Hall effect physics discoveries. Physica E. 2003;20:57.
[28] Du X, Skachko I, Duerr F, Luican A, Andrei EY. Fractional quantum Hall effect and insulating phase of Dirac electrons in graphene. Nature. 2009;462:192.
[29] Bolotin KI, Ghahari F, Shulman MD, Störmer HL, Kim P. Observation of the fractional quantum Hall effect in graphene. Nature. 2009;462:196.
[30] Abanin DA, Skachko I, Du X, Andrei EY, Levitov LS. Fractional quantum Hall effect in suspended graphene: transport coefficients and electron interaction strength. Phys. Rev. B. 2010;81:115410.
[31] Skachko I, Du X, Duerr F, Luican A, Abanin DA, Levitov LS, Andrei EY. Fractional quantum Hall effect

in suspended graphene probed with two-terminal measurements. Phil. Trans. R. Soc. A. 2010;368:5403.

[32] Zhang Y, Tan Y-W, Störmer HL, Kim F. Experimental observation of the quantum Hall effect and Berry's phase in graphene. Nature. 2005;438:201.

[33] MacClure JW. Diamagnetism of graphite. Phys. Rev. 1956;104:666.

[34] Papić Z, Goerbig MO, Regnault N. Atypical fractional quantum Hall effect in graphene at fillig factor 1/3. Phys. Rev. Lett. 2010;105:176802.

[35] Novoselov KS, Geim AK, Morozov SV, Jiang D, Katsnelson MI, Grigorieva IV, Dubonos SV, Firsov AA. Two-dimensional gas of massless Dirac fermions in graphene. Nature. 2005;438:197.

[36] Yannouleas C, Romanovsky I, Landman U. Edge and bulk components of lowest-Landau-level orbitals, correlated fractional quantum Hall effect incompressible states, and insulating behavior of finite graphene samples. Phys. Rev. B. 2010;82:125419.

[37] Dahal HP, Joglekar YN, Bedell KS, Balatsky AV. Absence of Wigner crystallization in graphene. Phys. Rev. B. 2006;74:233405.

[38] Bolotin KI, Sikes KJ, Jiang Z, Klima M, Fudenberg G, Hone J, Kim P, Störmer HL. Ultrahigh electron mobility in suspended graphene. Solid State Comm. 2008;146:351.

[39] Lee DS, Skákalová V, Weitz RT, von Klitzing K, Smet JH. Transconductance fluctuations as a probe for interaction-induced quantum Hall states in graphene. Phys. Rev. Lett. 2012;109:056602.

[40] Suen YW, Engel LW, Santos MB, Shayegan M, Tsui DC. Observation of a $\nu=1/2$ fractional quantum Hall state in a double-layer electron system. Phys. Rev. Lett. 1992;68:1379.

[41] Eisenstein JP, Boebinger GS, Pfeiffer LN, West KW, He S. New fractional quantum Hall state in double-layer two-dimensional electron systems. Phys. Rev. Lett. 1992;68:1383.

17

Modeling of metastable phase formation diagrams for sputtered thin films

Keke Chang, Denis Music, Moritz to Baben, Dennis Lange, Hamid Bolvardi and Jochen M. Schneider

Materials Chemistry, RWTH Aachen University, D-52074 Aachen, Germany

ABSTRACT

A method to model the metastable phase formation in the Cu–W system based on the critical surface diffusion distance has been developed. The driver for the formation of a second phase is the critical diffusion distance which is dependent on the solubility of W in Cu and on the solubility of Cu in W. Based on comparative theoretical and experimental data, we can describe the relationship between the solubilities and the critical diffusion distances in order to model the metastable phase formation. Metastable phase formation diagrams for Cu–W and Cu–V thin films are predicted and validated by combinatorial magnetron sputtering experiments. The correlative experimental and theoretical research strategy adopted here enables us to efficiently describe the relationship between the solubilities and the critical diffusion distances in order to model the metastable phase formation during magnetron sputtering.

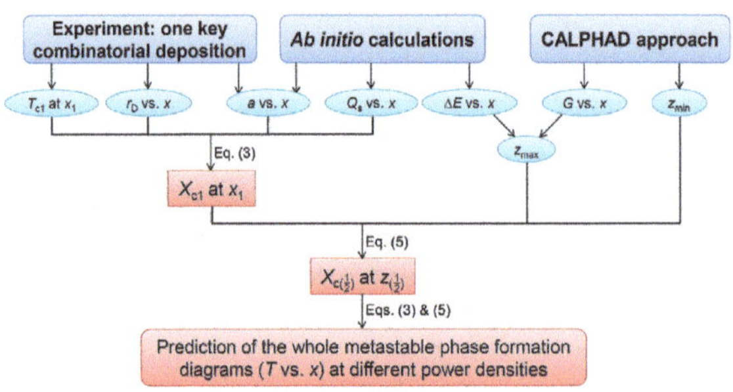

KEYWORDS

Combinatorial magnetron sputtering; metastable phase formation diagram; thin film growth; surface diffusion distance; Cu–W and Cu–V

CLASSIFICATION

40 Optical, magnetic and electronic device materials; 306 Thin film/Coatings; 300 Processing/Synthesis and Recycling; 401 1st principle calculations; 400 Modeling/Simulations; 403 CALPHAD/Phase field methods; 400 Modeling/Simulations; 307 Kinetics and energy/mass transport; 300 Processing/Synthesis and Recycling

1. Introduction

During energetic vapor phase condensation, for example by magnetron sputtering and cathodic arc deposition, the formation of metastable phases is often observed. Generally, this is attributed to kinetically limited growth processes that govern thin film structure evolution.[1–3] For thin films deposited at low temperatures, surface diffusion is the underlying physical mechanism that controls metastable phase formation.[1–3] The atomic mobility can be studied using the temporal dependence of the surface diffusion distance of an atom as given by Einstein [4]:

$$X = \sqrt{2D_s t} \tag{1}$$

where X is the diffusion distance, D_s is surface diffusivity and t is the time. Based on Equation (1), Cantor and Cahn [5] proposed the following equation to describe the deposition rate and temperature dependence of the diffusion distance during thin film deposition:

$$X = \sqrt{2\nu \frac{a}{r_D} \cdot a \cdot \exp\left(-\frac{Q_s}{2kT}\right)} \tag{2}$$

where ν is the vibrational frequency of surface atoms ($\sim 10^{13}$ s^{-1} [6]), a is the individual jump distance, r_D is the deposition rate, Q_s is the activation energy for surface diffusion, k is the Boltzmann constant and T is the substrate temperature during deposition. Saunders and Miodownik [7] also considered the effect of bulk diffusion and modified Equation (2) accordingly in order to investigate the metastable phase formation of $Cu_{0.57}Ag_{0.43}$, $Pd_{0.58}Rh_{0.42}$, $Cu_{0.885}Sn_{0.115}$ and $Cu_{0.895}Sn_{0.105}$ thin films. The surface diffusion distances were calculated as a function of temperature.[7] Guided by the experimental phase formation data, the critical surface diffusion distances

CONTACT Keke Chang chang@mch.rwth-aachen.de

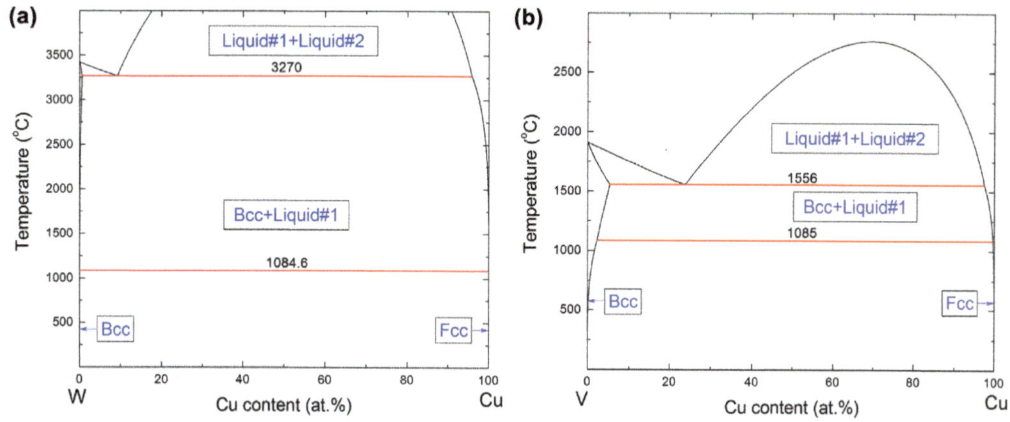

Figure 1. Stable Cu–W (a) and Cu–V [36] (b) phase diagrams calculated using the CALPHAD approach. The Gibbs energy expressions of all the phases are provided in Appendix A.

were estimated around an experimentally determined phase boundary.[7] Therefore, Saunders and Miodownik [7] have described the metastable phase formation for thin films at experimentally determined phase boundaries. However, no modeling attempt has been undertaken to predict metastable phase formation diagrams covering the whole composition range.

Metastable phase formation diagrams can be compiled based on experimental phase formation data of individual deposition experiments [8] or from combinatorial thin film synthesis experiments.[9–11] The CALPHAD (Computer Coupling of Phase Diagrams and Thermochemistry) approach provides thermodynamic input to study metastable phase formation [7] and has been applied to describe the metastable solubility limit in a system during thin film deposition, based on Gibbs energy versus composition (G vs. x) diagrams.[7,13–15] Using *ab initio* calculations, the metastable solubility limit and formation energy versus composition (E_f vs. x) can be estimated.[16–18] Together with experimental phase formation data, *ab initio* data has recently been employed to estimate the activation energy for surface diffusion during metastable phase formation.[19] It was demonstrated for sputtered Cu–W thin films that the activation energy data obtained [19] can be utilized to calculate the surface diffusion distance using Equation (2). Recently, attention has been given to immiscible Cu–W and Cu–V systems (for which phase diagrams are shown in Figure 1). The Cu–W thin films investigated by Vüllers and Spolenak [20] show a desirable property combination: W's high hardness and thermo-mechanical stability and Cu's excellent thermal conductivity and electrical performance, thereby allowing for possible applications including thermal management, high power and high voltage appliances. The Cu–V thin films feature excellent thermal stability, and have been applied as seed layers for Cu interconnects.[21–23]

The aims of the present work are (1) to develop a model with which to describe the relationship between solubilities and critical diffusion distances based on experimental phase formation data,[19] (2) to predict metastable Cu–W and Cu–V phase formation diagrams and (3) to validate these predictions using additional thin film growth and characterization experiments.

2. Experimental and theoretical methods

Cu–V thin films were synthesized by DC magnetron sputtering in an industrial CemeCon 800/9 system (CemeCon AG, Würselen, Germany), where the combinatorial approach [9–12] with split Cu and V targets was applied. The base pressure was less than 10^{-4} Pa and the argon partial pressure was 0.35 Pa during deposition. The distance between target and substrate was 9.5 cm. Si (100) and Al_2O_3 (0001) wafers were used as substrates. One set of thin films was deposited at a power density of 0.91 W·cm^{-2} and temperature of 240 °C, respectively. The power density is equal to the power applied to the target, divided by the target size (50 × 8.8 cm^2). The substrate temperature was measured using three thermocouples clamped to the substrate surface. More information about deposition setup can be found elsewhere. [19] Composition of the samples was analyzed with a scanning electron microscope (SEM; JEOL JSM 6480, JEOL Ltd., Tokyo, Japan) equipped with an EDAX2000 energy dispersive X-ray spectrometer (EDX, EDAX Inc., Mahwah, NJ, USA). The thin film thickness was measured by cross-sectional SEM and the deposition rate was calculated by dividing film thickness by deposition time. Structural analysis was carried out by means of X-ray diffraction (XRD) with a Bruker AXS D8 Discover General Area Diffraction Detector System (GADDS, Bruker AXS GmbH, Karlsruhe, Germany). The strain-free lattice parameters were determined using the $\sin^2\psi$ method. [24,25] These experimental data were then applied as the input in predicting metastable phase formation diagrams for the Cu–V system. Additional combinatorial sputtering processes were carried out at different power densities (0.91, 3.64 and 7.28 W·cm^{-2}) and different substrate temperatures (80–340 °C). This phase formation data was used to validate the predictions.

Lattice parameters and formation energetics of body centered cubic (bcc) and face centered cubic (fcc) Cu–V solid solutions were investigated using *ab initio* calculations utilizing the coherent potential approximation (CPA) and special quasi random structure (SQS) approaches. For CPA, the exact muffin tin orbitals (EMTO) formalism [26,27] based on Green's function [28] and full charge density [29] techniques was used. The generalized gradient approximation (GGA) [30] was applied. The core states were fixed and the total energy was converged within 10^{-7} Ry. For SQS, 16 atoms per unit cell of Cu_xW_{1-x} (x = 0.0625, 0.125, 0.1875, 0.25, 0.5, 0.75, 0.8125, 0.875 and 0.9375) for both bcc and fcc phases were used for calculations in the Vienna *ab initio* simulation package (VASP).[31] The valence electrons were explicitly treated by projector augmented plane-wave (PAW) potentials.[32] The GGA method was performed using Blöchl corrections for the total energy [33] with a plane-wave cutoff energy of 500 eV and a convergence criterion for the total energy of 0.01 meV. Integration in the Brillouin zone was done on appropriate k-points, which was determined after Monkhorst-Pack.[34] Migrations of Cu and V atoms in bcc (110) and fcc (111) surfaces were calculated using *ab initio* calculations in order to obtain surface diffusion activation barriers. A detailed description of the method can be found elsewhere.[19]

The CALPHAD approach was utilized to calculate stable phase diagrams of the Cu–W and Cu–V systems considering the phase equilibria and thermodynamic data from [35] and [36], respectively. FactSage software [37] was utilized for the calculations. Gibbs energy expressions for all of the phases are provided in Appendix A. While differing from [35], generally accepted Gibbs energy functions for pure Cu and W [38] were used in the present calculation. Therefore, the temperature of the invariant reaction (Liquid#2 = bcc + Liquid#1) in the Cu–W phase diagram is 3270 °C, compared to 3240 °C in [35]. As no experimental information is available for this reaction, the calculations in this work, as well as in [35], should therefore be regarded as hypothetical. Moreover, the Gibbs energies for the Cu–V system are identical to those employed in [36].

3. Results and discussion

3.1. A model to describe the relationship between solubilities and critical diffusion distance

Experimental phase formation data in the Cu–W thin films were obtained,[19] as shown in Figure 2(a). The maximum solid solubility limit of Cu in bcc-W is found to be approx. 78 at.%. It was suggested that as X reaches a critical diffusion distance (X_c), the formation of the second phase and hence decomposition is observed experimentally.[19] Equation (2) can thus be rewritten including X_c:

$$X_c = \sqrt{2\nu \frac{a}{r_{Dn}} \cdot a \cdot \exp\left(-\frac{Q_s}{2kT_c}\right)} \tag{3}$$

where T_c denotes the critical temperature for each Cu_xW_{1-x} composition at a certain deposition rate (r_{Dn}). The phase formation data (Figure 2(a)) are subsequently summarized in Figure 2(b). If the substrate temperature is less than T_c, the diffusion distance of atoms on bcc or fcc surfaces is smaller than X_c, meaning that the formation of thermodynamically stable phases is prevented by kinetic limitations in that the atomic mobility is insufficient. If the temperature increases to a value equal to or larger than T_c, atomic mobility is enhanced, causing decomposition of the metastable solid solutions into bcc + fcc phases. Thus, being able to quantify the critical diffusion distance is essential for modeling metastable phase formation diagrams. Using Equation (3), the composition dependence of X_c was obtained, as is shown in Figure 2(c). Activation energies (Q_s) derived from experimental data [19] were used in the calculations. It is interesting to note that all experiments result in a unified X_c vs composition behavior, suggesting that X_c is in fact a temperature-independent and deposition rate-independent property.

In order to model metastable phase formation (Figure 2(a)), it is essential to determine the composition dependence of X_c. In other words, if X_c for each x in Cu_xW_{1-x} is known, one can obtain the corresponding T_c using Equation (3) and the whole metastable phase formation diagram can be calculated. The mutual stable solid solubility at the growth temperature between Cu and W is negligible [35], as shown in the calculated Cu-W phase diagram (Figure 1(a)). However, metastable solid solubility (z) of Cu in bcc-W and W in fcc-Cu is obtained in magnetron sputtering experiments [19]. z is the composition of the boundary between single-phase and two-phase regions. The maximum metastable solubilities (z_{max}) for Cu in bcc-W and W in fcc-Cu at 110 °C and the power density of 0.92 W·cm^{-2} are about 78.0 at.% Cu and 22.0 at.% W [19], respectively. To compile a metastable phase formation model, the relationship of X_c and z needs to be determined, based on experimental data. The X_c vs. x diagram (Figure 2(c)) has been transformed into two diagrams (Figure 3(a) and (b)) describing X_c vs. z for the bcc and fcc phases, respectively. Each dataset shows a similar trend. Independent of crystal structure, the relationship of X_c and z can be fitted well using the following sigmoid function:

$$z = z_{min} + \frac{z_{max} - z_{min}}{1 + \left(\frac{X_c}{A}\right)^B}, \tag{4a}$$

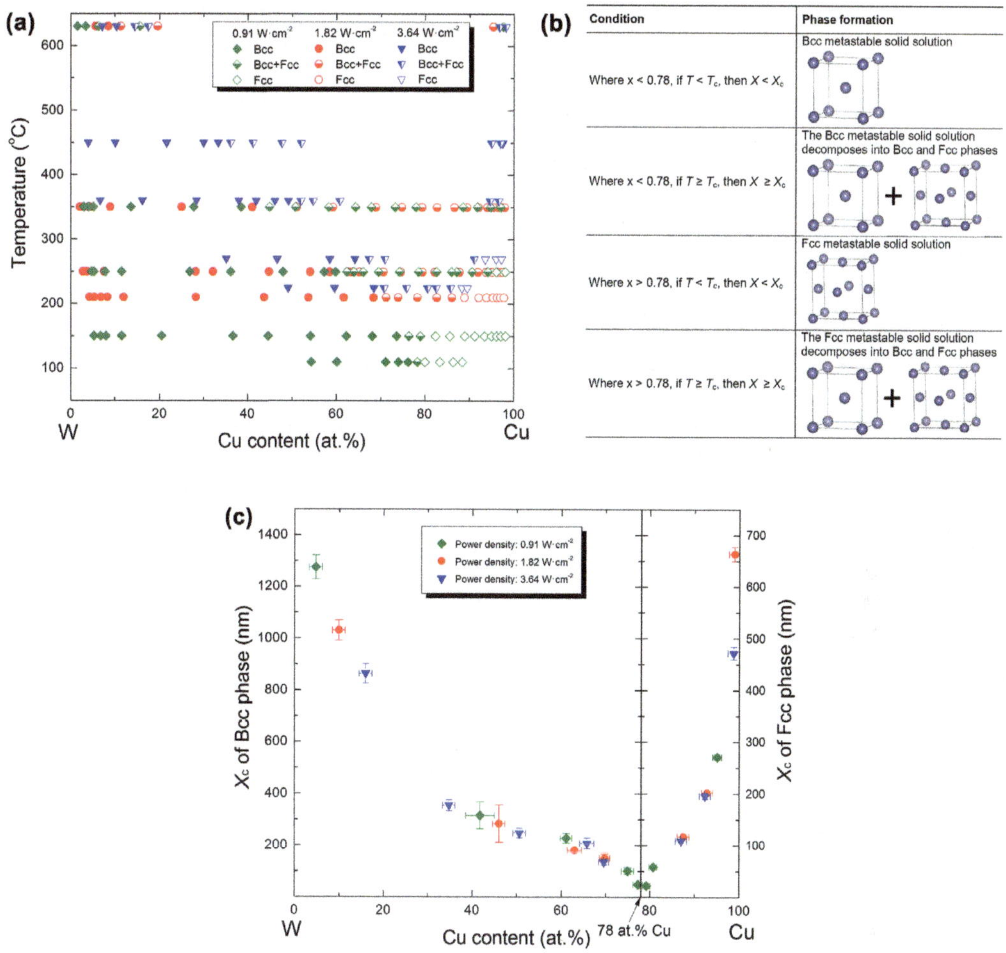

Figure 2. Metastable phase formation in Cu–W thin films: (a) experimental data of thin films grown at different power densities [19]; (b) structure evolution for Cu_xW_{1-x} thin films at a certain deposition rate (T_c: critical temperature; X_c: critical surface diffusion distance); and (c) composition dependence of X_c at the bcc and fcc surfaces.

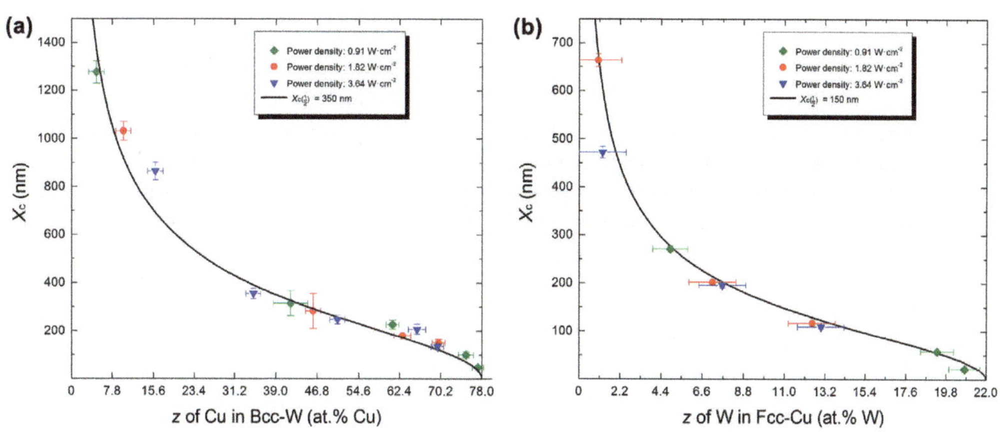

Figure 3. X_c vs. z plot: experimental data and fitted curves using Equation (5) for both bcc (a) and fcc (b) phases in the Cu–W system.

$$i.e.\, X_c = A \cdot \sqrt[B]{\frac{z_{\max} - z}{z - z_{\min}}}, \tag{4b}$$

where $z_{\max}$ denotes maximum metastable solid solubility and $z_{\min}$ denotes the stable solid solubility from the equilibrium phase diagrams (Figure 1(a) and (b)). A is

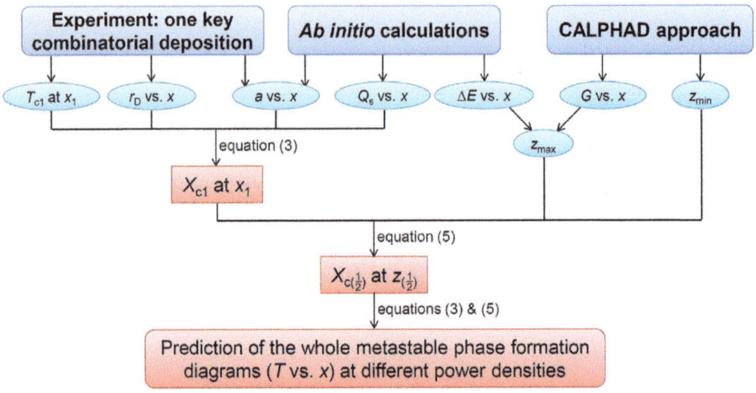

Figure 4. Flowchart of the present research strategy to predict metastable phase formation diagrams for sputtered thin films.

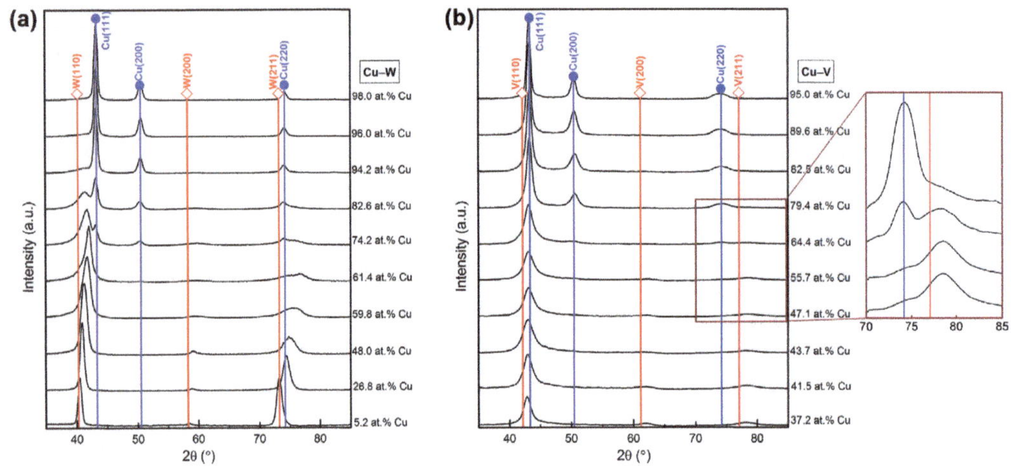

Figure 5. XRD profiles of (a) Cu–W thin films deposited at a temperature of 250 °C and power density of 0.91 W·cm^{-2}; (b) Cu–V thin films deposited at a temperature of 240 °C and power density of 0.91 W·cm^{-2}. Reference 2θ positions for the pure elements are taken from references [39–41].

the critical diffusion distance at half metastable solid solubility (when $z = \frac{1}{2}(z_{max} + z_{min})$, $X_c = A$) and will be denoted as $X_{c(\frac{1}{2})}$. In the Cu–W system, $z_{min} \approx 0$ for both bcc and fcc phases (see Figure 1(a)). According to experimental data [19], $z_{max} \approx 78$ at.% Cu for bcc phase and $z_{max} \approx 22$ at.% W for fcc phase. By fitting each dataset in Figure 3(a) and (b), the value of B is obtained ≈2 for both bcc and fcc phases and is henceforth assumed to be a constant. Different values of $X_{c(\frac{1}{2})}$ for the fitted curves can be obtained, and are shown in Figure 3. Therefore, Equation (4b) can be written in the following form:

$$X_c = X_{c(\frac{1}{2})} \cdot \sqrt{\frac{z_{max} - z}{z - z_{min}}} \qquad (5)$$

This model defines the surface diffusion required for atoms to drive second phase formation and reveals the relationship between solubilities and the critical diffusion distance. If $z = z_{max}$, meaning the phase is highly unstable, then $X_c = 0$, indicating that the atoms need only a minute diffusion distance in order to form a second phase. If $z = z_{min}$, then $X_c = +\infty$, denoting that the phase is thermodynamically stable and that the system is always in its equilibrium state.

3.2. A model based research strategy for predicting metastable phase formation diagrams

A combination of Equations (3) and (5) enables the prediction of metastable phase formation diagrams, based on a correlative experimental and theoretical research strategy. This method is delineated in the flowchart in Figure 4. One can calculate the diagram at a certain power density with input of the following variables: T_{c1}, r_D, a, Q_s, z_{max} and z_{min}. By varying the power density, r_D is affected and the metastable phase formation diagram changes accordingly. Assuming that r_D is directly proportional to power density, the diagrams at different power densities can be predicted. In Figure 4, T_{c1} denotes one critical temperature determined by experiments. r_D

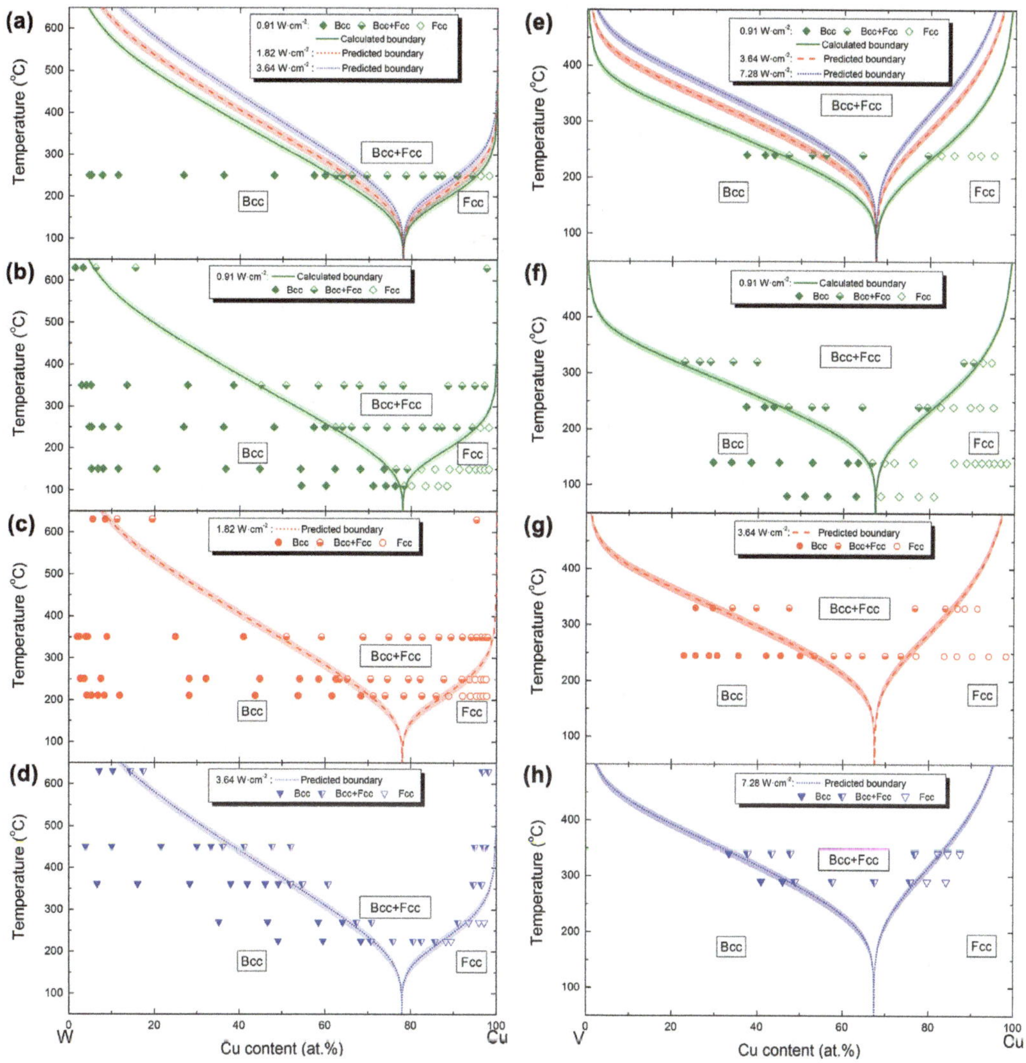

Figure 6. Metastable Cu–W phase formation diagrams: (a) calculated and predicted diagrams using experimental data [19] at a temperature of 250 °C and power density of 0.91 W·cm^{-2}; validation using experimental data [19] at power densities of (b) 0.91 W·cm^{-2}, (c) 1.82 W·cm^{-2} and (d) 3.64 W·cm^{-2}. The shaded phase boundaries indicate temperature fluctuations of ±5°C that occurred during thin film synthesis [19]; metastable Cu–V phase formation diagrams: (e) calculated and predicted diagrams using experimental data at a temperature 240 °C and power density of 0.91 W·cm^{-2}; validation using experimental data at power densities of (f) 0.91 W·cm^{-2}, (g) 3.64 W·cm^{-2} and (h) 7.28 W·cm^{-2}. The shaded phase boundaries indicate temperature fluctuations of ±7 °C that occurred during thin film synthesis.

and a can also be acquired using experiments. *Ab initio* calculations can be used to obtain a, Q_s and z_{max}, while the CALPHAD approach can be applied to describe both z_{max} and z_{min}.

Generally, one key combinatorial deposition together with the above-described calculations is needed to map metastable phase formation diagrams, provided the following necessary requirement is fulfilled: the experimental phase formation data has to cover the complete composition space, containing all of the phase boundaries to be modeled. For instance, one cannot obtain a critical temperature for the fcc phase based only on experimental data at 350 °C, because all of the data on the Cu-rich side exhibits bcc + fcc two phase structures (see Figure 2(a)).

3.3. *Modeling of the Cu–W system*

The phase formation data obtained from one deposition performed at a temperature of 250 °C and power density of 0.91 W·cm^{-2} have been selected to model the Cu–W system. XRD profiles of the Cu–W thin films are shown in Figure 5(a), where 2θ positions of pure Cu [39] and W [40] are added as references. Thin films with Cu concentrations below 59.8 at.% are composed of a bcc solid solution, while those with Cu concentrations above 96.0 at.% are composed of an fcc solid solution. The thin films with Cu concentration ranging from 61.4 to 94.2 at.% contain bcc and fcc phases. Therefore, T_{c1} is equal to 250 °C at the composition of ~60.6 and ~ 95.1 at.% Cu for the bcc and fcc phases, respectively. Previously,[19] a, z_{max} and Q_s have been described by *ab initio* calculations. z_{min} equaling zero was obtained here using the CALPHAD approach (Figure 1(a)). It is known that grain size is affected for example by the deposition rate and substrate temperature [42]. The model proposed here is based on the phase formation data only. Microstructural features such as grain size information are not included at present.

All the above data was considered, following the strategy outlined in Figure 4. A metastable Cu–W phase formation diagram, based on phase formation data obtained at a power density of 0.91 W·cm^{-2} covering the whole composition range, has been calculated. Based on this model, the position of the phase boundaries has been predicted for power densities of 1.82 and 3.64 W·cm^{-2}, as is shown in Figure 6(a). Temperature fluctuations of ± 5 °C that occurred during thin film synthesis [19] are represented in the phase formation diagrams as shaded phase boundaries. Additional experimental data [19] obtained at a power density of 0.91 W·cm^{-2} and substrate temperatures of 110, 150, 350 and 630°C are consistent with the model, as shown in Figure 6(b). The predicted phase formation diagrams for power densities of 1.82 and 3.64 W·cm^{-2} agree very well with the experimental phase formation data [19] depicted in Figure 6(c) and Figure 6(d), respectively. Therefore, the research strategy proposed here for modeling and predicting metastable phase formation has been validated for the Cu–W system. The evaluation criteria are the positions of the predicted metastable phase boundaries compared to the experimental data.[19]

Depending on the synthesis technology utilized and the deposition parameters employed, the formation of bcc [19,20,43–46] and fcc [19,20,43–51] solid solutions, amorphous [44–51] and a tetragonal phase [44], as well as mixtures thereof [19,20,43–51] are reported in the literature. The correlative experimental and theoretical research strategy proposed here for modeling the metastable phase formation during magnetron sputtering is clearly not limited to fcc and bcc solid solutions: Provided that the formation of metastable phases other than fcc and bcc solid solutions, for example amorphous phases, are observed experimentally and utilized as input data for the model, the formation of these metastable phases can be predicted.

3.4. Modeling of the Cu–V system

The same research strategy developed for and applied to the Cu–W system was also utilized to model and predict metastable phase formation within the Cu–V system. To that end, combinatorial deposition of Cu_xV_{1-x} thin films was performed at a temperature of 240 °C and power density of 0.91 W·cm^{-2}. XRD profiles of the Cu–V thin films are shown in Figure 5(b) and 2θ positions of pure Cu [39] and V [41] are added as references. A bcc solid solution is observed for Cu concentrations below 43.7 at.%, while a fcc solid solution forms at Cu concentrations above 82.5 at.%. Thin films with Cu concentrations ranging from 47.1 to 79.4 at.% comprise a mixture of bcc and fcc phases. Therefore, T_{c1} is equal to 240 °C at the composition of ~45.4 and ~81.0 at.% Cu for the bcc and fcc phases, respectively. r_D equaling $8.5 \cdot x + 0.5$ (Å/s) was acquired. As for the Cu–W system using the CALPHAD approach, z_{min} is found to be zero (Figure 1(b)). The *ab initio* results of a, z_{max} and Q_s for the Cu–V system are shown in Appendix B.

Following the research strategy specially validated for the Cu–W system (see Figure 4), metastable Cu–V phase formation diagrams were calculated, based on experimental phase formation data obtained at a power density of 0.91 W·cm^{-2} and a substrate temperature of 240 °C. This dataset fulfils the necessary requirement outlined above for modeling the metastable phase formation: the experimental data have to cover the complete composition space, thereby containing all phase boundaries to be modeled. On the basis of this model, phase formation diagrams at power densities of 3.64 and 7.28 W·cm^{-2} have been predicted, as shown in Figure 6(e). The shaded phase boundaries indicate temperature fluctuations of ±7 °C that occurred during thin film synthesis. Supplementary combinatorial depositions of Cu_xV_{1-x} thin films were carried out at a power density of 0.91 W·cm^{-2} and substrate temperatures of 80, 140 and 320 °C. The prediction based on the growth data obtained at 240 °C agrees very well with the experimental phase formation data obtained at substrate temperatures of 80, 140 and 320 °C, as shown in Figure 6(f). Further phase formation data obtained from depositions at power densities of 3.64 and 7.28 W·cm^{-2} are shown in Figure 6(g) and 6(h), respectively. These data also agree very well with the initial prediction. Hence, the predictive capability of the modeling strategy proposed here has been validated, as has the Cu–W system for the Cu–V system.

4. Conclusions

In this work, the following conclusions have been drawn:

(1) A model has been derived to describe the critical diffusion distance on the basis of experimental data for the Cu–W system, which is solubility dependent. It describes the surface diffusion required for atoms to drive the formation of a second phase.
(2) A strategy to predict metastable phase formation diagrams for sputtered thin films

based on one single combinatorial magnetron sputtering experiment, CALPHAD and *ab initio* calculations is proposed. The necessary requirement for the prediction is that the phase formation data cover the complete composition space containing all phase boundaries to be modeled.

(3) The predictive capabilities of the research strategy proposed here have been validated for both the Cu–W and Cu–V systems using additional experimental phase formation data, changing deposition temperature by up to 380 °C and deposition power by up to a factor of eight.

The correlative experimental and theoretical research strategy proposed here provides an efficient way to map metastable phase formation diagrams for sputtered thin films and hence expands the capabilities for future design of metastable thin film materials.

Acknowledgments

Ab initio calculations were performed with computing resources granted by JARA-HPC from RWTH Aachen University under project JARA0131. JMS gratefully acknowledges funding from the MPG fellow program.

Disclosure statement

No potential conflict of interest was reported by the authors.

Notes on Contributors

Keke Chang is a postdoctoral researcher at Materials Chemistry, RWTH Aachen University.

Denis Music is ab initio group leader at Materials Chemistry, RWTH Aachen University.

Moritz to Baben was a postdoctoral researcher at RWTH Aachen University. He currently works at GTT-Technologies, Germany.

Dennis Lange received his Master Degree at RWTH Aachen University and his Doctoral Degree at Ecole Polytechnique in Palaiseau (Île-de-France). He currently works as a project leader at the Herzogenrath R&D Center of Saint-Gobain, Germany.

Hamid Bolvardi received his doctoral degree at RWTH Aachen University. He currently works as a project manager R&D Tools at Oerlikon Surface Solutions AG, Liechtenstein.

Jochen M. Schneider, Ph.D., is Professor of Materials Chemistry at RWTH Aachen University, Germany. Formerly at Linköping University, Sweden, and Northwestern University, USA, Jochen received his Ph.D. degree in surface engineering from Hull University, UK. His research interest is the materials science of thin films grown by plasma-assisted vapor deposition. Jochen has been awarded the Sofya Kovalevskaya Prize by the Alexander von Humboldt Foundation and was named Fellow of AVS in 2013 and Max Planck Fellow and RWTH fellow in 2015.

References

[1] Petrov I, Barna PB, Hultman L, et al. Microstructural evolution during film growth. J. Vac. Sci. Technol. A. 2003;21:S117–28.

[2] Grovenor CRM, Hentzell HTG, Smith DA. The development of grain structure during growth of metallic films. *Acta Metall.* 1984;32:773–781.

[3] Barna PB, Adamik M. Fundamental structure forming phenomena of polycrystalline films and the structure zone models. Thin Solid Films. 1998;317:27–33.

[4] Einstein A. Elementare Theorie der Brownschen Bewegung. Z. Elektrochem. 1908;14:235–239.

[5] Cantor B, Cahn R. Metastable alloy phases by co-sputtering. *Acta Metall.* 1976;24:845–852.

[6] Ohring M. Materials science of thin films: deposition and structure. New York, NY: Academic; 2002.

[7] Saunders N, Miodownik AP. Phase formation in co-deposited metallic alloy thin films. J. Mater. Sci. 1987;22:629–637.

[8] Cremer R, Witthaut M, Neuschütz D Experimental determination of the metastable (Ti,Al)N phase diagram up to 700 °C. In: *Value addition metallurgy*. Warrendale: The Minerals, Metals & Materials Society; 1998. p. 249–258

[9] Cremer R, Neuschütz D. Optimization of (Ti, Al) N hard coatings by a combinatorial approach Int. J. Inorg. Mater. 2001;3:1181–1184.

[10] Gebhardt T, Music D, Takahashi T, et al. Combinatorial thin film materials science: From alloy discovery and optimization to alloy design. Thin Solid Films. 2012;520:5491–5499.

[11] Wang Q, Itaka K, Minami H, et al. Combinatorial pulsed laser deposition and thermoelectricity of (La1-xCax)VO3 composition-spread films Sci. Tech. Advan. Mater. 2004;5:543–547.

[12] Mardare AI, Ludwig A, Savan A, et al. Properties of anodic oxides grown on a hafnium–tantalum–titanium thin film library. Sci. Tech. Advan. Mater. 2014;15:015006.

[13] Saunders N, Miodownik AP. The use of free energy vs composition curves in the prediction of phase formation in codeposited alloy thin films. *CALPHAD*. 1985;9:283–290.

[14] Holleck H. Metastable coatings - prediction of composition and structure. Surf. Coat. Technol. 1988;36:151–159.

[15] Spencer P. Thermodynamic prediction of metastable coating structures in PVD process. Z. Metallkd. 2001;92:1145–1150.

[16] Mayrhofer PH, Music D, Schneider JM. Influence of the Al distribution on the structure, elastic properties, and phase stability of supersaturated $Ti_{1-x}Al_xN$. J. Appl. Phys. 2006;100:094906.

[17] Mayrhofer PH, Music D, Reeswinkel Th, et al. Structure, elastic properties and phase stability of $Cr_{1-x}Al_xN$. Acta Mater. 2008;56:2469–2475.

[18] Rovere F, Music D, Schneider JM, et al. Experimental and computational study on the effect of yttrium on the phase stability of sputtered Cr-Al-Y-N hard coatings. *Acta Mater.* 2010;58:2708–2715.

[19] Chang K, to Baben M, Music D, et al Estimation of the activation energy for surface diffusion during metastable phase formation. *Acta Mater*. 2015;98:135–140.
[20] Vüllers FTN, Spolenak R. From solid solutions to fully phase separated interpenetrating networks in sputter deposited "immiscible" W-Cu thin films. Acta Mater. 2015;99:213–227.
[21] Park J-H, Moon D-Y, Han D-S, et al. Self-forming barrier characteristics of Cu–V and Cu–Mn films for Cu interconnects. Thin Solid Films. 2013;547:141–145.
[22] Cao F, Wu GH, Jiang L-T, et al. Application of Cu–C and Cu–V alloys in barrier-less copper metallization. Vacuum. 2015;122:122–126.
[23] Park J-H, Kang M-S, Han D-S, et al. Effects of UV curing on the self-forming barrier process of Cu-V alloy films. Surf. Coat. Technol. 2015;276:254–259.
[24] Welzel U, Mittemeijer EJ. Diffraction stress analysis of macroscopically elastically anisotropic specimens: On the concepts of diffraction elastic constants and stress factors. J. Appl. Phys. 2003;93:9001–9011.
[25] Welzel U, Mittemeijer EJ. Diffraction stress analysis of elastic grain interaction in polycrystalline materials. Z. Kristallogr. 2007;222:160–173.
[26] Vitos L. Total-energy method based on the exact muffin-tin orbitals theory. Phys. Rev. B. 2001;64:014107.
[27] Andersen K, Jepsen O, Krier G. *Lecture on methods of electronic structure calculations*. Singapore: World Scientific; 1995. p. 63–124.
[28] Vitos L, Skriver HL, Johansson B, et al. Application of the exact muffin-tin orbitals theory: the spherical cell approximation. Comput. Mater. Sci. 2000;18:24–38.
[29] Vitos L, Kollár J, Skriver HL. Full charge-density scheme with a kinetic-energy correction: Application to ground-state properties of the 4d metals. Phys. Rev. B. 1997;55:13521.
[30] Perdew JP, Burke K, Ernzerhof M. Generalized gradient approximation made simple. Phys. Rev. Lett. 1996;77:3865–3868.
[31] Kresse G, Furthmüller J. Efficient iterative schemes for *ab initio* total-energy calculations using a plane-wave basis set. Phys. Rev. B. 1996;54:11169–11186.
[32] Kresse G, Joubert D. From ultrasoft pseudopotentials to the projector augmented-wave method. Phys. Rev. B. 1999;59:1758–1775.
[33] Blöchl PE. Projector augmented-wave method. Phys. Rev. B. 1994;50:17953–17979.
[34] Monkhorst HJ, Pack JD. Special points for brillouin-zone integrations. Phys. Rev. B. 1976;13:5188–5192.
[35] Subramanian PR, Laughlin DE. *The Cu-W (Copper-Tungsten) system phase diagrams of binary Tungsten alloys*. Calcutta: Indian Institute of Metals; 1991. p. 76–79.
[36] Zhao J, Du Y, Zhang L, et al. Thermodynamic reassessment of the Cu–V system supported by key experiments. Calphad. 2008;32:252–255.
[37] Bale CW, Bélisle E, Chartrand P, et al. FactSage thermochemical software and databases-recent developments. Calphad. 2009;33:295–311.
[38] Dinsdale AT. SGTE data for pure elements. *CALPHAD*. 1991;15:317–425.
[39] JCPDS card No. 4-836, International Center for Diffraction Data, JCPDF-ICDD.
[40] JCPDS card No. 4-806, International Center for Diffraction Data, JCPDF-ICDD.
[41] JCPDS card No. 22-1058, International Center for Diffraction Data, JCPDF-ICDD.
[42] Thompson CV. structure evolution during processing of polycrystalline films. Ann. Rev. Mater. Sci. 2000;30:159–190.
[43] Dirks AG, Van den Broek JJ. Metastable solid-solutions in vapor-deposited Cu–Cr, Cu–Mo, and Cu–W thin-films. J. Vac. Sci. Technol. A. 1985;3:2618–2622.
[44] Nastasi M, Saris FW, Hung LS, et al. Stability of amorphous Cu/Ta and Cu/W alloys. J. Appl. Phys. 1985;58:3052–3058.
[45] Rizzo HF, Massalski TB, Nastasi M. Metastable crystalline and amorphous structures formed in the Cu–W system by vapor deposition. Metall Trans A. 1993;24:1027–1037.
[46] Zong RL, Wen SP, Zeng F, et al. Nanoindentation studies of Cu–W alloy films prepared by magnetron sputtering. J. Alloys Compd. 2008;464:544–549.
[47] Engelhardt MA, Jaswal SS, Sellmyer DJ. Electronic-structure of amorphous and disordered crystalline Cu–W alloys of similar composition. Solid State Commun. 1990;75:663–665.
[48] Radić N, Gržeta B, Gracin D, et al. Preparation and structure of Cu–W thin films. Thin Solid Films. 1993;228:225–228.
[49] Gržeta B, Radić N, Gracin D, et al. Crystallization of Cu50W50 and Cu66W34 amorphous alloys. J. Non Cryst. Solids. 1994;170:101–104.
[50] Radić N, Stubicar M. Microhardness properties of Cu–W amorphous thin films. J. Mater. Sci. 1998;33:3401–3405.
[51] Zhou L, Wang M, Peng K, et al. Structure characteristic and its evolution of Cu–W films prepared by dual-target magnetron sputtering deposition. Trans. Nonferrous. Met. Soc. China. 2012;22:2700–2706.
[52] Straumanis ME, Yu LS. Lattice parameters, densities, expansion coefficients and perfection of structure of Cu and of Cu–In α phase. *Acta Cryst*. 1969;25:676–682
[53] Szalkowski FJ, Somorjai GA. The characterization of some vanadium (100) surface using LEED and AES. J. Chem. Phys. 1976;64:2985–2989.

Appendix A. Thermodynamic data of the Cu–W and Cu–V systems

The liquid, bcc and fcc phases in the Cu–W and Cu–V systems are described using an ideal solution model.[35,36] Gibbs energy expressions of the Cu–V system are identical to those in [36]. While differing from [35], Gibbs energies for the pure Cu and W are taken from a generally accepted SGTE compilation.[38] Interaction parameters of the liquid Cu_xW_{1-x} phase [36] are used, while those of bcc and fcc Cu_xW_{1-x} phases are adjusted on the basis of *ab initio* formation enthalpy data in a previous work.[19] Gibbs energy functions for all of the phases used to calculate the Cu–W and Cu–V phase diagrams (Figure 1) are listed below. Note that quantities are in J mol^{-1} and temperature (T) is in K.

(1) Gibbs energies of pure elements:

G(fcc,Cu) =

$-7770.458 + 130.485235 \cdot T - 24.112392 \cdot T \cdot \ln(T) - 0.00265684 \cdot T^2 + 1.29223 \cdot 10^{-7} \cdot T^3 + 52478 \cdot T^{-1}$ $(298.15\ K < T < 1357.77\ K)$

$-13542.026 + 183.803828 \cdot T - 31.38 \cdot T \cdot \ln(T) + 3.64167 \cdot 10^{29} \cdot T^{-9}$ $(1357.77\ K < T < 3200\ K)$

G(bcc, Cu) =

$G(fcc,Cu) + 4017 - 1.255 \cdot T$ $(298.15\ K < T < 3200\ K)$

G(Liquid, Cu) =

$G(fcc,Cu) + 12964.736 - 9.511904 \cdot T - 5.849 \cdot 10^{-21} \cdot T^7$ $(298.15\ K < T < 1357.77\ K)$

$G(fcc,Cu) + 13495.481 - 9.922344 \cdot T - 3.642 \cdot 10^{29} \cdot T^{-9}$ $(1357.77\ K < T < 3200\ K)$

G(bcc,W) =

$- 7646.311 + 130.4 \cdot T - 24.1 \cdot T \cdot \ln(T) - 0.001936 \cdot T^2 + 2.07 \cdot 10^{-7} \cdot T^3 - 5.33 \cdot 10^{-11} \cdot T^4 + 44500 \cdot T^{-1}$
$(298.15\ K < T < 3695\ K)$
$- 82868.801 + 389.362335 \cdot T - 54 \cdot T \cdot \ln(T) + 1.528621 \cdot 10^{33} \cdot T^{-9}$
$(3695\ K < T < 6000\ K)$
G(fcc,W) =
G(bcc,W) + 19300 + 0.63·T $(298.15\ K < T < 6000\ K)$
G(Liquid, W) =
G(bcc,W) + $52160.584 - 14.10999 \cdot T - 2.713468 \cdot 10^{-24} \cdot T^7$
$(298.15\ K < T < 3695\ K)$
G(bcc,W) + $52432.75 - 14.187335 \cdot T - 1.528621 \cdot 10^{33} \cdot T^{-9}$
$(3695\ K < T < 6000\ K)$
G(bcc,V) =
$- 7930.43 + 133.346053 \cdot T - 24.134 \cdot T \cdot \ln(T) - 0.003098 \cdot T^2 + 1.2175 \cdot 10^{-7} \cdot T^3 + 69460 \cdot T^{-1}$
$(298.15\ K < T < 790\ K)$
$- 7967.842 + 143.291093 \cdot T - 25.9 \cdot T \cdot \ln(T) + 6.25 \cdot 10^{-5} \cdot T^2 - 6.8 \cdot 10^{-7} \cdot T^3$
$(790\ K < T < 2183\ K)$
$- 41689.864 + 321.140783 \cdot T - 47.43 \cdot T \cdot \ln(T) + 6.44389 \cdot 10^{31} \cdot T^{-9}$
$(2183\ K < T < 4000\ K)$
G(fcc,V) =
G(bcc,V) + 7500 + 1.7·T $(298.15\ K < T < 6000\ K)$
G(Liquid, V) =
G(bcc,V) + $20764.117 - 9.455552 \cdot T - 5.19136 \cdot 10^{-22} \cdot T^7$
$(298.15\ K < T < 2183\ K)$
G(bcc,V) + $22072.354 - 10.0848 \cdot T - 6.44389 \cdot 10^{31} \cdot T^{-9}$
$(2183\ K < T < 4000\ K)$

(2) Gibbs energies of the Cu_xW_{1-x} phases in the Cu–W system.
G(fcc) =
$x \cdot$G(fcc,Cu) + $(1 - x) \cdot$G(fcc,W) + $R \cdot T \cdot [x \cdot \ln x + (1 - x) \cdot \ln (1 - x)] + x \cdot (1 - x) \cdot 120000$
G(bcc) =
$x \cdot$G(bcc,Cu) + $(1 - x) \cdot$G(bcc,W) + $R \cdot T \cdot [x \cdot \ln x + (1 - x) \cdot \ln (1 - x)] + x \cdot (1 - x) \cdot 130000$
G(Liquid) =
$x \cdot$G(Liquid,Cu) + $(1 - x) \cdot$G(Liquid,W) + $R \cdot T \cdot [x \cdot \ln x + (1 - x) \cdot \ln (1 - x)]$
$+ x \cdot (1 - x) \cdot (101000 - x \cdot 210000)$

(3) Gibbs energies of the Cu_xV_{1-x} phases in the Cu–V system:
G(fcc) =
$x \cdot$G(fcc,Cu) + $(1 - x) \cdot$G(fcc,V) + $R \cdot T \cdot [x \cdot \ln x + (1 - x) \cdot \ln (1 - x)] + x \cdot (1 - x) \cdot 53650$
G(bcc) =
$x \cdot$G(bcc,Cu) + $(1 - x) \cdot$G(bcc,V) + $R \cdot T \cdot [x \cdot \ln x + (1 - x) \cdot \ln (1 - x)] + x \cdot (1 - x) \cdot 42377.8$
G(Liquid) =
$x \cdot$G(Liquid,Cu) + $(1 - x) \cdot$G(Liquid,V) + $R \cdot T \cdot [x \cdot \ln x + (1 - x) \cdot \ln (1 - x)]$
$+ x \cdot (1 - x) \cdot (56400 - x \cdot 37000)$

Appendix B. *Ab initio* results of bcc and fcc phases in the Cu–V system

Ab initio calculated lattice parameters and formation energies of Cu–V phases using the CPA and SQS approaches are shown in Figure B1(a) and B1(b), respectively. In contrast to the experimental data (experimental lattice parameters of fcc-Cu and bcc-V are taken from [52,53]), CPA and SQS calculated lattice parameters exhibit a deviation of +0.84% – +4.42% and −1.08% – +2.37%, respectively, which is an acceptable agreement. Solid solubility limits of Cu in bcc-V calculated using CPA and SQS are 72.0 and 63.0 at.% Cu, respectively. An average value (i.e. 67.5 at.% Cu) is considered as z_{max} of Cu in bcc-V for modeling the Cu–V metastable phase formation diagrams. Figure B1(c) shows the *ab initio* obtained Q_s data, which decreases as the metastable solubility of Cu in bcc-V or V in fcc-Cu increases. This trend is expected and similar to that in the Cu–W system.[19]

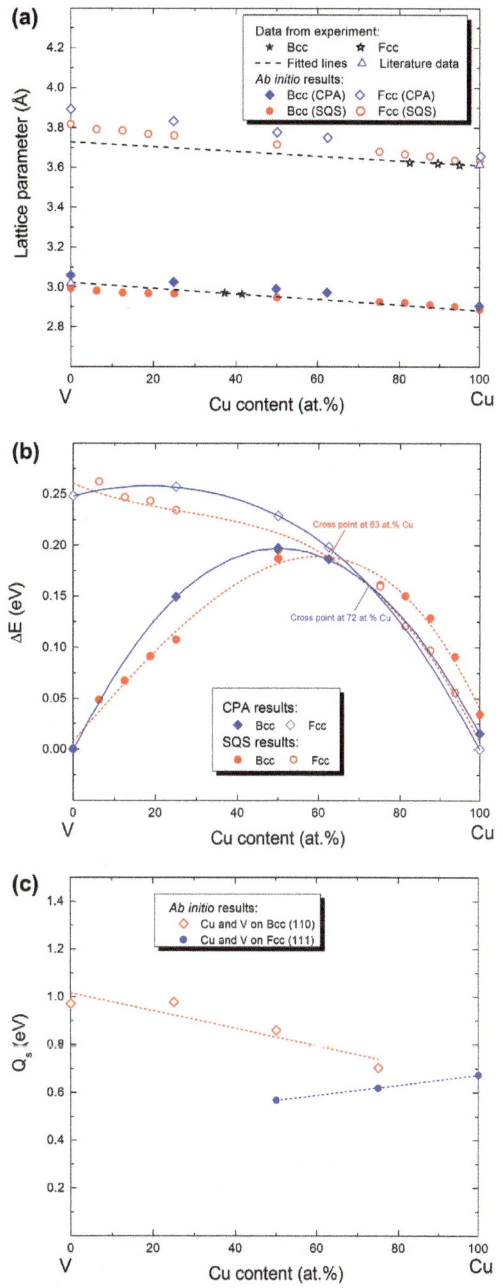

Figure B1. *Ab initio* results of Cu_xV_{1-x} as a function of composition: (a) lattice parameters (compared with experimental data), (b) formation energies (Δ*E*) of bcc and fcc phases, (c) activation barriers (Q_s) of Cu and V atoms at bcc (110) and fcc (111) surfaces. Experimental lattice parameters of fcc-Cu and bcc-V are taken from [52,53].

18

Magnetic microparticle-polydimethylsiloxane composite for reversible microchannel bonding

Chia-Wen Tsao and Yueh-Pu Lee

Department of Mechanical Engineering, National Central University, Taoyuan, ROC

ABSTRACT
In this study, an iron oxide magnetic microparticles and poly(dimethylsiloxane) (MMPs-PDMS) composite material was employed to demonstrate a simple high-strength reversible magnetic bonding method. This paper presents the casting of opaque-view (where optical inspection through the microchannels was impossible) and clear-view (where optical inspection through the microchannel was possible) MMPs-PDMS. The influence of the microchannel geometries on the casting of the opaque-view casting was limited, which is similar to standard PDMS casting. Clear-view casting performance was highly associated with the microchannel geometries. The effects of the microchannel layout and the gap between the PDMS cover layer and the micromold substrate were thoroughly investigated. Compared with the native PDMS bonding strength of 31 kPa, the MMPs-PDMS magnetic bonding experiments showed that the thin PDMS film with an MMPs-PDMS layer effectively reduced the surface roughness and enhanced MMPs-PDMS reversible magnetic bonding strength. A thin PDMS film-coated opaque-view MMPs-PDMS device exhibited the greatest bonding strength of 110 kPa, and a clear-view MMPs-PDMS device with a thin PDMS film attained a magnetic bonding strength of 81 kPa.

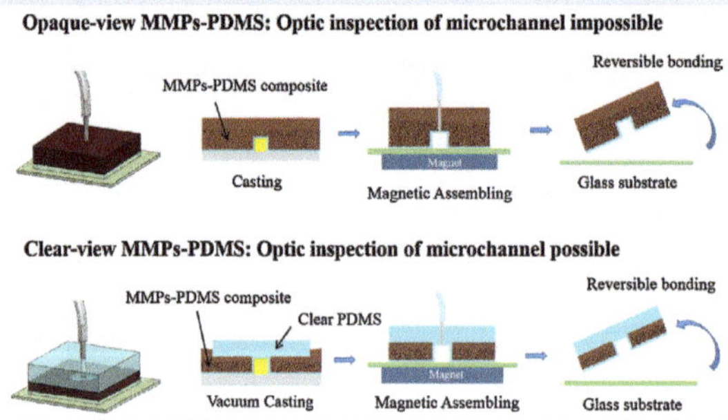

KEYWORDS
Microchannel; reversible bonding; magnetic microparticle; polydimethylsiloxane (PDMS)

1. Introduction

Microfluidic, biomedical, and lab-on-a-chip devices are used for rapid, high-throughput, high-automation, and high-integration analyses, which have gained prominence in the fields of chemistry [1,2] and biology [3–5]. Various materials, such as glass [6], silicon [7], and polymers [8] are used for manufacturing microfluidic devices. Recent trends in microfluidics technology are moving toward personalized diagnosis and point-of-care-testing applications. Polymer materials are an ideal material choice for these diagnostic applications because of the low cost and disposable advantages [9]. Polymers used in microfluidics can typically be divided into two major categories: thermoplastic and thermoset polymers. Thermoplastic polymers, such as polycarbonate, poly(methyl methacrylate), and cyclic olefin copolymer (COC), are commonly used in microfluidic devices because of their superior mechanical, optical, and chemical properties. Polydimethylsiloxane (PDMS) is the most common elastomer material used in microfluidics; because of its favorable biocompatibility and gas transmissivity, PDMS has been applied extensively in cell culture applications. In addition, the success of PDMS-based microfluidic valves and pumps developed by Quake [10] has advanced PDMS microfluidics to include various applications.

Thermoplastic and PDMS polymers can be fabricated through low-cost and high-fabrication throughput polymer replication processes. Thermoplastic polymers can be reshaped while being heated above a given glass transition temperature; therefore, thermoplastic microfluidic

CONTACT Chia-Wen Tsao cwtsao@ncu.edu.tw

devices are typically fabricated through hot embossing [8, 11] or inject molding [12]. PDMS is a thermoset material that is primarily used in casting for PDMS replication [13].

After polymer replication is completed, the polymer replication must be sealed using a different type of substrate layer to create an enclosed microfluidic channel. Developing an appropriate sealing method is critical for fabricating microfluidic devices. Thermoplastic polymers can be directly bonded through thermal bonding [14] by heating and pressing a pair of thermoplastic layers together. PDMS substrates can be directly attached to another PDMS substrate or glass substrate through the van der Waals force [15]. However, such microfluidic devices are prone to debonding or leaking during handling and several other types of high-pressure application. Consequently, high-strength polymer bonding methods have been developed to address these problems.

Thermoplastic bonding can be enhanced through solvent treatment [16–18] or surface modification [19,20] techniques. PDMS bonding is generally enhanced using O_2 plasma bonding [21,22], although plasma treatment is the most common bonding technique for PDMS because of its high bond strength and simple process. Nevertheless, plasma bonding is irreversible and the surface properties are typically altered after modification. A reversible bonding method enables fabrication of dismountable and reusable microfluidic devices, which is beneficial for biomolecules analysis, cell analysis, and other related fields. Although PDMS can reversibly bond to a smooth surface through the van der Waals force, the bond strength is typically below 35 kPa [15]. Accordingly, high-strength reversible PDMS bonding methods have been developed to resolve these problems. PDMS nanocapillaries can reversibly bond to gold or silicon substrates through electrostatic forces [23]. A bioinspired method that involves using gecko-foot-mimetic nanopillar arrays coated with a mussel-adhesive-protein-mimetic polymer has also been proposed [24]. Other reversible bonding techniques involve using a direct mechanical clamping glass-PDMS-glass device [25] or vacuum aspiration [26,27], both of which achieve a high bond strength (typically above 100 kPa), but additional working space is required for the mechanical clamps or vacuum pump. Additionally, a suction microchannel network and continuous vacuuming are required during microfluidic operation when using in the vacuum aspiration method, which limits the microchannel layout and its applications.

Reversible bonding through magnetic forces is an effective alternative bonding approach. Magnetic reversible bonding is achieved by placing a microfluidic device directly on a magnet and removing it from the magnet to dismount the microchannel. Rafat et al. [28] assembled iron slabs around a microchannel for magnetic reversible bonding. However, because of the presence of magnetic slabs, the microchannels were restricted to simple geometries. Rasponi et al. [29] presented another reversible magnetic bonding method in which iron micropowders were injected or plastered into micropowder layer cavities and then align-bonded to the microchannel. This bonding method required an additional micropowder layer and precise bonding alignment or micropowder injection during the fabrication process, rendering the fabrication process complex and making it difficult to create complex parallel multiplex microchannels. In addition, iron powders usually contain large particles (maximum grain size: 212 μm), which restricted the minimum microchannel width of the micropowder cavity layer. In their research demonstration, the minimum microchannel width in cavity layer is 800 μm. Thus, both the iron slab and micropowder approaches can provide magnetic sealing only near the major fluidic pathway instead of fully sealing each branch-microchannel boundary for complete sealing.

In this study, we first demonstrate the use of synthesized magnetic iron oxide (Fe_3O_4) microparticles and an MMPs-PDMS composite for reversible magnetic bonding applications. This paper presents the procedure for casting opaque-view (where the optical inspection of the microchannel was impossible) and clear-view (where the optical inspection of the microchannel was possible) MMP-PDMSs. The grain size of the synthesized MMPs is <27 μm, and the MMPs-PDMS composite directly fills the micromold cavity during the casting procedure, completely sealing each microchannel without requiring precision alignments and assembly during the casting procedure. Additionally, MMPs-PDMS casting and its effects on reversible magnetic bonding strength are evaluated in this paper in detail.

2. Experiment

2.1. Materials

P-type (100) silicon wafer with diameter of 10 cm was purchased from Summit-Tech Resource Corp. (Hsinchu, Taiwan). SU-8 3050 and SU-8 developer were purchased from MicroChem Corp. (Newton, MA, USA), and PDMS (Sylgard 184 silicone elastomer kit) was purchased from Dow Corning Corp. (Midland, MI, USA). Acetone and isopropanol were purchased from J.T Baker (Phillipsburg, NJ, USA). Surgical needles (SC20/15, LS20) were purchased from Instech Laboratories, Inc. (Plymouth Meeting, PA, USA). Medical grade tubing (Tygon S-50-HL) was purchased from Saint Gobain Performance Plastics Corp. (Aurora, OH, USA). Ammonia was purchased from Merck Co. (Darmstadt, Germany). Hydrochloric acid (HCl), iron (II) chloride tetrahydrate ($FeCl_2·4H_2O$), and iron (III) chloride hexahydrate ($FeCl_3·6H_2O$) were purchased from Sigma-Aldrich (St. Louis, MO, USA).

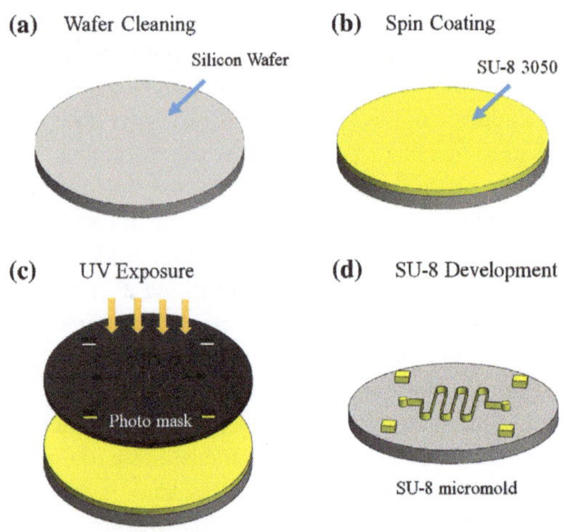

Figure 1. Schematic illustration of SU-8 micromold fabrication process: (a) wafer cleaning; (b) spin coat SU-8 photoresist; (c) UV exposure; and (d) development of SU-8 to create a micromold.

2.2. Iron oxide magnetic microparticle synthesis

Fe_3O_4 MMPs were fabricated using a magnetic nanoparticle synthesis procedure [30]; 2.0 g of $FeCl_2{\cdot}4H_2O$ and 5.4 g of $FeCl_3{\cdot}6H_2O$ were mixed with 25 ml of a 2 M HCl solution and degassed using ultrasonication. Subsequently, 25 ml of 25% (v/v) ammonia was added and stirred for 30 min at room temperature. A magnet was placed under the beaker to concentrate the Fe_3O_4 particles and completely rinsed with deionized water. The Fe_3O_4 was baked at 150°C for 4 h on a hotplate to remove the water from the MMPs in order to prevent bubbles forming during vacuum casting because of the MMPs, which would create voids after the MMPs-PDMS layer was cured. Finally, the dry clodded MMPs were finely ground using a mesh sieve to create MMPs 4–27 μm in size.

2.3. SU-8 micromold fabrication

The micromold was fabricated using SU-8 lithography process (Figure 1). First, a silicon wafer was cleaned with acetone, isopropanol, and deionized water in a sonicator, followed by drying with a N_2 gun and hot-baking at 130C for 15 min (Super-Nuova, Thermo Scientific, Inc., Waltham, MA, USA) to remove any surface moisture. Second, a two-step spinning process was performed at 500 rpm for 30 s and at 1000 rpm for 40 s (SPC-703, Yi Yang Co., Taoyuan, Taiwan) to coat a SU-3050 photoresist layer onto the silicon wafer, which was then soft-baked at 110C for 20 min. The photomask was aligned and the SU-8-coated wafer was exposed to UV radiation (AGL100 UV Light source, M&R Nano Technology Co., Taoyuan, Taiwan) for 90 s. Subsequently, the wafer was soft-baked at 95C for 5 min and then rinsed with the SU-8 developer to create a 100-μm-thick SU-8 micromold. The SU-8 coating steps (Figure 1(b)) were repeated to create a 200-μm-thick SU-8 micromold.

2.4. Magnetic microparticles-poly(dimethylsiloxane) composite preparation

The PDMS mixture was prepared by first mixing the PDMS base and a curing agent at a 10:1 weight ratio, and then fully mixing it with the MMPs powder at a weight ratio of 1:10, 1:5, and 2:5 to obtain the MMPs-PDMS composites. The MMPs-PDMS composite was degassed in a vacuum desiccator at a gauge pressure of <700 mm Hg (PC-250, Yeong-shin, Hsinchu, Taiwan) to remove the air bubbles from the composite.

2.5. Cover-PDMS layer preparation

The cover-PDMS layer was fabricated using a standard PDMS lithography process [31]. The vacuum-degassed 1:10 PDMS mixture was poured onto the silicon wafer by using a 3-mm-thick spacer and any excessive PDMS mixture was removed. Subsequently, the mixture was baked at 60°C for 3 h and cut-removed, yielding an optically clear PDMS as a cover layer for use in the clear-view MMPs-PDMS casting procedure.

3. Results and discussion

Figure 2 illustrates the opaque- and clear-view MMPs-PDMS layer casting procedures. The opaque-view MMPs-PDMS layer casting (Figure 2(a)) was fabricated using a standard PDMS process. In this process, the MMPs-PDMS composite was poured into the SU-8 micromold and cured on a hot plate at 60°C for 4 h to create the 3-mm-thick MMPs-PDMS layer. This process was simple and straightforward for creating an MMPs-embedded PDMS layer. However, the opaque-view MMPs-PDMS layer was completely covered with dark MMPs, which blocked the microchannel, rendering optical inspection impossible. This could introduce problems in some optical-based microfluidic applications.

Creating a clear top-view PDMS microchannel surrounded by the MMPs for reversible magnetic bonding purposes is a complex procedure. Multistep casting and bonding approaches are impractical because they require precise alignment and assembling. Cast-scraping, which involves removing the MMPs-PDMS mixture followed by casting another optically clear PDMS layer above the MMPs-PDMS layer, at first seems to be a feasible clear-view MMPs-PDMS fabrication method. However, using a thin MMPs-PDMS layer above the SU-8 micromold after the scraping step partially blocks the optical view in the microchannel after the PDMS has been cured. This is because of the behavior of the highly viscous MMPs-PDMS composite. Several adhesive microfluidic bonding techniques involve utilizing this viscous property to create a thin adhesive layer for adhesive bonding [32].

In this study, we propose a simple clear-view MMPs-PDMS casting procedure (Figure 2(b)) for creating an optically inspectable microchannel. First, the

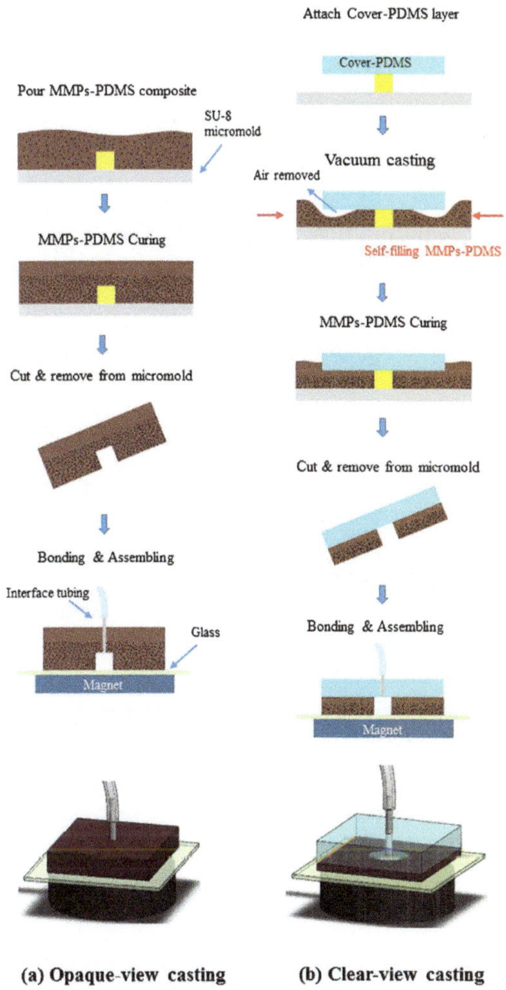

Figure 2. Schematic illustration of the (a) opaque-view and (b) clear-view MMPs-PDMS casting procedures.

3-mm-thick PDMS cover layer was attached to the SU-8 micromold, and the MMPs-PDMS composite was applied around the cover-PDMS layer. The micromold was placed in a vacuum desiccator with gauge pressure of <700 mm Hg and baked at 60C for 4 h. The MMPs-PDMS composite was adequately integrated with the PDMS cover layer, and no breakage was observed at the interface after the MMPs-PDMS casting. The MMPs-PDMS layers were pierced with 0.75-mm-diameter holes and surgical needles were inserted to form microchannel inlets and outlets. The MMPs-PDMS microchannel layer was attached to a glass substrate and assembled with a 5450 gauss magnet (diameter: 4 cm; thickness: 2 cm) as a reversible-bond microfluidic device.

Figure 3 shows consequent images of the opaque-view MMPs-PDMS vacuum-casting procedure, which involved removing air trapped from the intermediate layer and allowing the MMPs-PDMS composite to flow into the intermediate layer cavities to form the clear-view MMPs-embedded PDMS microchannel layer. As shown in Figure 3(a), when the vacuum was turned on, the air in the gap between the PDMS cover layer and the micromold was immediately removed. For simple open-cell geometries (e.g. straight microchannels or circular microfluidic chambers), the air can easily be removed and the gap in the MMPs-PDMS composite can be filled without trapping air. However, for several types of complex closed-cell geometries, such as the U-shaped serpentine microchannels, air becomes trapped in the U-turn sections of the microchannel (Figure 3(b)). However, because PDMS is a gas permeable material, the trapped air can gradually be removed through the PDMS membrane. When the vacuum was broken (Figure 3(c)), the MMPs-PDMS composite had completely self-filled the micromold cavities.

3.1. Microchannel geometry effects on MMPs-PDMS casting

In both the opaque- and clear-view MMPs-PDMS casting, the amount of MMPs embedded in the PDMS layer affected the reversible magnetic bonding strength, which depends on the concentration of the MMPs-PDMS composite. In addition, the layout of the microchannel may also affect the distribution of the MMPs embedded in the PDMS layer, as well as its reversible magnetic bonding performance. Therefore, this study investigated the MMPs-PDMS composite concentration and microchannel geometry effects to opaque- and clear-view MMPs-PDMS microchannel casting. Three

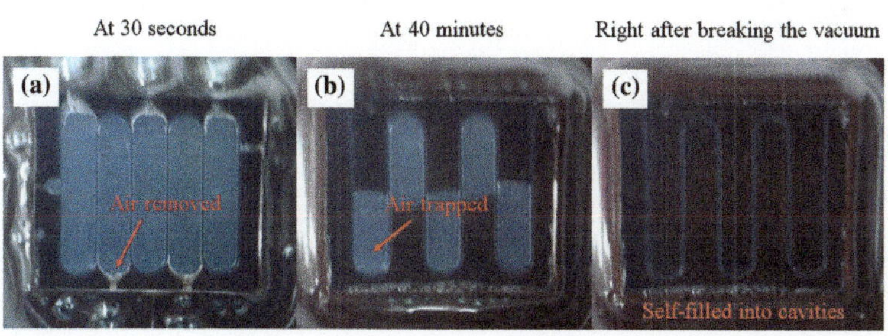

Figure 3. Vacuum casting steps in the serpentine clear-view casting process (a) at 30 s; (b) at 40 min; and (c) after breaking the vacuum.

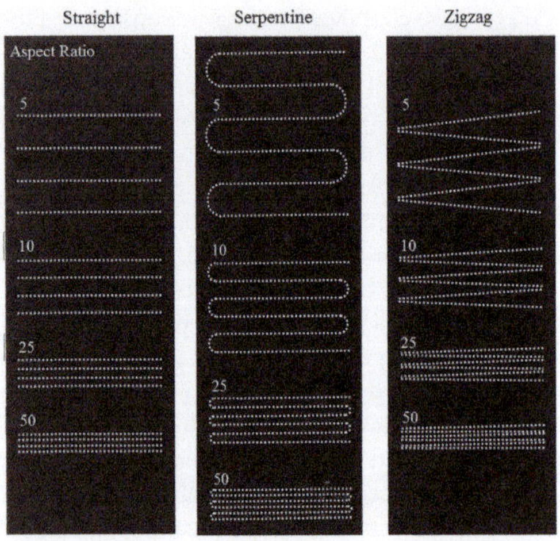

Figure 4. Opaque MMPs-PDMS casting of (a) straight; (b) serpentine; and (c) zigzag microchannels with aspect ratios of 5, 10, 25, and 50. White dot lines are used to indicate the microchannels because they cannot be clearly seen in the opaque-view MMPs-PDMS layer.

MMPs-PDMS composite concentrations of 1:10, 1:5, and 2:5 (MMPs:PDMS weight ratio) and three general microchannel geometries (i.e. straight, serpentine, and zigzag) were evaluated. The zag microchannel geometries were 100 μm wide and 10 mm long, with four length-to-space aspect ratios of 5 (space: 2 mm), 10 (space: 1 mm), 25 (space: 0.4 mm), and 50 (space: 0.2 mm).

3.1.1. *Opaque-view microchannel casting*

Opaque-view MMPs-PDMS casting is a simple procedure that is similar to standard PDMS casting. The MMPs-PDMS composite was successfully cast on a 100-μm-thick microchannel mold for all test geometry and concentration conditions. Figure 4 shows the opaque-view MMPs-PDMS straight, serpentine, and zigzag microchannel casting results with a 2:5 MMPs-PDMS concentration. Uniform MMPs were embedded in the PDMS layer for all geometry conditions.

3.1.2. *Clear-view straight microchannel casting*

The microchannel geometries were independent in the opaque-view casting because the MMPs-PDMS composite could move freely into the micromold cavities and remove air bubbles, such as in standard PDMS casting. However, the clear-view casting performance was closely related to the microchannel geometries. The geometrical constraints of the microchannel layout and gap between the PDMS cover layer and micromold substrate could result in trapped air bubbles or an irregular distribution of MMPs after MMPs-PDMS curing. Therefore, the clear-view straight microchannels were evaluated first; Figure 5 shows the casting results obtained from the various MMPs-PDMS composite concentrations.

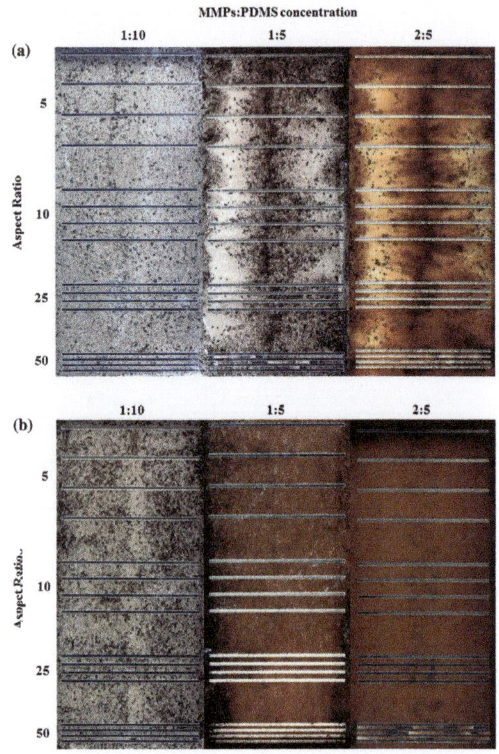

Figure 5. Clear-view MMPs-PDMS casting of (a) 100-μm-thick and (b) 200-μm-thick straight microchannels with composite concentrations of 1:10, 1:5, and 2:5 at aspect ratios of 5, 10, 25, and 50.

The MMPs-PDMS composite successfully filled the straight microchannel for the designs with a height of 100 μm (Figure 5(a)) and 200 μm (Figure 5(b)). The MMPs-PDMS layer became darker as the MMPs-PDMS composite became more concentrated, indicating that more MMPs were embedded in the PDMS layer at higher concentrations. No trapped air bubbles were observed in any condition in the straight microchannels; however, an irregular MMPs distribution was observed in the 100-μm-thick microchannel case (Figure 5(a)), where MMPs were more concentrated at the center of the microchannel. This occurred because the mixture was injected through the openings at both ends of the microchannel, merging in the center. When the gap between the PDMS cover layer and the micromold substrate was limited, MMPs accumulated and concentrated at the center. In the 200-μm-thick microchannel condition (Figure 5(b)), a regular distribution of MMPs was observed because the wider gap between the PDMS cover layer and SU-8 substrate allowed the MMPs to move more freely during the casting process. An irregular MMPs distribution was occasionally observed under the high aspect ratio conditions (i.e. 50). This irregular distribution can result in an uneven distribution of magnetic bonding

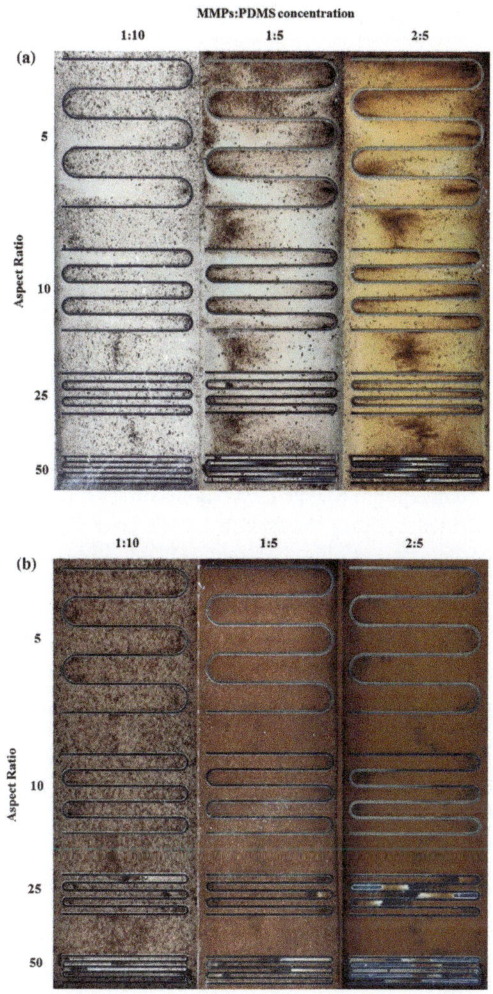

Figure 6. Clear-view MMPs-PDMS casting of (a) 100-µm-thick and (b) 200-µm-thick serpentine microchannels with composite concentrations of 1:10, 1:5, and 2:5 at aspect ratios of 5, 10, 25, and 50.

strength, causing local weak bonding, which can create local leakage during microfluidic operation.

3.1.3. Clear-view serpentine microchannel casting

In microfluidic applications, serpentine or U-shaped microchannels are a commonly used to minimize a device's footprint. Figure 6(a) shows the 100-µm-thick serpentine microchannel MMPs-PDMS casting results from this study. An irregular MMPs distribution was observed at all 1:10, 1:5, and 2:5 concentrations. During the vacuum casting procedure, the MMPs-PDMS composite filled the opening of the U-shaped cavities, causing the MMPs to accumulate at the U-shaped ends, filling the entire height of the 100-µm-thick microchannel. Compared with the 200-µm-thick microchannel condition (Figure 6(b)), the MMPs were more evenly distributed, indicating that the MMPs were less geometrically constrained during casting when taller microchannels were used.

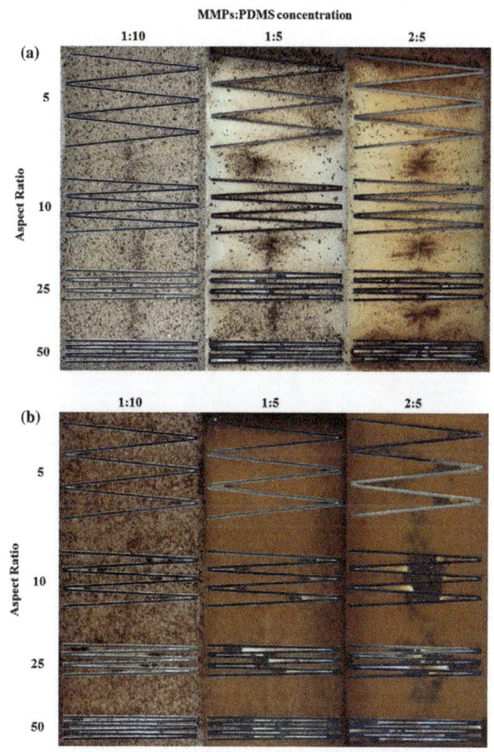

Figure 7. Clear-view MMPs-PDMS casting of (a) 100-µm-thick and (b) 200-µm-thick zigzag microchannels with composite concentrations of 1:10, 1:5, and 2:5 at aspect ratios of 5, 10, 25, and 50.

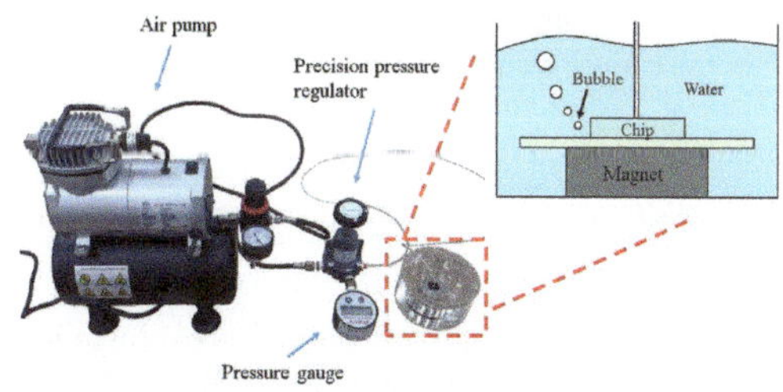

Figure 8. Schematic illustration of bonding experiment setup.

Trapped air was observed in the serpentine microchannel layout. During the casting process, air in the enclosed U-shaped pocket could not be expelled effectively, resulting in air bubbles forming in the high-aspect-ratio MMPs-PDMS layers after curing. As shown in Figure 6, trapped air was observed in the U-shaped section of the layers with aspect ratios of 25 and 50.

3.1.4. Clear-view zigzag microchannel casting

Microfluidic devices with zigzag microchannels feature sharp edges or turns in the microchannel design. The zigzag microchannel casting results (Figure 7)

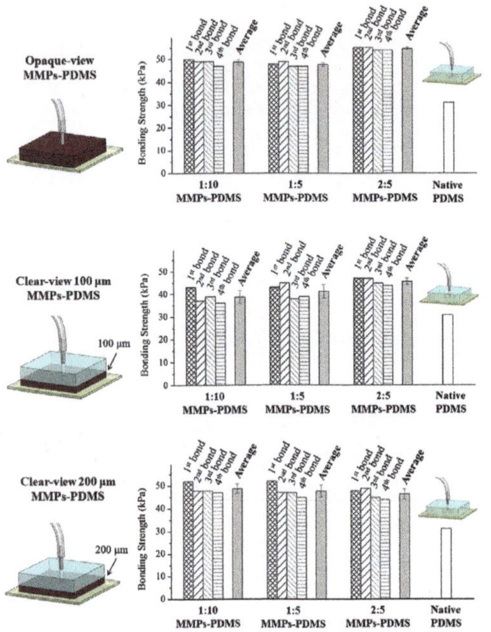

Figure 9. Bonding strength of the MMPs-PDMS microchannel.

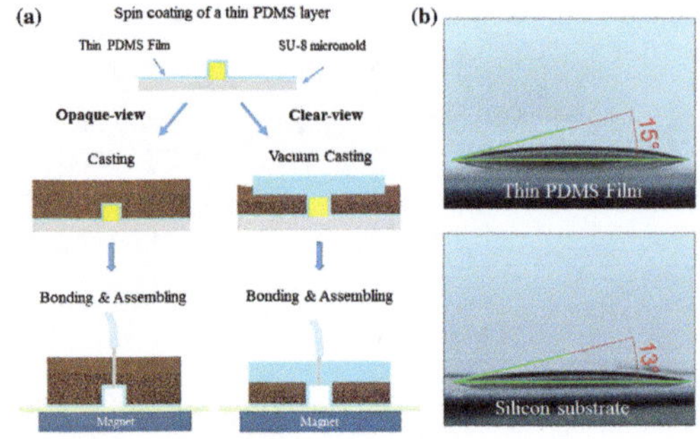

Figure 10. (a) MMPs-PDMS casting procedure with thin PDMS film. (b) Surface wettability of a 2:5 concentration MMPs-PDMS composite droplet on a thin PDMS film and on a silicon substrate.

are similar to those for the serpentine microchannel design; the MMPs tended to accumulate in the zigzag corners in the 100-μm-thick microchannel (Figure 7(a)), whereas the distribution was more even in the 200-μm-thick microchannel (Figure 7(b)). During the casting process, the sharp zigzag edges were more difficult to expel air from compared with the straight and U-shaped serpentine microchannels. Trapped air bubbles were more frequently observed in the high-aspect ratios of 25 and 50.

3.2. MMPs-PDMS reversible bonding

Figure 8 shows the reversible bonding test experiment setup. The magnetically assembled MMPs-PDMS microfluidic device was placed in a beaker filled with water. An air pump with a precision pressure regulator (Type 400, Control Air, Inc., Amherst, NH, USA) and pressure gauge (DPG-M2.5, Atlantis Ltd, Taipei, Taiwan) was connected to the MMPs-PDMS microfluidic device through a 5-mm-diameter enclosed microfluidic chamber designed for the bonding test. The air pump pressure was initially set at 0 kPa and increased by 2–3 kPa every 5 s. The values on the pressure gauge were recorded as the maximal bonding strength while air bubbles were leaking out of the microchannel. The MMPs-PDMS and glass surfaces were carefully cleaned and dried to ensure that no liquid film or surface contaminated the interface before each reversible bonding test.

Figure 9 shows the average and reversible bonding strength of the opaque- and clear-view MMPs-PDMS composites with concentrations of 1:10, 1:5, and 2:5. The maximum MMPs-PDMS concentration of 2:5 was selected because the MMPs-PDMS layer cracked at higher concentrations. The native PDMS-glass bonding strength was 31 kPa, which is comparable with another study (35 kPa) [15]. All bonding strength measurements of the opaque-view MMPs-PDMS were between 47 and 55 kPa. In the clear-view reversible bonding, the bonding strength was measured between 38 and 46 kPa for the 100-μm-thick microchannel and between 48 and 55 kPa for the 200-μm-thick microchannel. From the reversible bonding results (Bonds 1 to 4), as shown in Figure 9, the magnetic reversible bonding process was highly reversible because the MMPs-PDMS bonding strength was primarily created by the van der Waals adhesion from the MMPs-PDMS/glass interface and the magnetic force from the MMPs embedded in the MMPs-PDMS layer. However, we still observed an average 2.7% (maximum 13.9%) reduction trend, which might have been caused by surface damage or contamination at the bonding interface after each reversible bond, despite the surface cleaning and drying between each test.

According to the MMPs-PDMS bonding results (Figure 9), the bonding strength was enhanced only approximately 0.5-fold, and few correlations were observed between the bonding strength and MMPs concentrations. For example, for the 200-μm-thick clear-view reversible bonding (Figure 9), we observed a decrease in bonding strength when the MMPs concentration was increased. We attributed the weak correlations between the bonding strength and process parameters and the weak bonding strength enhancement to the rough surface at the MMPs-PDMS/glass substrate interface. Our MMPs-PDMS casting experiments revealed that the MMPs precipitated at the bottom layer during the vacuum casting procedure, and the MMPs accumulating in the bottom layer affected the flatness of the MMPs-PDMS surface. During the reversible MMPs-PDMS magnetic bonding process, the MMPs-PDMS layer was elastically deformed by magnetic forces and became attached to the glass substrate,

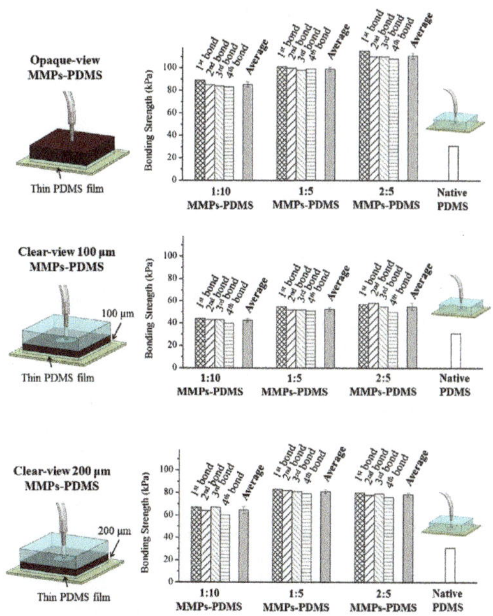

Figure 11. Bonding strength of MMPs-PDMS microchannel with thin PDMS film coating.

achieving conformity at the bonding interface. When the MMPs-PDMS surface was not adequately flat, the bonding strength from the magnetic force was reduced, resulting in an overall reduction in bonding strength.

To solve this problem, we spin-coated (at 1800 rpm for 40 s) a 50-μm-thick PDMS layer onto the SU-8 micromold substrate prior to MMPs-PDMS casting, thereby minimizing the roughness of the MMPs-PDMS bottom surface. Figure 10(a) shows the opaque- and clear-view MMPs-PDMS casting procedures with the thin PDMS films. The thin PDMS film-coated MMPs-PDMS casting, bonding, and assembling procedures are analogous to the previously described MMPs-PDMS casting procedure. The only difference is that a thin PDMS film was present at the bottom of the micromold during the casting procedure. Figure 10(b) shows the surface wettability of the 2:5 MMPs-PDMS composite droplet on a thin PDMS film and on a silicon substrate. The MMPs-PDMS composite contact angles were 15° on the PDMS thin film and 13° on the silicon substrate. These results show that the surface wettability of the silicon or PDMS film-coated silicon substrates had a limited effect during the casting procedure.

Figure 11 shows the average and reversible bonding strength of the MMPs-PDMS layers with the thin PDMS films. The opaque-view MMPs-PDMS bonding strength was measured at 85.3 ± 2.4 kPa (1:10), 99.0 ± 1.7 kPa (1:5), and 110.8 ± 2.8 kPa (2:5). The bond strength increased with the MMPs-PDMS composite concentration, and the maximum bonding strength enhancement was approximately 3.5-fold that of the bonding of native PDMS. For the clear-view bonding experiments, the bonding strength was 42.5 ± 1.6 kPa (1:10), 52.5 ± 1.6 kPa (1:5), and 55.0 ± 3.3 kPa (2:5) for the 100-μm-thick clear-view MMPs-PDMS layer and 64.5 ± 3.1 kPa (1:10), 81.0 ± 1.7 kPa (1:5), and 78.3 ± 1.6 kPa (2:5) for the 200-μm-thick clear-view MMPs-PDMS layer. The 200-μm-thick microchannels exhibited a greater enhancement in bonding strength compared with the 100-μm-thick microchannels because the active magnetic MMPs-PDMS layer was thicker and more uniformly distributed in the PDMS layer. The maximum bonding strength enhancement of the clear-view bonding process was approximately 2.8-fold that of the native PDMS bonding. Good bonding reversibility was also observed in the thin PDMS-coated MMPs-PDMS device. On average, the bonding strength exhibited a 2.4% reduction trend (maximum 10.4%) with each reversible bonding test.

The bonding strength of MMPs-PDMS can effectively be improved by using thin PDMS films. Figure 12 shows measurements taken with a surface profilometer (Dektak XT, Bruker, Billerica, MA, USA) of native PDMS (Figure 12(a)), a 2:5 MMPs-PDMS layer without the thin PDMS film (Figure 12(b)) and with the thin PDMS film (Figure 12(c)). The native PDMS surface was the smoothest surface (average roughness: 308 Å), and the 2:5 MMPs-PDMS layer exhibited a considerably rougher surface (average roughness: 16456 Å). With a thin PDMS film covering the 2:5 MMPs-PDMS layer, the surface became flatter (average roughness: 1397 Å). These results show

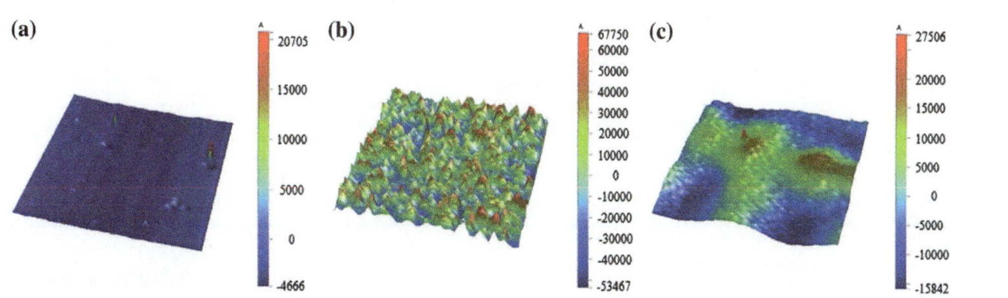

Figure 12. Surface roughness measurement of 1 × 1 mm (a) native PDMS, (b) 2:5 concentration MMPs-PDMS layer, and (c) 2:5 concentration MMPs-PDMS layer with thin PDMS film on the bottom.

that the surface roughness at the MMPs-PDMS/glass bonding interface was critical to the overall bonding strength. The MMPs-PDMS layer without the PDMS thin film (Figure 9) exhibited high surface roughness, which explains why the MMPs-PDMS layer without the thin PDMS film attained a bonding strength enhancement of only 50% compared with the layer with the thin film (in presence of magnetic forces). Although an additional thin PDMS film on the MMPs-PDMS surface can minimize the roughness effect and enhance the reversible MMPs-PDMS bonding strength, the effects from variation in the surface roughness persisted. The surface of the thin PDMS-coated MMPs-PDMS layer remained rougher than the native PDMS. Therefore, there is an optimal bonding strength for the 200-μm-thick 1:5 concentration clear-view MMPs-PDMS layer (81.0 kPa), and for the 2:5 concentration opaque-view MMPs-PDMS layer (110.8 kPa).

4. Conclusions

An Fe_3O_4 magnetic MMPs-PDMS composite was successfully used to demonstrate the effectiveness of a simple high-strength reversible magnetic bonding method. In this study, the proposed method was used to cast opaque-view (where an optical inspection is impossible) and clear-view (where an optical inspection is possible) MMPs-PDMS structures. The results show that microchannel geometries have limited influence on opaque-view casting because the MMPs-PDMS composite can freely fill the micromold cavities to remove air bubbles, which is similar to standard PDMS casting. On the other hand, clear-view casting performance was closely related to the microchannel geometries. The geometrical constraints of the microchannel layout as well as the gap between the PDMS cover layer and the micromold substrate caused air bubbles to become trapped or the MMPs distribution to become irregular at high-aspect ratios.

MMPs-PDMS magnetic bonding experiments showed that MMPs-PDMS composites facilitated a highly reversible bonding process because the MMPs-PDMS bonding strength was achieved through van der Waals adhesion and magnetic forces. A thin PDMS film-coated MMPs-PDMS layer can effectively reduce the surface roughness and enhance MMPs-PDMS magnetic bonding strength. The thin PDMS film-coated opaque-view MMPs-PDMS device attained the maximum bonding strength of 110 kPa in an MMPs-PDMS composite with a 2:5 concentration. For the clear-view MMPs-PDMS devices with a thin PDMS film, a reversible magnetic bonding strength as high as 80 kPa was achieved in the 1:5 and 2:5 MMPs-PDMS composites while providing an optically inspectable microchannel. We have successfully tested the MMPs-PDMS device in microfluidic cell culture application and the MMPs-PDMS layer may also be used as magnetic-based actuator as future work.

Disclosure statement

No potential conflict of interest was reported by the authors.

Funding

The authors thank the Ministry of Science and Technology, Taiwan, for financially supporting this project [grant numbers MOST 103-2221-E-008-027, MOST 104-2221-E-008-080].

Notes on contributors

Chia-Wen Tsao is Associate Professor in Department of Mechanical Engineering, in National Central University, Taoyuan, Taiwan. He obtained his Ph.D. degree in Department of Mechanical Engineering, University of Maryland at College Park, USA in 2008. After graduation, he join Department of Mechanical Engineering, National Central University, Taiwan as Assistant Professor in 2008. His research interest includes polymer and silicon micro/nano-fabrication, micro/nano-fluidics, mass spectrometry analysis and interfacing, and lab-on-chips devices.

Yueh-Pu Lee is master degree graduate school student in Department of Mechanical Engineering, in National Central University, Taoyuan, Taiwan. He received his master degree for National Central University in 2015. His research topic is polymer microfabrication process.

References

[1] Livak-Dahl E, Sinn I, Burns M. Microfluidic chemical analysis systems. Annu Rev Chem Biomol. 2011;2:325–353.
[2] Weibel DB, Whitesides GM. Applications of microfluidics in chemical biology. Curr Opin Chem Biol. 2006;10:584–591.
[3] Jakeway SC, de Mello AJ, Russell EL. Miniaturized total analysis systems for biological analysis. Fresen J Anal Chem. 2000;366:525–539.
[4] Alrifaiy A, Lindahl OA, Ramser K. Polymer-based microfluidic devices for pharmacy, biology and tissue engineering. Polymers-Basel. 2012;4:1349–1398.
[5] Nan L, Jiang ZD, Wei XY. Emerging microfluidic devices for cell lysis: a review. Lab Chip. 2014;14:1060–1073.
[6] Iliescu C. Microfluidics in glass: Technologies and applications. Inform Midem. 2006;36:204–211.
[7] Iliescu C, Taylor H, Avram M, et al. A practical guide for the fabrication of microfluidic devices using glass and silicon. Biomicrofluidics. 2012;6, 016505.
[8] Becker H, Locascio LE. Polymer microfluidic devices. Talanta. 2002;56:267–287.
[9] Fiorini GS, Chiu DT. Disposable microfluidic devices: fabrication, function, and application. BioTechniques. 2005;38:429–446.
[10] Unger MA, Chou HP, Thorsen T, et al. Monolithic microfabricated valves and pumps by multilayer soft lithography. Science. 2000;288:113–116.
[11] Jeon JS, Chung S, Kamm RD, et al. Hot embossing for fabrication of a microfluidic 3D cell culture platform. Biomed Microdevices. 2011;13:325–333.

[12] Attia UM, Marson S, Alcock JR. Micro-injection moulding of polymer microfluidic devices. Microfluid Nanofluid. 2009;7:1-28.

[13] Sia SK, Whitesides GM. Microfluidic devices fabricated in Poly(dimethylsiloxane) for biological studies. Electrophoresis. 2003;24:3563-3576.

[14] Tsao CW, DeVoe DL. Bonding of thermoplastic polymer microfluidics. Microfluid Nanofluid. 2009;6:1-16.

[15] McDonald JC, Whitesides GM. Poly(dimethylsiloxane) as a material for fabricating microfluidic devices. Accounts Chem Res. 2002;35:491-499.

[16] Lin CH, Chao CH, Lan CW. Low azeotropic solvent for bonding of PMMA microfluidic devices. Sensor Actuat B-Chem. 2007;121:698-705.

[17] Brown L, Koerner T, Horton JH, et al. Fabrication and characterization of poly(methylmethacrylate) microfluidic devices bonded using surface modifications and solvents. Lab Chip. 2006;6:66-73.

[18] Laher M, Hild S. A detailed micrometer scale investigation of the solvent bonding process for microfluidic chip fabrication. Rsc Adv. 2014;4:5371-5381.

[19] Tsao CW, Hromada L, Liu J, et al. Low temperature bonding of PMMA and COC microfluidic substrates using UV/ozone surface treatment. Lab Chip. 2007;7:499-505.

[20] Tennico YH, Koesdjojo MT, Kondo S, et al. Surface modification-assisted bonding of polymer-based microfluidic devices. Sensor Actuat B-Chem. 2010;143:799-804.

[21] Bhattacharya S, Datta A, Berg JM, Gangopadhyay S. Studies on surface wettability of poly(dimethyl) siloxane (PDMS) and glass under oxygen-plasma treatment and correlation with bond strength. J Microelectromech S. 2005;14:590-597.

[22] Eddings MA, Johnson MA, Gale BK. Determining the optimal PDMS-PDMS bonding technique for microfluidic devices. J Micromech Microeng. 2008;18, 067001.

[23] Kim P, Suh KY. Rigiflex, spontaneously wettable polymeric mold for forming reversibly bonded nanocapillaries. Langmuir. 2007;23:4549-4553.

[24] Lee H, Lee BP, Messersmith PB. A reversible wet/dry adhesive inspired by mussels and geckos. Nature. 2007;448:338-341.

[25] Chen Q, Li G, Nie Y, et al. Investigation and improvement of reversible microfluidic devices based on glass-PDMS-glass sandwich configuration. Microfluid Nanofluid. 2014;16:83-90.

[26] Le Berre M, Crozatier C, Casquillas GV, et al. Reversible assembling of microfluidic devices by aspiration. Microelectron Eng. 2006;83:1284-1287.

[27] Crozatier C, Le Berre M, Chen Y. Multi-colour micro-contact printing based on microfluidic network inking. Microelectron Eng. 2006;83:910-913.

[28] Rafat M, Raad DR, Rowat AC, et al. Fabrication of reversibly adhesive fluidic devices using magnetism. Lab Chip. 2009;9:3016-3019.

[29] Rasponi M, Piraino F, Sadr N, et al. Reliable magnetic reversible assembly of complex microfluidic devices: fabrication, characterization, and biological validation. Microfluid Nanofluid. 2011;10:1097-1107.

[30] Chen CT, Chen YC. Fe3O4/TiO2 core/shell nanoparticles as affinity probes for the analysis of phosphopeptides using TiO2 surface-assisted laser desorption/ionization mass spectrometry. Anal Chem. 2005;77:5912-5919.

[31] Xia YN, Whitesides GM. Soft Lithography. Angew Chem Int Edit. 1998;37:550-575.

[32] Dang F, Shinohara S, Tabata O, et al. Replica multichannel polymer chips with a network of sacrificial channels sealed by adhesive printing method. Lab Chip. 2005;5:472-478.

19

Room-temperature synthesis of three-dimensional porous ZnO@CuNi hybrid magnetic layers with photoluminescent and photocatalytic properties

Miguel Guerrero[a‡], Jin Zhang[a‡], Ainhoa Altube[b], Eva García-Lecina[b], Mònica Roldan[c], Maria Dolors Baró[a], Eva Pellicer[a] and Jordi Sort[d]

[a]Departament de Física, Facultat de Ciències, Universitat Autònoma de Barcelona, E-08193 Bellaterra, Spain
[b]Surfaces Division, IK4-CIDETEC, Paseo Miramón, 196, E-20009 San Sebastián, Spain
[c]Servei de Microscòpia, Universitat Autònoma de Barcelona, E-08193 Bellaterra, Spain
[d]Institució Catalana de Recerca i Estudis Avançats (ICREA) and Departament de Física, Universitat Autònoma de Barcelona, E-08193 Bellaterra, Spain

ABSTRACT
A facile synthetic approach to prepare porous ZnO@CuNi hybrid films is presented. Initially, magnetic CuNi porous layers (consisting of phase separated CuNi alloys) are successfully grown by electrodeposition at different current densities using H_2 bubbles as a dynamic template to generate the porosity. The porous CuNi alloys serve as parent scaffolds to be subsequently filled with a solution containing ZnO nanoparticles previously synthesized by sol-gel. The dispersed nanoparticles are deposited dropwise onto the CuNi frameworks and the solvent is left to evaporate while the nanoparticles impregnate the interior of the pores, rendering ZnO-coated CuNi 3D porous structures. No thermal annealing is required to obtain the porous films. The synthesized hybrid porous layers exhibit an interesting combination of tunable ferromagnetic and photoluminescent properties. In addition, the aqueous photocatalytic activity of the composite is studied under UV–visible light irradiation for the degradation of Rhodamine B. The proposed method represents a fast and inexpensive approach towards the implementation of devices based on metal-semiconductor porous systems, avoiding the use of post-synthesis heat treatment steps which could cause deleterious oxidation of the metallic counterpart, as well as collapse of the porous structure and loss of the ferromagnetic properties.

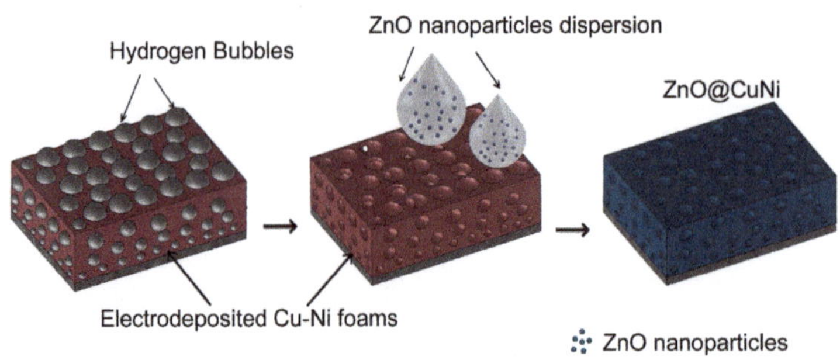

KEYWORDS
Porous nanocomposite; CuNi alloy; ZnO nanoparticles; ferromagnetism; photoluminescence

CLASSIFICATION
40 Optical, magnetic and electronic device materials; 102 Porous / Nanoporous / Nanostructured materials; 100 Materials; 103 Composites; 205 Catalyst / Photocatalyst / Photosynthesis; 200 Applications; 203 Magnetics / Spintronics / Superconductors

1. Introduction

The advances in the synthetic pathways to produce porous semiconducting materials (PSM) with controllable pore size and composition have boosted a wealth of applications in a number of different fields such as solar fuel production,[1] gas-sensing,[2,3] or environmental healing,[4,5] where materials with a high surface area are essential. Different PSM architectures including particles, rods, wires and films (the latter with porosity ranging from the macro- to the nanoscale) have been engineered in recent years.[6] Depending on the end application, PSMs have been manufactured either alone or in combination with other materials.

In general, the ability to use three-dimensional (3D) porous scaffolds as parent templates to be filled or coated with second phase materials (e.g. oxide semiconductors)

CONTACT Miguel Guerrero Miguel.Guerrero@uab.cat; Eva Pellicer Eva.Pellicer@uab.cat
‡These authors have equally contributed to this work.

has been shown to trigger new functionalities stemming from the synergy created between the host and the guest (or the matrix and the filler) components.[7] For example, in the case of metal-oxide ceramic composites, one can take advantage of the large hardness of the ceramic counterpart and the high ductility of the metal to design microstructures with optimized mechanical performance.[8,9] Magnetic exchange-interactions in certain metal-oxide nanocomposites have also triggered interesting applications in the field of nanomagnetism, as a means to enhance the coercivity of the resulting material.[10]

One of the main challenges to produce hybrid composite materials using polymeric or metallic porous scaffolds as templates, particularly when the second phase is intended to be an oxide, is that a high-temperature heat-treatment, often in the presence of an O_2-containing atmosphere, is required.[11] Such treatment can cause carburization of the polymer or oxidation of the parent metallic template, and can eventually provoke the collapse of the porous network. These issues need to be tackled when growing a metal oxide as the second phase by atomic layer deposition or colloidal templating (which typically involves subsequent heat-treatment in air).[11] Thus, new methodologies to fill or coat porous templates with suitable second-phase materials at temperatures close to room temperature are required.

Zinc oxide (ZnO), one of the most important n-type II-VI semiconductor materials with a wide band gap of 3.2 eV, possesses unique electrical and optical properties including good transparency, high electron mobility, and strong room-temperature luminescence.[12] Therefore, ZnO has recently attracted enormous attention due to its potential applications in integrated optoelectronics and laser technology[13] or photocatalysis.[14,15] Additionally, the piezo/pyroelectric properties of ZnO make it a smart candidate for sensors and other types of micro/nano-electromechanical systems (MEMS/NEMS).[16, 17] There are several synthetic techniques for the growth of ZnO-based materials. These include vapor deposition,[18] hydrothermal synthesis,[19] the sol-gel method,[20] or mechanochemical approaches.[21] Depending on the deposition process conditions, different structural, electrical and optical properties are obtained. Pulsed laser deposition,[22] chemical vapor deposition[23] or spray pyrolysis[24] can be used to prepare dense ZnO immobilized films. The synthesis of porous ZnO films is typically more challenging. Dip-coating methods of sol-gel precursors are receiving attention since they enable the low-cost deposition of ZnO films, eventually in porous form.[25, 26] Other approaches to synthesize porous metal oxide films on a substrate include spin-coating of sol-gel[27], template-assisted electrostatic spray deposition[28] and hydrothermal methods.[29] Both dip-coating and spin-coating of sol-gel mixtures show significant limitations when they are faced with large substrates and/or substrates with micrometer-sized features. Moreover, film thickness typically achieved in a single coating step is in the range of about 100–400 nm depending on coating conditions, whereas a film thicker than 1 μm can be obtained only by repeated coating/calcination cycles.[28] For the precise control of the film thickness, pore size and porosity degree, the use of pre-fabricated polymeric or metallic porous templates is often preferred. Nevertheless, in most cases, one or more post-synthesis heat treatment steps at temperatures ranging from 250°C to 900°C are necessary to convert the solution precursors into ZnO.[30, 31] As aforementioned, this can be a serious shortcoming when the porous scaffold is not intended to be eliminated but to remain unaffected (and hence keep the initial features like the composition, overall porosity and physical properties) after the heat treatment.

In this work, a novel approach to coat the interior of the pores of a CuNi 3D porous film scaffold with a ZnO shell at room temperature is presented. CuNi films are obtained by hydrogen-bubble assisted electrodeposition, which has proven to be a convenient technique to produce networked macroporous films of some compositions.[32] The Cu/Ni ratio is adjusted in order to deliver room-temperature ferromagnetic properties, which are not typically available in single-phase semiconductor materials. Thus, the obtained hybrid porous composites maintain the magnetic properties of the CuNi foams while exhibiting new functionalities (e.g. photoluminescence, photocatalytic activity) stemming from the presence of the ZnO semiconducting second phase. In particular, the aqueous photocatalytic activity of this composite system under UV-visible light irradiation for the degradation of Rhodamine B (RhB) is demonstrated. The multifunctional character of this composite material is possible because (i) the macro-porosity of the initial CuNi alloy foam is preserved and (ii) oxidation of the CuNi parent template is also avoided because of the low temperature used in all the synthetic steps.

2. Experimental section

2.1. Synthesis of the CuNi metallic foams

Two-phase CuNi metallic foams (MF) were successfully obtained by electrodeposition at different current densities on Ti/Si/Au substrates using H_2 as a dynamic template (Figure 1(a) and 1(b)). All chemicals were of analytical grade and were used as received without further purification. Electrolyte solution was prepared from ultrapure water. Si/Ti (25 nm)/Au (125 nm) substrates (working area of 0.25 cm^2) were employed as cathodes. Given the low roughness of the Au surface, surface conditioning (e.g. polishing) was not required. All depositions were carried out in

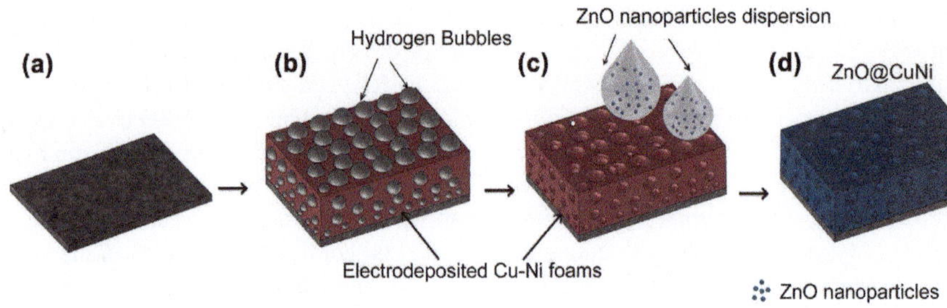

Figure 1. Schematic flow of the synthesis of the ZnO@CuNi hybrid porous layers. From (a) to (b) hydrogen bubble-assisted electrodeposition of porous CuNi MF and from (c) to (d) impregnation of the CuNi scaffold with ZnO NPs suspension.

a single-compartment, double-jacketed glass electrochemical cell. A platinum sheet served as the counter electrode, and a double junction Ag|AgCl 3 M KCl electrode (E = +0.210 V versus standard hydrogen electrode (SHE)) was utilized as the reference electrode. For CuNi MF electrodeposition, a constant current density of −1 A cm^{-2} and 300 s ($Cu_{80}Ni_{20}$) or −3 A cm^{-2} and 100 s ($Cu_{65}Ni_{35}$) was applied by using a PGSTAT302 N Autolab potentiostat/ galvanostat (Ecochemie, Utrecht, Netherlands). The deposition charge for the growth of CuNi MFs was kept constant in order to achieve a similar thickness (ca. 60 μm). The electrolyte consisted of $CuSO_4{\cdot}5H_2O$ (0.01 M), $NiSO_4{\cdot}6H_2O$ (0.15 M), H_2SO_4 (1 M), CH_3COOH (0.1 M), HCl (50 mM), and sodium citrate tribasic dehydrate (0.2 M). All deposition processes were carried out at 25°C and under stirring (800 rpm) using a magnetic stirrer bar.

2.2. Synthesis of the ZnO nanoparticles

Into a flask containing 20 ml of methanol were added 0.090 g of zinc acetate ($[Zn(CH_3COO)_2]{\cdot}2H_2O$ Aldrich, 98%) and 25 μl of deionized water. The solution was heated to 60°C under magnetic stirring. Later, a solution of 0.050 g of potassium hydroxide (KOH, Aldrich, 85%) and 20 ml of methanol heated to 60°C was dropwise added to the zinc solution. During this step, ZnO nanoparticles (NPs) were formed by sol-gel (ZnO NPs mean size = 7.3 nm). When the reaction was finished (24 h), the upper fraction of the reaction solution was withdrawn. Next, 40 ml of methanol were added to the solution and stirred for 15 min. Again, the upper fraction of the solution was discarded after 15 min. The methanol washing process was repeated twice. The isolated ZnO NPs were suspended in 40 ml of methanol, leading to a concentration of ca. 0.83 g ZnO l^{-1}.

2.3. Synthesis of the hybrid ZnO@CuNi porous film

As shown in Figure 1(c) and 1(d), 15 μl of the previously synthesized ZnO NPs suspension were dropwise added onto the surface of the CuNi MF. Next, the impregnated CuNi MF was placed inside a vacuum desiccator (0.5 mbar) for 120 s at room temperature. The impregnation process was consecutively repeated six times to yield ZnO@CuNi porous films with increasingly higher ZnO coverage. Samples were investigated after three and six consecutive impregnations (*middle* and *final* coverage, respectively). Samples were done in triplicate in order to ensure reproducible results.

2.4. Morphological and structural characterization

The morphology and structure of the porous films were studied by field emission scanning electron microscopy (FE-SEM), high resolution transmission electron microscopy (HRTEM) and X-ray diffraction (XRD). HRTEM characterization was carried out on a JEOL JEM-2011 microscope operated at 200 kV. For HRTEM observations, a portion of the film was scratched off with a metallic spatula and the powder released was dispersed in ethanol. The dispersion was then sonicated on a SELECTA Ultrasounds 230 V – 40 kHz apparatus (JP Selecta, Barcelona, Spain) for 10 s in order to break up big aggregates. Afterwards, one or two drops of the suspension were placed dropwise onto a holey carbon-coated Cu TEM grid. The chemical composition of the films was determined by energy dispersive X-ray spectroscopy (EDX). The stoichiometry of the Cu-Ni MFs is given in at%. XRD patterns were recorded on a Philips X'Pert diffractometer (Panalytical, Madrid, Spain) using the Cu K_α radiation, in the 20–90° 2θ range (0.03° step size, 8 s holding time). A full-pattern fitting procedure (Rietveld method, MAUD software)[33] was used to extract microstructural parameters from the XRD patterns.

2.5. Magnetic characterization

Room temperature hysteresis loops were collected at room temperature using a vibrating sample magnetometer (VSM) from Oxford Instruments (Abingdon), with a maximum applied magnetic field of 0.3 T. The field was applied parallel to the film plane.

2.6. *Confocal imaging*

The photoluminescence (PL) properties of the films were studied by confocal scanning laser microscopy (CSLM). The samples were mounted on Ibidi culture dishes (Ibidi GmbH, Martinsried, Germany). For 3D analysis, the samples were visualized with a TCS-SP5 CSLM microscope (Leica Microsystems CMS GmbH, Mannheim, Germany) using a Plan Apochromat (Leica, Mannhein, Germany) 10× / 0.4 (dry) objective. The ZnO@CuNi nanocomposites were excited with a blue diode laser (405 nm) and the autofluorescence was detected in the 420–580 nm range. Because the CuNi MF matrices are non-fluorescent, they were imaged using reflected light mode from an argon laser (488 nm) and subsequently detected in the 480–495 nm range. Stacks of 50 sections were collected at every 3 μm along the material's thickness. The three-dimensional images were processed by using Surpass Module in Imaris X64 v. 6.2.0 software (Bitplane; Zürich, Switzerland).

The emission signal of the specimens was excited at 405 nm with a blue diode laser, and its fluorescence intensity was recorded generating a lambda stack (69 images) with emission wavelength ranging from 425 to 750 nm, at 10 nm intervals. A set of 10 regions of interest (ROIs) of 400 μm^2 each was used to analyze the mean fluorescence intensity (MFI) of the samples in relation to emission wavelength. For each ROI, a graph plotting mean pixel intensity and emission wavelength of the lambda stack was generated.

Finally, photostability experiments were performed to monitor long time-lapse experiments. A defined region of interest (ROI: 620 μm × 620 μm) was photo-bleached at full laser power of 50 mW of blue diode 405 nm (100% power, 100% transmission). Fluorescence signal was detected in the range from 420 nm to 580 nm. Confocal time series were recorded with intervals of one frame / 850 ms for a period of 10 min. Data from all studies were analyzed using the LAS AF software 2.4.1 (Leica Microsystems) obtaining the variation of the MFI (normalized fluorescence values) as a function of irradiation time.

2.7. *Evaluation of photocatalytic activity*

The photocatalytic activity of the uncoated CuNi and hybrid ZnO@CuNi porous layers was evaluated by decolorization of a 5 ppm RhB aqueous solution (Alfa Aesar, Karlsruhe, Germany, 98%, without further treatment). The solutions were prepared by introducing the layers to be tested in 10 ml of the RhB solution. For each experiment, a blank RhB solution was used as a control and reference test. The reaction cells were placed in a SwiftCure IB irradiation cabin (Peschl Ultraviolet GmbH from Mainz, Germany) equipped with a mercury lamp. Cut-off filters were used to limit the wavelength radiation and to avoid direct photolysis of the dye (ISO 10678:2010 standard). The average light intensity used was 220 W and the wavelength ranged from 320 to 500 nm. The photocatalytic experiments were conducted under continuous magnetic stirring at a constant temperature of 29 °C.

After conditioning the suspensions for 60 min in dark to reach adsorption-desorption equilibrium, the light was turned on to initiate the reaction. Experiments were then carried out under ultraviolet-visible (UV-vis) irradiation for an overall time of 180 min. The red color of the solution faded gradually with time due to the decomposition of RhB. Aliquots (4 ml) were withdrawn regularly (at 0, 15, 30, 60, 120, and 180 min) from the reaction. The supernatant solutions were then tested with a UV-vis spectrophotometer (Shimadzu UV-1603, Shimadzu Corporation, Kyoto, Japan) by measuring absorption spectra of RhB (λ = 554 nm) as a function of the irradiation time. Following the measurement, the aliquot was immediately returned to the reaction vessel to continue the reaction. Thus, the final volume marginally decreased (approximately 5%). Photocatalytic activity of the catalysts was calculated as (C/C_0), where C_0 is the concentration of the test solution of RhB before irradiation and C is the concentration of RhB after UV-vis irradiation.

3. Results and discussion

3.1. *Morphological and structural characteristics of the porous layers*

XRD analyses were carried out to investigate the crystallographic structure of the prepared films. XRD patterns in the 30–60° 2θ range of $Cu_{65}Ni_{35}$ (fabricated with j = -3 A cm^{-2}) and $Cu_{80}Ni_{20}$ (fabricated with j = -1 A cm^{-2}) samples before the ZnO deposition are shown in Figure 2. All the samples show four diffraction peaks corresponding to Cu/Ni (111) and Cu/Ni (200) reflections (# and ×, respectively), proving the occurrence of two face-centered cubic (fcc) structures, one enriched in Cu and the other in Ni. The occurrence of two-phase CuNi films was anticipated considering the electrodeposition conditions. Namely, in order to favor intensive hydrogen co-evolution during electrodeposition, a highly acidic electrolyte was employed. In these conditions, citrate has negligible complexing capacity because it exists mostly in its protonated form. As a result, the co-deposition of Cu and Ni cations (and hence, the formation of a solid solution) is severely hindered, which results in heterogeneous nucleation (and hence, the occurrence of phase-separated deposits). From the magnetic viewpoint, formation of phase-separated CuNi films is desirable since room-temperature ferromagnetic properties are

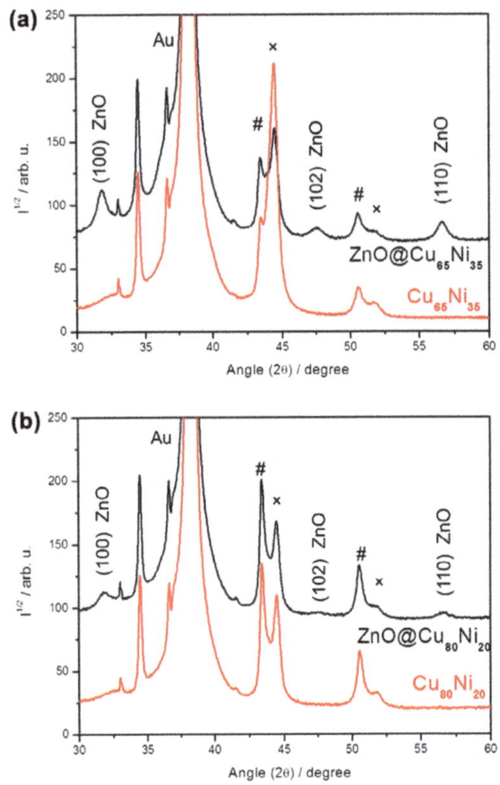

Figure 2. XRD patterns of (a) $Cu_{65}Ni_{35}$ (red) and ZnO@$Cu_{65}Ni_{35}$ (black) and (b) $Cu_{80}Ni_{20}$ (red) and ZnO@$Cu_{80}Ni_{20}$ (black). Peaks denoted by # and × belong to Cu-rich and Ni-rich phases, respectively. Two diffraction peaks at 33.06° and 34.36° belong to (100) Si and (100) α-Ti crystalline planes, respectively, of the substrate. Abbreviation: au = arbitrary units.

then accessible for almost any Ni content. If a single-phase alloy were formed, the ferromagnetic behavior would hold only for Ni percentages above 70 at%. Below this value, alloy layers would be paramagnetic.[34] The average crystallite size of the CuNi parent templates, determined using the MAUD software, ranges between 10 nm and 13 nm for both Cu-rich and Ni-rich phases. By applying the Bragg's and Vegard's laws to the (111) reflection, the amount of the Cu- and the Ni-rich phases was estimated to be 85 at% and 15 at%, respectively, for the $Cu_{80}Ni_{20}$ film, and 73 at% and 27 at%, respectively, for the $Cu_{65}Ni_{35}$ film. The precise stoichiometry of the Cu-rich and Ni-rich phases is given in the Supporting Information (Table S1). The XRD patterns of the hybrid ZnO@CuNi porous layers show, apart from the reflections coming from the metallic Cu-Ni, three additional diffraction peaks at 31.81°, 47.60°, and 56.69°, corresponding to (100), (102) and (110) crystalline planes, respectively, of ZnO in its wurtzite hexagonal phase (JCPDS 36–1451). These peaks are considerably broader, evidencing the nanocrystalline nature of the ZnO coating. Its crystallite size is around 7 nm, as estimated from the Rietveld refinements. Notice that the structural properties of the porous CuNi templates do not become altered after the coating with ZnO, as expected from the room temperature conditions used for ZnO deposition onto CuNi MF.

The evolution of surface morphology from the pristine CuNi MF film upon ZnO impregnation was studied by FE-SEM at different magnifications. As an example, representative FE-SEM images of the uncoated $Cu_{80}Ni_{20}$ MF and the corresponding film after being coated with ZnO (three and six impregnation steps or, in other words, *middle* and *final* coverages) are shown in Figure 3. At low magnification, a uniform distribution of 3D interconnected spherical pores of around 50 μm in diameter, was observed over the entire surface of the uncoated MF (Figure 3(a)). In close-up observations, the pore walls (about 20 μm in thickness) were found to be inherently porous (they are built up from numerous nanodendrites, see Figure S1), hence displaying a hierarchical porosity. The size of the 50 μm macropores directly correlates with the size of the H_2 bubbles. In the present work, acetic acid was added to the bath to prevent from bubble coalescence and, hence, formation of bigger pores. Instead, the porosity of the walls is likely due to the fast deposition rate. CuNi grains cannot stack in a dense manner (as would occur under lower deposition rates) and dendrites are formed. Due to the fractal architecture of such dendrites, numerous 'nano' voids are generated during growth.

Upon ZnO impregnation, the nanodendrites become less apparent and finally are no longer visible (Figure 3(b) and 3(c)). As a result, the pore walls show a smoother appearance. Nonetheless, the macro-porosity is preserved along the coating process. These macropores become smaller in size (from about 50 μm in $Cu_{80}Ni_{20}$ to about 30 μm in ZnO@$Cu_{80}Ni_{20}$) as expected from the increasingly thicker ZnO shell which conformally coats the CuNi network. Representative images of ZnO@$Cu_{65}Ni_{35}$ film are shown in Figure S2. In this case, the ZnO shell is less uniform than for ZnO@$Cu_{80}Ni_{20}$, which is in part due to the less regular morphology of the CuNi foams featuring larger Ni contents.[35] This is because there is strong dependence of the Ni content on the CuNi morphology. Cu deposition typically promotes the formation of a uniform foam-like architecture. With increasing the Ni content, pore interconnectivity worsens. EDX mapping images prove that Zn is well distributed among the CuNi MF (Figure S3). It is envisaged that the 3D interconnected macroporous structure of the ZnO@CuNi framework could be advantageous for catalytic applications. In particular, an enhanced surface diffusion of the reactant adatoms, ad-ions or molecules through the porous channels and subsequent desorption of the products could positively impact the kinetics of the reaction.

Figure 4 depicts HR-TEM images of the ZnO@$Cu_{80}Ni_{20}$ hybrid porous layer taken at the oxide-metal

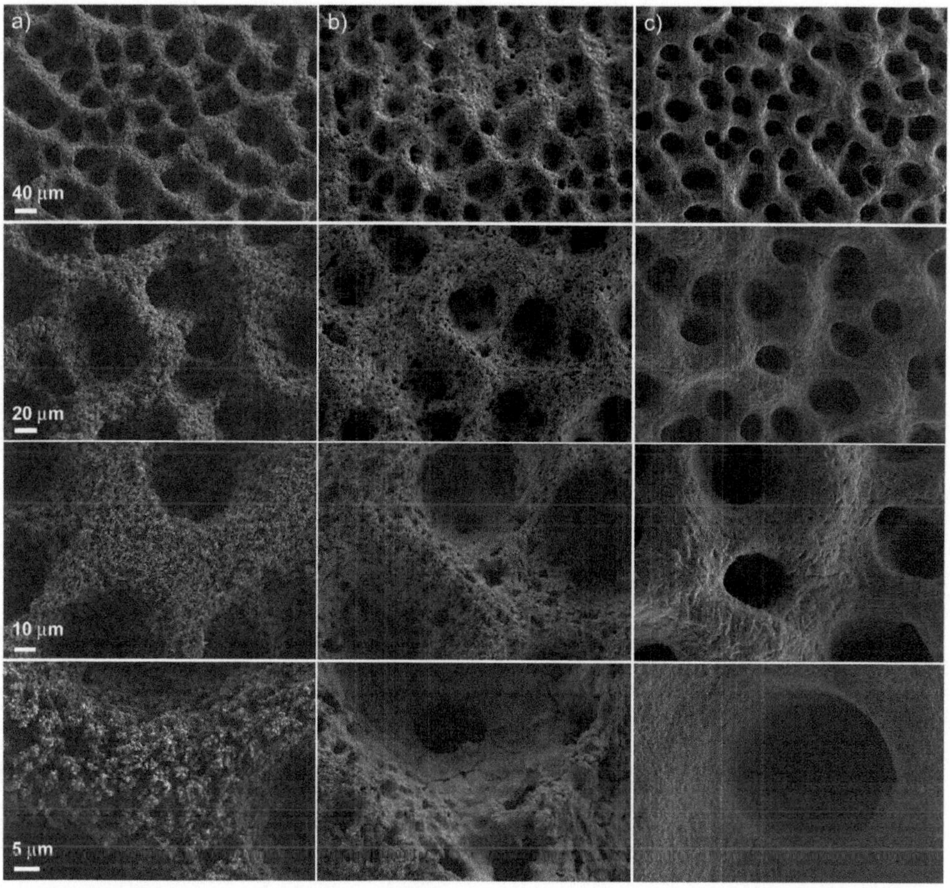

Figure 3. FE-SEM images of (a) the uncoated $Cu_{80}Ni_{20}$ MF; and the ZnO-coated $Cu_{80}Ni_{20}$ after (b) three consecutive impregnations (*middle* coverage) and (c) six consecutive impregnations (*final* coverage). In each column different magnifications of the three materials are shown.

edge. Figure 4(a) shows that the nanocomposite consists of a rather continuous shell of ZnO NPs well attached to the CuNi MF surface even after sonication of the sample (Figure S4). The selected area electron diffraction (SAED) pattern showed in Figure 4(b) displays spotty rings. Since the spot size was around 200 nm, the beam is actually embracing several small crystals with different orientations. This further confirms the nanocrystalline structure of the ZnO, in agreement with XRD analyses. Furthermore, in a zoomed detail of a single ZnO NP (Figure 4(c)) the lattice planes are sharply defined, thus confirming the crystallinity of the ZnO in the ZnO@$Cu_{80}Ni_{20}$ nanocomposite. Interestingly, this synthetic methodology preserves the morphology as well as the crystallinity of the as-prepared ZnO NPs (Figure 4(d)).

3.2. *Magnetic properties of the porous layers*

Well-defined hysteresis loops corresponding to the $Cu_{80}Ni_{20}$ MFs as well as ZnO@$Cu_{80}Ni_{20}$ are shown in Figure 5(a). It is known that CuNi single-phase alloys are ferromagnetic at room temperature for Ni

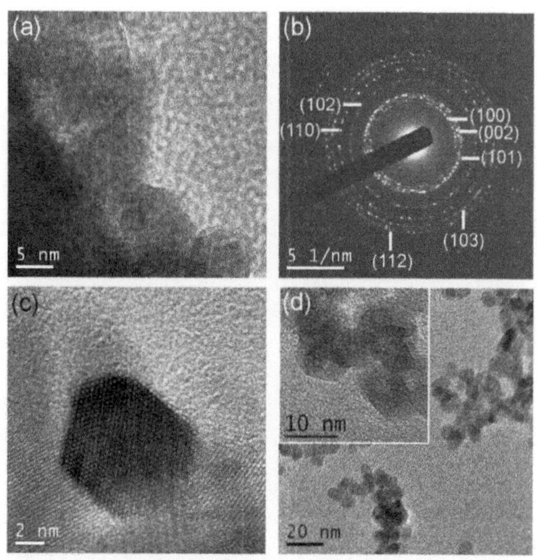

Figure 4. (a) High-resolution TEM image of ZnO@$Cu_{80}Ni_{20}$ hybrid porous layer; (b) the SAED pattern of the ZnO NPs stacked on the CuNi layer. (c) High-resolution TEM image of an individual ZnO NP. (d) TEM image of as-prepared ZnO nanoparticles; magnified in the inset.

contents above 70% at.[34, 36] Hence, the occurrence of ferromagnetism in spite of the low Ni content is a further indication that the CuNi MF are not single phase but, instead, Ni-rich clusters (ferromagnetic) co-exist together with a non-ferromagnetic Cu-rich matrix, in agreement with XRD results (Figure 2). To further explore the influence of the Ni content on the magnetic response of the layers, uncoated and ZnO-coated $Cu_{65}Ni_{35}$ films were also investigated. The $Cu_{65}Ni_{35}$ MF and its homologous ZnO@$Cu_{65}Ni_{35}$ porous film (Figure 5(b)) exhibit higher saturation magnetization (M_S) in comparison to $Cu_{80}Ni_{20}$ based films due to the higher Ni content. The hysteresis loops also reveal that the coercivity, H_C, tends to slightly decrease with the increase of the Ni content in the nanocomposites, in agreement with our previous work on uncoated CuNi porous films.[35] Remarkably, for both compositions the presence of the ZnO shell does not alter the magnetic response of the CuNi matrix as M_S remains the same before and after the ZnO coating. This indicates that the metallic character of the pristine CuNi MF is maintained during ZnO impregnation (i.e. no oxidation occurs). Hence, this synthetic approach can be of high interest in applications where it is necessary to retain the initial magnetic properties of the MF. Since the ZnO coating over the $Cu_{65}Ni_{35}$ MF was defective (Figure S1) or, at least, less conformal than over the $Cu_{80}Ni_{20}$ MF, the latter was selected for the subsequent analysis of the optical and catalytic properties.

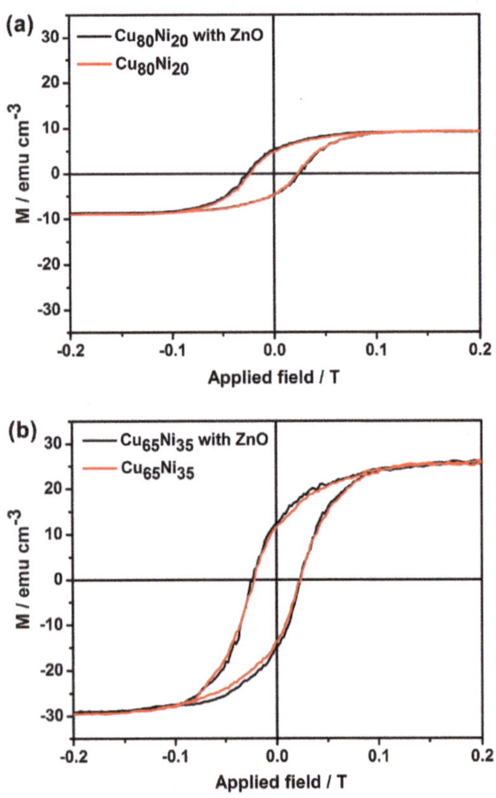

Figure 5. Room temperature hysteresis loops of the (a) $Cu_{80}Ni_{20}$ and ZnO@$Cu_{80}Ni_{20}$ and (b) $Cu_{65}Ni_{35}$ and ZnO@$Cu_{65}Ni_{35}$ samples.

3.3. Optical properties of the hybrid layers

3D image reconstruction and confocal methodologies offer great advantages for studying complex nanocomposites without mechanical distortions, providing valuable 3D information. Figure 6(a) shows the isosurface module of Imaris 3D photoluminescence (3D-PL, green) and structural (reflection mode, gray) image of the porous ZnO@$Cu_{80}Ni_{20}$ porous nanocomposite film. The full dataset was surveyed to determine the range of intensities (gray values) within the volume (0–255; black–white). An isovalue (intensity) was chosen and the software constructed a map of all the chosen values in the volume. After segmentation, isosurfacing produces simplified hollow objects (or 3D surfaces), which are accurate simplified representations of the original shape of the segmented structure. This surface model was extremely helpful to study the 3D shape and spatial relationship with different layers of the hybrid

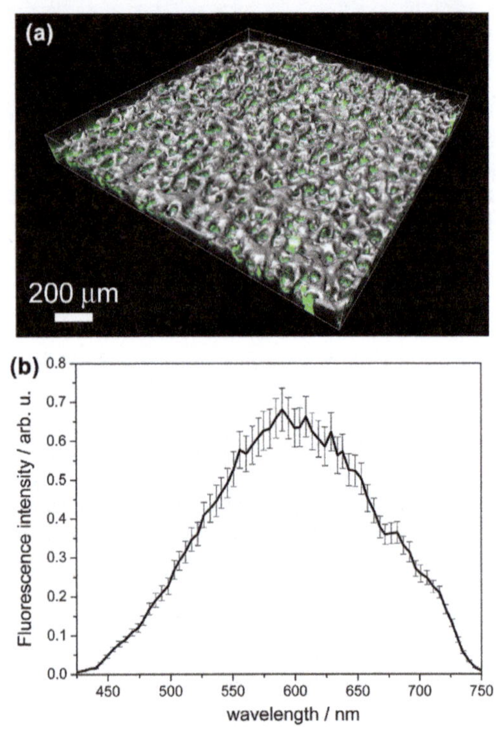

Figure 6. (a) 3D projection images of the hybrid ZnO@$Cu_{80}Ni_{20}$ nanocomposite isosurface view by Imaris v.6.2.0 software. The 3D representation was obtained from a 50-section stack in steps of 3 μm. CuNi MF shows as gray color (reflection mode) whereas the ZnO component shows as green color (fluorescence). (b) Fluorescence spectrum obtained with the Lambda scan module of CSLM at a 5 nm resolution. Abbreviation: au = arbitrary units.

nanocomposite. Notice that the green (arising from the ZnO) and the gray (arising from the CuNi) colors indicate a homogeneous distribution of ZnO within the CuNi matrix. Particularly, the PL response (Figure 6(b)) consists of broad-band emission which covers the entire visible range (in the wavelength interval 425–750 nm) and has a maximum of around 595 nm. In general, a broad green fluorescence in ZnO is attributed to oxygen defects.[37] Moreover, the PL signal is recovered across the entire thickness of the porous layer as can be observed in Figure 7. The intrinsic fluorescence follows the distribution of the reflection signal although it is more intense at 30 μm depth. A decay of the signal is observed at a higher depth. Nevertheless, this indicates that the ZnO is not only present at the outermost CuNi surface but it rather penetrates through the entire CuNi network and reaches the Au surface. Therefore, the obtained materials can be truly regarded as 3D porous composites. Interestingly, contrary to fluorescent dots, where a pronounced decrease in the PL intensity is observed during illumination with UV or visible light (photobleaching),[38] here only a decrease of 15% of PL intensity was observed after irradiation during 10 min (Figure S5). Due to this stability in the PL signal and the relatively large width of the observed PL peak, these novel ZnO@CuNi porous films could be appealing for optoelectronic devices.

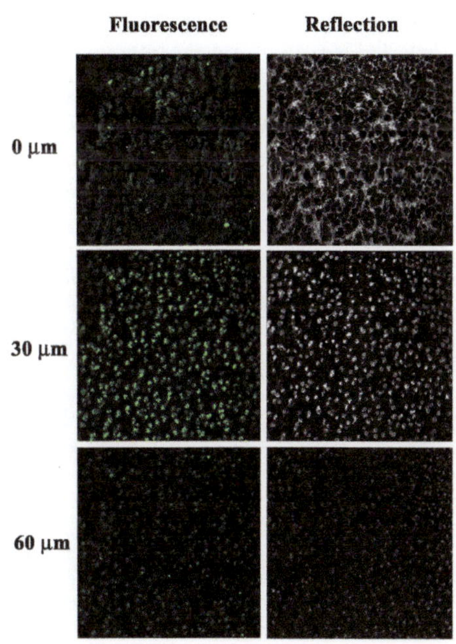

Figure 7. Fluorescence and reflection confocal micrographs (sample area: 621 μm x 621 μm) of the ZnO@$Cu_{80}Ni_{20}$ hybrid porous layer taken at different sections from the nanocomposite surface up to a total depth of 60 μm. Fluorescence microscopy was employed to visualize ZnO NP (in green). Reflection confocal microscopy was used to visualize the CuNi matrix (in gray).

3.4. Photocatalytic degradation of RhB dye

The photocatalytic performance of the ZnO@$Cu_{80}Ni_{20}$ porous film was evaluated by assessing the degradation rate of RhB under UV-vis irradiation ($320 < \lambda < 500$ nm) after the adsorption–desorption equilibrium was reached.

As can be observed in Figure 8, the control test without catalysts (blank) under UV-vis light irradiation showed that the photolysis of RhB is negligible. On the other hand, for the uncoated CuNi MF the concentration of RhB decreased very slightly: only 2% of the original organic matter was degraded after 3 h of exposure time. Regarding the ZnO-coated CuNi, it can be observed that the relative concentration (C/C_0) of RhB dye markedly decreased with the increase of exposition time. The final C/C_0 value was a function of the ZnO amount. The RhB concentration in solution decreases by 10% and 15% after 3 h when ZnO@$Cu_{80}Ni_{20}$ with *middle* and *final* coverages, respectively, were used. As expected, the photoactivity of hybrid porous ZnO@CuNi thin films is higher than that of bare CuNi MF, and it increases with ZnO content. Furthermore, the activity of a ZnO-coated non-porous electroplated CuNi flat dense film is much lower. This puts forward the importance of a large specific surface area for efficient dye removal and suggests that the macropore sidewalls do contribute to the photocatalytic activity. According to the literature,[39] RhB dye degradation mechanism may proceed according to (1–5):

$$ZnO + h\nu \rightarrow e^-_{cb} + h^+_{vb} \quad (1)$$

$$O_2 + e^- \rightarrow O_2^{\cdot -} + H^+ \rightarrow HOO^{\cdot} \quad (2)$$

$$e^- + HOO^{\cdot} + H^+ \rightarrow H_2O_2 \quad (3)$$

$$H_2O_2 + e^- \rightarrow OH^{\cdot} + OH^- \quad (4)$$

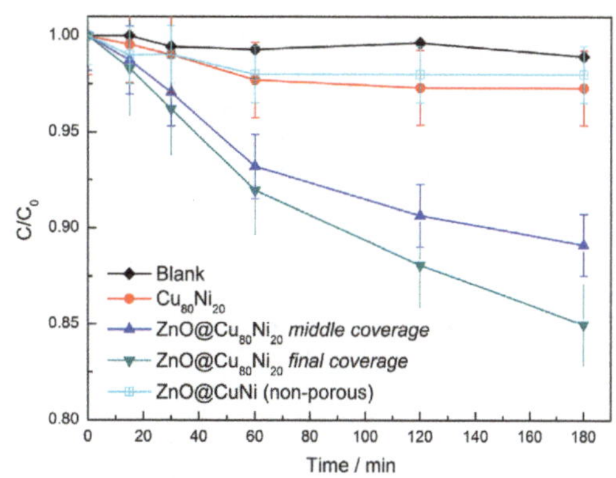

Figure 8. C/C_0 plot of the degradation of RhB in the blank, in the presence of $Cu_{80}Ni_{20}$ MF, and in the presence of ZnO@$Cu_{80}Ni_{20}$ photocatalysts with *middle* and *final* coverages. The RhB degradation profile for ZnO-coated non-porous Cu-rich flat layer is also shown for comparison.

$$RhB + O_2/O_2^{\cdot-}/HOO^{\cdot}/OH^{-} \rightarrow \text{mineralized product} \quad (5)$$

Upon illumination, the generation of electron–hole (e_{cb}^- and h_{vb}^+) pairs between conduction (CB) and valence bands (VB) of ZnO NPs occurs (Equation (1)). The generated e^- are then scavenged by the dissolved O_2 to generate superoxide anion $O_2^{\cdot-}$ (Equation (2)), which, in turn, generate H_2O_2, OH ions and radicals (Equations (3) and (4)). All the generated reactive species are able to degrade RhB and mineralize it (Equation (5)).

Previous conditioning under dark conditions for 60 min in the presence of the ZnO@$Cu_{80}Ni_{20}$ did not bring any change in the RhB concentration, indicating that UV-vis irradiation is indispensable for dye degradation. Upon several consecutive degradation tests under irradiation conditions, a decrease of the photocatalytic efficiency was observed (Figure S6), likely due to gradual deactivation of ZnO caused by the adsorption of species present in solution.[40] No apparent detachment of the ZnO NPs from the CuNi MF was observed though. The solution remained fully transparent (eventual turbidity arising from detached ZnO NPs was not noticed) and SEM images of the CuNi MF after the photocatalytic tests did not disclose appreciable changes in morphology. Compared to the literature dealing with ZnO-based materials, the photocatalytic activity of the here-developed hybrid composites is not particularly high. Different strategies could help improve the photocatalytic performance: through doping of ZnO with suitable cations/anions or by further increasing the specific surface area of the CuNi MF (e.g. via the creation of homogeneously distributed nanopores) and, hence, of the ZnO coating. Moreover, diffusion of the RhB dye through the 3D macroporous network is perhaps not fully efficient. Indeed, taking into account the mean pore diameter (several micrometers), it is expected that a non-negligible fraction of the inner part of the film contributes to the photocatalytic activity. If the pores were much narrower, diffusion of RhB through the pores would have been probably more challenging. Nevertheless, detailed studies are required to evaluate the efficiency of dye diffusion within the 3D network. Although the photocatalytic tests with ZnO-coated non-porous CuNi films do indicate that porosity is definitely beneficial, further work should be done in order to take full advantage of the macroporous architecture of the material. This also includes optimizing the way the material is irradiated so that the inner ZnO shells fully contribute to the catalytic response.

4. Conclusions

In summary, a fast and inexpensive synthetic approach to obtain hybrid porous ZnO@CuNi films has been presented. First, two-phase CuNi porous magnetic films were grown by means of electrodeposition. Then, a suspension containing ZnO nanoparticles was dropwise added onto the CuNi framework, coating not only its outermost external surface but also the inner pores, as revealed by CLSM imaging, hence rendering truly 3D porous ZnO@CuNi hybrid films. This strategy overcomes the problems associated with high temperature post-synthesis heat treatments that can be deleterious from both morphological and compositional viewpoints. Indeed, thanks to the lack of annealing treatments, the inherent macro-porosity of the initial CuNi MFs as well as their ferromagnetic properties are preserved after coating with ZnO. The obtained ZnO@CuNi hybrid porous layers show some activity for the aqueous degradation of Rhodamine B under UV-vis light irradiation. When supported onto substrates, these materials could be handled as classical heterogeneous catalysts. Compared to ZnO-based powders, tedious recycling steps such as filtration and centrifugation are not required. If released from the substrate, the composite films could eventually be wirelessly manipulated using external magnetic fields, whilst being both photoluminescent and photocatalytic upon irradiation under UV-vis light. This would be possible due to the combination of the various properties of the different phases comprising the hybrid composites. The reported synthetic strategy could also be extrapolated to other types of materials, paving the way for the preparation of hybrid materials in general, suitable for multifunctional solid-state devices.

Acknowledgements

This work has been funded by the Generalitat de Catalunya [project 2014-SGR-1015] and the Spanish Ministerio de Economía y Competitividad (MINECO) [projects MAT2014-57960-C3-1-R and MAT2014-57960-C3-3-R]. Partial funding from the European Research Council [Consolidator Grant, project number 648454, SPIN-PORICS] is acknowledged. E. Pellicer is grateful to MINECO for the 'Ramón y Cajal' contract [RYC-2012-10839]. J. Zhang is grateful to the China Scholarship Council (CSC) for the PhD grant. M. Guerrero acknowledges the support of the Secretary for Universities and Research of the Government of Catalonia and the COFUND Programme of the Marie Curie Actions of the 7th R&D Framework Programme of the European Union for the 'Beatriu de Pinos' contract [2013 BP-B 00077].

Disclosure statement

No potential conflict of interest was reported by the authors.

References

[1] Jiangtian L, Nianqiang W. Semiconductor-based photocatalysts and photoelectrochemical cells for solar fuel generation: A review. Catal. Sci. Technol. 2015;5:1360–1384.

[2] Wagner T, Haffer S, Weinberger C, et al. Mesoporous materials as gas sensors. Chem. Soc. Rev. 2013;42:4036–4053.

[3] Prim A, Pellicer E, Rossinyol E, et al. A novel mesoporous CaO-loaded In_2O_3 material for CO_2 sensing. Adv. Funct. Mater. 2007;17:2957–2963.

[4] Hu X, Li G, Yu JC. Design, fabrication, and modification of nanostructured semiconductor materials for environmental and energy applications. Langmuir. 2010;26:3031–3039.
[5] Zarei-Chaleshtori M, Correa V, López N, et al. Synthesis and evaluation of porous semiconductor hexaniobate nanotubes for photolysis of organic dyes in wastewater. Catalysts. 2014;4:346–355.
[6] Lin J, Heo Y-U, Nattestad A, et al. 3D hierarchical rutile TiO_2 and metal-free organic sensitizer producing dye-sensitized solar cells 8.6% conversion efficiency. Sci. Rep. 2014;4:5769, 1–8.
[7] Pellicer E, Cabo M, López-Ortega A, et al. Controlled 3D-coating of the pores of highly ordered mesoporous antiferromagnetic Co_3O_4 replicas with ferrimagnetic $FexCo_{3-x}O_4$ nanolayers. Nanoscale. 2013;5:5561–5567.
[8] Rodriguez-Suarez T, Bartolomé JF, Moya JS. Mechanical and tribological properties of ceramic/metal composites: A review of phenomena spanning from the nanometer to the nicrometer length scale. J. Eur. Ceram. Soc. 2012;32:3887–3898.
[9] Menéndez E, Salazar-Alvarez G, Zhilyaev AP, et al. Cold consolidation of metal–ceramic nanocomposite powders with large ceramic fractions. Adv. Funct. Mater. 2008;18:3293–3298.
[10] Sort J, Nogués J, Amils X, et al. Room-temperature coercivity enhancement in mechanically alloyed antiferromagnetic-ferromagnetic powders. Appl. Phys. Lett. 1999;75:3177–3179.
[11] Velev OD, Kaler EW. Structured porous materials via colloidal crystal templating: From inorganic oxides to metals. Adv. Mater. 2000;12:531–534.
[12] Kołodziejczak-Radzimska A, Jesionowski T. Zinc oxide-from synthesis to application: A review. Materials. 2014;7:2833–2881.
[13] Nunomura S, Che X, Forrest SR. Charge trapping in mixed organic donor–acceptor semiconductor thin films. Adv. Mater. 2014;26:7555–7560.
[14] Zhang X, Qin J, Xue Y, et al. Effect of aspect ratio and surface defects on the photocatalytic activity of ZnO nanorods. Sci. Rep. 2014;4:1–8.
[15] Tian C, Zhang Q, Wu A, et al. Cost-effective large-scale synthesis of ZnO photocatalyst with excellent performance for dye photodegradation. Chem. Commun. 2012;48:2858–2860.
[16] Huang MH, Mao S, Feick H, et al. Room-temperature ultraviolet nanowire nanolasers. Science. 2001;292:1897–1899.
[17] Bashami RM, Hameed A, Aslam M, et al. The suitability of ZnO film coated glassy carbon electrode for the sensitive detection of 4-Nitrophenol in aqueous medium. Anal. Methods. 2015;7:1794–1801.
[18] Shi J, Grutzik S, Wang X. Zn cluster drifting effect for the formation of ZnO 3D nanoarchitecture. ACS Nano. 2009;3:1594–1602.
[19] Xu X-H, Zhang M, Wu J-F, et al. Hydrothermal synthesis of ZnO crystallite with multi-morphology on zinc foil and crystal growth mechanism. Chem. Res. Chin. Univ. 2011;32:1703–1708.
[20] Fang H, Wu W, Huang Z. Photocatalytic performance of ZnO prepared by sol-gel method with assistance of CTAB. Adv. Mat. Res. 2014;881:256–2559.
[21] Zhou J, Zhang M, Zhu Y. Preparation of visible light-driven $g\text{-}C_3N_4$@ZnO hybrid photocatalyst via mechanochemistry. Phys. Chem. Chem. Phys. 2014;16:17627–17633.
[22] Zhang Y, Guo T, Luo Y-D, et al. Tunable ferromagnetic behaviors observed in highly orientated Co-doped ZnO thin films by the bandgap engineering. J. Am. Ceram. Soc. 2013;96:361–364.
[23] Chen S, Carraro G, Barreca D, et al. Growth and electro-optical properties of Ga-doped ZnO films prepared by aerosol assisted chemical vapour deposition. Thin Solid Films. 2015;584:316–319.
[24] Vijayalakshmi S, Venkataraj S, Subramanian M, et al. Physical properties of Zinc doped Tin Oxide films prepared by Spray Pyrolysis Technique. J. Phys. D: Appl. Phys. 2008;41:1–7.
[25] Poongodi G, Kumar R-M, Jayavel R. Structural, optical and visible light photocatalytic properties of Nanocrystalline Nd doped ZnO thin films prepared by Spin Coating Method. Ceram. Int. 2015;41:4169–4175.
[26] Nam G, Yoon H, Kim B, et al. Effect of Co doping concentration on structural properties and optical parameters of Co-doped ZnO thin films by Sol-gel Dip-coating Method. J. Nanosci. Nanotechnol. 2014;14:8544–8548.
[27] Redel E, Petrov S, Dag Ö, et al. Green Nanochemistry-Metal Oxide nanoparticles and porous thin films from bare metal powders. Small. 2012;8:68–72.
[28] Sokolov S, Paul B, Ortel E, et al. Template-assisted electrostatic spray deposotion as a new route to mesoporous, macroporous, and hierarchically porous oxide films. Langmuir. 2011;27:1972–1977.
[29] Zhang R, Liu J, Guo H, et al. Rational synthesis of three-dimensional porous $ZnCo_2O_4$ film with nanowire walls via simple hydrothermal method. Materials Letters. 2014;115:208–211.
[30] Eswar KA, Rouhi J, Husairi HF, et al. Annealing heat treatment of ZnO nanoparticles grown on porous Si substrate using Spin-coating Method. Adv. Mater. Sci. Eng. 2014;796759:1–6.
[31] Znaidi L. Sol–gel-deposited ZnO thin films: A review. Mater. Sci. Eng. B-Adv. Funct. Solid-State Mater. 2010;174:18–30.
[32] Plowman BJ, Jones LA, Bhargava SK. Building with bubbles: The formation of high surface area honeycomb-like films via hydrogen bubble templated electrodeposition. Chem. Commun. 2015;51:4331–4346.
[33] Young RA. The Rietveld method; Union of crystallography. Oxford, U.K.: Oxford University Press; 1995.
[34] Bakonyi I, Tóth-Kádár E, Tóth J, et al. Magnetic and electrical transport properties of electrodeposited Ni-Cu alloys and multilayers. J. Phys. Condens. Matter. 1999;11:963–975.
[35] Zhang J, Baró MD, Pellicer E, et al. Electrodeposition of magnetic, superhydrophobic, non-stick, two-phase Cu–Ni foam films and their enhanced performance for hydrogen evolution reaction in alkaline water media. Nanoscale. 2014;6:12490–12499.
[36] Varea A, Pané S, Gerstl S, et al. Ordered arrays of ferromagnetic, compositionally-graded $Cu_{1-x}Ni_x$ alloy nanopillars prepared by template-assisted electrodeposition. J. Mater. Chem. C. 2013;1:7215–7221.
[37] Djurisic AB, Choy WCH, Roy VAL, et al. Photoluminescence and electron paramagnetic resonance of ZnO tetrapod structure. Adv. Funct. Mater. 2004;14: 856–864.

[38] Moyo T, Maruyama K, Endo H. Photodarkening and photobleaching of CdS microclusters grown in Zeolites. J. Phys. Condens. Matter. 1992;4:5653–5664.
[39] Rahman QI, Ahmad M, Misra SK, et al. Effective photocatalytic degradation of rhodamine B dye by ZnO nanoparticles. Mater. Lett. 2013;91:170–174.
[40] Sun W-J, Li J, Mele G, et al. Enhanced photocatalytic degradation of rhodamine B by surface modification of ZnO with copper (II) porphyrin under both UV–vis and visible light irradiation. J. Mol. Catal. A: Chem. 2013;366:84–91.

20

Study of {332}<113> twinning in a multilayered Ti-10Mo-xFe (x = 1–3) alloy by ECCI and EBSD

Ivan Gutierrez-Urrutia[a], Cheng-Lin Li[a], Satoshi Emura[a], Xiaohua Min[b] and Koichi Tsuchiya[a,c]

[a]National Institute for Materials Science, Tsukuba, Ibaraki 305-0047, Japan
[b]School of Materials Science and Engineering, Dalian University of Technology, Dalian 116024, PR China
[c]Graduate School of Pure and Applied Sciences, University of Tsukuba, Ibaraki 305-0047, Japan

ABSTRACT

We have investigated the propagation of {332}<113> twins in a multilayered Ti-10Mo-xFe (x = 1–3) alloy fabricated by multi-pass hot rolling. The material contains a macroscopic Fe-graded structure (about 130 μm width) between 1 and 3 wt% Fe in the direction perpendicular to rolling. We observe strong influence of the Fe-graded structure in the twin propagation behavior. The propagation of {332}<113> twins that are nucleated in Fe-lean regions (~1 wt% Fe) is interrupted in the grain interiors at a specific Fe content, namely, about 2 wt% Fe. We ascribe this effect to the role of Fe content in solid solution on the stress for twin propagation. The interruption of twins in the grain interiors results in the development of characteristic dislocation configurations such as highly dense dislocation walls (HDDWs) associated to strain localization phenomena. The nucleation and propagation of these dislocation configurations is ascribed to the underlying plastic accommodation mechanisms of the stress field at the twin tips. We find that the crystallographic alignment of HDDWs is determined by the stress field at the twin tips and the deformation texture. The excellent plastic accommodation at the interrupted twin tips allows attaining the good ductility of the present material (total elongation of 28%).

PROPAGATION OF {332}[113] TWINS IN A MULTILAYERED Ti-10Mo-xFe (x = 1-3) ALLOY

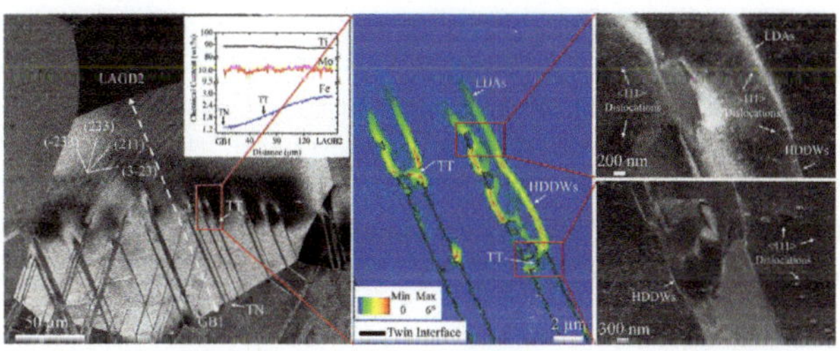

BSE AND EPMA KAM-EBSD ECCI

KEYWORDS
Dislocations; plasticity; interface; multi-layered materials

CLASSIFICATION
10 Engineering and Structural materials; 106 Metallic materials; 100 Materials; 303 Mechanical / Physical processing; 300 Processing / Synthesis and Recycling; 503 TEM, STEM, SEM; 500 Characterization

1. Introduction

In the last decade, β-Ti-Mo alloys have received considerable attraction as structural materials in several technological areas such as aerospace, automotive and biomedical industries. These alloys contain outstanding combination of high specific strength, low elastic modulus and excellent bio-compatibility [1–6] Several processing techniques such as rolling, drawing and forging are typically used to optimize the mechanical behavior of these alloys [1–6] Recently, Min et al. [7] have processed a multilayered β-Ti-10Mo-xFe (x = 1–3 wt%) alloy by multi-pass hot rolling that exhibits superior combination of strength and ductility (yield strength, YS, of 800 MPa; total elongation, TE, of 28%), compared to those of the individual alloys. This material contains a macroscopic gradient in Fe that tailors the combination of specific deformation mechanisms, namely, <111> dislocation slip and {332}<113> deformation twinning that results in superior mechanical behavior.

The activation of deformation modes in β-Ti-Mo alloys is associated to the stability of β-Ti (bcc phase) against the β-α′ and β-α″ stress-induced phase transformation [8–11] Whereas <111> slip has been intensively studied in bcc metals [12–15] the details of {332}<113> twinning mode are still unclear [16–18] Specifically, two main models

CONTACT Ivan Gutierrez-Urrutia gutierrezurrutia.ivan@nims.go.jp have been proposed for {332}<113> twinning, namely, shear and shuffle-type mechanisms,[16,17,19] and a pole dislocation-type model involving partial dislocations.[18,20] The first approaches suggest that twinning takes place by a complex atomic movement that involves atomic shearing and shuffling by specific crystallographic criteria. The dislocation-type model suggests that <111> perfect dislocations are dissociated into <113> dislocations after characteristic dislocation interactions. The subsequent revolution of <113> dislocations around the dislocation node by the pole mechanism results in the growth of the twin plate. The twin models described above provide relevant details on the crystallography and nucleation mechanism of {332}<113> twinning. However, twin propagation also plays significant role on the strain hardening behavior and fracture mode of polycrystalline alloys. The influence of microstructure on twin propagation is complex mainly due to the large number of parameters involved.[21–30] In particular, several works have addressed the influence of grain size [21,27,29,30] and twin–twin interactions [25] on twin propagation in several bcc alloys.

Novel β-Ti-Mo alloys such as multilayered materials contain another interesting microstructural feature that may also influence twin propagation phenomena, namely, chemical graded structure. The propagation of a twin plate through a composition gradient may be different to that in homogeneous alloys due to solute effects on twinning mode [16,17] and twin propagation stress,[21,29] which are currently unclear. The aim of this work is to investigate the propagation of {332}<113> deformation twins in a multilayered β-Ti-10Mo-xFe (x = 1–3) alloy that contains a macroscopic composition gradient of Fe. Microstructure characterization was performed by electron probe microanalysis (EPMA), electron backscatter diffraction (EBSD) and electron channeling contrast imaging (ECCI). The material was fabricated by multi-pass hot rolling and contains a Fe-graded structure ranged between 1 and 3 wt% of about 130 μm width. We observe a strong influence of the Fe-graded structure in the twin propagation behavior. Specifically, the propagation of {332}<113> twins is interrupted at a certain Fe content, namely, about 2 wt% Fe. We ascribe this effect to the role of Fe content on the stress for twin propagation. The occurrence of interrupted growth twins in the grain interiors results in the development of specific dislocation configurations that are ascribed to the underlying plastic accommodation mechanism at the twin tips. We claim that this effect contributes to the good ductility of the multilayered Ti-Mo-Fe alloy (total elongation of 28%).

2. Experimental details

A multi-layered stacked material consisting of 10 alternative layers of Ti-10Mo-1Fe and Ti-10Mo-3Fe alloys (wt%) was used in this study. To generate a macroscopic Fe-graded structure, the material was processed by multi-pass hot rolling at 1000 °C and subsequent solution treatment at 900°C for 1 h.[7] According to the current melting method (cold crucible levitation melting) and thermomechanical conditions, the oxygen content is negligible (<0.1 wt%).[31] Tensile samples were machined out from the multilayered material and from individual alloys, namely, Ti-10Mo-1Fe, Ti-10Mo-2Fe and Ti-10Mo-3Fe. The tensile samples had 18 mm gage length, 4 mm gage width and 2.5 mm gage thickness. Tensile axis was parallel to the rolling direction (RD). Complete and interrupted tensile tests to a true strain of 4% were performed on an Instron 5581 tensile machine at room temperature and at an initial strain rate of 2.8×10^{-4} s^{-1}. Deformation structure was characterized by two Scanning Electron Microscopy (SEM) techniques, namely ECCI and EBSD in a Sigma Zeiss field emission gun scanning electron microscope (Carl Zeiss AG, Oberkochen, Germany)(FEG-SEM) equipped with a TSL (TSL, EDAX, NJ, USA) Orientation Imaging Microscopy (OIM) EBSD system. The rolling direction (RD)–normal direction (ND) section of the tensile sample was evaluated. High-resolution EBSD maps with a step size of 50 nm were taken at 15 kV acceleration voltage. The deformation structure of the grains mapped by EBSD was subsequently examined by ECCI under controlled diffraction conditions as described in [32,33]. ECCI images were obtained with optimum channeling contrast by tilting the matrix crystal into the Bragg condition in a "two-beam" mode using a high intensity diffraction vector g_{111}. ECCI observations were carried out at a working distance of 4–5 mm and at 9 kV acceleration voltage by using an objective aperture of 120 μm. Elemental mapping was performed by EPMA on a JEOL JXA-8900F system operated at 15 kV.

3. Results

Figure 1 shows the deformation microstructure (a) and the elemental distribution (b) of the multilayered material tensile deformed to 0.04 true strain. The elemental map shows that at the present thermomechanical conditions, the material contains a macroscopic gradient of Fe ranged between 1 and 3 wt% Fe that is distributed in the form of bands parallel to RD, while Ti and Mo are homogeneously distributed within the Fe-graded structure [7]. These bands are about 130 μm thick (green to yellow to red colored bands in Figure 1(b)) and are stacked between the original Ti-10Mo-1Fe and Ti-10Mo-3Fe layers, which are about 50 μm thick (3 wt% Fe: pink layer; 1 wt% Fe: blue layer). The width of the Fe-graded structure is determined by the thermomechanical conditions. At the present conditions and considering the diffusion coefficient of Fe in β-Ti (60×10^{-13} m^2 s^{-1}),[34] a diffusion length of Fe in β-Ti of about 50 μm is obtained. Using these data, the estimated length of the Fe concentration profile between two layers is about 100 μm, which roughly agrees with the experimental

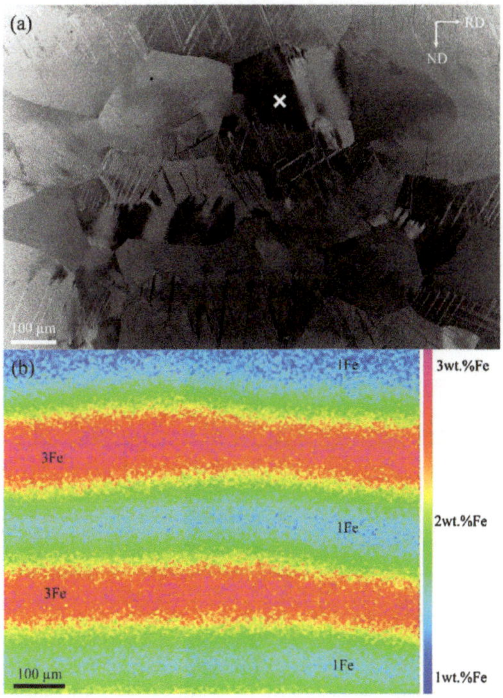

Figure 1. (a) ECCI image of the deformation structure of the multilayered Ti-10Mo-xFe (x = 1–3) material tensile deformed to 0.04 true strain. The white cross indicates the grain analyzed in Figure 2. (b) EPMA map of Fe distribution of the grain indicated by the white cross in (a).

value of 130 µm. Figure 1(a) shows that the deformation structure is formed by {332}<113> deformation twins and <111> slip traces. This structure is similar to that reported in conventional β-Ti-Mo alloys.[14,35–38] However, in the present multilayered Ti-Mo-Fe alloy the propagation of deformation twins is interrupted within the grain interiors at the same location, i.e. at about 2 wt% Fe. This effect results in the confinement of twin plates within areas of about 100 µm long (average grain size is about 250 µm). This finding indicates that the Fe-chemical graded structure has a significant influence on the propagation of {332}<113> deformation twins in the present Ti-Mo-Fe alloy.

Figure 2(a) shows an example of the deformation structure of a grain oriented close to the [1 15 16] crystallographic direction (the location of the grain is indicated by the cross in Figure 1(a)). This ECCI image was obtained by tilting the sample in the channeling condition with g_{111} excited so that the crystal matrix is in an out-of-Bragg condition and hence it appears as bright background, whereas the twin crystal plates are close to the Bragg condition and accordingly, they appear dark. The grain contains two active twin systems that correspond to the twin systems with the highest Schmid factors (SFs), namely, (−233)[311] with SF = 0.41, and (233)[−311] with SF = 0.36. These twins are nucleated at GB1 (TN: twin nucleation in Figure 2(a)) and are interrupted at an area containing about 1.8 wt% Fe (TT: twin tip in Figure 2(a)), i.e. they propagate through the grain interior a finite length of about 60 µm. Elemental mapping reveals that the grain contains a Fe gradient of about of about 0.01 wt% Fe/µm which is 160 µm long. Specifically, Fe content varies between 1.2 wt% at the grain boundary GB1 and 2.8 wt% at the low angle grain boundary LAGB2. On the other hand, Mo content is relatively homogeneous and ranges between 9.8 and 10.2 wt%. Detailed analysis of the EPMA data, Figure 2(b), did not reveal the occurrence of any chemical interface within the composition gradient (line analysis along the dashed line in Figure 2(a)).

Figures 3 and 4 show the deformation structure around two interrupted twin tips of the grain shown in Figure 2. Specifically, Figures 3(a) and (b) correspond to an IPF-EBSD map along the tensile direction and a kernel average misorientation (KAM)-EBSD map (first neighbors), respectively. These EBSD maps reveal that the deformation structure is mainly formed by paired dislocation layers (indicated by yellow lines in Figure 3(a)) and few small twin plates of about 1 µm long and 0.5 µm width. The twin system of these small twins is the same as that of the primary twins, namely, (−233) [311]. The dislocation layers are nucleated at the edges of the twin tips (TT in Figures 3 and 4) and propagate into the matrix for a length of about 5 µm. The KAM-EBSD map of Figure 3(b) shows that these dislocation configurations are misoriented between 3 and 15° from the crystal matrix. The misorientation angle θ decreases with the distance from the twin tip, namely, from $\theta = 10–15°$ in the vicinity of the twin tips to $\theta \sim 3°$ at 5 µm from the twin tips.

Details of the dislocation configurations around the twin tips are revealed by the ECCI images of Figure 4, which were taken with g_{111} excited close to Bragg condition. Channeling contrast of dislocation configurations reveals that the paired dislocation layers correspond to highly dense dislocation walls (HDDWs).[39,40] These dislocation configurations are imaged by ECCI at the current channeling conditions as dense dislocation arrays which appear as bright straight compact layers; see Figure 4(b). HDDWs are characteristic of fcc metals with pronounced planar slip such as Al-Mg, Fe-Mn-C and Fe-Mn-Al-C alloys.[39,41,42] We observe that HDDWs tend to align almost parallel to the (−233) twin habit plane trace; see Figure 4(a). Specifically, they lay along three slip systems, namely, (112), (10−1) and (110), which are slightly deviated from the (−233) twin habit plane (about 2–5°). The KAM-EBSD map of Figure 3(b) reveals that HDDWs are formed by a high density of geometrically necessary dislocations (GNDs) that enables the accommodation of matrix misorientations of up to 15° in the vicinity of the twin tips. The dislocation density of HDDWs is not uniform, as reflected in the

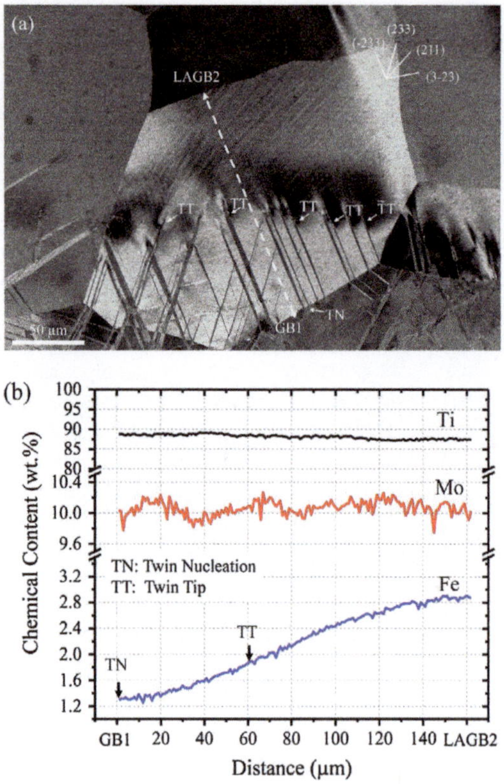

Figure 2. (a) ECCI image of the deformation structure of a typical grain. The grain is oriented close to the [1 15 16] crystallographic orientation. (b) Chemical line analysis performed by EPMA along the dashed line of the grain shown in (a). TN: twin nucleation; TT: twin tip; GB: grain boundary; LAGB: low angle grain boundary.

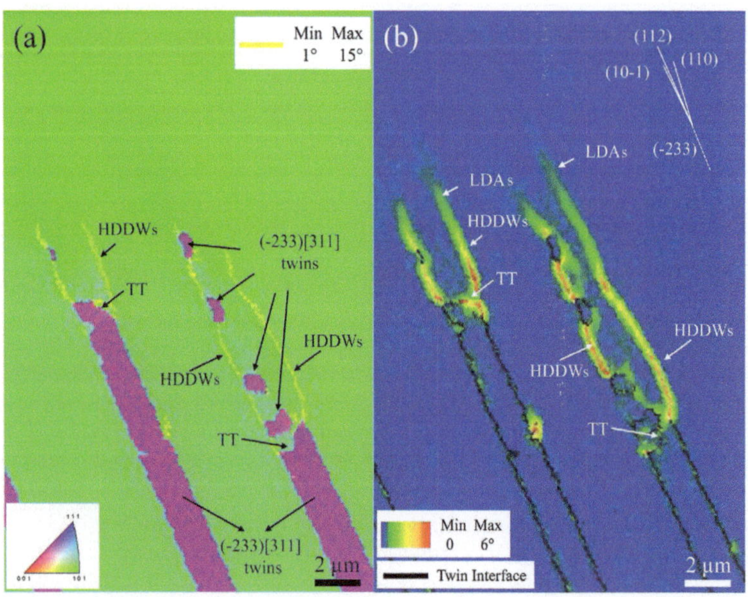

Figure 3. EBSD maps of the deformation structure around two interrupted twins inside the grain shown in Figure 2(a). (a) IPF-EBSD map along the tensile direction. Yellow line represents low angle grain boundaries (misorientation between 1 and 15°); (b) KAM-EBSD map. Black line indicates twin interfaces. HDDWs: highly dense dislocation wall; LDAs: loose dislocation arrangements; TT: twin tip.

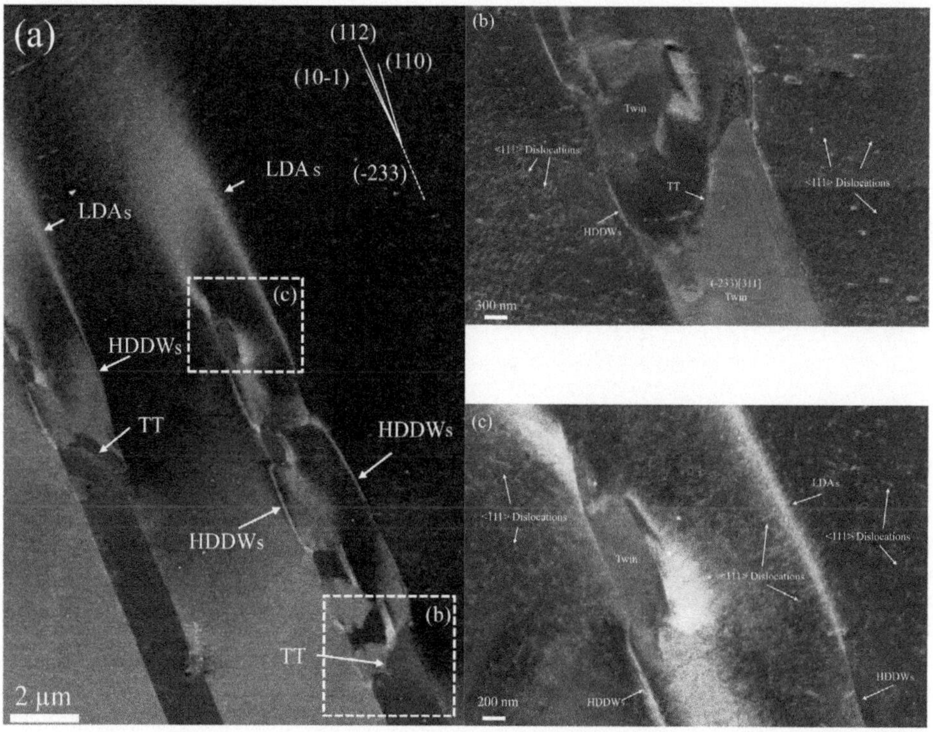

Figure 4. ECCI images of the deformation structure of the twins shown in Figure 3. Images were taken with g_{111} excited close to Bragg condition. (b) and (c) reveal details of the dislocation configuration of the areas indicated by dashed boxes in (a). HDDWs: highly dense dislocation wall; LDAs: loose dislocation arrangements.TT: twin tip.

decrease of the misorientation angle (θ) with the distance from the twin tip. The propagation of HDDWs in the matrix is limited to a distance of about 2–5 μm. At this point, ECCI reveals that the dislocation substructure evolves from planar configurations to loose dislocation arrangements (LDAs) that propagate about 5 μm further into the matrix; see Figure 4(c). These dislocation configurations are imaged by ECCI at the current channeling conditions as wide loose bright layers. They are visualized in the corresponding KAM map (Figure 3(b)) as localized orientation gradients with $\theta \sim 1–3°$. Figure 4(c) also reveals that the interior of the crystal delimited by HDDWs/LDAs contains high density of <111>-type dislocations gliding on {110} and {112} slip planes (dislocations are imaged in Figure 4(c) as bright sharp lines over a dark matrix background).

4. Discussion

4.1. *Propagation of twins*

Homophase interfaces such as grain boundaries, annealing twins, and mechanical twins play a significant role on the propagation of mechanical twins in bcc metals.[21,29,43] In particular, grain boundaries and twin interfaces can act as strong barriers on twin propagation and, depending on the crystallographic orientation of the crystals involved, may even completely hinder twin propagation. The propagation of twins in the present chemical graded Ti-10Mo-xFe (x = 1–3) alloy is interrupted in the grain interiors at a specific Fe content, namely, about 2 wt% Fe. EPMA and high-resolution EBSD mapping do not reveal the formation of any intragranular homophase interface such as diffuse chemical interface, low angle grain boundary or annealing twin. On the other hand, in the present Ti-Mo-Fe alloy and under the current thermomechanical conditions, the role of precipitates such as ω-omega phase on the precipitate-twin interaction can be ruled out due to the small volume fraction (<0.01) and their shearable nature.[15,44] These findings indicate that Fe content in solid solution is the main microstructural parameter controlling twin propagation in the present chemical graded Ti-Mo-Fe alloy.

In order to clarify the role of Fe content on twin propagation, we have estimated the twinning stress, σ_{tw}, of the Ti-10Mo-xFe (x = 1–3) system. We define σ_{tw} as the macroscopic stress at which twins are detected. According to dislocation-based models of twinning in bcc metals,[21,29,30,43] the experimental twinning stress corresponds to the stress to grow an existing twin nuclei. Specifically, the twinning stress of three alloys, namely Ti-10Mo-1Fe, Ti-10Mo-2Fe and Ti-10Mo-3Fe, was

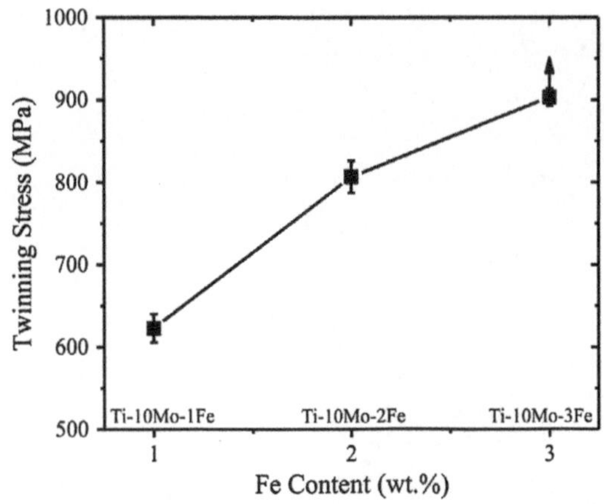

Figure 5. Effect of the Fe content on the twinning stress in the Ti-10Mo alloy system. Error bars indicate experimental deviation.

determined by tensile loading experiments. Deformation structure was examined by ECCI and EBSD. The alloys contained similar average grain sizes of ~250 μm in order to rule out grain size effects on twinning stress. The alloys with low Fe content, i.e. Ti-10Mo-1Fe and Ti-10Mo-2Fe, contained {332}<113> twins readily at yielding. For these alloys we thus consider twin stress as the yield stress. On the other hand, deformation twins were not visible in the Ti-10Mo-3Fe alloy at any investigated strain level. Accordingly, we can only estimate a lower bound of the twinning stress for this alloy system. The experimental values of the twin stress with respect to Fe content are plotted in Figure 5. This figure shows that twinning stress scales with Fe content. We observe that additions of Fe higher than 1 wt% promote a significant increase of the twinning stress, namely, 2 wt% Fe increases the twinning stress about 205 MPa, and 3 wt% Fe increases the twinning stress at least 330 MPa. These findings indicate that the stress to propagate a twin plate that is nucleated in the Fe-lean region (~1 wt% Fe) and propagates through the Fe-graded structure increases significantly with the twin propagation length. At the current deformation conditions, the macroscopic applied stress is about 830 MPa, which is slightly higher than the twinning stress for the Ti-10Mo-2Fe alloy (σ_{tw} = 780 MPa). This result suggests that twins nucleated in the Fe-lean region (~1 wt% Fe) are able to propagate within the interior of a grain up to an area containing about 2 wt% Fe, which is close to the experimental value of 1.8 wt% Fe, see Figure 2(b). This difference can be explained as follows. The twinning stress is given by two contributions, namely, one term that is associated to the critical resolved shear stress ($\tau_{critical}$) to grow a twin, and another term (τ_{GS}) that accounts for the role of grain boundaries on twin propagation,[21,29,30,43] i.e.:

$$s_{tw} = M(t_{critical} + t_{GS}) \quad [1]$$

where M is the Taylor factor. $\tau_{critical}$ is determined by the corresponding twinning model. Its mathematical expression is still unknown. τ_{GS} is commonly considered as a Hall–Petch type relation ($\tau_{GS} = K/D^{1/2}$) or Frank–Read type stress ($\tau_{GS} = Gb/D$), where K is a constant, D is the average grain size, G is the shear modulus and b is the magnitude of Burgers vector.[21,23,29,30] The experimental twinning stress values of Figure 5 were obtained in polycrystalline alloys. As Eq. (1) shows, grain boundary effects on twin propagation (τ_{GS}) are inherently estimated in such experiments. However, these effects are not relevant for the present case where we investigate the propagation of twins within a grain interior free of interfaces, similar to that occurring in single crystals. Accordingly, we envisage that the calculated twin propagation lengths by using σ_{tw} of individual polycrystalline alloys are slightly overestimated, which agrees with our findings.

4.2. Plastic accommodation of twin tips

The main relevant feature of the deformation structure associated to the plastic accommodation of the twin tips is the formation of specific dislocation configurations such as HDDWs and LDAs around the twin tips. HDDWs are characteristic of strain localization phenomena associated to intense dislocation activity on single slip or texture softening.[39,45–48] They are formed along highly stressed crystallographic planes and typically contain high density of GNDs. In the present case, we observe that HDDWs are nucleated at the edges of the twin tips and propagate into the matrix for about 2–5 μm. At this point, HDDWs evolve into loose dislocation arrangements, i.e. they lose their planar confinement and become less defined dislocation configurations.

The formation of HDDWs at the edges of the twin tips suggests that these locations contain high stress concentrations, and hence are the main sources of plasticity. This observation agrees with the geometry of the stress field around twin tips calculated by the dislocation theory.[49–53] In particular, the grain analyzed here, which is oriented close to the [1 15 16] crystallographic orientation, contains three slip planes where HDDWs are propagated, namely, (112), (10–1) and (110). These slip planes are slightly deviated to the trace of the (–233) twin habit plane (~2–5°). As these slip planes do not follow the Schmid law with respect to the macroscopic tensile stress state, we suggest that these planes correspond to the highly stressed slip planes with respect to the microscopic stress field around the twin tip.

The evolution of HDDWs into loose dislocation arrangements can be explained as follows. ECCI images of the deformation structure around the twin tips show a high density of <111>-type dislocations within the crystal delimited by the paired HDDWs; see Figures 4(b) and (c). Slip trace analysis reveals that they glide on {112} and {011} slip planes. In the current grain, <111> pencil glide rotates the tensile axis toward [11–1] direction [54]. This crystal rotation can be described as a single rotation around the [101] axis. This means that the intense dislocation activity around the twin tip results in lattice rotations mainly around [101] axis. As a result, the crystal delimited by the paired HDDWs is just rotated several degrees (5–15° from EBSD) around [101] axis with respect to the crystal matrix away from HDDWs. This finding indicates that HDDWs can be considered as simple dislocation tilt boundaries. The stable configuration of dislocation tilt boundaries requires a specific dislocation spacing that is determined by the local stress field.[52,53] Perturbations of the stress field may destabilize such tilt boundaries that subsequently evolve into less defined arrangements. We suggest that the evolution of well-defined dislocation arrangements such as HDDWs to LDAs is associated to the decrease of the intensity of the stress field with the distance from the twin tip. It is worth commenting that due to the bcc structure, the grains with stable crystallographic orientations after tensile loading (orientations along the line between <100>//tensile axis and <110>//tensile axis) contain several {110} and {112} crystallographic slip planes that are slighted deviated from the {233} twin habit planes. Accordingly, in the present bcc Ti-Mo-Fe alloy deformed to tensile loading, we can expect that the HDDWs nucleated at twin tips tend to be aligned to the twin habit plane traces.

The occurrence of interrupted growth twins in the grain interiors of the present multilayered Ti-10Mo-xFe (x = 1–3) alloy can play a significant role on the mechanical behavior due to the development of localized stress gradients that are typically formed around twin tips. These in-grain stress concentrations may result in a high activity of dislocation plasticity, nucleation of twins or even lead to intragranular failure according to the underlying stress relaxation mechanism.[21,26,29,55,56]. In the present case, these stress concentrations are relaxed by dislocation plasticity on {112}<111> and {110}<111> slip systems, and {332}<113> twinning. In particular, our observations reveal that the accommodation of twin tips is mainly performed by dislocation plasticity, as reflected by the low density of twin plates around the twin tips and the high dislocation activity, mainly in the form of HDDWs. Deformation twinning is then not a relevant relaxation mechanism in the present multilayered material. We conclude that the excellent plastic accommodation of the twin tips has significant contribution to the good ductility of the multilayered Ti-Mo-Fe alloy (elongation of 28%).

5. Conclusions

We have investigated the propagation of {332}<113> twins in a multilayered Ti-10Mo-xFe (x = 1–3) alloy fabricated by multi-pass hot rolling. The material contains a macroscopic Fe-graded structure (about 130 μm width) between 1 and 3 wt% Fe in the direction perpendicular to rolling. The following conclusions can be drawn:

The Fe-graded structure has a significant influence on the twin propagation behavior. The propagation of {332}<113> twins that are nucleated in Fe-lean regions (~1 wt% Fe) is interrupted in the grain interiors at a specific Fe content, namely, about 2 wt% Fe. We ascribe this effect to the strong influence of Fe content on the stress for twin propagation, which has been experimentally validated.

Stress accommodation of the twin tips is mainly performed by dislocation plasticity on {112}<111> and {110}<111> slip systems. Deformation twinning is not a relevant relaxation mechanism of the stress concentrations around the twin tips in the present multilayered material.

The high dislocation activity around the twin tips leads to the formation of HDDWs, which are typically associated to strain localization phenomena. These dislocation configurations nucleate at the edges of the twin tips and propagate on highly stressed crystallographic planes. These dislocation configurations are not stable and evolve into loose dislocation arrangements. We ascribe this effect to the stress field around the twin tips. Due to the bcc structure and the deformation texture, HDDWs tend to be aligned to the twin habit plane traces.

The excellent plastic accommodation of the interrupted twin tips in the grain interiors results in the good ductility (total elongation of 28%) of the present multilayered Ti-10Mo-xFe alloy (x = 1–3).

References

[1] Niinomi M. Mechanical biocompatibilities of titanium alloys for biomedical applications. J Mech Beh Biom Met. 2008;1:30–42.

[2] Al-Zain Y, Sato Y, Kim HY, Hosoda H, Nam TH, Miyazaki S. Room temperature aging behavior of Ti-Nb-Mo-based superelastic alloys. Acta Mater. 2012;60:2437–2447.

[3] Al-Zain Y, Kim HY, Hosoda H, Nam TH, Miyazaki S. Shape memory properties of Ti-Nb-Mo biomedical alloys. Acta Mater. 2010;58:4212–4223.

[4] Geetha M, Singh AK, Asokamani R, Gogia AK. Ti based biomaterials, the ultimate choice for orthopaedic implants - a review. Prog Mater Sci. 2009;54:397–425.

[5] Banerjee D, Williams JC. Perspectives on titanium science and technology. Acta Mater. 2013;61:844–879.

[6] Clément N, Lenain A, Jacques PJ. Mechanical property optimization via microstructural control of new metastable beta titanium alloys. JOM. 2007;59:50–53.

[7] Min XH, Emura S, Meng FQ, Mi GB, Tsuchiya K. Mechanical twinning and dislocation slip multilayered dislocation microstructures in b-type Ti-Mo base alloy. Scripta Mater. 2015;102:79–82.

[8] Hida M, Sukedai E, Henmi C, Sakaue K, Terauchi H. Stress induced products and ductility due to lattice instability of β phase single crystals of Ti-Mo alloys. Acta Metall. 1982;30:1471–1479.

[9] Ohyama H, Nishimura T. Effects of alloying elements on deformation mode in Ti-V based β titanium alloy system. ISIJ Int. 1995;35:927–936.

[10] Ling FW, Starke EA, Lefevre BG. Deformation behavior and texture development in beta Ti-V alloys. Metall Trans. 1975;5:179–187.

[11] Hanada S, Takemura A, Izumi O. The mode of plastic deformation of β Ti-V alloys. Trans Jpn Inst Met. 1982;23:507–517.

[12] Christian JW. Some surprising features of the plastic deformation of body-centered cubic metals and alloys. Metall Trans A. 1983;14A:1237–1256.

[13] Taylor G. Thermally-activated deformation of BCC metals and alloys. Prog Mater Sci. 1992;36:29–61.

[14] Lee SH, Hagihara K, Nakano T. Microstructural and orientation dependence of the plastic deformation behavior in β-type Ti-15Mo-5Zr-3Al alloy single crystals. Metall Mater Trans A. 2012;43:1588–1597.

[15] Gysler A, Lütjering G, Gerold V. Deformation behavior of age-hardened Ti-Mo alloys. Acta Metall. 1974;22:901–909.

[16] Kawabata T, Kawasaki S, Izumi O. Mechanical properties of TiNbTa single crystals at cryogenic temperatures. Acta Mater. 1998;46:2705–2715.

[17] Tobe H, Kim HY, Inamura T, Hosoda H, Miyazaki S. Origin of 332 twinning in metastable β-Ti alloys. Acta Mater. 2014;64:345–355.

[18] Rusakov GM, Litvinov AV, Litvinov VS. Deformation twinning of titanium β-alloys of transition class. Met Sci Heat Treat. 2006;48:244–251.

[19] Bilby BA, Crocker AG. The theory of the crystallography of deformation twinning. Proc R Soc A. 1965;288:240–255.

[20] Cottrell AH, Bilby BA. A mechanism for the growth of deformation twins in crystals. Phil Mag. 1951;42:573–581.

[21] Christian JW, Mahajan S. Deformation twinning. Prog Mater Sci. 1995;39:1–157.

[22] Steinmetz DR, Jäpel T, Wietbrock B, Eisenlohr P, Gutierrez-Urrutia I, Saeed-Akbari A, Hickel T, Roters F, Raabe D. Revealing the strain hardening behavior of twinning induced plasticity steels through a dislocation density- and twin evolution-based constitutive model: theory, simulations, experiments. Acta Mater. 2013;61:494–510.

[23] Gutierrez-Urrutia I, Zaefferer S, Raabe D. The effect of grain size and grain orientation on deformation twinning in a Fe–22 wt.% Mn–0.6 wt.% C TWIP steel. Mater Sci Eng A. 2010;527:3552–3560.

[24] Fernandez A, Jerusalem A, Gutierrez-Urrutia I, Perez-Prado MT. Three-dimensional investigation of grain boundary–twin interactions in a Mg AZ31 alloy by electron backscatter diffraction and continuum modeling. Acta Mater. 2013;61:7679–7692.

[25] Mahajan S. Accommodation at deformation twins in bcc crystals. Metall Trans A. 1981;12:379–386.

[26] Eisenlohr A, Gutierrez-Urrutia I, Raabe D. Adiabatic temperature increase associated with deformation twinning and dislocation plasticity. Acta Mater. 2012;60:3994–4004.

[27] Hull D. Effect of grain size and temperature in slip, twinning and fracture in 3% silicon iron. Acta Metall. 1961;9:191–204.

[28] Worthington PJ, Smith E. Slip, twinning and fracture in polycrystalline 3% silicon iron. Acta Metall. 1966;14:35–41.

[29] Mahajan S, Williams DF. Deformation twinning in metals and alloys. Int Metall Rev. 1973;18:43–61.

[30] Meyers MA, Vöhringer O, Lubarda VA. The onset of twinning in metals: A constitutive description. Acta Mater. 2001;49:4025–4039.

[31] Min XH, Tsuzaki K, Emura S, Tsuchiya K. Enhancement of uniform elongation in high strength Ti-Mo based alloys by combination of deformation modes. Mater Sci Eng A. 2011;528:4569–4578.

[32] Gutierrez-Urrutia I, Zaefferer S, Raabe D. Electron channeling contrast imaging of twins and dislocations in twinning-induced plasticity steels under controlled diffraction conditions in a scanning electron microscope. Scripta Mater. 2009;61:737–740.

[33] Gutierrez-Urrutia I, Zaefferer S, Raabe D. Coupling of electron channeling with EBSD: toward the quantitative characterization of deformation structures in the SEM. JOM. 2013;65:1229–1236.

[34] Collings EW. The physical metallurgy of titanium alloys. New York, NY: American Society for Metals; 1984.

[35] Kawabata T, Kawasaki S, Izumi O. Mechanical properties of TiNbTa single crystals at cryogenic temperatures. Acta Mater. 1998;46:2705–2715.

[36] Hanada S, Izumi O. Deformation of metastable beta Ti-15Mo-5Zr alloy single crystals. Metall Trans A. 1980;11:1447–1452.

[37] Sun F, Zhang JY, Marteleur M, Brozek C, Rauch EF, Veron M, Vermaut P, Jacques PJ, Prima F. A new titanium alloy with a combination of high strength, high strain hardening and improved ductility. Scripta Mater. 2015;94:17–20.

[38] Hanada S, Ozeki M, Izumi O. Deformation characteristics in β phase Ti-Nb alloys Metall Trans A. 1985;16:789–795.

[39] Gutierrez-Urrutia I, Raabe D. Multistage strain hardening through dislocation substructure and twinning in a high strength and ductile weight-reduced Fe-Mn-Al-C steel. Acta Mater. 2012;60:5791–5802.

[40] Gutierrez-Urrutia I, Raabe D. Microbanding mechanism in a Fe-22Mn-0.6C (wt.%) high-Mn twinning induced plasticity steel. Scripta Mater. 2013;69:53–56.

[41] Hughes DA. Microstructural evolution in a non-cell forming metal: Al-Mg. Acta Metall Mater. 1993;41:1421–1430.
[42] Kuhlmann-Wilsdorf D. Theory of plastic deformation:-properties of low energy dislocation structures. Mater Sci Eng A. 1989;113:1–41.
[43] Narita N, Takamura JI. Deformation twinning in f.c.c. and b.c.c. metals. In: Nabarro FRN, editor. Dislocations in solids. Vol. 9. Dislocations and disclinations. Amsterdam: North-Holland. 1992, pp. 135–190.
[44] Banerjee S, Naik UM. Plastic instability in an omega forming Ti-15%Mo alloy. Acta Mater. 1996;44:3667–3677.
[45] Chen QZ, Ngan AHW, Duggan BJ. Microstructure evolution in an interstitial-free steel during cold rolling at low strain levels. Proc R Soc Lond A. 2003;459:1661–1685.
[46] Chen QZ, Duggan BJ. On cells and microbands formed in an interstitial-free steel during cold rolling at low to medium reductions. Metall Mater Trans A. 2004;35:3423–3430.
[47] Shen K, Duggan BJ. Microbands and crystal orientation metastability in cold rolled interstitial-free steel. Acta Mater. 2007;55:1137–1144.
[48] Poddar D, Cizek P, Beladi H, Hodgson PD. The evolution of microbands and their interaction with NbC precipitates during hot deformation of a Fe-30Ni-Nb model austenitic steel. Acta Mater. 2015;99:347–362.
[49] Hu SY, Henager CH, Chen LQ. Simulations of stress-induced twinning and de-twinning: A phase field model. Acta Mater. 2010;58:6554–6564.
[50] Kamat SV, Hirth JP, Müllner P. The effect of stress on the shape of a blocked deformation twin. Phil Mag A. 1996;73:669–680.
[51] Barnett M, Setty M, Siska F. Estimating critical stresses required for twin growth in a magnesium alloy. Metall Mater Trans A. 2013;44:2962–2969.
[52] Nabarro FRN. Theory of crystal dislocations. Oxford: Oxford University Press; 1967.
[53] Hirth JP, Lothe L. Theory of dislocations. New-York: McGraw-Hill; 1968.
[54] Hosford WF. Mechanical behavior of materials. Cambridge (UK): Cambridge University Press; 2010.
[55] Priestner R, Leslie WC. Nucleation of deformation twins at slip plane intersections in B.C.C. metals. Phil Mag. 1965;11:895–916.
[56] Bieler TR, Eisenlohr P, Zhang C, Phukan HJ, Crimp MA. Grain boundaries and interfaces in slip transfer. Curr Opin Solid State Mater Sci. 2014;18:221–226.

Investigation of microstructure in additive manufactured Inconel 625 by spatially resolved neutron transmission spectroscopy

Anton S. Tremsin[a], Yan Gao[b], Laura C. Dial[b], Francesco Grazzi[c] and Takenao Shinohara[d]

[a]Space Sciences Laboratory, University of California at Berkeley, Berkeley, CA, USA
[b]General Electric Global Research Center, Niskayuna, NY, USA
[c]Consiglio Nazionale delle Ricerche, Istituto Sistemi Complessi (CNR-ISC), Sesto Fiorentino (FI), Italy
[d]Japan Atomic Energy Agency Tokai-mura, Naka-gun Ibaraki, Japan

ABSTRACT

Non-destructive testing techniques based on neutron imaging and diffraction can provide information on the internal structure of relatively thick metal samples (up to several cm), which are opaque to other conventional non-destructive methods. Spatially resolved neutron transmission spectroscopy is an extension of traditional neutron radiography, where multiple images are acquired simultaneously, each corresponding to a narrow range of energy. The analysis of transmission spectra enables studies of bulk microstructures at the spatial resolution comparable to the detector pixel. In this study we demonstrate the possibility of imaging (with ~100 μm resolution) distribution of some microstructure properties, such as residual strain, texture, voids and impurities in Inconel 625 samples manufactured with an additive manufacturing method called direct metal laser melting (DMLM). Although this imaging technique can be implemented only in a few large-scale facilities, it can be a valuable tool for optimization of additive manufacturing techniques and materials and for correlating bulk microstructure properties to manufacturing process parameters. In addition, the experimental strain distribution can help validate finite element models which many industries use to predict the residual stress distributions in additive manufactured components.

KEYWORDS
Non-destructive testing; additive manufacturing; microstructure; neutron imaging

CLASSIFICATION
10 Engineering and Structural materials; 106 Metallic materials; 303 Mechanical/Physical processing; 504 X-ray/ Neutron diffraction and scattering

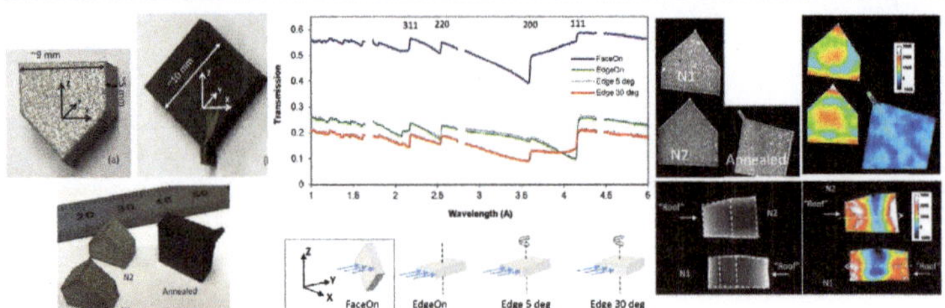

Inconel 625 additive manufactured samples

Neutron transmission spectra measured for different sample orientations

Neutron transmission images and strain maps reconstructed from measured spectra

1. Introduction

The additive manufacturing (AM) technology recently progressed from the technique mostly used for rapid prototyping to production of complicated engineering components consisting not only of plastics but also of metals and metallic alloys [1–4]. In many cases additive manufacturing can produce parts that cannot be made with conventional machining technology. Additive manufacturing is fundamentally different than many traditional, subtractive manufacturing processes. In direct metal laser melting (DMLM) or selective laser melting (SLM) processes, substantial differences in microstructure are observed as compared to traditional manufacturing methods. In these processes where samples are grown layer by layer, the microstructure is shaped by the interaction of materials within the layer and between subsequent layers as complex thermal processing cycles are applied. Residual stress distribution is among the most important microstructural variables to be investigated and understood by additive manufacturers in the industry. Also the presence of processing defects (e.g. micropores or impurities) can introduce local strength

CONTACT Anton S. Tremsin ast@ssl.berkeley.edu

deficiencies. Despite all the knowledge and optimization of the manufacturing tools there is a substantial variability between different machines.[4] At present, additive manufactured components may still be inferior to conventionally fabricated ones in overall performance. Therefore understanding the links between the microstructure, composition, and processing parameters is very important for the progress of this technology.[1–5]

In this article we demonstrate the possibility of studying the microstructure of Inconel 625 materials made by additive manufacturing with energy-resolved neutron imaging. This neutron technique can provide microstructure and composition information on relatively large components, which cannot be obtained by conventional characterization techniques, such as optical and electron microscopies, and X-ray diffraction. The unique capability of neutrons to penetrate thick metal samples (e.g. several centimeters of steel, titanium and nickel-based materials) can reveal the bulk properties of samples without alteration of sample integrity. Conventional neutron radiography and tomography are widely used to reveal the internal structure of complicated objects.[6,7] including additive manufacturing samples.[8] Addition of spectroscopic information in each pixel of the detector [9,10] also enables studies of material microstructures, such as texture,[11–14] strain,[15–17] grain orientation,[18] and, in some cases, bulk elemental composition.[19,20] This paper presents the results of the non-destructive study of Inconel 625 samples, manufactured by the DMLM method.[1] The results of these experiments demonstrate the capability of achieving ~100 μm spatial resolution for investigating the bulk microstructure properties of metal AM samples.

2. Methods

The samples studied in this experiment were manufactured by using an SLM 250 additive manufacturing machine (SLM Solutions GmbH, Lübeck, Germany). Two 9 × 9 × 5 mm samples, N1 and N2, were prepared using similar processing parameters (Figure 1). A third additional sample (as shown in Figure 1(c)) was built in the same way and was heat treated at 1050°C in argon for 2 h and air cooled prior to measurement to relieve as-built stresses in the part. This sample was used as an unstrained reference for strain calculation as described below. The photographs of the samples are shown in Figure 1, with growth direction along the Z-axis. During the experiment, neutrons were propagated along both the Y-axis (referred to as face-on orientation) and X-axis (edge-on orientation). There are

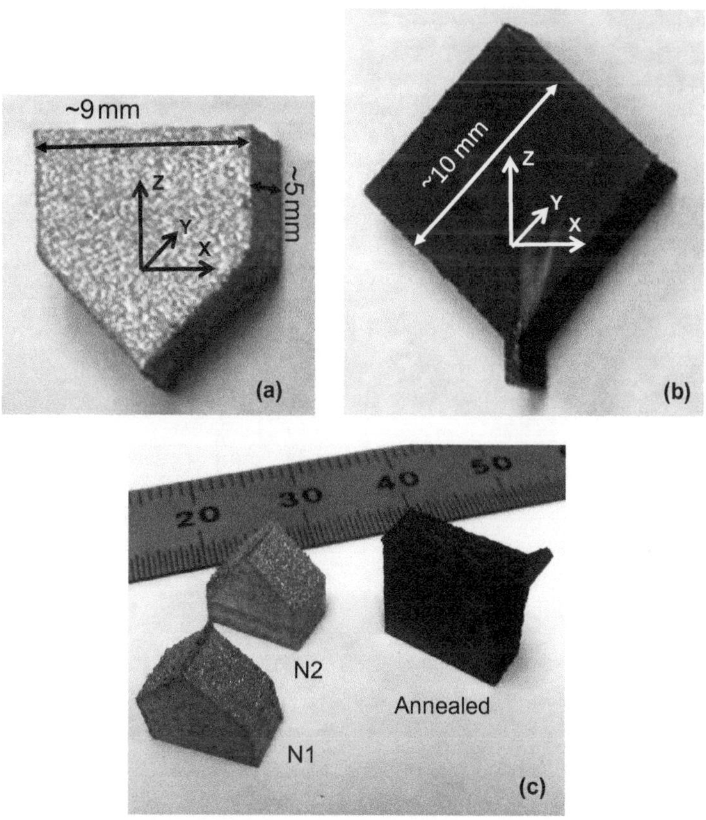

Figure 1. Photograph of Inconel 625 additive manufacturing samples investigated in the present study. The samples were measured in two orthogonal orientations: with neutrons propagating along the Y-axis and along the X-axis. The sample build direction was along the Z-axis. The sample shown in (b) was annealed for 2 h in argon at 1050°C.

several methods to measure spatially resolved neutron transmission spectrum. Velocity selectors [21] and double crystal monochromators [22,23] can be utilized at continuous neutron sources. The energy resolution ΔE/E in these setups is ~10–15% for the former and ~3% for the latter. The transmission spectra in these experiments can be taken by scanning the neutron energy and recording consecutive images at each energy. In some cases that energy resolution is sufficient for the studies of microstructure in metal samples, where the presence of Bragg edge scattering provides the contrast in the transmission images. However, measurement of strain is limited by the energy resolution of these scanning techniques. It has been demonstrated that imaging of strain distribution can be done with better accuracy, down to ~100 microstrain [15–17] if neutron spectra are measured with 10× better energy resolution of ~0.3%, as in the case of pulsed neutron sources. It is important to note that the neuron transmission spectrum over a broad energy range can be obtained simultaneously in one measurement at pulsed neutron sources. Although studies involving measurements of neutron transmission spectra at spallation sources have been widely used to date in a single pixel configuration, it is only recently that spatially resolved neutron transmission has been possible due to the development of event counting neutron detectors with multiple pixels, such as 10 × 10 detector with 2 × 2 mm^2 pixels [24], 512 × 512 detector with 55 × 55 μm^2 pixels [25,26] and event-encoding detectors [27,28] with sub-mm spatial resolution.

The experiments reported in this paper were conducted at the Noboru beamline [29] of Japanese Spallation Neutron Source, operated at 25 Hz frequency and 250 kW power during the measurements. The neutron pulses are produced by spallation initiated by two 100 ns proton pulses, separated by 600 ns. After moderation the spallation neutrons propagated towards the sample and detector over ~14.5 m distance, Figure 2. The energy of each neutron can be calculated from its time of flight (TOF) providing the detection system can record the time of neutron arrival relative to the source trigger. A neutron counting MCP/Timepix detector [25,26] was used for the measurement of the transmission spectra in each pixel. Unique neutron sensitive microchannel plates for this detector were provided by Nova Scientific, Inc (Sturbridge, MA, USA). This type of detector is capable of registering multiple (>10^5) simultaneous events with 55 μm spatial resolution and 20–500 ns timing resolution, depending on the energy of the neutron. A dataset consisting of several thousand images, each corresponding to a particular neutron energy, was acquired in each measurement. Thus 262144 spectra, from the detector consisting of 512 × 512 pixels, were acquired simultaneously, enabling spatially resolved studies of bulk microstructure. The data were acquired with the energy binning of ΔE/E ~0.05–0.1%. The strain was reconstructed from datasets integrated over several hours, in order to acquire sufficient number of neutron counts in each energy bin and at each pixel of the 2D detector.

2.1. Reconstruction of strain distribution across the samples

The high degree of penetration into many metallic materials makes neutrons unique probes of internal microstructure in these samples, which can be as thick as a few centimeters (and even tens of centimeters for the case of aluminum). Measurements of neutron transmission spectra in each 55 μm pixel, enabled recently by advances in neutron instrumentation, allow the analysis of microstructure with a high spatial resolution. In the transmission configuration the measured spectra apparently provides information on the microstructure averaged

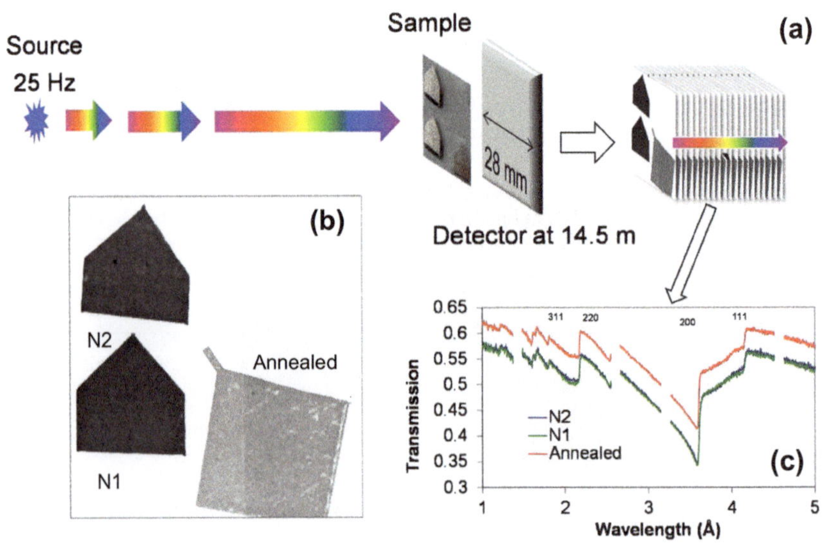

Figure 2. (a) Schematic diagram of measurement setup: neutrons with a wide energy range are produced by a pulsed source. Sample is placed in front of detector at 14.5 m distance from the source. A set of images (several thousands), each corresponding to a certain range of neutron energy/wavelength is acquired simultaneously. The full spectrum (summed over the entire energy range) transmission image is shown in (b). Transmission spectra (c) can be obtained from selected areas of the image.

over the entire sample thickness, except for some special cases such as imaging of a diffracted signal from a single crystal or grain, which can be specific to a certain crystal/grain and is independent from the surrounding materials. Therefore the stain reconstructed in the present study represents the value averaged along the path of neutrons within the sample. However, that averaged strain value can be reconstructed within each 55 μm pixel, providing there is sufficient number of neutrons acquired in such a small area and maps of residual strain distribution can be obtained across the entire sample in one measurement.

The presence of sharp variation in transmission spectra, often referred to as Bragg edges, caused by diffraction of incident neutrons, enables determination of crystal lattice strain, as suggested in [30] and demonstrated in subsequent experimental studies.[15–17] The wavelength λ_{hkl} at which neutrons scatter to angle ϑ from a set of planes with interplanar distance d_{hkl} is determined by the Bragg equation:

$$\lambda_{hkl} = 2d_{hkl} \sin \vartheta \tag{1}$$

where the Miller indices (*hkl*) [31] define a specific set of lattice planes in a crystalline sample. With ϑ reaching $\pi/2$ neutrons are scattered backward and there is no scattering possible at longer wavelength for that set of planes, which results in a sharp increase of neutron transmission. Thus measurement of a Bragg edge wavelength is directly related to the interplanar distance d_{hkl} and can be used for determination of strain from the equation:

$$\varepsilon = (d - d_0)/d_0 = (\lambda - \lambda_0)/\lambda_0. \tag{2}$$

where d_0 is the interplanar distance of the unstrained material. In this study we assumed that the strains were relieved in the annealed sample and consequently the averaged value of d_{hkl} of that sample was taken as the unstrained interplanar distance d_0 for a particular Bragg edge.

The energy resolution of our experimental setup is determined by the width of the moderated neutron pulse, typically ΔE/E ~ 0.2–0.4% for the wavelengths of interest, limiting reconstructed strain values to 2000–4000 microstrain resolution. However, the position of the Bragg edge, as in case of diffraction techniques, can be determined with a better accuracy by fitting an analytical function to measured spectra,[15–17,30] improving the resolution to ~100 microstrain. The fitting of an analytical function with five fitted parameters (λ_e – the wavelength of the Bragg edge; σ, τ, C_1 and C_2 – the width, asymmetry, offset and height of the edge, respectively) [15,16,30]

$$T(\lambda)|_{\lambda_0,\sigma,\tau,C_1,C_2} = C_1 + C_2\left[erfc\left(\frac{\lambda_e - \lambda}{\sqrt{2}\sigma}\right) - \exp\left(\frac{\lambda_e - \lambda}{\tau} + \frac{\sigma^2}{2\tau^2}\right) * erfc\left(\frac{\lambda_e - \lambda}{\sqrt{2}\sigma} + \frac{\sigma}{\sqrt{2}\tau}\right)\right] \tag{3}$$

can be performed for the spectra acquired for each 55 × 55 μm² pixel of the dataset. However, due to the limited neutron flux, the measured spectrum in such a small area is noisy even after several hours acquisition time. In order to reduce the statistical noise we combined the spectra from neighboring pixels, typically in the area from 0.5 × 0.5 to 1.5 × 1.5 mm², and repeated that process ~250,000 times covering all 55 μm pixels of the dataset. The spectral averaging obviously would smooth sharp variations of reconstructed strain with "running average", but still allows for the averaged strain reconstruction within each 55 μm pixel.

3. Results

3.1. Full spectrum radiography

The conventional white-spectrum neutron radiography (summation over all images in a dataset) of the samples is shown in Figure 3. The contrast in these images is due to the density and/or sample composition variation instead of crystallographic (including texture) differences. The white spots in these images are most likely due to voids and micropores, which in the case of the annealed sample are elongated perpendicular to the growth axis. The black spots are likely caused by impurities containing elements with larger neutron attenuation. The contrast in these images is greatly enhanced so that non-uniformities, even at a very small fraction of the sample volume, can be observed. No substantial difference between the annealed and the as-printed samples is observed based on the results shown in Figure 3.

3.2. Energy resolved neutron imaging

The energy-resolved imaging data enables extraction of the transmission spectra from small areas within a sample. The transmission spectra shown in Figure 4 exhibit sharp variations of transmission at a specific set of wavelengths, characteristic of the fcc crystal structure of Inconel 625. The transmission of the annealed sample is higher due to the smaller thickness of the sample along the Y-axis. There is no Bragg scattering above ~4.14 Å (the (1 1 1) edge wavelength), corresponding to the largest interplanar distance of ~2.07 Å in these samples. The images acquired in a narrow-energy range around each Bragg edge are shown in Figure 5. There are some weak features along the X-axis, in the growth plane, on the image around (1 1 1) edge, while the image at (2 0 0) edge reveals the vertical stripes, which are parallel to the growth direction. All these features are most likely related to the difference in texture or anisotropy of grain orientation. The lack of large contrast variation in these images indicates that these samples are uniform in texture and no gross crystal structural defects are present. The transmission image acquired with wavelengths above the last Bragg edge (4.5 to 7.5 Å, Figure 6) reveals the presence of features most likely caused by impurities

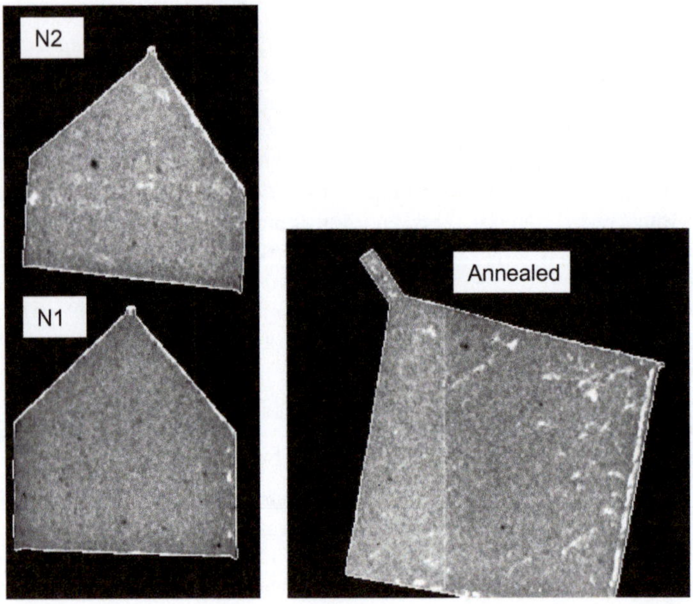

Figure 3. Neutron transmission images obtained by summation over the entire energy range (full spectrum transmission). The features seen within the samples are due to the presence of inclusions or voids, as confirmed by the images shown in Figure 7. The vertical line seen in the annealed sample is an artifact of detector calibration (the border between the two chips of the readout is visible).

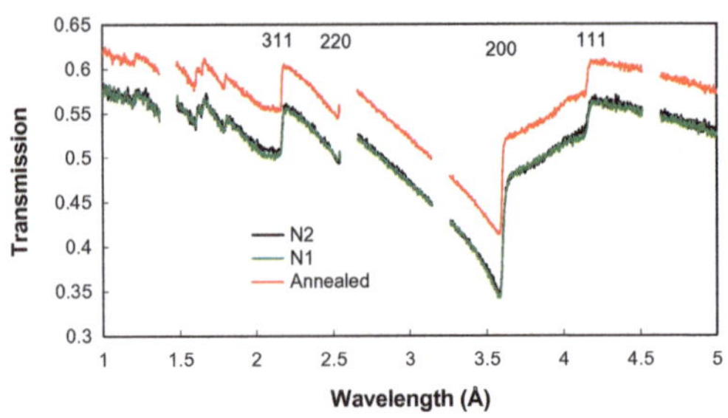

Figure 4. The measured transmission spectra of three samples with neutron beam along the Y-axis of the sample. Sharp increase of transmission is due to Bragg scattering at the corresponding crystallographic planes, marked with Miller indices. The gaps in the spectrum are due to the detector deadtime required for the data readout.

and voids and not related to crystallographic properties as no scattering exists at these wavelengths.

3.3. Strain and microstructure along Y-axis reconstructed by Bragg edge analysis

The three Bragg edges used in the reconstruction of strain and texture are shown in Figure 7. The strain was calculated from Equation (2) for the measured edge wavelength, parameter λ_e in Equation (3), while variation of texture was reconstructed from parameter σ (the width), C_1 (the offset) and C_2 (the height) of the Bragg edges obtained by fitting to the spectra acquired around each pixel according to Equation (3). These reconstructed parameters obviously represent the values averaged over the sample thickness as measured transmission spectra is affected by absorption and scattering along the neutron path within the entire sample. The noise of measured data for (1 1 1) edge is the highest among three edges due to the amplitude of that edge being the smallest, while the strongest edge was (2 0 0). The nonlinear least squares fitting was implemented in the Bragg edge analysis. The fitting was performed for each 55 × 55 μm pixel of our dataset, thus requiring ~60,000 fits to be performed for each reconstructed image of parameters representing crystallographic properties of the samples, e.g. Figure 8. The results of the analysis for the reconstructed Bragg edge position are shown in Figure 8(a). There is substantial difference between the annealed and as-manufactured samples, as well as within the as-manufactured samples. The averaged λ_e values across the annealed sample was used as λ_0 of the unstrained material, which was used to calculate the strain values shown in Figure 8(b). Substantial tensile residual strains are present in the non-annealed samples, with the highest strain concentration at the bottom (~1500 με) and at the tip (~2000 με) of the samples, as well as in the area around the center. The distribution of measured λ_e and strain values is shown in Figure 9, while the cross sections through the strain maps are shown in Figure 10. Figure 9 shows histograms of lattice spacing and strain distribution based on pixelated data over the entire samples. Note the distribution of λ_e is narrower for the annealed sample than for the as-built samples, which supports using post-printing annealing to produce an unstrained sample for strain analysis. Figure 9(b) shows that the strain distributions for both N1 and N2 are similar, about twice broader than the

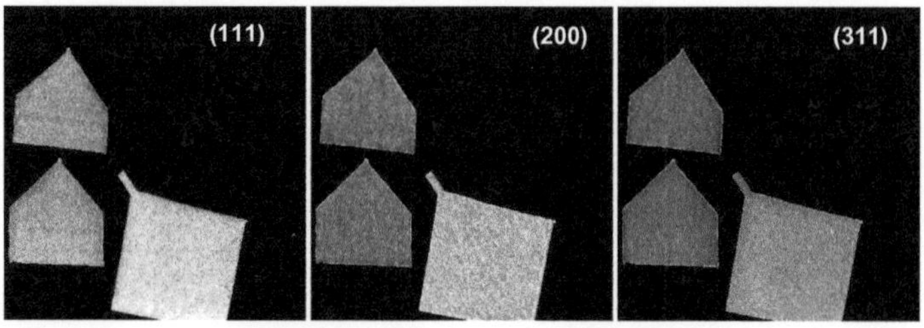

Figure 5. Neutron transmission images obtained at the narrow range of wavelength (0.1–0.2 Å) near the specific Bragg edges. The contrast is mostly due to the variation of crystallographic properties of the samples, e.g. texture.

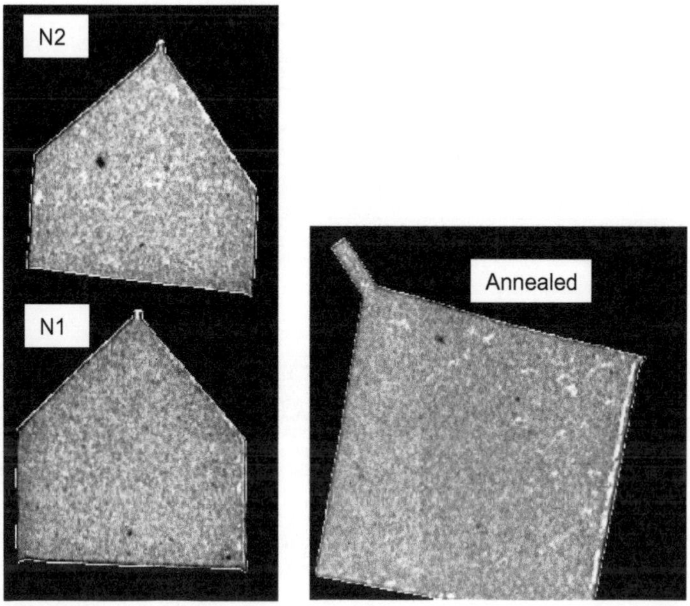

Figure 6. Neutron transmission image obtained by summation of images acquired for the wavelengths past the last Bragg edge (4.5 to 7.5 Å). The contrast is due to density/composition variation from either inclusions or voids, instead of crystallographic properties as no coherent scattering is possible at these wavelengths.

annealed sample, and both are in tensile, with maximum tensile strain up to ~1700 microstrain. Note that the measured strains are averaged over the sample thickness (5.2 mm along the Y-axis), and the transmission Bragg edge analysis shown here is sensitive *only* to the strain along the direction of neutron propagation (the Y-axis in this sample orientation).

Figure 10 shows the strain profiles along two specific directions, along the build direction Z and along the growth plane X. Each point represents strain, calculated for the spectrum integrated over 30 × 30 pixels (~1.6 × 1.6 mm^2 area). The strain variations of the annealed sample is much smaller than those for N1 and N2, especially along the build direction (Figure 10(b)).

The maps of fitted parameters σ and C_2 are shown in Figure 11. The width of the Bragg edge (Figure 11(a)) is a manifestation of grain size and intergranular defects distribution with respect to the neutron beam direction. For a single crystal sample which ideally is a single grain with a unique orientation, the measured spectrum shows a sharp dip as the Bragg scattering occurs at a single wavelength for a specific lattice plane. For a non-textured small grain polycrystalline sample for which grains are randomly oriented, the measured transmission has a smooth and gradual decrease from shorter wavelength towards the Bragg edge, appearing at the wavelength, defined by scattering from those grains whose lattice planes are perpendicular to the incident beam. For a relatively large size set of grains and for grains containing internal defects, the shape of the Bragg edge will appear different from these two extreme cases, and therefore neutron transmission spectra may be used as a qualitative measure of the degree of microstructure. A full spectrum Rietveld-type analysis is more appropriate for microstructural analysis and is not performed in this study as data analysis tools are still under development. The Bragg edge width σ, on the other hand, is a measure

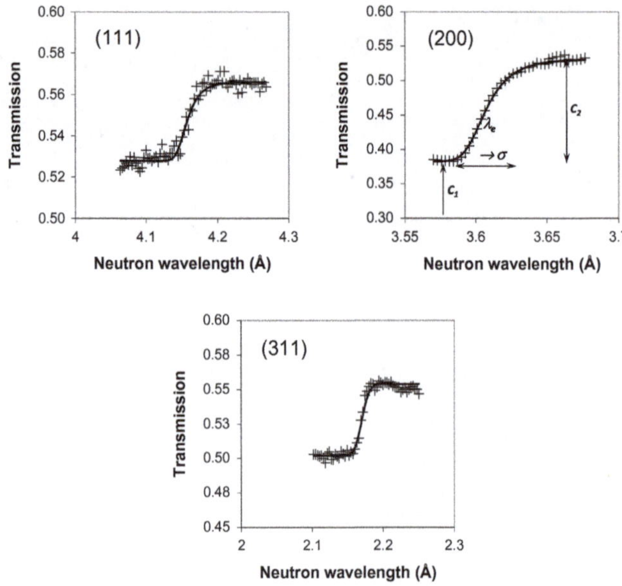

Figure 7. The measured Bragg edges for sample N1 (markers) and fitted curves. The spectra are from pixels summed over an area of several mm wide to improve counting statistics.

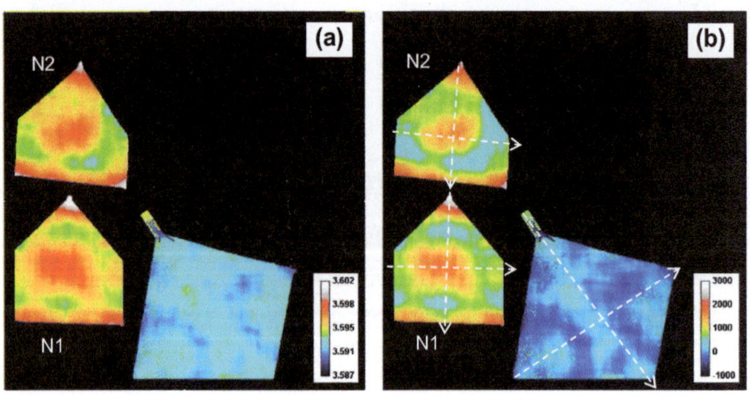

Figure 8. (a) Map of (2 0 0) Bragg edge wavelengths λ_e for the transmission spectra acquired along the direction perpendicular to the face of the samples (along the Y-axis as shown in Figure 1). λ_e values are obtained by fitting Equation (3) to the measured transmission spectra. Around 60,000 fits are performed for that image. The value in each pixel is based on fitting of combined spectra from a 30 × 30 pixels area (or 1.65 × 1.65 mm) around this pixel. The color scale indicates the values of the λ_e edge position. (b) Map of strain values (integrated along the Y-axis) calculated from (a) assuming the unstrained value λ_0 of 3.592 Å (average value measured from the annealed sample). The color scale indicates strain values in microstrain. The dashed lines indicate the position of cross sections through the strain map shown in Figure 10.

of edge sharpness and is related to the mosaic spread of lattice plane orientation perpendicular to the beam. As the curve fitting method (shown in Figure 7) covers only a limited wavelength range, σ doesn't measure the overall microstructure of the sample. In Figure 11(a), the annealed sample shows overall narrower edge width than as-built samples, which probably means that annealing has reduced internal defects and increased the size of grains which are nearly aligned to the direction of the Y-axis. The consistency of edge width between N1 and N2 is indicative of the robustness of the curve fitting technique, as the mosaic spreads of grain orientations for as-built samples are expected to be similar. The maps of fitted height and the offset parameters of the Bragg edge (parameter C_2 of Equation (3)) reveal the presence of some horizontal (along the growth planes) and vertical (along the growth direction) lines, caused by variation of the sample microstructure, namely the presence of referred grain orientations in this case. The height of the Bragg edge (parameter C_2) gets smaller with the presence of preferred grain orientations,

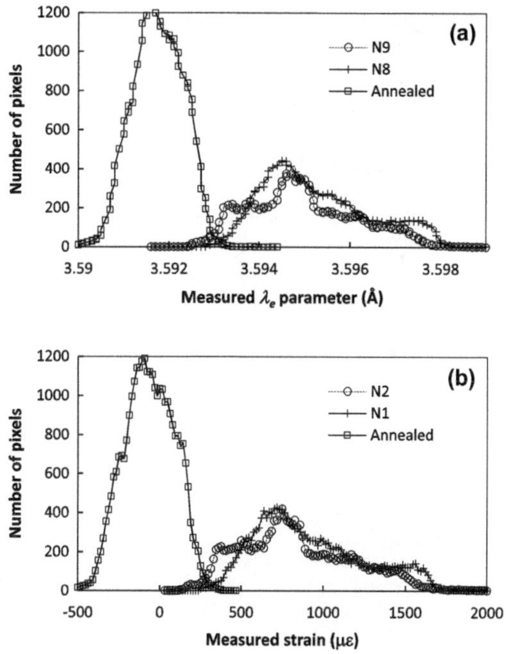

Figure 9. Histograms of measured λ_e parameter (2× of the interplanar lattice spacing) and strain at (2 0 0) Bragg edge: (a) distribution of measured λ_e values; (b) distribution of the strain values along the Y-axis.

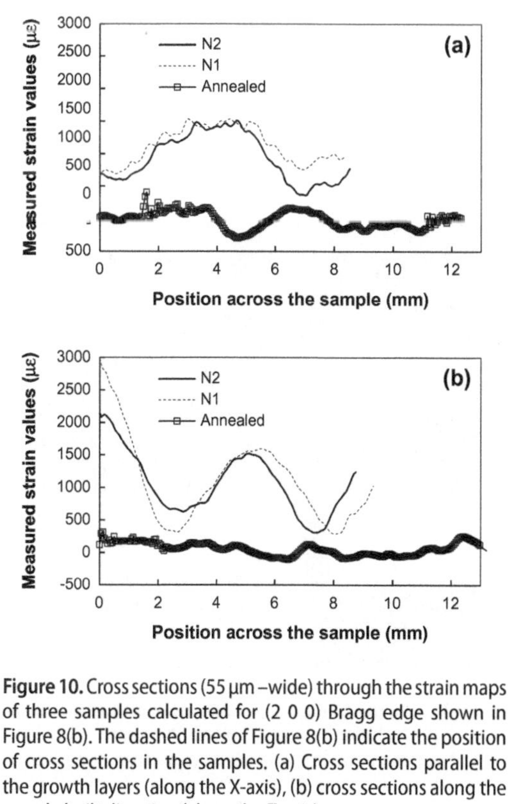

Figure 10. Cross sections (55 μm –wide) through the strain maps of three samples calculated for (2 0 0) Bragg edge shown in Figure 8(b). The dashed lines of Figure 8(b) indicate the position of cross sections in the samples. (a) Cross sections parallel to the growth layers (along the X-axis), (b) cross sections along the sample built direction (along the Z-axis).

as seen in Figure 11(b). The striations along Z-axis seen in these images indicate that grain orientations have some mm-scale correlation between the layers along the build direction. These features are not seen in the narrow-energy image (Figure 5) acquired around the (2 0 0) Bragg edge, demonstrating the increased sensitivity of fitted data to the microstructure variation. The higher energy resolution of the pulsed neutron source is crucial for revealing such microstructural non-uniformities, which may not be possible with continuous neutron sources.

The strain maps obtained from the Bragg edge analysis of (3 1 1) and (1 1 1) edges (Figure 12) show different residual strain values for different crystallographic planes. For example, the strain reconstructed from the (1 1 1) edge for sample N2 shows the compressive strains at the bottom of the sample, and nearly zero-strain values at the bottom of sample N1. Actually the strains from different edges are not necessarily the same as they are from different grains. For example, the (2 0 0) edge is formed by Bragg scattering from a set of grains whose (2 0 0) planes are perpendicular to the neutron beam, while the (1 1 1) edge is formed by a *different* set of grains whose (1 1 1) planes are perpendicular to the neutron beam. These two sets of grains with different orientations may have experienced different local stresses. The intergranular interaction may put one grain in tension and its neighboring grains in compression. Even when both grains are under the same stress, their strain responses will depend on Miller indices. Moreover there is a strong grain orientation level and distribution for different reflections as observed in Figures 11(b) and 13. For a Ni base alloy like Inconel 625, the (1 1 1) planes are harder to deform than (2 0 0) planes. In other words, under the same stress, one would expect a larger strain from the (2 0 0) edge than the (1 1 1) edge. Considering the multiplicity and the distribution of the grains along the three (1 1 1), (2 0 0) and (3 1 1) directions, for an estimation of the average strain in the sample it is more reliable to base it on the one exhibiting the higher spherical symmetry, as is the (3 1 1) reflection. The stresses experienced by certain grains result from a combination of thermal cycles and geometric constraints from the preceding layer onto which the new layer is formed and from their neighboring grains. Qualitatively, it is feasible to visualize that during heating or cooling of the manufacturing process, the thermal expansion can induce different strain responses from different grains depending upon their orientations; and due to their different elastic constants, some grains may be under tension and some under compression. The total stress might be balanced or cancelled out when stresses from all grains are taken into account.

Figure 13 shows the distributions of edge height (coefficient C_2) for the (3 1 1) Bragg edge. Distinct horizontal features parallel to the growth planes are apparent for all

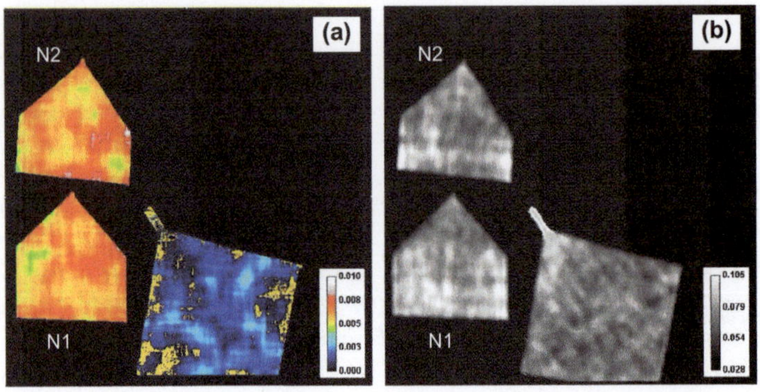

Figure 11. The maps of (2 0 0) Bragg edge parameters obtained by fitting Equation (3) to the measured data. (a) The width parameter σ (Å) calculated with 1.65 × 1.65 mm (30 × 30 pixels) running average over the area where spectra are obtained. (b) The height of the edge C_2 calculated with 0.55 × 0.55 mm (10 × 10 pixels) running average of the spectra.

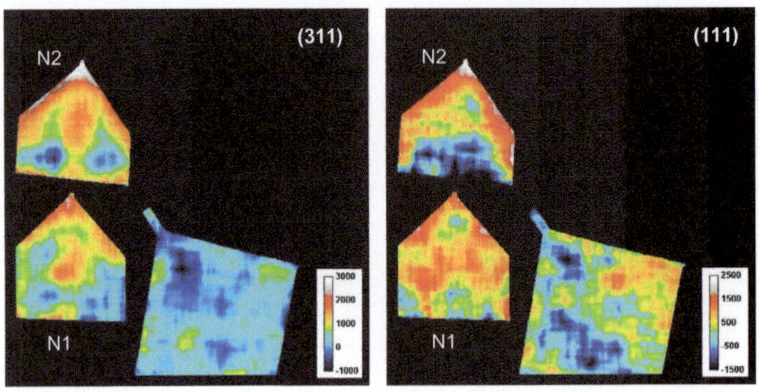

Figure 12. Strain maps along Y-axis measured for the (3 1 1) and (1 1 1) Bragg edges. Averaged spectra in the 1.65 × 1.65 mm area are used as in Figure 8, but accuracy of reconstructed strain for these edges is lower as these Bragg edges are not as strong as (2 0 0), leading to noisier fitted results. The average λ_0 values from the annealed sample, 2.1655 Å and 4.1415 Å, respectively, were used as unstrained values for calculating the (3 1 1) and (1 1 1) strain maps.

three samples, with the most prominent line present in the middle of samples N1 and N2. Their connection to the manufacturing process is being investigated.

3.4. Strain and microstructure along X-axis (along the growth planes)

To understand the strain distribution in the orthogonal direction, the measurements were also performed with samples positioned in the edge-on orientation relative to the neutron beam. For the annealed sample, the neutron beam was at 45° with respect to the X-axis (i.e. perpendicular to one of the edges of the sample), while samples N1 and N2 were positioned such that neutrons propagate nearly parallel to their growth planes, along the X-axis. The white-beam neutron transmission radiography shown in Figure 14 exhibits no strong features. Not even pores or impurities which are seen in Figure 3 are visible in this sample orientation. The neutron transmission integrated across the entire thickness (~9 mm) of the sample might have canceled out the effect of small pores or impurities seen in Figure 3. However, the

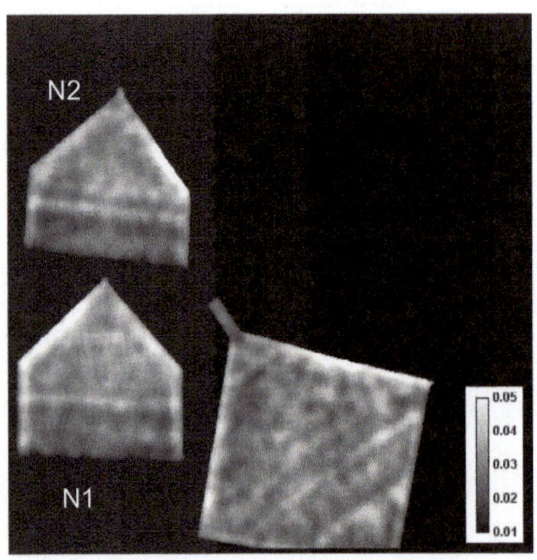

Figure 13. The maps of (3 1 1) Bragg edge height (parameter C_2) obtained by fitting Equation (3) to the measured data. Fitted map of parameter σ is not shown as it was too noisy due to lower neutron counting statistics for that edge. The height of the edge C_2 is calculated with 0.55 × 0.55 mm running average of the spectra.

spectra shown in Figure 15 reveal substantial differences in the bulk microstructure of these samples. The (2 0 0) Bragg edge is nearly absent in the spectra measured for the annealed sample. At the same time, the sharpness of (1 1 1) edge for the annealed sample is substantially higher than that of samples N1 and N2. The small height of the (2 0 0) edge and sharpness of the (1 1 1) edge could be interpreted as that the build direction is along (1 0 0), which is the diagonal of this sample, which would result in large population of (1 1 1) and small populations of (2 0 0) planes perpendicular to the beam. Note the angle between (1 0 0) and (1 1 1) is 54.74° and this is very close to the incident beam direction. This is also consistent with the data from N1 and N2, as both samples exhibit a sharp (2 0 0) edge: this is because the neutron beam was along the X-axis and there would be a large population of grains with (2 0 0) planes perpendicular to the beam if the build direction is along (1 0 0). For the same reason, there would be a small population of (1 1 1) planes perpendicular to the neutron beam for N1 and N2; as a result, the (1 1 1) edge is nearly missing for N1.

The strain map and the map of the (2 0 0) Bragg edge height parameter C_2 for the samples N1 and N2, shown in Figure 16, indicate that the tensile strain is present at the top and at the bottom of these samples, similar to the strains averaged along Y-axis. However, the middle section of these samples did not show tensile strain along

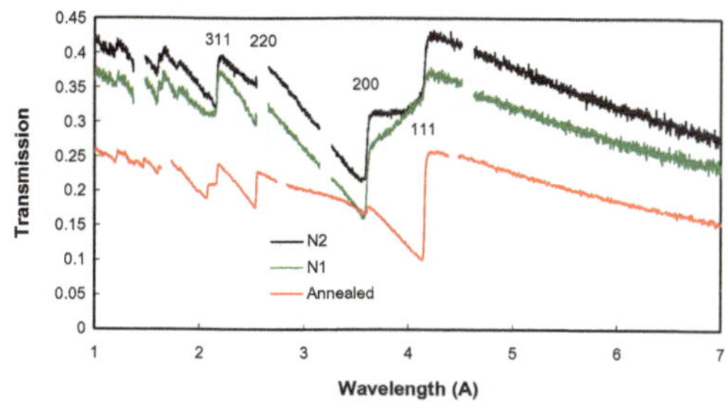

Figure 15. The measured transmission spectra, combined data from all the pixels within a sample, with neutron beam propagating along the growth plane (the X-axis) for samples N1 and N2 and at 45° relative to the growth plane for the annealed sample. The gaps in the spectrum are due to the detector deadtime required for the data readout. The pronounced presence of texture in sample N1 is evident by the slow transmission increase between (2 0 0) and (1 1 1) edges.

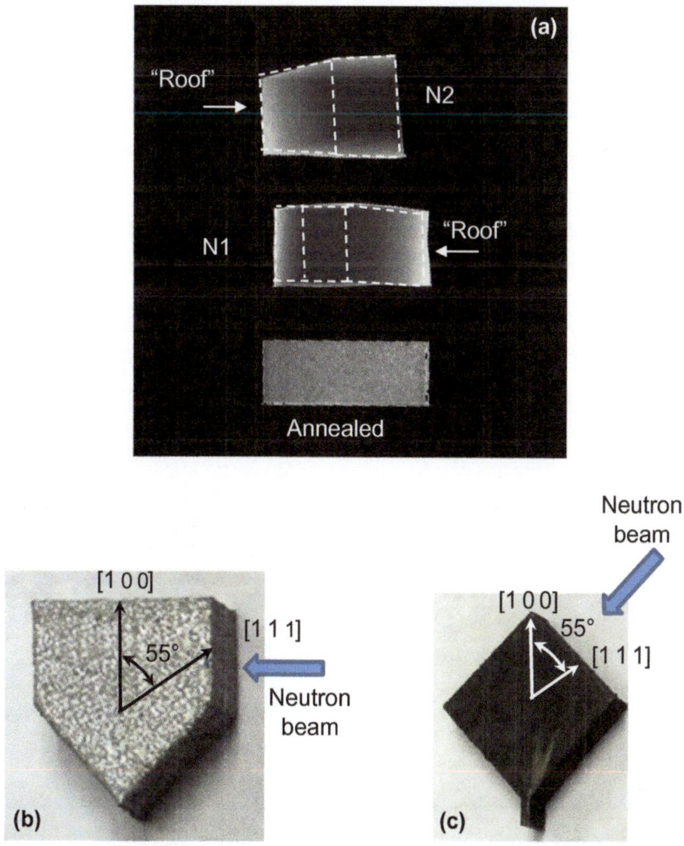

Figure 14. (a) Full spectrum transmission images obtained when the neutron beam entered the sample edge on. Samples N1 and N2 were oriented with neutrons travelling along the growth plane (along the X-axis, as shown in (b)), while for the annealed sample neutron beam was approximately perpendicular to an edge, or at 45° relative to the growth plane (as shown in (c)).

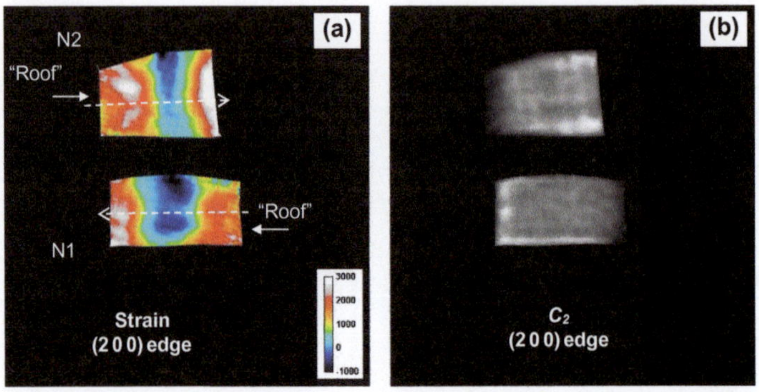

Figure 16. (a) The map of averaged strain along the X-axis calculated by fitting to measured neutron transmission spectra of (2 0 0) Bragg edge. (b) The edge height parameter C_2 measured when samples N1 and N2 were mounted with neutrons travelling along the growth plane (X-axis). The strain was obtained with 1.65 × 1.65 mm running average of the spectra, while the parameter C_2 was obtained with 0.55 × 0.55 mm averaging of the spectra. The dashed lines in (a) indicate the location of cross section through the strain map shown in Figure 17.

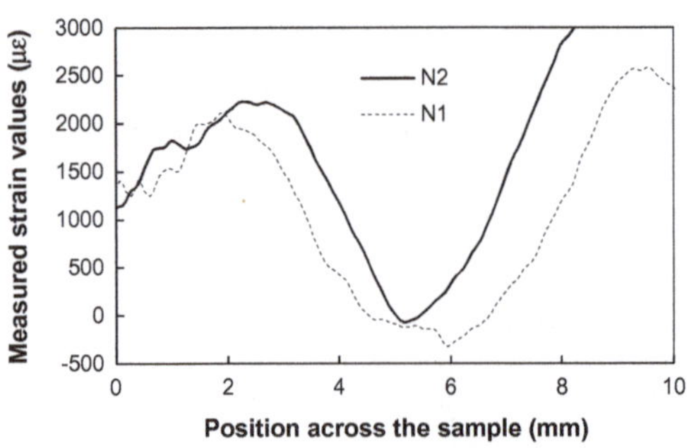

Figure 17. Cross sections (55 μm wide) through the X-axis strain maps for samples N1 and N2 calculated for (2 0 0) Bragg edge shown in Figure 16(a).

the X-axis, but rather small compressive strain values, as shown in Figure 17. There are also some features parallel to the growth direction seen in Figure 16(b), confirming the above-mentioned correlation of grain orientations between subsequent layers. However, that correlation is not preserved along the Y-axis as the layers along that axis are formed by separate passes during sample manufacturing.

To study the variation of microstructure dependence on sample orientation, the annealed sample was measured in the edge-on orientation with 0, 5 and 30° rotations around the Y-axis, Figure 18. The transmission spectra of the annealed sample showed substantial variation with sample rotation, indicating

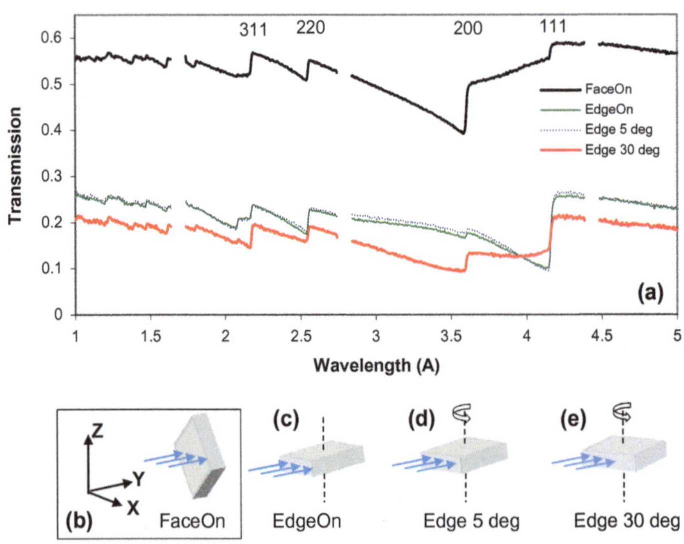

Figure 18. The transmission spectra of the annealed sample measured at different sample orientations relative to the incoming beam. The sample build direction is along the Z-axis (b). The sample orientations are as follows: (b) Face-on – with neutron incidence along Y-axis, (c) Edge on – 45° relative to the sample growth plane (X-axis); (d) and (e) the sample was rotated by 5 and 30° around the Y-axis, respectively.

the presence of strong texture of (1 0 0) as build direction. The height of (2 0 0) edge, for example, is larger for the neurons propagating along Y-axis and X-axis (orientations (b) and (e) in Figure 18), but is smaller when neutrons propagate at 45° relative to the X-axis (orientations (c) and (d)). At the same time, the (1 1 1) edge is the strongest for the orientations (c) and (d) and weakest for orientation (b). All these observations can be explained by the assumption that the sample is (1 0 0) textured in the build direction.

4. Conclusions

The results of our experiments with Inconel 625 samples, manufactured by a direct metal laser melting method, demonstrate the possibilities of using energy-resolved neutron imaging to study bulk microstructure properties of additive manufactured components. The low attenuation of neutrons for various metals provides the possibility to study thick samples non-destructively, when it is impossible with other characterization techniques. The contrast in energy-resolved neutron imaging due to neutron coherent scattering can reveal unique crystallographic properties of the samples, most importantly the distribution of internal strains and stresses (averaged over the neutron path through the sample), the grain size and intra-granular defects distribution, and the presence of texture or preferred orientation of crystalline grains. The spatial resolution of this imaging technique is currently determined by the resolution of event counting detectors (55 μm in our case), the neutron beam intensity and the data integration time. The distribution of macroscopic (~100 μm) defects such as voids and impurities can also be imaged through the entire volume of the samples along with strain and microstructure. At the same time the elemental distribution within the sample may be obtained with resonance absorption imaging,[19,20] providing the elements of interest have reasonably high resonance absorption cross section, such as Ta, W and Hf which are commonly in Ni-base alloys. As very high neutron flux is essential for this measurement, such measurements cannot be accomplished by using portable neutron sources which may be desirable for process or quality control. This technique obviously cannot be applied for real-time feedback control of manufacturing processes as it can be implemented only in large scale neutron beamline facilities. However, it can be an additional tool in the optimization of specific additive manufacturing techniques and may be used to correlate spatial variation of microstructures with processing parameters of a specific manufacturing method and materials composition.

Acknowledgements

The detector used in these experiments was developed in collaboration between UC Berkeley and Nova Scientific, Inc (Sturbridge, MA, USA) and the readout electronics developed within the Medipix collaboration. The authors would like to acknowledge generous donation of Vertex 5 FPGA, Vivado design suite and DK-K7-CONN-G connectivity kit by Xilinx Inc. of San Jose, California through their Xilinx University Program.

Disclosure statement

No potential conflict of interest was reported by the authors.

Funding

This work was supported in part by the US Department of Energy [STTR grant numbers DE-FG02-07ER86322, DE-FG02-08ER86353 and DE-SC0009657].

References

[1] Gu DD, Meiners W, Wissenbach K, et al. Laser additive manufacturing of metallic components: materials, processes and mechanisms. Int Mater Rev. 2012;57:133–164.

[2] Vaezi M, Seitz H, Yang SF. A review on 3D micro-additive manufacturing technologies. Int J Adv Manuf Technol. 2013;67:1721–1754.

[3] Frazier WE. Metal additive manufacturing: a review. J Mater Eng Perform. 2014;23:1917–1928.

[4] Herderick, E. Additive manufacturing of metals: a review. 2011 proceedings of MS&T 11; 2011 October 16–20; Columbus, Ohio; 2011.

[5] Dinda GP, Dasgupta AK, Mazumder J. Laser aided direct metal deposition of Inconel 625 superalloy: microstructural evolution and thermal stability. Mater Sci Eng A. 2009;509:98–104.

[6] Bilheux HZ, McGreevy R, Anderson IS, editors. Neutron Imaging and Applications. Springer Science+Business Media, LLC; 2009

[7] Strobl M, Manke I, Kardjilov N, et al. Advances in neutron radiography and tomography. J Phys D: Appl Phys. 2009;42:243001.

[8] Watkins T, Bilheux H, An K, et al. Neutron characterization for additive manufacturing. Adv Mat & Proc. 2013;171:23–27.

[9] Kockelmann W, Frei G, Lehmann EH, et al. Energy-selective neutron transmission imaging at a pulsed source *Nucl.* Instrum Methods Phys Res A. 2007;578:421–434.

[10] Tremsin AS, Vogel SC, Mocko M, et al. Energy resolved neutron radiography at LANSCE pulsed neutron facility. Neutron News. 2013;24:28–32.

[11] Santisteban JR, Edwards L, Stelmukh V. Characterization of textured materials by TOF transmission. Physica B. 2006;385-386:636–638.

[12] Sato H, Takada O, Iwase K, et al. Imaging of a spatial distribution of preferred orientation of crystallites by pulsed neutron Bragg edge transmission. J Phys: Conf Ser. 2010;251:012070.

[13] Lehmann EH, Frei G, Vontobel P, et al. The energy-selective option in neutron imaging. Nucl Instrum Methods Phys Res A. 2009;603:429–438.

[14] Tremsin AS, Kockelmann W, Zhang Shu-Yan, et al. Investigation of microstructure within metal welds by energy resolved neutron imaging subm. Phy Proc. Forthcoming 2015

[15] Steuwer A, Withers PJ, Santisteban JR, et al. Bragg edge determination for accurate lattice parameter and elastic strain measurement. Phys Stat Sol A. 2001;185(2):221–230.

[16] Tremsin AS, McPhate JB, Steuwer A, et al. High-resolution strain mapping through time-of-flight neutron transmission diffraction with a microchannel plate neutron counting detector. Strain: Intern. J. Experim. Mechanics. 2012;48. 296–305.

[17] Tremsin AS, McPhate JB, Vallerga JV, et al. High-resolution strain mapping through time-of-flight neutron transmission diffraction. Mat Sci Forum. 2014;772:9–13.

[18] Dehoff RR, Kirka MM, Sames WJ, et al. Site specific control of crystallographic grain orientation through electron beam additive manufacturing. Mat Sci Tech. 2015;31:931–938.

[19] Tremsin AS, Vogel SC, Mocko M, et al. Non-destructive studies of fuel rodlets by neutron resonance absorption radiography and thermal neutron radiography. J Nucl Mater. 2013;440:633–646.

[20] Tremsin AS, Shinohara T, Kai T, et al. Neutron resonance transmission spectroscopy with high spatial and energy resolution at J-PARC pulsed neutron source Nucl Instr Meth A. 2014;746:47–58.

[21] Wagner V, Friedrich H and Wille P. Performance of a high- tech neutron velocity selector. Phys B: Cond Matter. 1992;180–181:938–940

[22] Treimer W, Strobl M, Kardjilov N, et al. A wave length tunable device for neutron radiography and tomography Appl Phys Lett. 2006;89:203504.

[23] Schulz M, Böni P, Calzada E, et al. Energy-dependent neutron imaging with a double crystal monochromator at the ANTARES facility at FRM II Nucl Instr Meth A. 2009;605:33–35.

[24] Gorini G, Perelli-Cippo E, Tardocchi M, et al. The resonant detector and its application to epithermal neutron spectroscopy. Nucl Instr Methods Phys. Res A. 2004;529:293–300.

[25] Tremsin AS. High resolution neutron counting detectors with microchannel plates and their applications in neutron radiography, diffraction and resonance absorption imaging. Neutron News. 2012;23:35–38.

[26] Tremsin AS, Vallerga JV, McPhate JB, et al. Optimization of high count rate event counting detector with Microchannel Plates and quad Timepix readout. Nucl Instrum Methods Phys Res A. 2015;787:20–25.

[27] Parker JD, Harada M, Hattori K, et al. Applications of a micro-pixel chamber (μPIC) based, time-resolved neutron imaging detector at pulsed neutron beams. J Phys Conf Ser. 2014;502:012048.

[28] Kiyanagi Y, Kamiyama T, Kino K, et al. Pulsed neutron imaging using 2-dimensional position sensitive detectors J. Instrum, JINST. 2014;9:C07012.

[29] Harada M, Oikawa K, Kasugai Y, et al. Shielding design of a neutron beam line "NOBORU" at JSNS/J-PARC. Pro Nucl Sci Techn. 2011;1:94–97.

[30] Vogel. A Rietveld-Approach for the Analysis of Neutron Time-of-Flight Transmission Data (Ph.D. thesis). Germany: Kiel University;2000.

[31] Ashcroft NW, Mermin ND. Solid state physics. New York, NY: Harcourt; 1976.

Permissions

All chapters in this book were first published in STAM, by Taylor & Francis Online; hereby published with permission under the Creative Commons Attribution License or equivalent. Every chapter published in this book has been scrutinized by our experts. Their significance has been extensively debated. The topics covered herein carry significant findings which will fuel the growth of the discipline. They may even be implemented as practical applications or may be referred to as a beginning point for another development.

The contributors of this book come from diverse backgrounds, making this book a truly international effort. This book will bring forth new frontiers with its revolutionizing research information and detailed analysis of the nascent developments around the world.

We would like to thank all the contributing authors for lending their expertise to make the book truly unique. They have played a crucial role in the development of this book. Without their invaluable contributions this book wouldn't have been possible. They have made vital efforts to compile up to date information on the varied aspects of this subject to make this book a valuable addition to the collection of many professionals and students.

This book was conceptualized with the vision of imparting up-to-date information and advanced data in this field. To ensure the same, a matchless editorial board was set up. Every individual on the board went through rigorous rounds of assessment to prove their worth. After which they invested a large part of their time researching and compiling the most relevant data for our readers.

The editorial board has been involved in producing this book since its inception. They have spent rigorous hours researching and exploring the diverse topics which have resulted in the successful publishing of this book. They have passed on their knowledge of decades through this book. To expedite this challenging task, the publisher supported the team at every step. A small team of assistant editors was also appointed to further simplify the editing procedure and attain best results for the readers.

Apart from the editorial board, the designing team has also invested a significant amount of their time in understanding the subject and creating the most relevant covers. They scrutinized every image to scout for the most suitable representation of the subject and create an appropriate cover for the book.

The publishing team has been an ardent support to the editorial, designing and production team. Their endless efforts to recruit the best for this project, has resulted in the accomplishment of this book. They are a veteran in the field of academics and their pool of knowledge is as vast as their experience in printing. Their expertise and guidance has proved useful at every step. Their uncompromising quality standards have made this book an exceptional effort. Their encouragement from time to time has been an inspiration for everyone.

The publisher and the editorial board hope that this book will prove to be a valuable piece of knowledge for researchers, students, practitioners and scholars across the globe.

List of Contributors

Kazuhiko Ishihara, Weixin Chen, Yihua Liu, Yuriko Tsukamoto and Yuuki Inouea
Department of Materials Engineering, School of Engineering, The University of Tokyo, Tokyo, Japan
Department of Bioengineering, School of Engineering, The University of Tokyo, Tokyo, Japan

Muhammad Farooq Khan, Ghazanfar Nazir, Volodymyr M. lermolenko and Jonghwa Eom
Department of Physics & Astronomy and Graphene Research Institute, Sejong University, Seoul 05006, Korea

Xia Yu
School of Medical Instrument and Food Engineering, University of Shanghai for Science and Technology, Shanghai, P.R. China

Yufang Zhu
School of Materials Science and Engineering, University of Shanghai for Science and Technology, Shanghai, P.R. China

Elena Filatova, Aleksei Konashuk, Yuri Petrov, Evgeny Ubyivovk, Andrei Selivanov and Victor Drozd
aInstitute of Physics, St Petersburg State University, Ul'yanovskaya Str. 1, Peterhof, 198504, St Petersburg, Russia

Andrey Sokolov
Institute Nanometre Optics and Technology (FG-INT), Helmholtz-Zentrum Berlin für Materialien und Energie GmbH, Albert Einstein Str.15, 12489, Berlin, Germany

Carmine Urso
Dipartimento di Chimica, Università degli Studi di Bari, Via Orabona 4, 70126 Bari, Italy

Angela Agostiano, Michela Corricelli and Elisabetta Fanizza
Dipartimento di Chimica, Università degli Studi di Bari, Via Orabona 4, 70126 Bari, Italy
Istituto per i Processi Chimico Fisici IPCF Consiglio Nazionale delle Ricerche CNR, Via Orabona 4, 70126 Bari, Italy

Marinella Striccoli, Nicoletta Depalo, Annamaria Panniello and M. Lucia Curri
Istituto per i Processi Chimico Fisici IPCF Consiglio Nazionale delle Ricerche CNR, Via Orabona 4, 70126 Bari, Italy

Nunzio Denora, Valentino Laquintana
Dipartimento di Farmacia - Scienze del Farmaco, Università degli Studi di Bari, Via Orabona 4, 70126 Bari, Italy

R. Maria Iacobazzi
Dipartimento di Farmacia - Scienze del Farmaco, Università degli Studi di Bari, Via Orabona 4, 70126 Bari, Italy
Istituto tumori IRCC S Giovanni Paolo II, Bari, Italy

Meng Li and Jialin Shi
State Key Laboratory of Robotics, Shenyang Institute of Automation, Chinese Academy of Sciences, Shenyang 110016, PR China
University of the Chinese Academy of Sciences, Beijing 100049, PR China

Lianqing Liu, Peng Yu and Yuechao Wang
State Key Laboratory of Robotics, Shenyang Institute of Automation, Chinese Academy of Sciences, Shenyang 110016, PR China

Ning Xi
State Key Laboratory of Robotics, Shenyang Institute of Automation, Chinese Academy of Sciences, Shenyang 110016, PR China
Department of Electrical and Computer Engineering, Michigan State University, East Lansing, Michigan 48824, PR USA

Iván Gutierrez-Urrutia
Microstructure Physics and Alloy Design, Max-Planck-Institut fűr Eisenforschung, Max-Planck Str. 1, D-40237 Dűsseldorf, Germany
Structural Materials Unit, Research Center for Strategic Materials, National Institute for Materials Science, Tsukuba, Ibaraki 305-0047, Japan

Fady Archie and Dierk Raabe
Microstructure Physics and Alloy Design, Max-Planck-Institut fűr Eisenforschung, Max-Planck Str. 1, D-40237 Dűsseldorf, Germany

Feng-Kai Yan and Nai-Rong Tao
Institute of Metal Research, Chinese Academy of Sciences, Shenyang National Laboratory for Materials Science, Shenyang 110016, PR China

Ke Lu
Institute of Metal Research, Chinese Academy of Sciences, Shenyang National Laboratory for Materials Science, Shenyang 110016, PR China
Herbert Gleiter Institute of Nanoscience, Nanjing University of Science & Technology, Nanjing 210094, PR China

Muhammad Boota
Faculty of Science and Technology, MESA+ Institute for Nanotechnology, University of Twente, P.O. Box 217, 7500AE Enschede, the Netherlands
Engineering Department, University of Rome "ROMA TRE", Via della Vasca Navale 79, 00146 Rome, Italy

Matthijn Dekkers
SolMates BV, Drienerlolaan 5, Building 6, 7522NB Enschede, the Netherlands

Giulia Lanzara
Engineering Department, University of Rome "ROMA TRE", Via della Vasca Navale 79, 00146 Rome, Italy

Minh D. Nguyen
Faculty of Science and Technology, MESA+ Institute for Nanotechnology, University of Twente, P.O. Box 217, 7500AE Enschede, the Netherlands
SolMates BV, Drienerlolaan 5, Building 6, 7522NB Enschede, the Netherlands

Kurt H. Vergeer, Gertjan Koster, Guus Rijnders and Evert P. Houwman
Faculty of Science and Technology, MESA+ Institute for Nanotechnology, University of Twente, P.O. Box 217, 7500AE Enschede, the Netherlands

Thomas Lutz, Paul E. Barclay, Wolfgang Tittel and Lucile Veissier
Department of Physics & Astronomy, Institute for Quantum Science and Technology, University of Calgary, Calgary Alberta T2N 1N4, Canada

Charles W. Thiel, Philip J. T. Woodburn and Rufus L. Cone
Department of Physics, Montana State University, Bozeman, MT 59717, USA

Jun Kang and Myeong-Hoon Lee
Division of Marine Engineering, Korea Maritime and Ocean University, Busan 606-791, Republic of Korea

Hye-min Kim
Graduate School of Engineering, Nagoya University, Nagoya 464-8603, Japan

Nagahiro Saito
Department of Material Science and Engineering, Nagoya University, Furo-cho, Chikusa-ku, Nagoya 464 8603, Japan

Dong Wang and Kaiwen Guan
State Key Laboratory of Coordination Chemistry, School of Chemistry and Chemical Engineering, Nanjing University, Nanjing 210023, P.R. China
State Key Laboratory of Pollution Control and Resource Reuse, School of the Environment, Nanjing University, Nanjing 210046, P.R. China

Zhiping Bai
State Key Laboratory of Coordination Chemistry, School of Chemistry and Chemical Engineering, Nanjing University, Nanjing 210023, P.R. China

Fuqiang Liu
State Key Laboratory of Pollution Control and Resource Reuse, School of the Environment, Nanjing University, Nanjing 210046, P.R. China

Matthew R. Skorski, Jake M. Esenther and Matthew R. Hartings
Department of Chemistry, American University, 4400 Massachusetts Ave., NW, Washington, DC, 20016, USA

Zeeshan Ahmed
Thermodynamic Metrology Group, Sensor Science Division, Physical Measurement Laboratory, National Institute of Standards and Technology, Gaithersburg, MD 20899, USA

Abigail E. Miller
Department of Chemistry, American University, 4400 Massachusetts Ave., NW, Washington, DC, 20016, USA
Food and Drug Administration, Washington, DC, USA

Daniele Scopece, Daniele Passerone and Alex Dommann
nanotech@surfaces laboratory, Empa, Swiss Federal Laboratories for Materials Science and Technology, Überlandstrasse 129, CH-8600, Dübendorf, Switzerland

Max Döbeli
Ion Beam Physics, ETH Zürich, Otto-Stern-Weg 5, CH-8093, Zürich, Switzerland

Jürgen Ramm and Beno Widrig
Oerlikon Surface Solutions AG , Oerlikon Balzers, Iramali 18, 9469 Balzers, Liechtenstein

Ulrich Müller and Xavier Maeder
Laboratory for Nanoscale Materials Science, Empa, Swiss Federal Laboratories for Materials Science and Technology, Überlandstrasse 129, CH-8600 Dübendorf, Switzerland

Antonia Neels
Center for X-ray Analytics, Empa, Swiss Federal Laboratories for Materials Science and Technology, Überlandstrasse 129, CH-8600 Dübendorf, Switzerland

Yin Liu, Yuxuan Gong, Nathan P. Mellott, Bu Wang and Yiquan Wu
New York State College of Ceramics, Alfred University, Alfred, NY , USA

Haitao Ye
School of Engineering and Applied Science, Aston University, Birmingham, UK

Dudekula Althaf Basha, Hidetoshi Somekawa and Alok Singh
Structural Materials Unit, National Institute for Materials Science, Sengen 1-2-1,Tsukuba 305-0047, Japan

Koichi Tsuchiya and Takashi Miyawaki
Structural Materials Unit, National Institute for Materials Science, Sengen 1-2-1,Tsukuba 305-0047, Japan
Graduate School of Pure and Applied Sciences,University of Tsukuba, Tsukuba, Ibaraki305-8577, Japan

JulianM. Rosalie
Erich Schmid Institute of Materials Science,Austrian Academy of Sciences, Austria

Janusz Jacak and Lucjan Jacak
Institute of Physics,Wrocław University of Technology, Wyb.Wyspiańskiego 27, 50-370, Wrocław, Poland

Keke Chang, Denis Music, Moritz to Baben, Dennis Lange, Hamid Bolvardi and Jochen M. Schneider
Materials Chemistry, RWTH Aachen University, D-52074 Aachen, Germany

Chia-Wen Tsao and Yueh-Pu Lee
Department of Mechanical Engineering, National Central University, Taoyuan, ROC

Miguel Guerrero, Jin Zhang, Maria Dolors Baró and Eva Pellicer
Departament de Física, Facultat de Ciències, Universitat Autònoma de Barcelona, E-08193 Bellaterra, Spain

Ainhoa Altube and Eva García-Lecina
Surfaces Division, IK4-CIDETEC, Paseo Miramón, 196, E-20009 San Sebastián, Spain

Mònica Roldan
Servei de Microscòpia, Universitat Autònoma de Barcelona, E-08193 Bellaterra, Spain

Jordi Sort
Institució Catalana de Recerca i Estudis Avançats (ICREA) and Departament de Física, Universitat Autònoma de Barcelona, E-08193 Bellaterra, Spain

Ivan Gutierrez-Urrutia, Cheng-Lin Li and Satoshi Emura
National Institute for Materials Science, Tsukuba, Ibaraki 305-0047, Japan

Xiaohua Min
School of Materials Science and Engineering, Dalian University of Technology, Dalian 116024, PR China

Koichi Tsuchiya
National Institute for Materials Science, Tsukuba, Ibaraki 305-0047, Japan
Graduate School of Pure and Applied Sciences, University of Tsukuba, Ibaraki 305-0047, Japan

Anton S. Tremsin
Space Sciences Laboratory, University of California at Berkeley, Berkeley, CA , USA

Yan Gao and Laura C. Dial
General Electric Global Research Center, Niskayuna, NY , USA

Francesco Grazzi
Consiglio Nazionale delle Ricerche, Istituto Sistemi Complessi (CNR-ISC), Sesto Fiorentino (FI), Italy

Takenao Shinohara
Japan Atomic Energy Agency Tokai-mura, Naka-gun Ibaraki, Japan

Index